The Automotive Body

Mechanical Engineering Series

Frederick F. Ling
Editor-in-Chief

The Mechanical Engineering Series features graduate texts and research monographs to address the need for information in contemporary mechanical engineering, including areas of concentration of applied mechanics, biomechanics, computational mechanics, dynamical systems and control, energetics, mechanics of materials, processing, production systems, thermal science, and tribology.

For other titles published in this series, go to
http://www.springer.com/1161

Lorenzo Morello • Lorenzo Rosti Rossini
Giuseppe Pia • Andrea Tonoli

The Automotive Body

Volume II: System Design

Lorenzo Morello
Via Bey 5B
10090 Villarbasse
Italy
E-mail: lormorello@gmail.com

Giuseppe Pia
via Filadelfia 237/8 B
10137 Torino
Italy
E-mail: giuseppe.pia@libero.it

Lorenzo Rosti Rossini
via Canova 9
20145 Milan
Italy
E-mail: lorenzo_rosti@hotmail.com

Andrea Tonoli
via Oronte Nota 55
10051 Avigliana (TO)
Italy
E-mail: andrea.tonoli@polito.it

ISSN 0941-5122
ISBN 978-94-007-9220-3 ISBN 978-94-007-0516-6 (eBook)
DOI 10.1007/978-94-007-0516-6
Springer Dordrecht Heidelberg London New York
Softcover re-print of the Hardcover 1st edition 2011

The accuracy and completeness of information provided in this book are not guaranteed to produce any particular results. Therefore, the Authors and the Publisher will not be liable for any direct or indirect loss or damages incurred from any use of the information contained in the book.

Cover design: eStudio Calamar S.L.

Printed on acid-free paper

Springer is part of Springer Science+Business Media (www.springer.com)

Contents

7
Introduction to Volume II

The purpose of Volume II is to explain the links which exist between satisfying the needs of the customer (either driver or passenger) and the specifications for vehicle design, and between the specifications for vehicle system and components.

For this study a complete vehicle system must be considered, including, according to the nature of functions that will be discussed, more component classes than considered in Volume I, and, sometimes, also part of the chassis and the powertrain.

Since the first element of the chain of elements to be taken into consideration is the human being, it is appropriate to consider physiology issues to some extent in order to better understand the needs to be satisfied and the control parameters to be evaluated.

The Chapter 8, the first in this Volume, is dedicated to body requirements and functions. An introductory framework regarding statistic vehicle usage in Europe is provided, followed by an analysis of vehicle functions, with particular reference to those functions that are more conditioned by body design: Ergonomics and internal space utilization, thermal comfort, acoustic and vibration comfort, structural integrity and passive safety.

Although specific aspects of marketing are beyond the scope of this text, some examples are provided as a reference regarding the procedures applied to define technical quantitative specifications on the basis of customer's needs, usually rated qualitatively. Bearing in mind their correlation with vehicle system specifications, the European regulations relevant to the car body and its components are also explained, with particular reference to vehicle active and passive safety.

L. Morello et al.: The Automotive Body, Vol. 2: System Design, MES, pp. 1–2.
springerlink.com

The following chapter, dedicated to Ergonomics and packaging, describes the most important issues to be considered in terms of occupants well-being inside passenger compartment: seating comfort, direct and indirect visibility, the space necessary to install mechanical components into the vehicle will be particularly considered. The analysis of these issues allows to the primary design criteria of internal space and the lay-out of seats and controls to be identified.

Chapter 10 addresses the topic of thermal comfort and includes an introduction to the human body physiology thermal comfort conditions and the main parameters of relevance; the metabolic activity is also examined as function of ambient conditions and the influence on comfort, taking into account the thermal exchange. A thermal balance can therefore be evaluated of relevance with respect to the design of the air conditioning system. After summarizing the main components of the HVAC system, already described in Volume I, the chapter concludes with design and testing criteria description.

Starting from human sensitivity to noise and vibration, the main issues influencing comfort are considered in the next chapter. The main sources of dynamic excitation are examined, including those which generate noise and vibration inside the vehicle including tires and powertrain, and outside the vehicle such as road shocks and aerodynamic noise. The main parameters needed to describe the dynamic behavior of the body are also introduced, together with the technical solutions proposed to reduce the transmission of excitations to the body.

The main objective of the following chapter concerning structural integrity is to explain the role of the body primary components and the baseline criteria for their design. Quasi-static reference loads acting on the body during normal service life are introduced. Some additional concepts, appropriate to understand the effect of bending and torsion deformation on a car body, are also discussed in terms of the contribution of the framework. The chapter concludes by describing the options available and the critical issues to be considered when modeling a car body using the finite elements technique.

The last chapter is dedicated to the subject of passive safety, opening with an introduction to biomechanics and the criteria used to evaluate the severity of injuries consequent to an accident. This very broad topic is limited here to empirical relationships and the acceptability limits proposed by governments regulations to limit the severity of injuries during accidents. The need to guarantee a high level of protection to mitigate the consequence of accidents is a major issue that directly conditions body design. The objective of the explanation included is to describe the most common solutions and design criteria in use to protect car occupants during an accident. Simplified mathematical models are introduced to describe restraint systems functions and the structural behavior of deformable parts of the body involved in the crash. The high reliability of results obtained from finite elements method, also for the prediction of the large displacements expected as a result of a crash, justifies the outline included of the computer codes applied to model plastic behavior.

8
Functions and Specifications

Knowledge of how the vehicle is constructed and manufactured and of how its components should be designed, which are the subjects of Volume I, and of how they should be integrated into the system, the scope of Volume II, does not complete the entire picture with regard to how to conceive and develop a car which will meet with commercial success since in practice many of the functions are not directly related to purely technical aspects.

Such knowledge is essential for obtaining a set of assigned targets for the product. However the vehicle is a mature product meaning that fundamental characteristics are almost always standardized and technical excellence is now considered to be a 'must' than a topic for advertisement. Without underestimating the fundamental importance of this knowledge and the resulting product targets, it is appropriate to recognize that the success of a product is mostly dependent on how well these targets are able to interpret the customer's needs.

The combination of vehicle technical objectives and an overall description of its architecture comprise what is universally known, with particular reference to the case of the car, as product *concept* or *conceptual design.*

The concept is the starting point for the development of a car and its production tools. It can be expressed with a sketch or by using a three dimensional simplified model, suitable for illustrating its appearance and its main functions. Its aesthetic appearance must also be addressed because it must be coherent with the expectations of potential customers. This visual documentation must be accompanied by an exhaustive quantitative definition of the technical and economical characteristics which demonstrate congruence and feasibility.

Concept creation is a truly creative process, where its leadership is usually assigned to a marketing specialist, but where meeting its target successfully must

L. Morello et al.: The Automotive Body, Vol. 2: System Design, MES, pp. 3–125.
springerlink.com

include contributions from every discipline involved in product development. In fact, the characteristics of the concept must be derived from a good understanding of customer needs and of how much the customer is willing to spend for their satisfaction; even if it is true that this knowledge is the speciality of marketing experts, an innovative contribution essential for success is required from all involved who are competent regarding relevant product functions.

Defining a concept means, therefore:

- to describe a product in terms of technical functions and specifications;
- to determine the product configuration and to choose its main components;
- to identify character, personality, feelings and other traits this new product will offer to the customer.

Each car manufacturer emphasizes different aspects of the product concept and determines, consequently, its characteristics and potential success from the outset of the development process. The most obvious aspects may be defined, for example, according to the categories of convenience, luxury and sportsmanship.

The central issue of the concept definition process is to obtain involvement throughout the company; the concept is partly driven by objective and measurable facts, the job of technicians, and partly by insight that will be contributed by marketing experts and others involved with sufficient experience to contribute creatively.

Those in charge of detailed product design, component specification, styling, production means development, sales and service must also be involved since their expertise will condition customer satisfaction.

Nevertheless, strong leadership by marketing experts is necessary: While ignoring the any of the operations listed above during the concept development can cause significant inconvenience, it is also true that excessive involvement of many can cause premature conflicts and compromises, leading to product characteristics which may be 'flat' or even trivial.

A new vehicle cannot be simply the extrapolation of a previous perception of customer needs, as assessed by the popularity of existing products; very often, successful cars have been born out of a response to needs that were unexpressed until the time of product launch.

To underline this point, it is worthwhile recalling the first launch of sport utility vehicles, coupe-cabriolets, minivans and a series of other product innovations that met with commercial success which initially may have been unexpected by other vehicle manufacturers.

When defining a new concept it is important to proceed according to the following steps:

- Focus on customer needs;
- Identify latent or hidden needs, in addition to those demonstrated by existing products;

- Develop a full comprehension of customer needs.

At this stage it is important to distinguish between customer needs and product specification.

Customer needs are independent of the product under development and are therefore not bound to the concept. Needs must be identified and focused on without necessarily knowing if or how they may be satisfied.

Product specifications depend, on the other hand, on the defined product architecture and the kind of components that have been defined; their linkage with customer needs can be understood only by knowing and interpreting the success of similar products already in the market.

If a good channel of communication is established between the customers and those people in charge of product development, a sound knowledge of customer needs results.

Customers are identified through the following methods:

- Market surveys;
- Direct interviews;
- Product use monitoring.

These methods can also be applied to develop a measuring system to quantify the appraisal of the customer; this subject will be discussed in a separate, dedicated section.

The purpose of this chapter is not to attempt to present the complete knowledge of these marketing techniques, but rather to provide the future vehicle engineer with a different view which complements the technical perspective in order to help enhanced interaction with marketing experts or customers.

In the following different topics will be analyzed with particular reference to those aspects which have a major impact on the automotive body and on the body components design.

Transportation statistics

Transportation statistics represent the starting point for knowledge of vehicle use; these statistics make reference to Italy and to the European Union, as far as passenger traffic and the transportation of goods are concerned. In particular, the volume breakdown between different means of transportation will be analyzed in order to determine the average expected lifetime of vehicles.

Readers from different countries could take these data into consideration and adapt this method of analysis appropriately.

Vehicle functions

This chapter addresses vehicle system design which involves the definition of the technical characteristics of the vehicle that determines the level of satisfaction of customer needs.

Since the field of our study is limited to the body, system design focuses on packaging, ergonomics comfort and safety.

Because this subject is broad, only outline examples are used to suggest the methods of correlation of objective technical characteristics with customer needs which, being subjective, are only expressed qualitatively.

Vehicle mission definition, as a description of the expected events during the useful life of the vehicle, will be addressed; these data are necessary to forecast the resistance in terms of aging of a body and define its scheduled maintenance program.

Regulations

Legislation and regulations imposed by governments significantly influence vehicles in general and, therefore, the characteristics of the body. These rules evolve continually and within the European Union are now standardized by general directives with which the technical regulations of the member states must comply; again, European legislation is proposed as a reference, noting that other states have introduced legislation that is quite similar in many cases. Issues which are subject to existing regulations and are of direct relevance to body design typically include passive safety, active safety systems, pollution and emissions, energy consumption and recycling.

Moreover many other technical organizations, supported by consumer associations and specialist magazines, influence product design with a power equal to legislation, an issue that must also be taken into account in accordance with their importance.

8.1 Transportation Statistics

The data reported in this chapter were extracted from institutional documents of ANFIA, ACEA, ISTAT and Eurostat.

ANFIA (Associazione Nazionale Fra le Industrie Automobilistiche), the Italian national association of automotive manufacturers, was established in 1912 and is spokesman for its associates, on all issues (from technical, economic, fiscal and legislative to qualitative and statistical) regarding the mobility of people and goods.

Among its objectives, ANFIA has the task of gathering data and information, providing official statistical data for this sector of industry.

ANFIA publishes an annual report entitled Autoincifre (Figures of the Automobile), which is one of the fundamental references for statistical data on motoring in Italy and Europe. Much of the data collected in this report comes also from PRA (Pubblico Registro Automobilistico), the public vehicle registry managed by ACI, the association of Italian motorists.

More information can be found at:

www.anfia.it.

In the European Union, the role of ACEA (Association des Constucteurs Européen d'Automobile) is similar to that of ANFIA in Italy; the 13 major vehicle manufacturers with headquarters in Europe are associated with ACEA (BMW Group, DAF Trucks NV, Daimler AG, FIAT S.p.A, Ford of Europe GmbH, General Motors Europe AG, MAN Nutzfahrzeuge AG, Dr. Ing. h. c. F. Porsche AG, PSA Peugeot Citroën, Renault SA, Scania AB, Volkswagen AG, AB Volvo).

This association represents European manufacturers in the European Union under a wide spectrum of activities, setting up research groups, supporting the manufacturers with objective data and creating new legislative proposals in the fields of mobility, safety and environmental protection.

More information can be found at:

www.acea.be.

ISTAT (Istituto nazionale di STATistica) is the Italian government institution for statistics. Established in Italy in 1926, ISTAT is the main producer of official statistics for citizens and decision takers in the public sector. It works in full autonomy with continuous interactions with the academic and scientific world.

This institution is directly involved in collecting European statistics (according to regulation R 322) and gathering data according to the fundamental rules of impartiality, reliability, efficiency, privacy and transparency.

More information can be found at:

www.istat.it.

Eurostat is the statistical office of the European Union. Its mission is to provide the statistics from corresponding national services. The European Statistic Service (ESS) adopts similar methods, allowing it to obtain comparable data. This service was established in 1953.

These data, accessible to the public, concern:

- key indicators of Union policies;
- general and national statistics;
- economy and finance;
- population and social conditions;
- industry, commerce and services;
- agriculture and fisheries;
- commerce with foreign nations;
- transportation;
- environment and energy;

- science and technology.

  More information can be found at:

epp.eurostat.cec.eu.int.

A further source of information within the European Union derives from the public documents of the different Directorate Generals (DGs) which, for the European Union, are the equivalent term for Department or Ministry.

Among the DGs with particular relevance to the topics addressed here are DG Environment, DG Energy and DG Transport. Many papers and pamphlets on their related subjects can be downloaded free of charge from:

bookshop.europa.eu.

Since all data on these issues rapidly become obsolete, those interested are invited to consult regularly the mentioned public sites, which allow access to the original archives.

During these years of rapid expansion of the European Union, it has become particularly difficult to set up homogeneous historical series. Therefore as regards the data presented, reference is made either to E15, the group of 15 states which constituted the EU until few years ago, or to E25, following the recent expansion to 25 states.

8.1.1 Traffic Volume

Traffic volumes are conventionally measured as the product of transported units times the distance covered by such transportation; therefore:

- passenger traffic is measured in passengers per kilometer [pass×km];
- goods traffic is measured in metric tons per kilometer [t×km].

It should be pointed out that the metric ton, equivalent to 1,000 kg, is a unit of mass; in any case, of relevance is the quantity of transported material, therefore mass and not weight.

Nevertheless, the habit of considering the kilogram as a unit of weight and not of mass persists and therefore we sometimes see statements that *traffic volume has the same dimensions as energy*, which is only correct if the kilogram is assumed to be a unit of weight.

It is also true that if we assume a value for the acceleration of gravity and we know the vehicle coefficient of resistance, traffic volume is proportional to the energy expended to overcome motion resistance relative to the payload.

This could also apply to passenger traffic by substituting the number of passengers with the corresponding mass (conventionally 70 kg per passenger).

These considerations do not take into account the altitude difference between origin and destination, or any speed variations along the route, which are, instead, relevant for determining motion resistance and primary energy consumption.

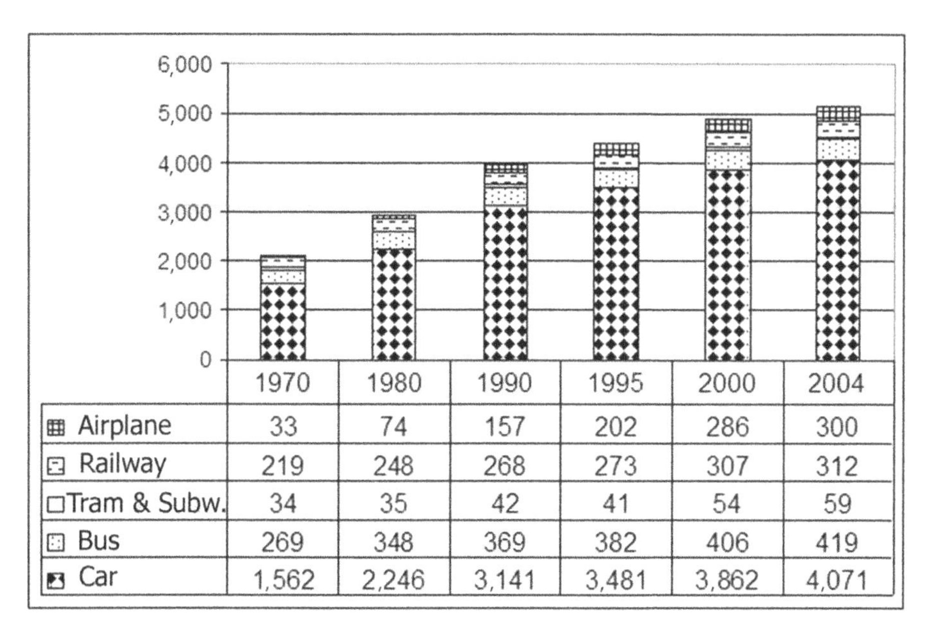

	1970	1980	1990	1995	2000	2004
Airplane	33	74	157	202	286	300
Railway	219	248	268	273	307	312
Tram & Subw.	34	35	42	41	54	59
Bus	269	348	369	382	406	419
Car	1,562	2,246	3,141	3,481	3,862	4,071

Fig. 8.1. Passenger traffic volume in the European Union, from 1970 to 2001 (in billions of pass×km), broken down by main vehicle types: airplanes, railroads, urban railroads, including subways, buses and cars (Source: Eurostat).

Passengers transportation

Fig. 8.1 reports passenger traffic volume in the European Union from 1970 to 2004, broken down according to the primary passenger transportation vehicles, such as cars, buses, urban railways with subways, trains and airplanes.

Cars definitely predominate over other means of transportation; car traffic represents in 2004 about 79% of the total, and traffic on tires (cars and buses) is about 88%; this breakdown has varied little over this period.

The total volume increased about 4% yearly during the first twenty years considered in this diagram; successively the growth slowed down to approximately zero in the last years available.

Air transportation with its continuous development is an exception; the last figure of the series on air transportation is extrapolated from the increase in transported passengers.

A similar table is reported for Italy in Fig. 8.2.

The situation for Italy is not so different from that of Europe as a whole; in this case, car traffic represented about 80% of the total in 2004 and traffic on tires (car and bus) was about 94% in the same year. This percentage has slightly increased over recent years, mainly due to the reduction in railroad traffic.

The total traffic volume increased more than the average of the European Union during the last years considered. Air transportation also increased during this period more than the average.

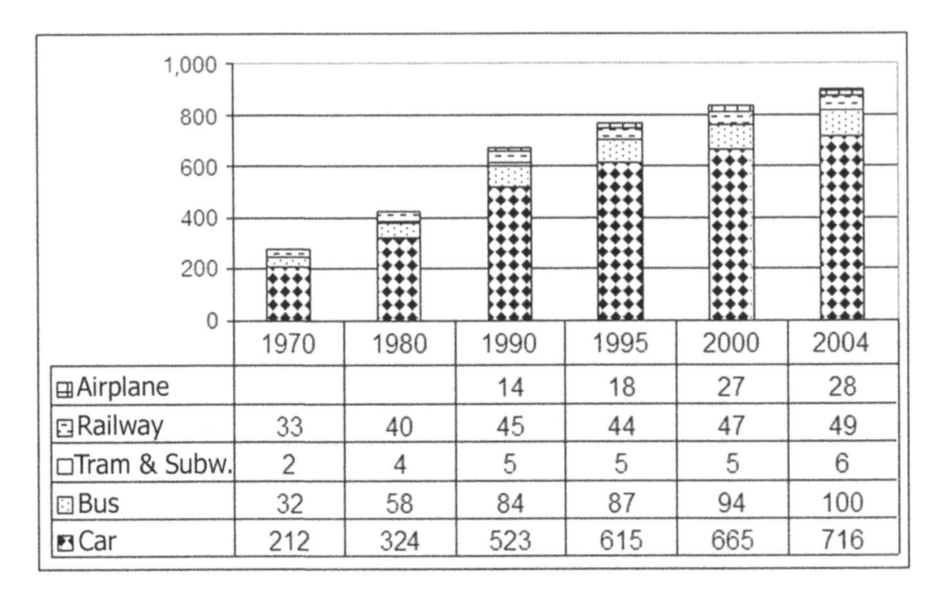

	1970	1980	1990	1995	2000	2004
Airplane			14	18	27	28
Railway	33	40	45	44	47	49
Tram & Subw.	2	4	5	5	5	6
Bus	32	58	84	87	94	100
Car	212	324	523	615	665	716

Fig. 8.2. Passenger traffic volume in Italy, from 1970 to 2001 (in billions of pass×km), broken down by main vehicle types: airplanes, railroads, urban railroads, including subways, buses and cars (Source: ANFIA).

In Italy (source ISTAT) traffic volume is well correlated with Gross Domestic Product.

The total ground transportation system made use of a network of about 6,500 km of toll motorways, more than 46,000 km of national roads, about 120,000 km of country roads and about 20,000 km of railroads, interconnecting 8,100 communities, 146 harbours, 101 airports and many railroad stations.

About 43 million of vehicles were in service, in addition to ships, trains and airplanes, serving approximately 57 million residents with an annual average of distance travelled totalling about 15,000 km.

Transportation of goods

Fig. 8.3 shows the volume of transportation of goods in Europe from 1970 to 2001, broken down according to the main vector types; in this case, road, rail, inland and sea navigation, and pipeline transportation are considered. The years considered here are different from those of passengers traffic because sampling methods are different.

Here again road transportation is predominant, accounting for 45% of the total in the last years of this period, starting from a percentage of 35% in 1970. The role of railroads has been reduced from 20% in 1970 to 8% in 2001. The

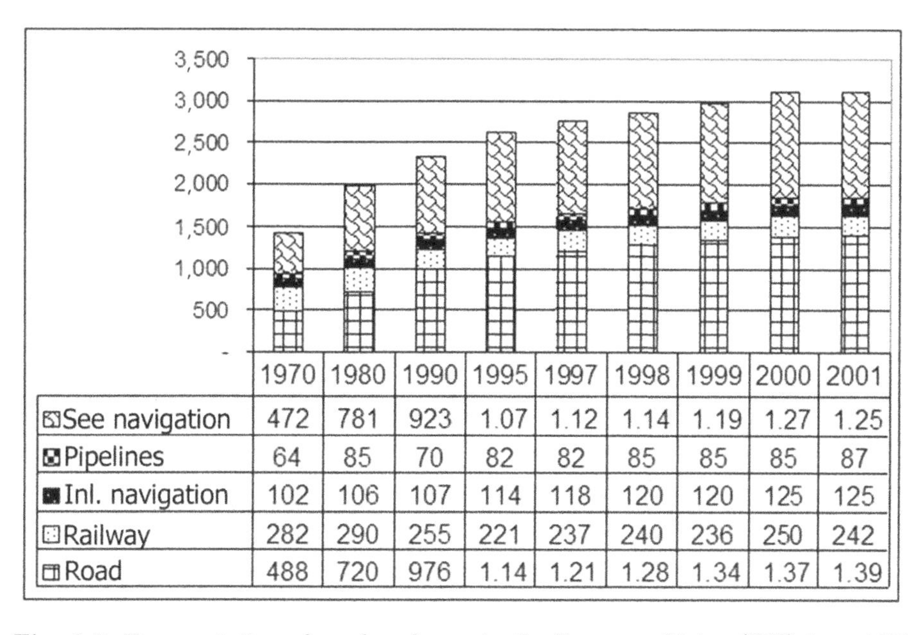

	1970	1980	1990	1995	1997	1998	1999	2000	2001
See navigation	472	781	923	1.07	1.12	1.14	1.19	1.27	1.25
Pipelines	64	85	70	82	82	85	85	85	87
Inl. navigation	102	106	107	114	118	120	120	125	125
Railway	282	290	255	221	237	240	236	250	242
Road	488	720	976	1.14	1.21	1.28	1.34	1.37	1.39

Fig. 8.3. Transportation of goods volumes in the European Union (E15) from 1970 to 2001 (in billion of t×km, broken down according to the different kinds of carrier; road, railroad, inland navigation, oil pipes and sea navigation are considered (Source: Eurostat).

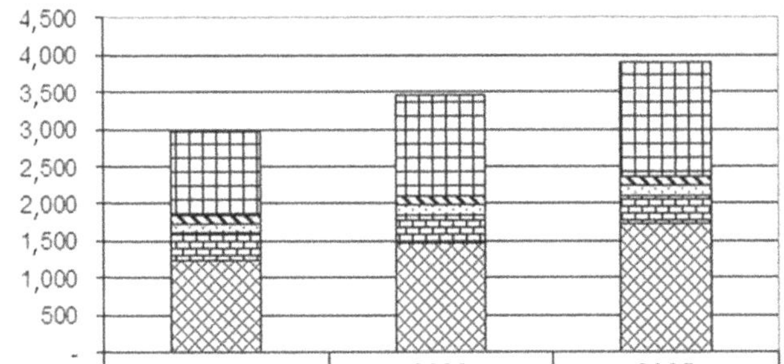

	1995	2000	2005
Aviation	2	2	2
See navigation	1,133	1,345	1,525
Pipelines	112	124	131
Inl. navigation	117	130	129
Railroad	358	374	392
Road	1,250	1,487	1,724

Fig. 8.4. Transportation of goods volumes in the European Union (E25) from 1995 to 2005 (in billion of t×km, broken down according to the different kinds of carrier (Source: Eurostat).

contribution of sea navigation is relevant, considering the higher average distance travelled.

To gain insight into the reasons behind the recent trend, consider Fig. 8.4 a short time series on goods transportation with respect to E25 where 10 new states have recently been added.

The increase in volumes is about 31% from 1995 to 2005; the volume of traffic in 2005 corresponds to 1 t of goods per 23 km for each of the citizens of the Union. The increase is due primarily to road and sea transportation taking up respectively 38% and 35% of the total.

Fig. reports a similar table for Italy.

Road transportation plays a respectively more important role in Italy than the European Union, carrying 89% of the total in the last year considered, starting from a share of 70% in 1970. In a similar way, the railroad share has reduced from an initial 16% to 6% in 2001. The contribution of sea navigation is not relevant, since the data includes domestic transportation only.

In the most recent years, all developed countries have recorded continuous growth in transportation demand. Factors stimulating this growth have been many (economical and fiscal integration, market globalization) and seem likely to last over the medium term.

The most stimulating factor for Italy has been the process of European economic integration, resulting in the free transfer of goods in the Union. The

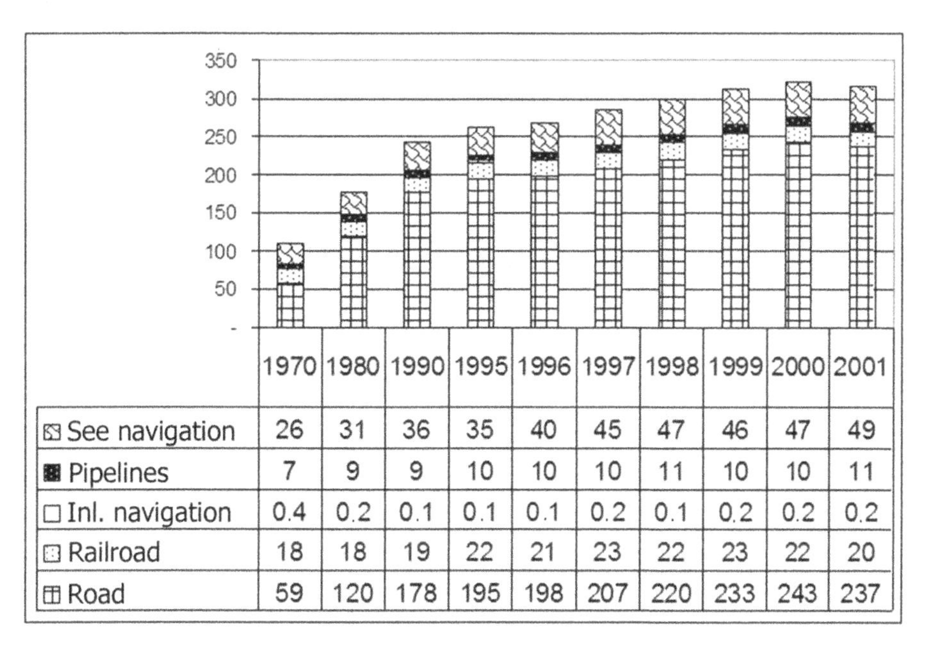

	1970	1980	1990	1995	1996	1997	1998	1999	2000	2001
See navigation	26	31	36	35	40	45	47	46	47	49
Pipelines	7	9	9	10	10	10	11	10	10	11
Inl. navigation	0.4	0.2	0.1	0.1	0.1	0.2	0.1	0.2	0.2	0.2
Railroad	18	18	19	22	21	23	22	23	22	20
Road	59	120	178	195	198	207	220	233	243	237

Fig. 8.5. Goods transportation volumes in Italy from 1970 to 2001 (in billion of t×km, broken down in terms of the different carriers; road, railroad, inland navigation, oil pipelines and sea navigation are considered (Source: ANFIA).

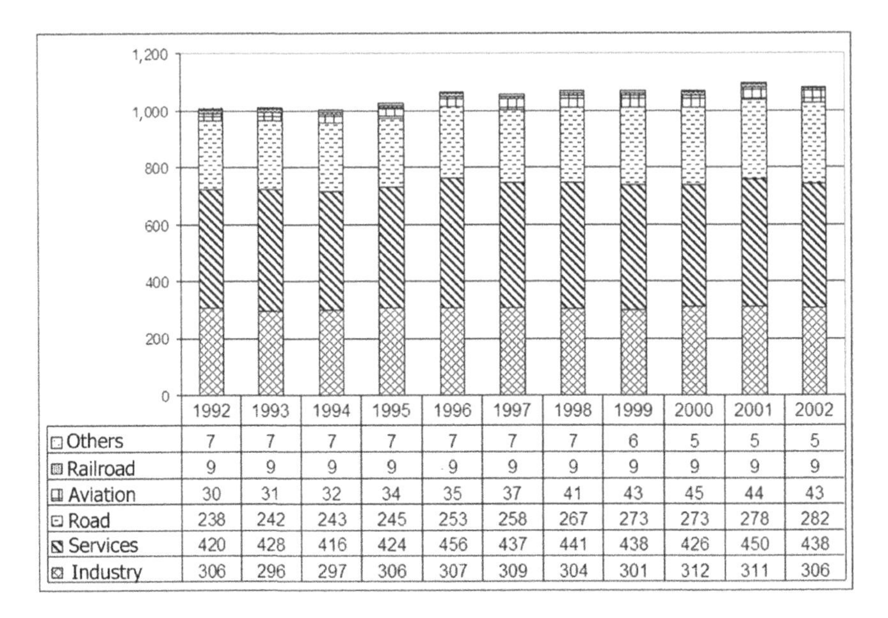

	1992	1993	1994	1995	1996	1997	1998	1999	2000	2001	2002
Others	7	7	7	7	7	7	7	6	5	5	5
Railroad	9	9	9	9	9	9	9	9	9	9	9
Aviation	30	31	32	34	35	37	41	43	45	44	43
Road	238	242	243	245	253	258	267	273	273	278	282
Services	420	428	416	424	456	437	441	438	426	450	438
Industry	306	296	297	306	307	309	304	301	312	311	306

Fig. 8.6. Energy consumption in the European Union (E25) for most important transportation systems and final applications; consumption is measured in million of TEP(Source: Eurostat).

introduction of the Union currency and the prospect of a further enlargement of the European Union portend a continuation of this trend in the future.

8.1.2 Energy Consumption

Energy consumption is usually measured in Tons of Equivalent Petroleum quantity [TEP], corresponding conventionally to 41.87 GJ or 11.63 MWh; these values define the equivalent quantity of heat that is yielded by burning a ton of oil of average quality.

This unit is also used to measure energy from sources other than oil, evaluated in terms of the energy cost for their production.

For instance, railroad transportation uses a combination of electric energy and oil refinery products; electric energy itself is produced partly in thermal power stations using oil products or natural gas and partly in hydroelectric power stations. Other contributions can come from nuclear energy, geothermal energy or wind power.

Each contribution is converted to an oil value, considering production losses and thermal equivalence.

Fig. 8.6 displays a time series of energy consumption in Europe (E25) for the most important means of transportation and other final applications.

The energy consumption of the transportation system is about 32% of the total; this share can be broken into:

- 2.7% for railroad transportation;
- 83.2% for road transportation;
- 12.7% for air transportation and
- 1.7% for remaining means, including inland navigation.

This last figure includes not only river, lake and channel navigation, but also maritime navigation in the European Union. The figure therefore includes sea navigation; this correction is particularly important for Italy because its extensive coastline.

The energy used for sea navigation, the so-called bunkered quantity at the sailing harbour, is partially used for transportation to countries outside the European Union; it is treated conventionally as an oil export. In 2004 this quantity was estimated as 48.4 Mtep, about 14% of total transportation consumption.

The transportation system relies mainly on oil products; railroad transportation uses diesel fuel for 30% of the total energy consumption and a notable part of electric energy comes from oil combustion as well.

Road transportation uses primarily oil refinery products; Italy and Holland are an exception consuming respectively 9% and 7% liquefied petroleum gas for traction (1999); at this time the contribution of coal and natural gas is negligible. This situation may remain nominally unchanged over the near future; total consumption has shown a leveling in recent years.

In Italy, road transportation seems to follow a different trend, as shown in Fig. 8.7, which concerns the consumption of oil products for ground transportation.

The following Fig. 8.8 shows the share between diesel fuel and gasoline.

The relative growth of diesel fuel over gasoline is evident; this trend is partly justified by the different retail price of the two fuels and partly by the more efficient combustion of diesel engines. It is important also to remember that quantities are measured by mass units, but customers pay by volume; for the same volume, diesel fuel contains 12% more energy than gasoline.

Using the available data, it is possible to compare the energy consumption of different means of transportation; If *energy efficiency* E_e is defined as the amount of energy necessary to perform a unit of traffic volume and, as a common indicator, the goods traffic unit [t×km] is used to summarize with a single measurement goods and passengers transportation. Assumed an average mass of 70 kg for each passenger, including the transported baggage, the following expression holds:

$$E_e = \frac{C_e}{V_t}$$

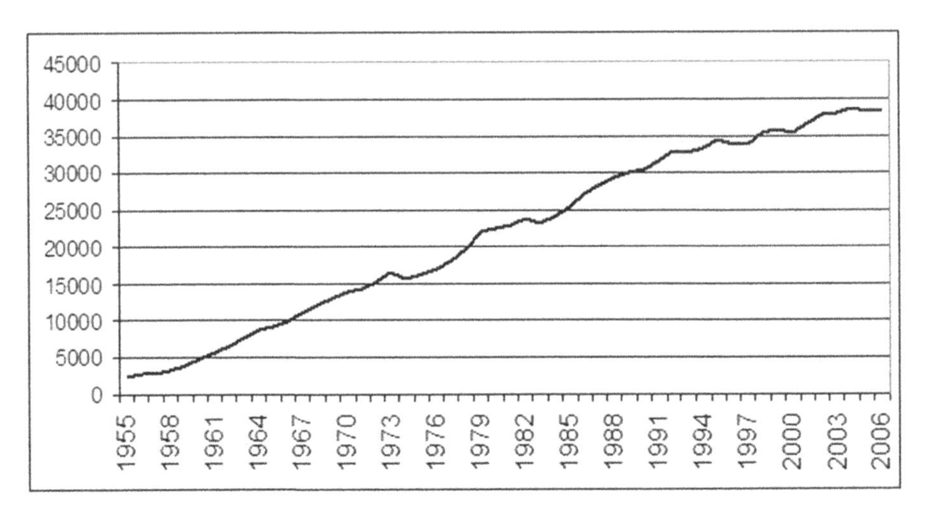

Fig. 8.7. Total oil product consumption for ground transportation in Italy; the quantity in Ktep includes gasoline, diesel fuel and lubricants, the latter accounting for about 1% of the total (Source: ANFIA).

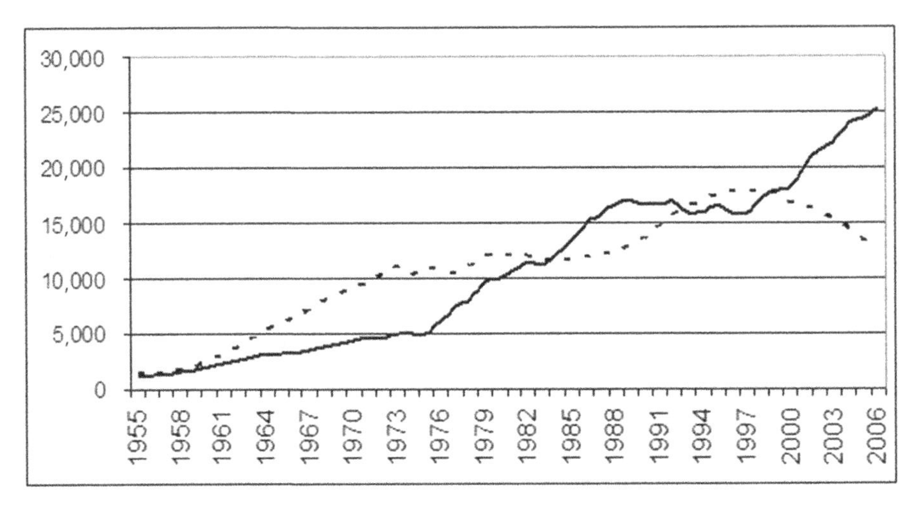

Fig. 8.8. Gasoline and diesel fuel used in Italy by ground transportation, measured in Ktep (Source: ANFIA).

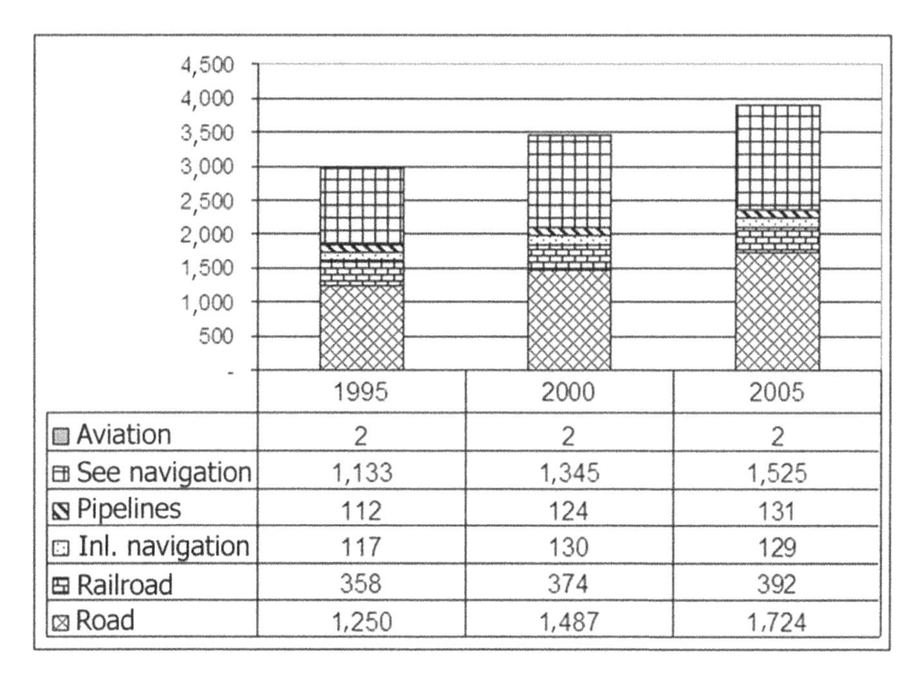

	1995	2000	2005
Aviation	2	2	2
See navigation	1,133	1,345	1,525
Pipelines	112	124	131
Inl. navigation	117	130	129
Railroad	358	374	392
Road	1,250	1,487	1,724

Fig. 8.9. Total volume of traffic in million of t×km for the European Union (E15), for the different transportation modes; the last column shows the percentage variation (Source: Eurostat).

where:

- E_e is the specific energy consumption,
- C_e is the total annual energy consumption of a given transportation mean,
- V_t is the volume of traffic over the same period of time.

This parameter does not take into account mean transportation speed, vehicle stops, traffic environment, including their influence in a bottom line figure.

Accepting this questionable equivalence, we obtain the time series of Fig. 8.9. The last column of the table reports the percentage variation in the period of time.

The second time series in Fig. 8.10 shows the specific energy consumption E_e measured in gep/t×km (grams of equivalent petroleum per unit of traffic); the inverse of this parameter could be considered to be proportional to an efficiency.

The values shown here display an increase over time of about 12% for road transportation and 16% for air transportation, covering a period of about ten years.

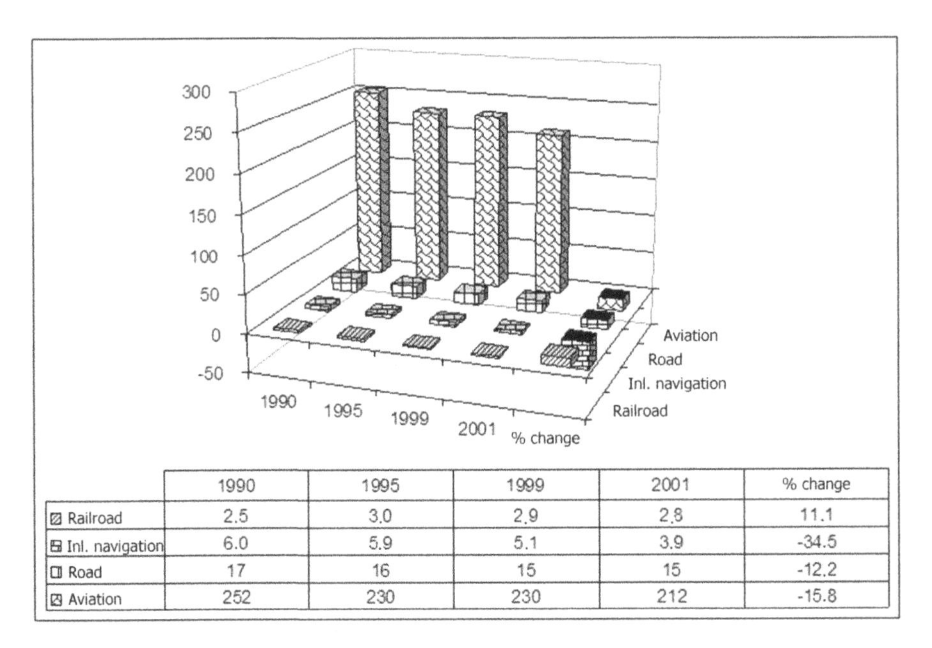

	1990	1995	1999	2001	% change
Railroad	2.5	3.0	2.9	2.8	11.1
Inl. navigation	6.0	5.9	5.1	3.9	-34.5
Road	17	16	15	15	-12.2
Aviation	252	230	230	212	-15.8

Fig. 8.10. Specific energy consumption, measured in gep/t×km for the E15 group; the last column reports the percentage variations.

8.1.3 Operating Fleet

Quantity

Vehicles owned by naturalized or legal residents of Europe totalled about 215 million in 2002; they comprise the so-called vehicle operating fleet.

Fig. 8.11 shows a time series for private vehicles, mainly cars; while Fig. 8.12 shows public service vehicles, including commercial vehicles, light, medium and heavy duty trucks and buses.

The figures from the year 2000 regarding total traffic volume are also available (source: Eurostat):

- the railway fleet included 40,000 engines and approx. 76,000 cars for passenger transportation and 500,000 freight cars;
- the navigation fleet, included about 15,000 vessels;
- the air fleet, included about 4,900 airplanes.

The private car fleet is predominant; about 469 cars for every 1,000 citizens were in circulation in 2000. The fleet growth in thirty years has been approx. 184%, with a yearly growth rate of around 3.5%; this growth has slowed but has not stopped.

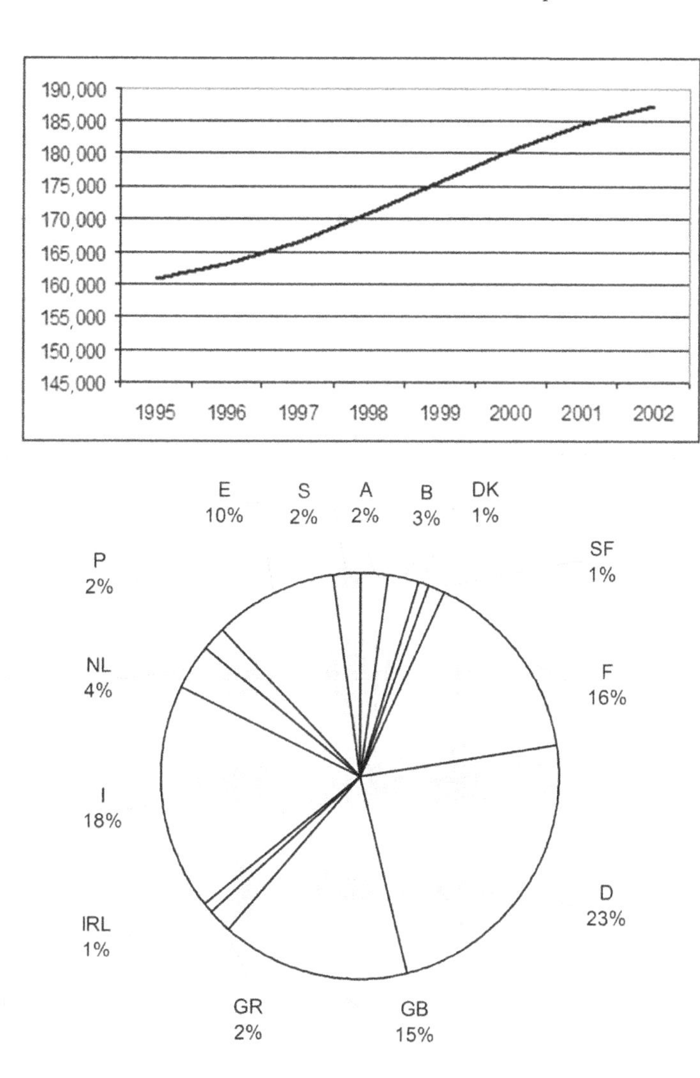

Fig. 8.11. Time series of private cars in the European Union (E15), in thousands; the lower pie chart shows the breakdown of the 2002 figures into the 15 considered countries, identified according to the international licence plate (Source: ACEA).

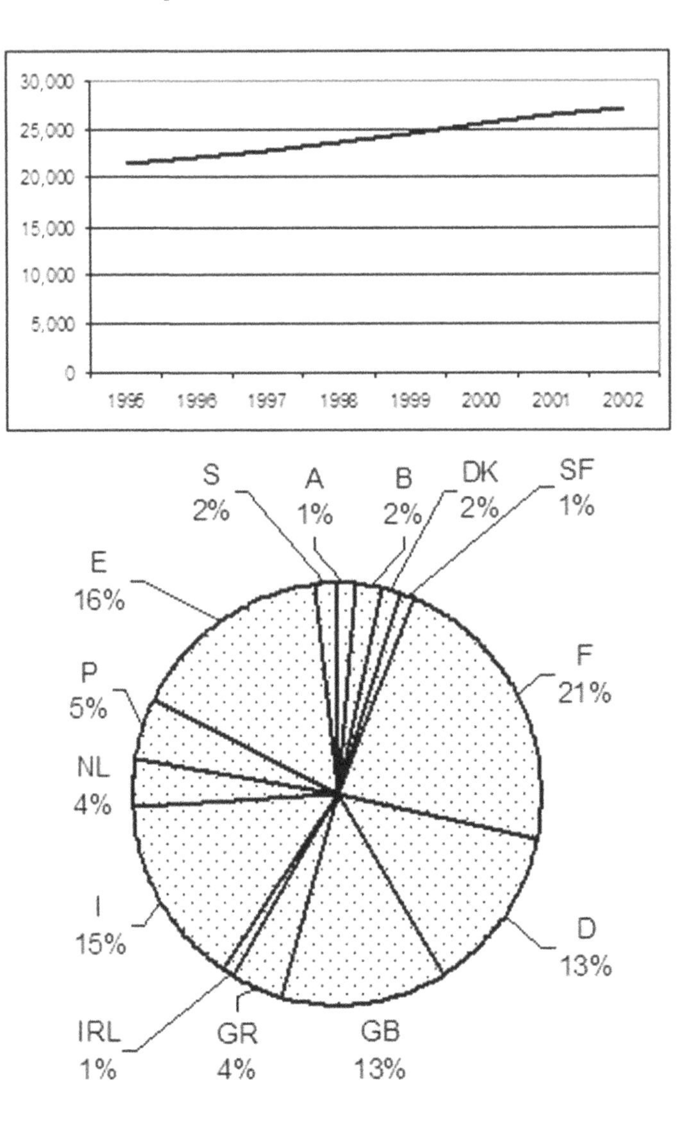

Fig. 8.12. Time series of public service vehicles in the European Union (E15), in thousands; the lower cake diagram shows the breakdown of the 2002 figures into the 15 considered countries, identified according to the international licence plate (Source: ACEA).

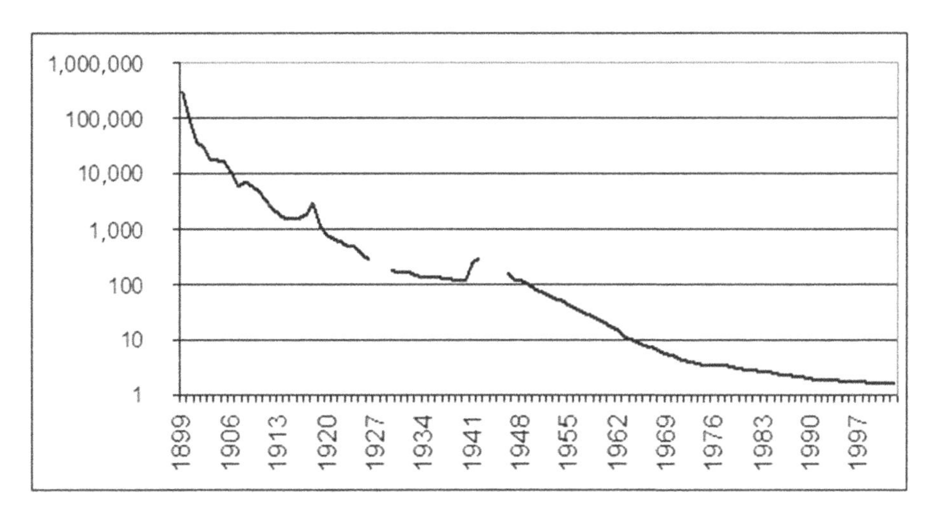

Fig. 8.13. Citizens per car in Italy; this index has been decreasing continuously over time, with the exception of two discontinuities (not shown) at the time of the two world wars, in 1915 - 18 and 1939 - 45 (Source: ANFIA).

In the United States, car density has reached 750 cars per 1,000 citizens and is now steady; statistics show, in fact, that new car sales largely keep pace with written off units.

Although it is not inevitable for the European Union that car density will reach US levels, the EU fleet is still growing and countries with faster growing economies exhibit higher rates of increase, such as Greece with 9.2%, Portugal with 7.3%, and Spain with 6.9%, while countries with a more mature economy show lower rates of growth, such as Denmark with 1.8%, and Sweden with 1.9%.

The highest car density in the year 2000 was reached in Luxemburg with 616 cars/1,000 person (corresponding to 1.62 persons per car), Italy with 563 cars/1,000 person and Germany with 522 cars/1,000 person.

Fig. 8.13 shows a time series of the ratio of citizens per car in Italy; this diagram, if compiled from the beginning of the motoring era, would have shown a figure of 300,000 citizens per car in 1899; from that time on the index decrease was continuous, except during the two world wars in 1914 - 1918 and 1939 - 1945, when the total fleet decreased.

In 2003 this index has decreased to 1.5 citizens/vehicle.

Characteristics

To better understand the composition of the car fleet, consider the histogram of Fig. 8.14, showing the breakdown of cars registered from 1995 to 2004 according to different market segments and body types.

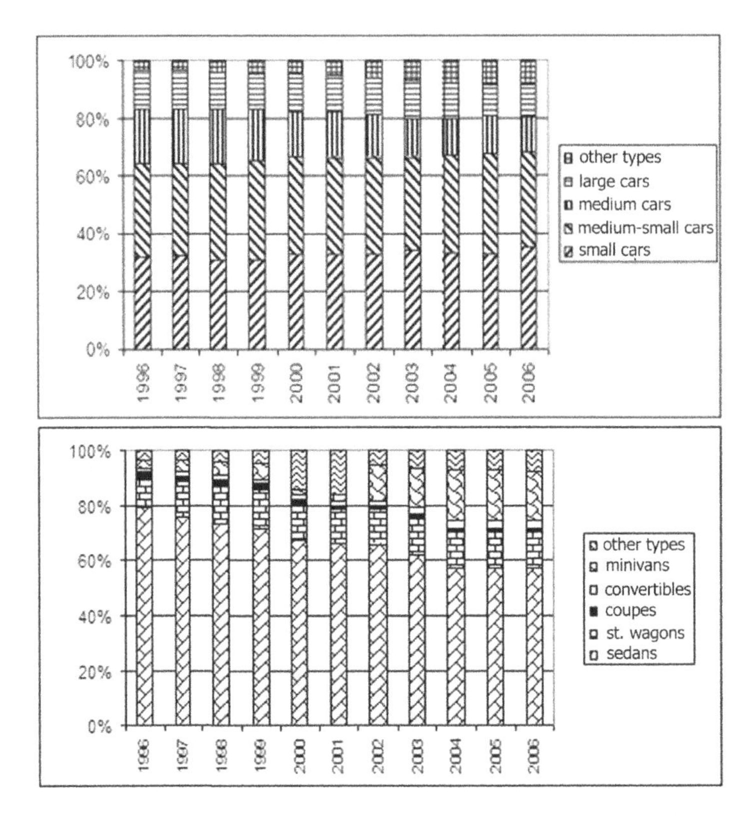

Fig. 8.14. Car types registered from 1995 to 2004, classified according to market segments and body types; large, medium, medium small and small cars are defined, according to their length: larger than 4.5 m, 4 m, 3.5 m or equal or lower than 3.5 m (Source: ACEA).

Table 8.1. Road vehicles of the Italian operating fleet by age (Source: ISTAT).

Age	Cars	Buses	Trucks
0	2,033,296	3,819	222,443
1	2,541,933	6,056	260,116
2	2,518,499	5,381	243,297
3	2,391,709	5,485	197,700
4	2,399,014	4,569	173,021
5	2,381,000	3,936	136,940
6	1,667,344	3,409	142,841
7	1,619,341	2,610	136,149
8	1,533,972	1,898	108,402
9	1,497,088	1,866	102,005
10	1,993,366	2,852	134,100
>10	11,068,915	49,519	1,563,077
N.A.	60,376	316	9,761
Total	33,706,153	91,716	3,429,852

Large, medium, medium small and small cars are considered, defined according to an overall length of more than 4.5 m, more than 4 m, more than 3.5 m and less or equal to 3.5 m.

These segments show no substantial variation; in terms of type, a constant decline of sedans can be noticed, with a simultaneous growth in what were once considered niche segments, in particular minivans.

The percentage of four wheel driven vehicles has increased from 2.6% in 1996, to 7.7% in 2007.

Diesel engines, introduced into mass production after the second world war, have suffered from fiscal and regulatory intervention; at this time their fleet share is about 24%, while their market share is 44% (Source: ANFIA).

Also the average engine displacement of cars registered in Europe has slowly increased from 1.66 l in 1996, to 1.73 l in 2006, as the average rated power, from 65 kW to 83 kW in the same period of time (Source: ACEA). The displacement increase is partly due to diesel share increase.

By analyzing and elaborating the characteristics of fleets, some information on expected life can be gained; this task is particularly difficult because much of the relevant data referring to the past are missing or are not comparable with present information.

As an example it is appropriate to consider Tab. 8.1, where the Italian fleet is classified according to vehicle categories and their registration age. In Italy, the average age of the running fleet is rather high: 32.8% of cars were more than 10 years old in 2002. The percentage for trucks reaches 46.1%.

We observe also that the weight of cars and trucks more than 10 years old has increased as compared with previous years, because in previous years purchases have benefited from incentives favouring newer, less polluting vehicles. Newer cars (less than one year of age) have moved from 7.2% in the year 2000 to 6.0%

in 2002. A different trend appears in trucks, where newer vehicles went from 5.1% in 2000 to 6.5% in 2002.

The ANFIA data, covering more age classes, provide an estimate of an average age of 8.9 years for cars in the year 2002. Analyzing these data, cars with gasoline engines appear to be the older category (9.4 years), but this must be weighed against the relatively recent success of diesel engines.

In the European Union the average age of cars is approximately 8 years; 70% of the fleet is younger than 10 years (source: ACEA).

For industrial vehicles in Italy, the average age is about 10.5 years.

The estimate of vehicle life expectancy is hard to predict; but barring major changes, an expected life of 15 years for cars and 20 years for industrial vehicles would be a reasonable forecast.

The expected evolution of regulations on emissions and passive safety would tend to promote a shorter vehicle life, increasing fleet obsolescence. Macroeconomics should not be forgotten.

The average distance covered by a vehicle in a year can be estimated by dividing traffic volume by the number of operating vehicles and seats offered; in reality, not many vehicles operate at maximum capacity.

For example, in the European Union cars deliver a traffic volume of 3.779 Gpass×km (see Fig. 8.1) with a working fleet of 187,400,000 units (see Fig. 8.11); by crediting each car with five places, we would obtain about 4,000 km/year. However the so called *occupation factor* should be taken into account; it is defined as the ratio between occupied and available places; statistical surveys measured a mere 26.5% for this value, reducing total occupation to only 1.33 passengers per car.

The average distance covered yearly is therefore about 15,000 km/year (Source: ACEA).

A reasonable estimate for the life expectancy of a car, therefore, should be close to 200,000 km.

Following the same process for other vehicle categories, we obtain:

- more than 400,000 km for buses;
- more than 800,000 km for long haul trucks.

8.1.4 Infrastructures

Even when compared with the United States or Japan, the European Union has a very dense transportation network including roads, railways, pipelines and inland navigation routes with respect to the populated surface area.

The recent Union enlargement in 2004, primarily eastward, has given the Union a continental character that requires the development of new means of communication.

Considering as road network only motorways, a total extension of about 300,000 km is obtained for total infrastructure, according to the data shown in Fig. 8.15.

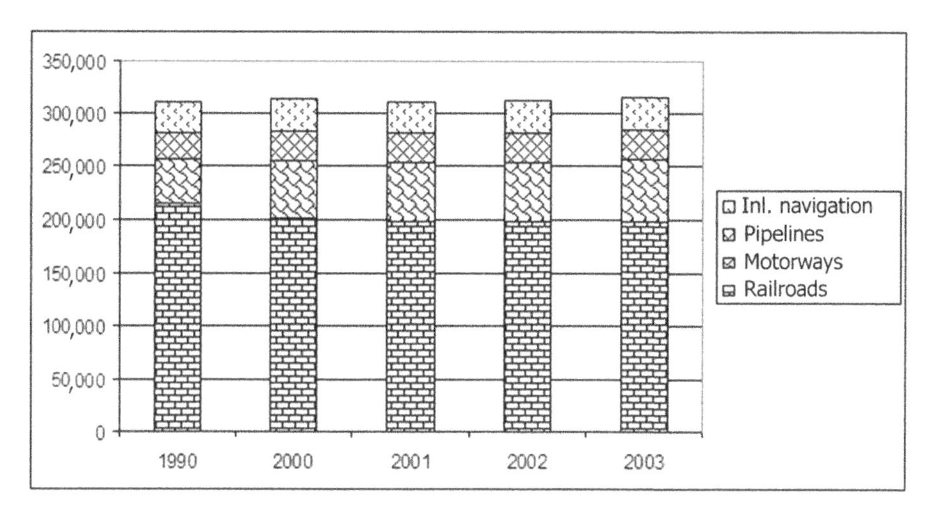

Fig. 8.15. Extension in [km] of the terrestrial transportation network of the European Union (E25), including motorways, railways and pipelines; inland navigation routes refer to E15 only (Source: Eurostat).

It should be noted that data refer to E25, but for inland navigation routes that evaluated for the only E15 group.

If national and regional roads are included, the total in 1990 should be implemented by 3,960,000 km and that of 2003 by 4,820,000.

With respect to the total network growth of about 20%, primarily motorways contribute with 41% while the railway network has reduced by 8% in the same period.

This figure materializes the conclusion of a long lasting reorganization to adapt railways to international and medium distance traffic for personal mobility and long distance transportation of goods. In fact the decreasing trend exhibited a turnaround already in 2001. New growth is still slowed down by many problems which still arised concerning the utilization of vehicles outside of the country of origin.

In this context the construction of railroads to remove heavy road vehicles from crossing the Alps or Pyrenees have to be considered, in the way which is now possible to cross the Channel with high-speed trains.

The European high-speed train network was born in France in 1990 initial covering about 700 km; the Trans Europe Network (TEN) reached 2,800 km in 2003 with lines connecting Belgium, Germany, Spain and Italy.

Inland navigation routes include rivers, lakes and artificial canals which can be navigated by ships of at least 50 t of payload used primarily for goods transportation. These routes offer a great potential for development, particularly in Rhine-Danube and Elbe-Oder basins, requiring significant investment to build harbours and interchanges with road and sea transportation systems, particularly in eastern Europe.

The skies of Europe are among the most congested ones, including a large number of skyways.

It is not particularly appropriate to count skyways in terms of available infrastructure; despite of the existence of sky corridors that comprise a virtual transportation network, many other routes are used with flexibility in accordance with traffic needs.

In this case, airport count is usually taken as an indicator; these figures regard the E25 group in 2004:

- 370 airports with a traffic volume exceeding 15,000 passengers per years, of which
- 255 airports exceeded 150,000 passengers per year and
- 112 large airports exceeded 1,500,000 passengers per year.

8.1.5 Social Impact

As already seen, transportation has a strong bearing on daily life. Every morning transport services in the European Union have to move more than 150 million people to their working place and return them later to their homes in the evening, as well as serving longer routes; in addition, approximately 50 million t of freight are transported every day.

Considering passenger traffic only, each citizen travels approximately 12,700 km per year, using all available means of transportation; as a consequence, transportation is highly relevant to how people live.

In the following paragraphs the following will be considered:

- the accidents attributable to the use of transportation means;
- the emissions of the primary pollutant products;
- the jobs offered by this economic sector;
- the tax revenue generated by the transportation system.

As far as energy consumption is concerned, the previous section deals with this topic. Mainly motor vehicles are considered, being the main area of interest, although some reference data for other means of transport are also reported.

Accidents

Like all activities involving humans, road transportation involves risks and the number of accidents caused by the use of motor vehicles is remarkable in all countries of the world. Their economic and human cost is high and the objective of increasing vehicle safety is generally considered to be a top priority from social and technical perspectives.

Table 8.2. Death risk for different causes, referring to the USA population in 2002.

Cause	Total number of fatalities	Percentage
All causes	2,403,351	100
Heart troubles	936,923	39.0
Cancer	553,091	23.0
Accidents (total)	97,900	4.1
Motor vehicles	43,354	1.8
Generic accidents	17,437	0.73
Falls	13,322	0.55
Poisoning	12.757	0,53
Drowning	3,842	0.16
Burns	3,377	0.14

To understand the extent of such damage it is useful to refer to statistics on the causes of death in the United States; these data are similar to those in any other developed country.

Tab. 8.2 shows the figures relating to 2002. The number of fatalities attributed to road transport is higher than those caused by all remaining modes of transportation and represents 44% of all fatalities from accidental causes.[1]

In the European Union, transportation accidents caused 41,500 fatalities in the year 2000; 98% of these were due to road accidents. For people younger than 45, road accidents are the leading cause of death overall.

Fig. 8.16 summarizes this worrying situation.

Total fatalities are decreasing, despite the increase in traffic volume; this result is attributable to better driving education, improved infrastructures and enhanced vehicle passive safety also due to the increasing severity of relevant regulations.

61% of deaths in road accidents corresponds to the occupants of cars, 29% to users of two-wheelers; these figures refer to the occupants of the vehicles involved in the accident. 18% of fatalities regard pedestrians while 82% vehicle occupants.

The average mortality rate was about 90 deaths for each million residents ($0.9{\cdot}10^{-4}$); Italy and Austria were close to this average; the lowest value was found in the Netherlands ($0.46{\cdot}10^{-4}$), with the highest in Lithuania ($2.23{\cdot}10^{-4}$).

A survey in Sweden, the Netherlands and United Kingdom, the countries with the lowest percentage of fatalities, has revealed that the primary reasons for the lower numbers are:

- between 15 % and 20 %, vehicles passive safety increase;
- between 15 % and 20 %, seat belts usage;
- between 15 % and 20 %, alcohol enforcement laws;
- between 30 % and 40 %, improvements in safety of vulnerable users;

[1] Source: http:\\www.the-eggman.com/writings/death_stats.html

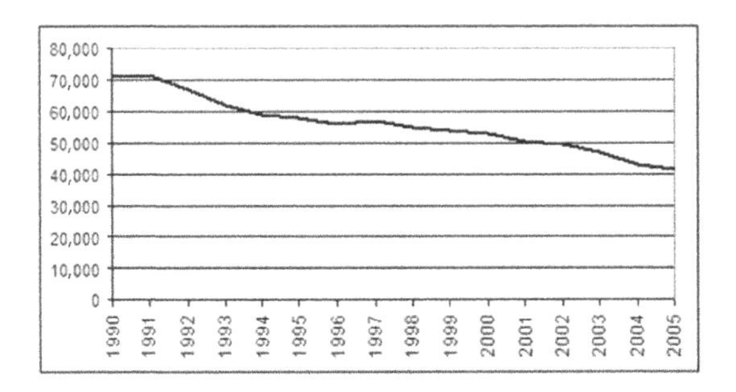

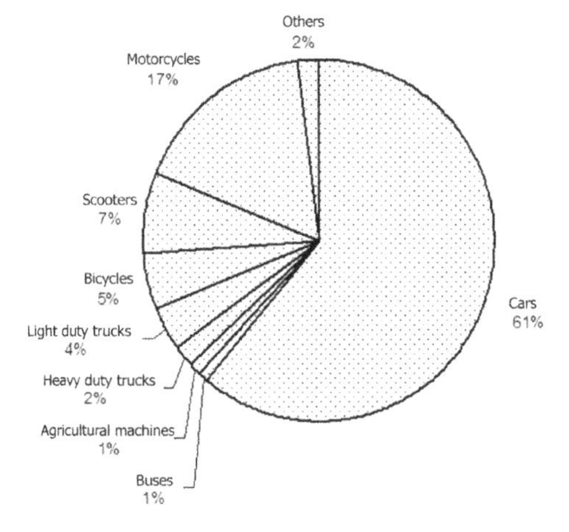

Fig. 8.16. Time series of fatalities caused by road accidents in the European Union (E 25 in 2004); the pie chart (E 15) shows the contributions of the different vehicles (Source: Eurostat).

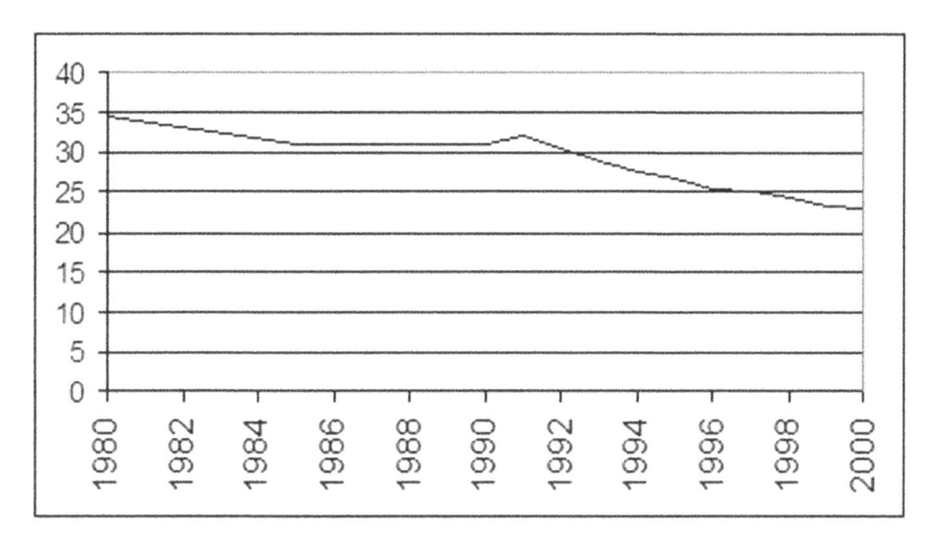

Fig. 8.17. Time series of the ratio between deaths and non fatal injuries due to road accidents in the European Union (Source: Eurostat).

- between 5 % and 10 %, infrastructure improvements;
- between 7 % and 18 %, education and communication.

Referring fatalities to different passenger traffic volumes, the following mortality rates can be determined:

- $10 \cdot 10^{-9}$ deaths/pass×km, for road transportation;
- $3 \cdot 10^{-9}$ deaths/pass×km, for railway transportation;
- $0.27 \cdot 10^{-9}$ deaths/pass×km, for air transportation.

Air transportation poses an additional challenge with regard to such an analysis; for example, one might consider accidents occurring inside the borders of the European Union, or accidents involving European airlines both inside and outside the EU. Just European citizens might be considered or, alternatively, any person involved.

If we consider that most accidents occur near airports, the different counting policies yield different conclusions. Moreover, these accidents, fortunately few, fluctuate over time and are difficult to average meaningfully.

The figure reported refers to all accidents occurring in 1999 within the borders of the European Union.

Returning to road transportation, accident severity has also decreased, as we can conclude by examining Fig. 8.17 which shows the time series of the ratios between deaths and non fatal injuries.

Emissions

The main pollutants emitted by the combustion of oil refinery products, in general, and by road traffic, in particular — all of which have been demonstrated to be harmful to public health — are the following:

- carbon monoxide (CO);
- nitrogen oxides (NOx);
- non-methane organic compounds (NMOC);
- particulate matter (PM).

In recent times other gases have been added to the list; these are not directly harmful, but contribute to creating the so-called greenhouse effect. They are known, therefore, as greenhouse gases (GHG).

Carbon monoxide is a flavorless, colorless and poisonous gas; if exchanged with blood hemoglobin, in the lungs, it reduces the quantity of oxygen delivered to body organs and tissues.

A significant quantity of CO emission is produced during the combustion in the engine of gasoline or diesel and, therefore, by cars; all combustion processes of organic fuels that lack oxygen in their composition contribute to the production of CO. Corresponding there are many contributors including all vehicles (cars, trucks, motorcycles, etc.) with gasoline or diesel engines, incinerators and heating and electricity generating appliances in homes.

Fig. 8.18 shows a CO breakdown by source as estimated for the European Union in the year 2000.

These values are constantly decreasing because of the conversion to natural gas of many wood-heating furnaces, and due to the increasingly stringent regulations on vehicle emissions that reduced the allowed limits, for example, for gasoline engine cars from 4.05 g/km, in 1992, to 1 g/km in 2005; the introduction of catalysts in 1992 had already reduced CO emissions by a factor of ten.

Nitrogen oxides (NO_x) are made by mixing NO and NO_2 and are the result of the combination of atmospheric nitrogen and oxygen due to combustion processes at high temperature and pressure; in general, the more efficient the combustion process, the higher the rate of nitrogen oxide formation. For this reason fuel consumption and CO_2 emission reductions conflict with reduced emissions of NO_x.

A second major source of NO_x is nitrate salts used in agriculture, which produce acids emitting nitrogen in the presence of water.

Nitrogen dioxide (NO_2) irritates the lungs and can reduce resistance to infection, causing increased risk of bronchitis and pneumonia.

Contributions to this pollutant are many, as shown in Fig. 8.19, again based on the year 2000.

It should be remarked that NO_x, together with NMOC are also precursors of complex chemical reaction leading to the formation of ozone (O_3) into the low altitude atmosphere, proven to be noxious to human health.

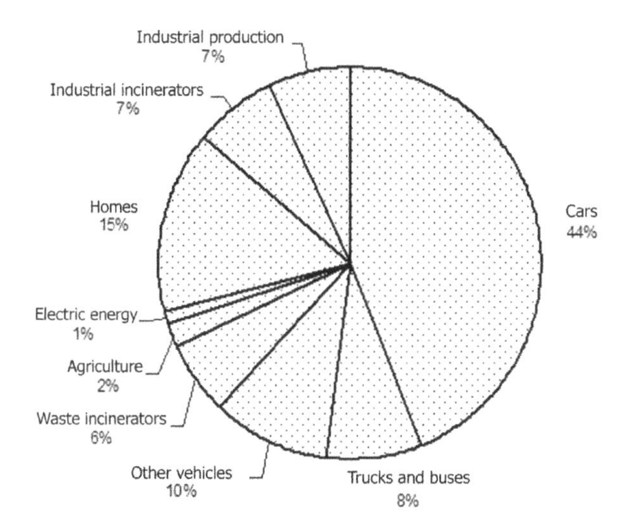

Fig. 8.18. CO breakdown by source for the European Union for the year 2000 (Source: ACEA).

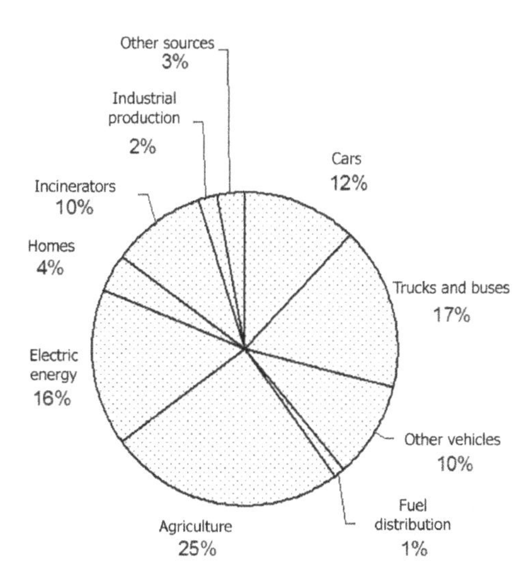

Fig. 8.19. NO_x breakdown by source for the European Union in the year 2000 (Source: ACEA).

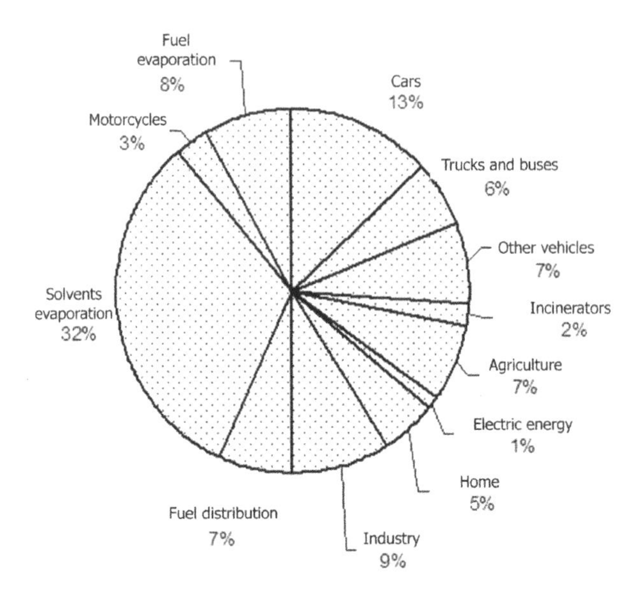

Fig. 8.20. NMHC breakdown by source, in the European Union in the year 2000 (Source: ACEA).

Anthropogenic sources of this pollutant are many; also in this case, a clear decreasing trend exists. The evolution of vehicle regulations has reduced NO_x limits from 0.78 g/km in 1992, to 0.25 g/km in 2005.

Fig. 8.20 shows a similar diagram for NMOC; the evaporation of fuels and solvents is a major contributor.

NMOC also follows a decreasing trend; vehicle regulations have resulted in reduced levels from 0.66 g/km to 0.10 g/km for gasoline engines, and from 0.2 g/km to 0.05 g/km for diesel engines in the period 1992 to 2005.

Particulate matter is a mix of particles of different size that is harmful to health; it is also damaging to exposed materials and can reduce visibility. It is usually classified according to the average diameter of particles involved; the smaller the particle, the greater the risk for human health. The most noxious particulates are those suspended in the atmosphere which precipitate very slowly; extended exposure to these particles affects breathing, can worsen existing pulmonary diseases and increases cancer risk.

PM-10 indicates particles smaller than 10 μm, while PM-2,5 refers to sizes smaller than 2,5 μm.

Apart from combustion products, particles also contain dust, ash, smoke and airborne droplets. If not washed away by rain or artificial means, powders on the ground can again become airborne due to natural wind or passing vehicles.

Fig. 8.21 shows breakdown of the main sources of PM-10.

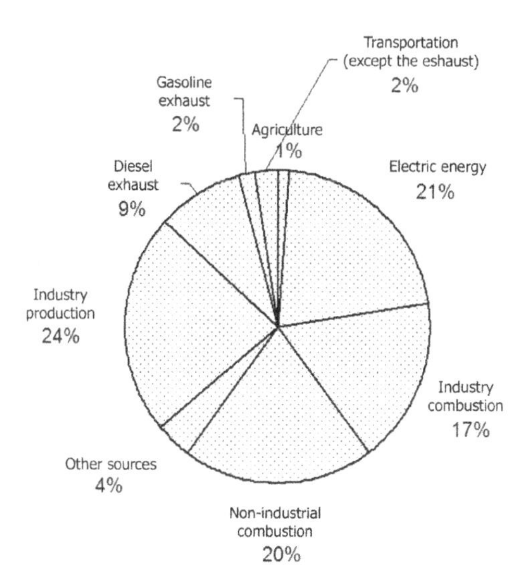

Fig. 8.21. PM-10 particulate breakdown by source in the European Union in the year 2000 (Source: Auto-Oil II).

Greenhouse gases (GHG) include six chemical compounds that were identified in the final document of the Kyoto protocol; these are: carbon dioxide (CO_2), methane (CH_4), nitrogen dioxide (NO_2), chlorofluorocarbons (HFC), perfluorocarbons (PFC) and SF_6.

Each of these gases, if diffused into the atmosphere, limit infrared radiation, contributing to an increase in the average temperature of the atmosphere. They are measured according to their heating potential, which is reported as CO_2 equivalent; their quantity is multiplied by weights p_i, which express the carbon dioxide equivalent in global warming potential.

The weights are the following: $p_{CO_2} = 1$, $p_{CH_4} = 21$, $p_{NO_2} = 310$, $p_{SF_6} =$ 23,900. HFC and PFC include two large families of different gases, each with its own weight.

The pie chart showing the GHG breakdown by source is shown in Fig. 8.22[2].

GHG emissions are strongly correlated to the size of the population: France, Germany, Italy and the United Kingdom account for more than 50% of the total for Europe. These emissions can be reduced by avoiding burning fossil fuels.

Air pollution is a phenomenon preceding the development of the automobile, especially in urban environments and has been declining for many years. However

[2] These data are unfortunately not consistent with the others, because they refer to the European Union as extended to 25 countries.

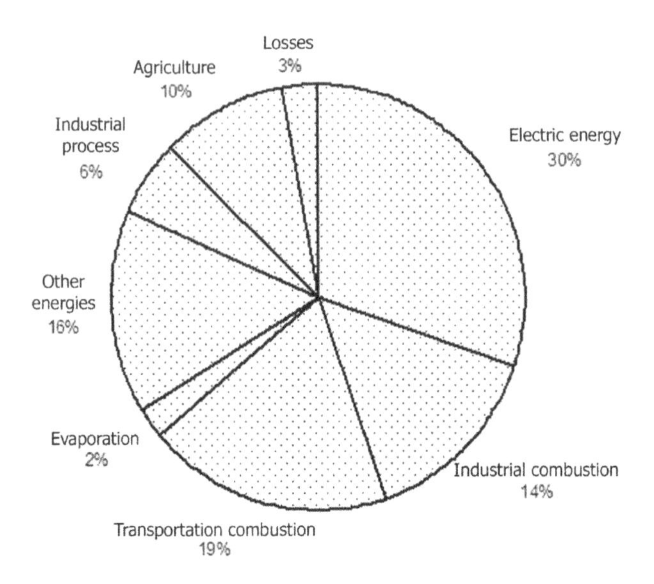

Fig. 8.22. Greenhouse gases broken down by source, measured as CO_2 equivalents (Source: Eurostat).

public perception of the issues has increased, particularly over recent years, due more to politics and the evolution of legislation.

Fig. 8.23 shows an interesting diagram reporting SO_2 and smoke evolution recorded in London during the last four centuries; the period up to 1920 is an estimate based on coal consumption, while figures after that date correspond to measurements.

Similar diagrams are available for all major cities; it is interesting to note that, whereas urban pollution decreased considerably in the second half of the past century and is now lower than at any time in the last 400 years, towns of more recent development show curves that are still increasing.

Economic figures

Limiting the figures to road transportation only, ACEA manufacturers produced about 18,600,000 vehicles in 2006, 18 million of these registered in the European Union and the remainder exported outside the EU.

This production activity accounts for 2,300,000 jobs at vehicle and parts manufacturers; moreover related activities including distribution, service, parts and fuels distribution, car rentals, insurance, waste disposal, driving schools, dedicated press and infrastructure management, create jobs for 12,300,000 people.

The yearly sales of vehicles alone reached 560 billion euro in the same year; investments corresponded to approximately 7% of sales, with about 4% spent on

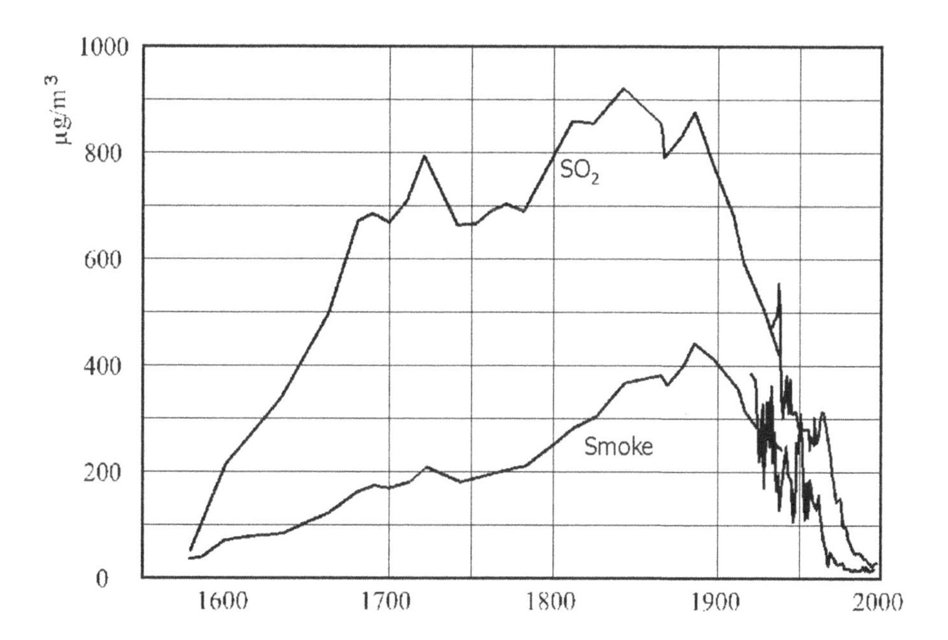

Fig. 8.23. Diagram of SO_2 concentration and smoke, as functions of time, in London.

research and development. The positive contribution to the trade balance was 41.6 billion euro.

Taxation on sales and property transfers, on vehicle ownership, on tolls and petroleum products was 360 billion euro in 2003, about 3.5% of the European Union GNP (8.2% of total tax revenue).

8.2 Vehicle Functions

8.2.1 System Design

The goal of a system approach to vehicle design (*system design)* is to define the technical specifications of each component, in such a way that the vehicle as a whole performs its functions according to assigned procedures and objectives.

Technical specifications means a set of physical measurements that define each part completely, without the use of detailed drawings; sometimes technical specifications are a complement to a conventional detailed drawing, but can be used as a substitute.

Detailed drawings are needed solely to explain how a part must be manufactured, while technical specifications explain the function of a part, how its performance can be measured together with the acceptable values for this performance. Clearly, coherence between detailed drawings and technical specifications is necessary, but only the latter guarantee that the performance targets can be reached.

Technical specifications can, therefore, be the starting point for designing an assigned part while providing a concise way to describe a part by means of how its functions must be performed.

Considering a body door, for example; its functions can be divided into three categories:

- aesthetics;
- ergonomics;
- passive safety.

Taking into account the third category only, just the function of protecting occupants from injuries after lateral crashes could be considered.

If, of the possible lateral crashes scenarios, collision against a pole of assigned size is considered, a technical specification for the anti-intrusion bar could take into account the plastic deformation work during a crash of this kind, requiring the minimum to be reached in a given deformation space.

Although this specification does not consider how the anti-intrusion bar will be designed or its material, it does represent a necessary logical step before starting to design this component.

The system approach to design enables projects, even the most elaborate, to be undertaken by assigning activities to teams working in parallel, each with comprehensible objectives that can be checked autonomously and finalized with respect to the overall performance of the vehicle. The system approach also allows a project to be developed, using standard components produced by suppliers; these components may require development specifically for the purpose, or selected from a catalogue.

Finally, system design is the initial phase of each project, when the feasibility of reaching the assigned targets is verified; this phase is usually called *feasibility study.* The technical specifications of the main components are part of the concept documentation.

At this point it is appropriate to consider how to assign and measure overall vehicle performance and functions.

Performance and functions cannot be defined absolutely, being conditioned by the customer's expectations of the product which depend on both objective and subjective parameters. Furthermore these expectations are conditioned by product alternatives which exist on the market when the product will be sold, and these alternatives are rarely known in detail when the project is initially launched.

To highlight the complexity of this problem, it is again appropriate to consider an example, in this case the air conditioning system. Some functions of primary importance are easily identified. It can be assumed for instance that the objective of the climate system is to allow the passenger compartment to reach the assigned comfort temperature in a given time, independently of the current outside temperature.

In reality this time (cool-down or warm-up time) is only an approximate parameter; ideal comfort temperatures should not be reached too rapidly, since passenger discomfort may result due to too strong a flow of air on the skin or high temperature gradients in the air.

Heat flux on passengers appears still to be a matter of empirical judgement, further complicating understanding of the phenomenon.

In a luxury car, it might be mandatory to obtain comfort conditions automatically, requiring the passenger to adjust only the temperature control. A few years ago the design of a multi-zone climate system for medium or small cars would have been largely empirical and may have seemed largely a waste of time and money. Nevertheless this feature is beginning to appear widely and may become standard in the near future.

Looking into the future, on the next generation medium cars, automatic abatement of external odor may become indispensable for example; if this were the case, the abatement efficiency of various odorous reference substances would need to be defined and measured.

In this context, it should be remembered that too ambitious a specification could increase the product cost in ways that could prove to be unrecoverable from the final selling price.

Likewise, there are other functions that, although of secondary importance, cannot be neglected and which may occasionally affect the performance of other systems. As such, design goals could include:

- minimize additional fuel consumption due to the climate system;
- minimize additional power required from the engine, especially during sudden acceleration;
- minimize the noise of air flux into the passenger compartment;
- control humidity to avoid fogging of the windshield or side windows, etc.

Thus even in an apparently simple case, it can be difficult to identify functions and specifications and this identification should be done only after a careful study of customer expectations, which in part can be defined *a priori* but in part depend on the competitor's products.

It is also difficult to define the boundaries of the climate system, or to identify which components influence the climate system by their behavior. Indeed, following a traditional approach, the climate system would be limited to:

- the heat exchange group, including the heater and the evaporator, the air channels and nozzles that direct the conditioned air to the passenger compartment, a part of the body system;
- the compressor, a part of the engine system;

- the electronic control box, a part of the electric system.

However, to improve our control over climate system functions, this list should be expanded to include, for example, windows and the windshield, which are responsible for an important part of the radiated heat; door panels, responsible for part of the transmitted heat; seats, as they influence human body heat exchange; and gearbox transmission ratios for their influence on engine warm-up.

Hence it can be seen that the boundaries of the technical specifications which influence certain functions are wide and exceed the boundaries of the components dedicated to this function.

We suggest that a correct approach to system design should include the following steps at least:

1. define the functions performed by the system.

2. define the parameters that best measure those functions and the target values they should reach to obtain customer satisfaction.

3. define which components are part of the system and influence the achievement of the target values.

4. identify any other system functions which may compete with those of point 1.

5. establish a set of technical specifications for each component which are coherent with the system function target values.

Therefore system engineering implies the study of components that are normally classified under different automotive engineering disciplines; in the example of the climate system, components are included that are part of the body, the powertrain and the electric system; these components are also located in different sub-assemblies of the car.

The body system design approach described in this book is similar including, for example, acoustic comfort; for a correct system approach, it is necessary to consider not only how noise is transmitted to the passenger compartment (part of body structure design) and how it may be reduced by using insulation and anti-vibration materials (part of interiors design), but also how sound is generated by the engine and transmission (part of powertrain engineering) and by wheels and suspensions (part of chassis engineering).

Since engineering subjects are, by their nature, interdisciplinary, system engineering must likewise go beyond the boundaries of its individual subjects. A traditional, topological approach, on the other hand, classifies and studies vehicles according to three main subsystems:

- The chassis, the group of components dedicated to vehicle path control, such as transmission, suspension, brakes, wheels and steering mechanism, with their dedicated supporting structures.

- The powertrain, the group of components dedicated to traction power generation, such as engine, fuel supply, intake and exhaust plants.

- The body, the structure supporting all other components including the passenger and payload compartment.

This book, dedicated to body system design, will consider all functions that are primary to this system; however, while studying a function such as comfort, some aspects of the engine and the chassis will be considered and, in terms of automatic system controls, the issues interfacing with the electric and electronic system will be addressed.

System design is necessarily general since only the baseline specifications of the components included in the system are considered; design details of these components are left to specialists respectively. This approach is useful to engineers and anyone involved in the new vehicle development process.

8.2.2 Functions Perceived by Customers

Let us consider all functions performed by the vehicle, with particular reference to automobiles.

Vehicle functions can be defined as the categories by which the customer rates vehicle performance.

A complete list of functions, probably to be expanded in the future, can include the following:

- Appearance.

- Available space.

- Ergonomics.

- Climate comfort.

- Dynamic comfort.

- Dynamic performance.

- Handling.

- Safety.

- Resistance to age.

Each function can be explained through a certain set of *requirements*, which are qualitative and quantitative attributes that the vehicle must possess in order to perform each function correctly; these requirements are described shortly.

The first listed function (not ranked by priority) is appearance, the ability to appeal to the customer; even though designing appealing bodies is beyond the scope of this book, salient requirements involve the body, in terms of shape, volume, materials and details.

These requirements do not seem to have a major impact on traditional tasks assigned to engineering during product development; instead, they must be taken into the account appropriately when assigning priorities or taking decisions.

The creative activities of body engineers, production engineers and style architects must be integrated correctly in such a way that no technical requirement could sacrifice the aesthetic appearance of the car. It has been demonstrated that appearance is one of the most relevant features determining the commercial success of a car.

Roominess, or, from the designer's point of view, the use of space, is important because it embodies the primary objective of carrying people and goods.

Customers do not expect unlimited room, but associate interior space with the class of the car; what is important is how much the space available, albeit limited, can be rationally used. Since the room used by car components is unavailable for this use, component lay-out should be designed to limit as much as possible intrusion into the passenger compartment. This explains how so much space was dedicated to structure bulk and component lay-out in the first volume.

Other important requirements also hinge on body design, such as roominess and the availability of space to organize small objects; another important requirement is adaptability (tilting or removable seats), in order to enable the customer to change the car interior to suit different transportation needs.

Car ergonomics can be defined as the ability to minimize the physical activity required by a given operation while using the car. Within this function, we usually include the pleasure of driving the car, including the many sensations the customer feels while driving.

The requirements of this function again involve the car body and include:

- the ease of entering and exiting the car for driver and passengers, of opening and closing doors, the glove compartment, hood, trunk, etc.;
- the ease of identifying and operating the most important controls with minimal reach;
- the comfort of the driver's posture;
- the ease of loading and unloading the transported goods.

Requirements regarding controls operation, such as the steering wheel, gearbox, clutch and brakes are, instead, normally considered when designing the chassis; they include the effort needed to operate a control, the geometry of motion of levers and their position inside the body, with reference to human anatomy.

Secondary controls, like for example door handles, studied in body engineering, do not imply major effort but require a particular feed-back to the operator confirming the accomplishment of the desired manoeuvre.

In the case of climate control, the related requirements affect body design and, partially, the engine.

The dynamic comfort function is evaluated by the ability to suppress all noise and vibration disturbances from outside (road pavement and other vehicles) and from inside (engine operation and component vibration).

The related requirements involve almost all vehicle components which participate as sources and potential transmitters of such disturbances.

Noise and vibrations also transmit information useful for both driver and passengers. A totally silent vehicle, without vibrations, could even prove to be dangerous, as has been demonstrated by experience with active noise suppression. In addition, noise may be considered to be a characteristic feature of specific types of car, as, for example, sports cars. So the target is not total suppression but an acoustic environment coherent with customer expectations.

Filtering of components vibrations is a task usually assigned to the body system, while filtering noise and vibrations from the road is usually assigned to the tires and suspensions.

Filtering powertrain noise (engine operation) involves powertrain suspension and the intake and exhaust system.

Unbalance specifications are assigned to any potential source of vibration.

The dynamic performance function includes requirements that are easy to measure, such as top speed, gradeability, acceleration and pick-up. Requirements that are more difficult to evaluate meaningfully include drivability and fuel economy.

Requirements involve all vehicle components from the engine to the body; the body contributes with its aerodynamic performance and weight.

A correct allocation of the weight target requirement for each part of the vehicle, even if weight is not directly appreciated by customers, is one of the most important tasks of preliminary design.

Handling functions are usually defined as the ability of the vehicle to follow driver inputs on the controls, when modifying car speed or trajectory; these controls include separate or combined operations on the steering wheel, brakes and accelerator.

Handling function requirements involve not only the suspensions, tires, steering mechanism and brakes, but also engine and transmission. Overall properties of inertia (mass and momentum) have vital importance for this function.

The safety function is usually classified in three ways:

- preventive safety, such as the ability of the vehicle to keep the driver constantly updated on corrective maneuvers to be undertaken; a typical example of this category includes outside visibility, visibility of the main instruments (i.e. speedometer, outside thermometer, etc.) and car trim variations;
- active safety, such as the ability of the vehicle to react to driver inputs with a response that should be immediate, stable and proportional to the action, while avoiding obstacles and dangerous situations;

- passive safety, such as the ability to mitigate, when a collision is unavoidable, the severity of injuries to car occupants, pedestrians or passengers of other cars involved in the collision.

By definition safety cannot be total, but requirements should be established for the most statistically relevant situations; homologation requirements are an important part of this approach, together with the technical policies of the manufacturer.

In the passive safety category, repair cost limitations following low speed collisions have recently been added.

Safety involves all main vehicle components; the body is particularly involved in preventive (inside and outside visibility, lights) and passive safety (structures, passive and active restraint systems, component lay-out, surface materials and finishing).

The chassis must comply with all active safety requirements for suspension, brakes and tires, and with passive safety requirements, such as intrusion into the passenger compartment following a collision.

The engine system is involved in passive safety as far as fuel spills following a crash and consequent fire hazards are concerned.

The aging resistance function is the ability of vehicle system and components to maintain their functions unchanged or restrict their degradation with age within acceptable limits; reliability is a requirement of this function.

Aging resistance involves all vehicle components and, obviously, all body components.

8.2.3 Technical Specifications

Each vehicle function can be described through a coherent set of measurable requirements; compliance with these guarantees customer satisfaction with the vehicle.

These requirements determine the technical specifications of all components.

Each part or subassembly could be fully defined by an engineering drawing, containing all relevant geometric dimensions and materials.

In reality, the detailed knowledge of complex components is irrelevant to the car manufacturer, more interested in performance than details. A very detailed drawing cannot always guarantee complete fulfillment of the desired performance and the technical competencies may be insufficient to understand complex details.

Technical specifications solve this problem by providing global and concise information only; it is therefore necessary to establish, for a certain component, what is relevant for system function. A technical specification should list:

- the physical properties that describe the requested requirements for the component;
- the conditions under which those properties must be measured;

- the values (with allowed tolerance) of the properties in order to obtain the desired system performance.

These technical specifications, together with simple outline drawings, represent the only technical documentation needed by car manufacturers for managing their relationship with component suppliers; in this case a single piece of paper could include all the necessary information.

The component manufacturer's point of view is necessarily different, since technical documents to produce the needed part consistently must be created. Indeed, some second tier suppliers often produce other parts that will be integrated into the final subassembly. The first tier supplier must therefore use the technical specifications to best advantage.

The example of weather strips may be useful to clarify this point.

This kind of gasket is used in doors to make the passenger compartment tight with respect to water, air and dust coming from the outside; their elasticity must compensate for dimensions variations due to body deformation and production tolerance.

Adequate contact forces must be guaranteed within an ample field of gap dimensions; technical specifications will primarily address this issue. A minimum value for contact pressure will be considered, for the maximum displacements that are allowed during the life of the body; in addition, an allowable tolerance must be determined.

While determining these values, it should be remembered that this pressure affects also the load to close the door; minimum pressure allowed will be a trade-off between passenger compartment tightness and the ergonomic issues involved in closing a door.

Other requirements are set to useful life and endurance, for instance, how much contact pressure can decay due to the effect of aging.

The only information necessary to define its geometry, on the side of the car manufacturer, is the space allowed in the gap between door and body side.

The common characteristics of the information reported by the technical specification are that:

- they are correlated to functions to be obtained on the vehicle (tightness, closure load, etc.).
- their measurement can be made by the supplier to check its design and fabrication process, without requiring a direct knowledge of the vehicle on which they will be installed.

Other information such as, for example, the chemical composition of the elastomer applied to the weather strip, are not relevant to system operation and form part of the proprietary know-how of the supplier.

In conclusion, technical specifications define the attended performance, not the details that allow this performance to be obtained. On the other hand, specifications should not be too superficial; for example, it is necessary to avoid providing a supplier specifications on road durability without referring to the driving conditions and the typical trip in question; a good specification should enable the supplier to evaluate the results of his effort independently.

Continuing with the weather strip example, it is clear that the technical documents available to the supplier will be much more detailed than those used by the car manufacturer such as the technical specifications; the supplier will have available a complete set of drawings of the part, including detailed dimensions, metallic core description, materials, fixtures and production set up, etc. The design tools of the supplier will be able to correlate these parameters with the pressure performance on the vehicle system, which is the most relevant issue of the technical specifications.

Some details, such metallic core inserted in the weather strip for clamping to the car body — the performance of which is not only dependent on reference dimensions but also on the manufacturing process in the steel mill — should be described, for a second tier supplier, by specifications including only dimensions and clamping force desired.

Technical specifications represent a universal simplified language, allowing different industrial organizations such as final manufacturer and supplier to cooperate in reaching the same objective, the satisfaction of the final customer.

The same logic can be usefully applied within a company, particularly a vehicle manufacturer, to integrate the activities of different departments.

Although there is no conceptual obstacle to developing each component from scratch, it is always useful, before taking this decision, to clarify which function the component performs on the vehicle, how it can be quantified, and from which values the objective of satisfying the customer is obtained.

In this way it is possible to manage, with relative simplicity, complex activities involving a number of people, breaking down each objective into sub objectives that are measurable and understandable by the different parties involved.

8.2.4 Body System Design

As already seen, the vehicle functions pertaining primarily to the car body are:

- Ergonomics and packaging;
- Thermal comfort;
- Acoustic and vibration comfort;
- Structural integrity and
- Passive safety.

Correspondingly the objective of the methods explained in this book is to design body components that satisfy the above functions at the vehicle system level.

A specific chapter is dedicated to each of these functions, starting from the physiological requirements of the human body and developing through related design methods.

Explained is how to verify which function an assigned vehicle is able to perform, identifying also which components condition those functions, but not how these components must be specified in order to perform the functions at the desired level; this problem can only be addressed *a posteriori*, although an *a priori* approach would be advantageous.

This qualification could apply to all design courses; if designing means to define a product that does not yet exist, what is actually taught is how to verify whether an already defined product is able to perform an assigned function. The designer's job is, therefore, to assume an hypothesis and to verify the results that can be achieved; a deviation from the objective will guide towards defining a different hypothesis that will again be verified. The designer will be more efficient, if the first approximation hypothesis is close to correct; in any case, design will remain a trial and error process.

A technical specification definition is further complicated by the fact that the final judgments on the product will be issued by the customer and not by the designer, and customer judgments are sometimes difficult to express concretely, being influenced by immeasurable parameters and alternative offers on the market that may be unknown at the beginning of the development process.

Technical specifications are developed and determined through different strategies, according to a process that can be divided into two parts, called *target setting* and *target deployment*. The target setting phase consists in setting objectives for each of the functions perceived by the customer; this task is more successful if subjective judgments are avoided and only objective measurements are used.

If this requirement appears relatively straightforward with respect to functions such as top speed, acceleration, and gradeability, it will be difficult for functions such as ergonomics where subjective feelings come into play. In the following paragraphs, it will be shown how subjective feelings can be transformed into objective measurements and how just a few figures can summarize functions including multiple requirements.

In the next phase of target deployment, as a first step, vehicle subsystems according to function are identified and their specifications tentatively set; the specifications adequacy to the targets will be verified, correcting any errors in the specification.

These verifications may be performed using mathematical models of the vehicle and in some cases also by building and testing simplified prototypes (*mule cars*) that allow complex subsystems to be verified.

8.3 Requirements Measurement

So far it has been seen that required is the translation into physical, measurable parameters of how the expectations of the customer are satisfied for a given product: the car body, in our case.

The problem to be solved, at this point, is to determine how these parameters can be measured.

Processes and methods involved in solving the problem are mostly statistical analyses concerning customer's behavior regarding the car.

Although these analyses usually form part of the science of marketing, it is appropriate to introduce engineers to these aspects considering the importance of a correct requirements definition while aiming to enhance the future interaction between experts of different disciplines.

To avoid explanations concerning general purpose methods outside of the scope of this book, the presentation is limited to a few cases involving different but typical and recurring problems regarding:

- How to translate subjective judgments of a customers sample into engineering quantitative requirements.
- How to rate competing products through their capacity to exceed regulation specifications.
- How to derive components requirements from experimental tests made on the entire system.

In this presentation some concept are presented that will be better explained in the following chapters; thus the reader is encouraged to limit, for the time being, their interest to understanding these different approaches.

8.3.1 Translation of Subjective Judgments into Measurable Parameters

Here two different cases are considered; the first regards the secondary controls of the vehicle (on board information system), as far as their ergonomics and visibility is concerned; the second concerns postural comfort of the seat. Certainly these cases are far from being exhaustive but do offer two approaches that can be applied also to different circumstances.

Secondary controls ergonomics

This section will focus on the task of controlling a vehicle and concern cognitive workload measurement, the parameter selected to evaluate the overall ergonomic value of the driver's place, with particular reference to secondary controls and to on board information systems.

The approach taken follows the research of Toffetti A., Nodari E., Zoldan C., Rambaldini A.[1].

Table 8.3. Hierarchical model of vehicle control proposed by Michon. Driving can be described as an activity including tasks of different complexity which involve different knowledge and ability.

LEVEL	REACTION TIME [s]	TASK	BEHAVIOR BASE
STRATEGIC	100	Communication Navigation Diagnosis Supervision Identification	KNOWLEDGE
TACTICAL	10	Classification Ricognition Execution	RULE
OPERATIONAL	0,1	Detection Reaction Stabilization	ABILITY

The driver interacts with the vehicle, using primary controls, as steering wheel, gear stick and pedals and secondary controls as appliances switches, navigation systems switches, etc. In addition, he or she acquires information about vehicle status by checking the on board instruments and interpreting what is conveyed from the outside through eyesight, hearing and sense of touch. In this way, the driver and the vehicle can be considered to be a dynamic system in a changing environment, in relation with traffic, weather and lighting conditions.

One of the references applied to describe the task of driving is the hierarchical model proposed by Michon (1985) that is outlined in Tab. 8.3. Driving a vehicle can be described as an activity where different tasks are performed by the driver involving different levels of difficulty, knowledge and ability.

The human mind works according with two different modes, that can be generally classified as falling between two extremes: the automatic and the conscious mode.

The conscious mode involves taking decisions and performing critical analyses about the results of actions performed and modifying them according to a predefined reference model. Instead the automatic mode consists in performing quick actions, almost unconsciously, leaving the mind available to perform other activities.

In general, driving implies both modes; three different behavior models are identified in Tab. 8.3.

- *Ability based behavior.* Ability based actions refer to tasks that are performed automatically, because they are the result of rules learned in the past and firmly rooted. Important feature of this behavior is a very short reaction time to external stimuli: for example, in front of a red stop light, the brake pedal is immediately depressed to stop the vehicle; for this reason operations relevant to safety should involve behaviors based upon ability exclusively. This category includes detection of a critical situation, the consequent reaction, stabilizing the response of the vehicle, and avoiding an obstacle.

- *Rule based behavior.* Rule based actions refer to tasks that are governed by rules deriving from a preceding experience or specific training. This behavior comes into play when the previous behavior is not applicable. The person in charge of the action examines and interprets the current situation and chooses the rule that better matches the problem (for instance, if there are many car in line on the same lane, a rule can be applied that provides that the car is driven along a less crowded lane. This category includes the detection of a specific situation or the execution of a complex maneuver.
- *Knowledge based behavior.* Knowledge based actions involve reasoning, inference, judgment and evaluation. If previous behaviors cannot solve the problem, processing the knowledge deriving from previous experience is necessary. This happens when a new situation is encountered and when available rules are not sufficient. This category includes the use of a navigation system, the detection of non normal situations, and the use of communication systems.

From the first to the last behavior type, we pass from a purely physical engagement to situations that involve an increasing involvement of brain activity.

Due to product evolution, current vehicles require an ever-decreasing load of physical activities with an increasing load of brain activities with respect to the vehicles of the past. In addition, the information arriving from the outside environment has increased significantly, as has the need to use this information appropriately in order to respond to increased expectations and requirements in terms of vehicle performance, with a wide range of communication tools, such as radios, portable phones, electronic diaries, navigation systems having been introduced also on cars in recent years.

Often the mental resources of the driver have to be shared between concurrent tasks that can interfere each other. A rational design of the driver's space must address the target of avoiding an unacceptable usage of these mental resources.

Cognitive workload measurement can be assumed to be an evaluation method of ergonomics of controls and, more in general, of driving environment, and corresponds to the effort required to an individual to perform a task with an acceptable level of performance.

In the case of the driver, an appropriate level of acceptance corresponds to being able to control the vehicle safely, in an outside environment which is randomly changing.

Since mental resources are limited, it is fundamentally important to study how they are deployed and how adequate human-machine interfaces can enhance their usage.

On one hand, when free resources are available, they can be used for tasks other than ordinary driving, such as trip planning, on board facilities usage, interaction with other passengers or emergency maneuvers. On the other hand, the threshold of cognitive overload is said to be reached when the driver is no longer able to accomplish to all tasks because there are no more resources available. This threshold depends on the characteristics of the driver and his/her psychological and emotional status.

Sometimes, when the driver is operating in a too restful environment (as in a motorway trip, with little traffic and low speed limits) the induced mental workload is so low that an opposite threshold of cognitive underload is reached, with the risk of loss of concentration and drowsiness.

The cognitive load is increased when the driver is traveling in critical conditions such as in fog or high density traffic.

When secondary tasks are performed, as for example a phone call, in addition to the cognitive workload induced by the primary task of driving, another load is demanded.

If these secondary tasks are too complex, the driver may be brought close to cognitive overload.

Many methods have been developed to evaluate cognitive load which can be classified according to three categories:

- *Performance based measurements*; the rate of success in the accomplishment of primary tasks (e.g. driving a car on a path traced with rubber cones) and secondary tasks (e.g. using the on board information system) is evaluated. In particular the rate of success of the primary task can be evaluated objectively from driving parameters (e.g. the number of rubber cones knocked over or the lateral acceleration) while the secondary task can be used to manipulate the effort requested to the driver.

- *Subjective rating*; standard evaluation check lists can be developed to estimate the cognitive workload in an interview of the driver after the test.

- *Physiologic parameters measurements*; the utilization of physiologic parameters is sometimes preferred. This is because non intrusive instruments are now available that can be applied without altering the operation environment, such as heart beat, eye movement, an enlargement of the pupil which can be well correlated to cognitive workload and can be easily measured using a videocamera.

The application of this method is particularly appropriate while developing secondary controls that interfere minimally with the primary driving task.

During the preliminary design phase (when the control has not been identified yet) it is helpful to outline the mental model that the driver can have of this control; this is the functional representation of this control on the part of the driver, including expected functions, lay-out, how and which information should be displayed. This initial phase is important; bearing in mind the driver's expectations while designing will contribute to reduce cognitive work load, because the new control will avoid using knowledge-based behaviors in favor of less resource-consuming ability-based behaviors.

Simplified or virtual prototypes can be developed at this time.

A group of selected users could test these prototypes with a drive simulator in a laboratory, trying to perform suggested secondary tasks in the meantime.

Drivers reactions are carefully recorded to select the best solution according to the proposed criteria.

A full scale prototype is finally built and tested on a real car by potential customers in order to refine the design further.

In both tests, normal videocameras record eye positions during driving. A valuable evaluation parameter is how long eyesight is diverted from the road path in order to read information or identify the positions of the new control under evaluation.

An empirical and subjective evaluation can be transformed into an objective quantitative measurement.

Seat comfort

This paragraph is based on the research of Demontis S. and Giacoletto M.[2].

Many features of the seat such as thermal and dynamic properties, and the shape of the cushion and backrest, have a major influence on perceived comfort; the definition of a comfortable seat requires the knowledge of different parameters and their mutual interaction.

Traditionally such issues are investigated in clinic tests involving potential customers in the development process. During these tests, a consistent number of potential customers of cars in the same segment class as that under development is required to rate alternative prototypes using questionnaires that are developed for this purpose.

The inconvenience of this procedure relates to the fact that customers can evaluate only seats in a very simplified environment because real prototype cars are not yet available; if real prototypes are used, the cost of the test will increase and the suggested modifications will be too late or too expensive.

Therefore a mathematical model is considered to be important in order to reduce clinic tests or reduce at least the number of unsuccessful tests.

It is better to occasionally perform a large set of tests on a sufficient sample of vehicles available on the market to help develop design criteria, rather than correcting errors on a product close to production start.

In many studies it has been demonstrated that the subjective comfort evaluation depends on the duration of the test.

Limiting the scope of this study to small duration tests that report the initial response of the customer to a car in the dealer show room, the following parameters are included in the evaluation:

- Appearance.
- Ease of adjusting seat position.

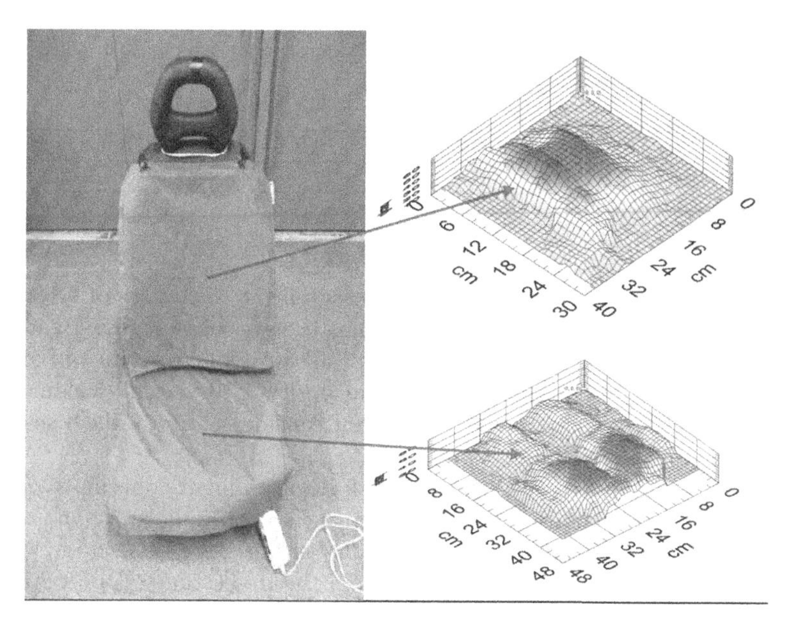

Fig. 8.24. On left: the pressure sensitive film on its seat. On right: example of *BPD* diagrams at the contact between seat and occupant..

- Postural comfort.
- Filling softness.
- Upholstery.

When considering just postural comfort, the purpose of this research is to correlate the customer's subjective rating with measurable parameters having physical meaning, that could be addressed using any of the available modelling tools.

In this case many authors consider the so-called *body pressure distribution* (*BPD*) between the seat and the occupant contacting surfaces to be a parameter which is significant in terms of postural comfort.

32 different subjects have been considered in this analysis in order to evaluate 6 different seats, with cloth upholstery, for economy cars; for the sake of homogeneity only seats with mechanical adjustment of longitudinal position and backrest inclination have been considered.

The acquisition of *BPD* is performed using pressure sensitive films, able of measuring contact pressure according to a reticule of 48x42 sensor points; the low thickness of the film (about .1 mm) does not significantly influence occupants feel.

Table 8.4. Evaluation weights of cushion and back rest, postural comfort, softness and upholstery on overall rating.

	POSTURAL COMFORT	SOFTNESS	UPHOLSTERY
CUSHION	0.60	0.55	0.55
BACK REST	0.40	0.45	0.45

Fig. 8.24 illustrates the results of this measurement; the effect of error measurements and the time to stabilize output has been accurately investigated.

BPD measurement has been performed for all available subjects and also for a SAE dummy, used to determine the H point position, as we will explain in the next chapter. Usual comfort angles have been considered to establish seat rest inclination and leg angle value.

Subjective ratings have been collected with a questionnaire developed for this purpose, including questions regarding postural comfort, upholstery and the remaining parameters described above. Also an overall score has been requested.

The questions set have been developed from the free comments of a group of different potential users, collected during interviews as usual.

The test protocol provides for the following steps:

- Information concerning the purpose of the test.
- Seat self-adjustment to the best position.
- Seat rating.
- *BPD* measurement.
- Subject anthropometric measurement.

BPD acquisition areas have been grouped according to different criteria, as explained in a dedicated section in the following chapter; these areas are particularly significant with respect to the anatomy of the human body, consisting in ischial, popliteal, lateral, lumbar and sacral zones. The basic idea is to identify parameters and measures identifying the solutions with the highest scores.

A linear regression method is applied to all measures to identify the relationships existing between identified parameters.

The correlation factor of 76%, between the parameters that were identified and the final score, indicates that more parameter should exist that explain the customer's judgments; nevertheless such a correlation factor can be considered acceptable.

Accepting this processing scheme, it appears that the overall score is influenced by the static comfort for 50% and by remaining parameters for 25%.

The evaluation weights of the considered parameters are reported in Tab. 8.4, demonstrating a slightly higher importance of the cushion in comparison to the back rest.

For each of these correlations, the correlation factor is reasonably satisfactory, exceeding .95.

Considering the different pressure distribution between body and seat, the following qualitative conclusions can be drawn:

- The higher the postural comfort rating, the lower the popliteal BPD with the cushion and the lower the dorsal BPD with the back rest; instead the lumbar BPD can be increased.
- The seat is judged to be 'soft' if the popliteal BPD is reduced, accepting a consequent increase in lateral and ischiatic areas.
- The upholstery rating is higher if the ratio between the peak and mean ratios of popliteal and lateral areas is also high.

This kind of approach enables a seat to be designed by means of parameters that can be calculated with conventional mathematical models, anticipating the results of customer clinic tests on paper.

Nevertheless it necessary to bear in mind the purely empirical nature of this approach and the fact that the rating is influenced by the availability of comparison items; this issue suggests the need to occasionally repeat research of this type in order to update with respect to the state of the art.

This case will be expanded upon further in the following chapter.

8.3.2 Euro NCAP Rating

The protection of car occupants from potential injuries in the event of a crash, which forms part of the passive safety requirements, is subject to legislation as far as its minimum acceptable level is concerned; however many manufacturers have introduced new models over recent years that claim levels of protection that exceed the legal requirements, determining a new market trend that regards high levels of passive safety to be an important factor of competition between brands.

Correspondingly a new issue has emerged regarding how to objectively measure the actual protection level of a car in a rigorous and comparable manner; obviously this measurement cannot be made directly by the customer.

Since 1970, on the basis of research conducted under the ESV (Experimental Safety Vehicle) project in the USA, many European governments and organizations have been cooperated to develop procedures and test instruments to enable the passive safety of a car to be assessed and classified. In particular the EEVC (European Experimental Vehicle Committee) has been established for this purpose.

Only since the start of the 1990's has a new passive safety test, which is accepted by all members, been made available that describes a full scale test for safety measurement during frontal and lateral crashes; successively a test for the passive safety of pedestrians was also developed.

At that time, European legislation required just a crash test against rigid barrier, made with a prototype without dummies in order to measure acceleration levels; only the steering column displacement was measured after the test to demonstrate compliance with regulations.

In the same years USA already requested biomechanical limits to be respected by dummies, and the NHTSA (National Highway and Traffic Safety Administration) set up a survey for evaluating a large number of cars bought from conventional dealerships. The objective of this survey was to measure the effect of crashes going beyond law compliance, in order to provide potential customers with additional information concerning the respective safety margins. This program was called NCAP (New Car assessment Program).

In Europe, the ADAC (Allgemeiner Deutscher Automobil Club), the association of German car owners and the magazine 'Auto Motor und Sport' started to perform offset crashes against rigid barriers in order to inform customers about the safety of cars in tests which are considered to be more realistic than the centred crash required by law.

At the same time, using funds of the Department of Transportation, consumer associations in the UK performed test programs of crashes against deformable barriers for the same purpose. In 1994, TRL (the Transport Research Laboratory in the UK) proposed that the Department of Transportation start a new program, again called NCAP, at a UK national level [3], the intention being to extend this test across Europe subsequently; in fact the program was initiated successively.

Although the official position of the association of car manufacturers to this proposal was negative, the reaction of individual car manufacturers proved to be highly constructive, with each striving to obtain positive ratings bearing in mind the commercial implications; in fact, despite not being enforced by law, successful achievement of positive ratings has been included in all development programs over recent years.

This program was officially started in the United Kingdom in 1995 and, in the same year, contact with the European Union initiated with a view to extending the program across the EU. SNRA (Swedish National Road Administration), FIA (Federation Internationale de l'Automobile) and International Testing were the first organizations to join and sponsor the program with the first results presented in 1997. In the same year, the Volvo S 40 was the first to receive 4 Star score in what had become known as the Euro NCAP program.

Other state organizations joined the program later and the European Union endorsed this program immediately after.

The Euro NCAP test procedure transformed tests that were initially designed for homologation into a tool suitable for evaluation by performing homologation tests with more severe control parameters; it summarizes in one index only the overall result, without losing single tests results.

---

[3] Test of English cars are specific, because of the right hand position of the drivers seat.

Cars are crashed in an offset impact against a deformable barrier and in a lateral crash. The lateral crash against pole was introduced in 2000 together with a new test to evaluate the danger of injuries for pedestrians hit during a crash.

Test and evaluation procedures, the criteria for assigning partial and overall scores, and the results of cars tested to date are available on the official website:

www.euroncap.com,

where more detailed information can also be found; here only a short summary is presented.

Euro NCAP assigns a number of stars to represent the overall level of protection provided by a car in front and lateral crashes; this star number depends on partial scores obtained in these tests: A maximum score of 34 points can be obtained in these tests, where 16 are available for each of the two test categories.

These scores evaluate whether current standards are met and if an excellent protection has been reached.

A partial score is assigned to each body region of the dummy according to specific criteria. This score is corrected afterwards in consideration of the dummy motion during the crash that is recorded with a high speed camera; this correction takes into account how small variations in dummy position or size could affect results.

Also the mechanical behavior of the body structure is evaluated following the crash, with particular reference to wheel wells, A pillars, pedals and steering wheel deformations.

The corrected score is presented in a diagram as in Fig. 8.25 with different colors assigned to different body regions (not represented in the picture). This procedure is followed for the driver and the passenger occupying the front seat, for both front and lateral impacts. A different color code is assigned to the front region of the car potentially involved in impacts with pedestrians.

From this presentation an overall score is derived for driver and passenger, while a separate score is assigned for pedestrians.

A similar procedure is applied to assess protection levels of the rear seat area using dummies representing children of 18 months and 3 years of age respectively occupying the prescribed rear safety seat.

The evaluation scale for each of the tests presents the maximum value for excellent performance to the minimum value for law compliance; below this limit no scores are available.

To supply more details about this method, only front impact with a speed of 64 km/h. is considered, bearing in mind that all details are available in the website cited.

After the crash, the following body regions are evaluated.

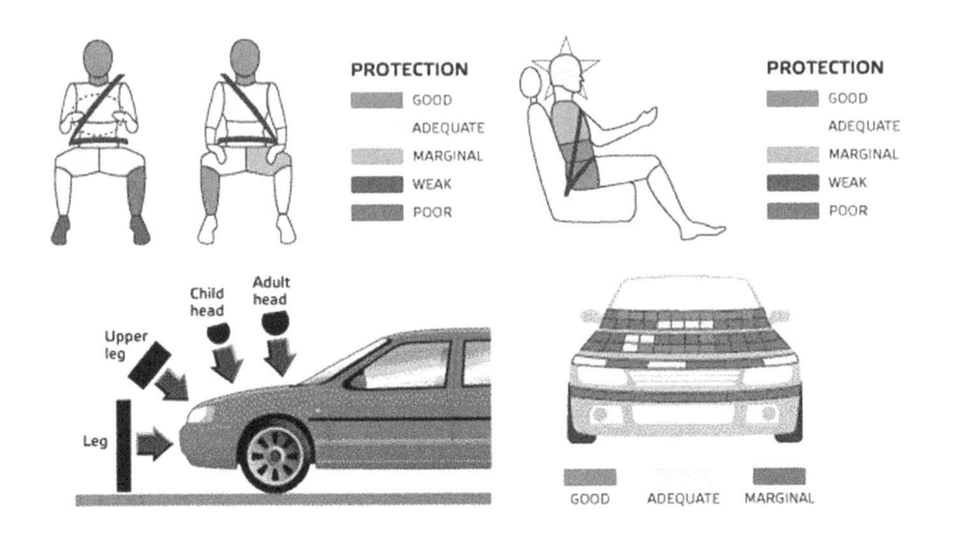

Fig. 8.25. Examples of biomechanical results evaluation of Euro NCAP tests. At top: front impact protection for driver and passenger; at right: lateral impact against pole protection; below: pedestrian protection.

Head

If air-bags are provided, 4 points are assigned if there is no direct contact of the head with rigid parts; The Euro NCAP procedure identifies a direct contact when visible contact traces are present on the head dummy skin after the test or if the deceleration of the head is equal or exceeds 80 g.

If there is a direct contact, the maximum performance is $HIC_{36} = 650$, while the minimum for acceptance is $HIC_{36} = 1{,}000$[4].

If air-bags are not available and the value of $HIC_{36} = 1{,}000$ is overcome, the kind of impact must be assessed using a honeycomb face profile to identify the most critical impact area which are usually the joints of the steering wheel ring with its spokes; in this case, the maximum performance is a deceleration of 80 g (no more than 65 g, for 3 ms), while the minimum performance corresponds to 120 g (no more than 80 g, for 3 ms).

In any case the maximum attributed score is 2, meaning that in this test a car without air-bags can only achieve a best score which corresponds to half the maximum available.

Neck

The best performance is reached when:

---

[4] *HIC*, Head Injury Criterion is correlated to head deceleration; it will be explained in the last chapter.

- Shear force is less than 1.9 kN, or less than 1.2 kN, for 25÷35 ms, or 1.1 kN for 45 ms.
- Tension force is less then 2.7 kN, or less than 2.3 kN, for 35 ms, or 1.1 kN for 60 ms.
- Extension is less then 42 Nm.

The worst allowed performance is:

- Shear force is less than 3.1 kN, or less than 1.5 kN, for 25÷35 ms, or 1.1 kN for 45 ms.
- Tension force is less then 3.3 kN, or less than 2.9 kN, for 35 ms, or 1.1 kN for 60 ms.
- Extension is less then 57 Nm.

Thorax

The best performance is reached when:

- Compression is less then 22 mm.
- Viscous criterion is less then .5 m/s.

The worst allowed performance is:

- Compression is less then 50 mm.
- Viscous criterion is less then 1.0 m/s.

Knee, femur and pelvis

The best performance is reached when:

- Femur compression is less than 3.8 kN.
- Knee articulation compression is less than 6 mm.

The worst allowed performance is:

- Femur compression is less than 9.07 kN, or 7.56 kN for more than 10 ms.
- Knee articulation compression is less than 15 mm.

Leg

The best performance is reached when:

- Tibia index is less than 0.4.
- Tibia compression is less than 2 kN.

The worst allowed performance is:

- Tibia index is less than 1.3.
- Tibia compression is less than 8 kN.

Foot and ankle

The best performance is reached when:

- Pedal displacement is less than 100 mm.

The worst allowed performance is:

- Pedal displacement less than 200 mm.

Displacement is measured for each pedal, paying attention to not apply loads: In cases in which the pedals are designed to separate from their articulations during crash, the existing displacement will not be taken into account if separation occurred during the crash and the separated pedals do not oppose significant force to their motion. If pedals are designed to move forward during crash, only the final position will be taken into account.

Score

Notice that values associated with worst allowed performance given above correspond to homologation limits. In future, a limit associated with wheel well displacement could also be set.

Each of the body regions is assigned one point for maximum performance; if more than one evaluation criterion is available for the same body region, the lowest point is assigned as overall evaluation.

The generated total score can be modified in cases in which worse behavior may occur for slightly different impacts or for occupants of different size.

If during its motion consequent to the impact, the center of gravity of the head is in any position outside the volume occupied by the air-bag after its deployment, the contact between head and air-bag is judged to be unstable and the total score is reduced by one point.

The same reduction is applied if the air-bag protection is compromised by a rupture of the steering wheel or by interference with solid bodies.

A penalty is applied to the thorax score if the A pillar displacement, measured at a height 100 mm over the lowest point of the front window, is judged to be excessive. An acceptable figure for this displacement is 100 mm or less to incur no penalty, whereas a displacement exceeding 200 mm would incur a penalty of two points.

One point penalty is applied when the structural integrity of the passengers compartment looks to be compromised; this situation is determined on the basis of whether one of the following occurs:

- Rupture of one door hinge or lock, unless doors are firmly kept in position by the shape of their frame.
- Body panels buckling.
- Separation of anti-intrusion beams.

- Doors which cannot be opened by hand.

Other penalties may be applied for similar situations and for doors that open as a consequence of the impact, incurring the risk of occupants being ejected out of the passenger compartment.

The visual presentation of the overall result is made using a dummy sketch, with its different body regions being colored according to the score obtained, as can be seen in Fig. 8.25.

The color code is the following:

- Green is associated to 4.00 points.
- Yellow is assigned between 2.67 and 3.99 points.
- Orange is assigned between 1.33 and 2.66 points.
- Brown is assigned between 0.01 and 1.32 points.
- Red is associated to 0 point.

A similar treatment is made for the result of the side impact test, that is summed with that of front impact; if the pole impact obtains the maximum score, a green star is added over the head of the dummy sketch; the color of this star becomes yellow or white if results are marginal or negative.

Finally, the result of the pedestrian impact test is displayed with a rectangle divided into colored squares; the color of each square is green for 2.00 points, yellow between .01 and 1.99 points, red for .0 point, as can be seen again on the same figure.

The overall result of front and lateral crash tests are summarized by taking into account the worst results of driver and passenger for each body region.

Body regions are joined in the following groups:

- Head and neck.
- Thorax.
- Right and left knees, femurs and basin.
- Right and left legs and feet.

To each group the lowest score obtained by the associated regions is assigned for both front and lateral impact. For pole impact, only the head is considered.

The total maximum score is 34 points, since each of the four considered regions can obtain a maximum of 4 points in each of the two front and lateral tests plus a maximum of 2 points in the pole test.

Seat belt signals, if applied, can obtain up to 3 additional points, since 3 additional points can be added for outstanding innovative safety devices not yet stipulated by the regulations.

Since the impact areas considered in the pedestrian impact test are 18, a maximum of 36 points can also be assigned for this rating category.

Occupants and pedestrian tests are summarized by stars.

The occupant protection test summary scores are as follows:

- 5 stars with an overall score between 33 and 40 points, with no less than 13 points in any single test.
- 4 stars with an overall score between 25 and 32 points, with no less than 9 points in any single test.
- 3 stars with an overall score between 17 and 24 points, with no less than 5 points in any single test.
- 2 stars with an overall score between 9 and 16 points, with no less than 2 points in any single test.
- 1 star with an overall score between 1 and 8 points.
- No star for 0 points.

Result reliability

A study was carried out by Lie A. and Tingvall C. in order to determine the correlation, if any, between Euro NCAP scores and the outcome of real road accidents which occurred on roads in Sweden. This study does not yet take into account pedestrian and child tests scores.

The analysis technique applied consisted in considering accidents between pairs of cars; an injury risk is evaluated by comparing the actual injuries of occupants of one car with those of the other car.

Results are corrected to take into account the different mass of cars included in the same pair since it is accepted that mass directly influences crash severity.

According to this research, results are grouped into four classes:

- Injuries of occupants of both cars (cases count x_1).
- Injuries of occupants of the car under consideration, but not in its counterpart (cases count x_2).
- No injury to occupants of the car under consideration, but only in its counterpart (cases count x_3).
- No injury to occupants or either cars (cases count x_4).

The injury risk of one car occupants, the ratio between reported and potential injuries, is given by the ratio:

$$R_1 = \frac{x_1 + x_2}{x_1 + x_3}.$$

Table 8.5. Injury risk (R_1: overall risk; $R_{1,gf}$: severe and fatal injury risk) as function of Euro NCAP score (P) and number of assigned stars (S). M_s is the average mass of cars under exam, while M_c is the average mass of their counterpart.

S	$R_{1,gf}$	R_1	M_s [kg]	M_c [kg]	P
n.v.	0,92	0,99	1332	1287	n.v.
2	0,92	1,10	1260	1288	13,09
3	0,61	0,91	1450	1297	21,06
4	0,65	0,96	1362	1304	25,98

If $R_1 = 1$, the car under examination and the opponent car have the same injury risk; if $R_1 < 1$, the car under examination has a lower risk than its opponent, and vice versa.

If a difference of mass between the two cars in the pair exists, a higher speed variation in the lighter car will occur whereas the opposite occurs in the heaviest car; therefore a mass difference plays a double role in terms of increasing and decreasing injury risks depending on the car.

Previous research aiming to address this issue determined that the risk is decreased by 7% for each tenth of a metric ton added to the average mass of the cars in the same pair.

On this basis a corrected risk can be defined by the following formula:

$$R_c = R_1 \frac{1.07^{\frac{M_s - M_m}{100}}}{1.07^{\frac{M_c - M_m}{100}}},$$

where M_s, M_c, M_m are respectively the mass of vehicle under study, of its counterpart and the average mass.

All cases of accidents between couples of car reported to the police between 1994 and 1999 were considered, if correctly identified; injuries were classified into three categories: light, serious and fatal. A total of 1,779 cases with serious and fatal injuries were identified and a further 12,214 with light injuries.

However no 5 stars cars were available because at that time the existing Euro NCAP rules considered 4 stars to be a maximum. Identified injuries were detected on:

- 8,460 cars without rating;
- 1,534 on 2 stars cars;
- 1,866 on 3 stars cars;
- 354 on 4 stars cars.

Tab. 8.5 summarizes results, where R_1 is the overall risk and $R_{1,gf}$ is the sum of severe and fatal injury risk. The Euro NCAP score is P and the number of assigned stars is S. Instead M_s is the average mass of the principal cars considered, while M_c is the average mass of their counterpart; no data correction according to the proposed formula has been applied.

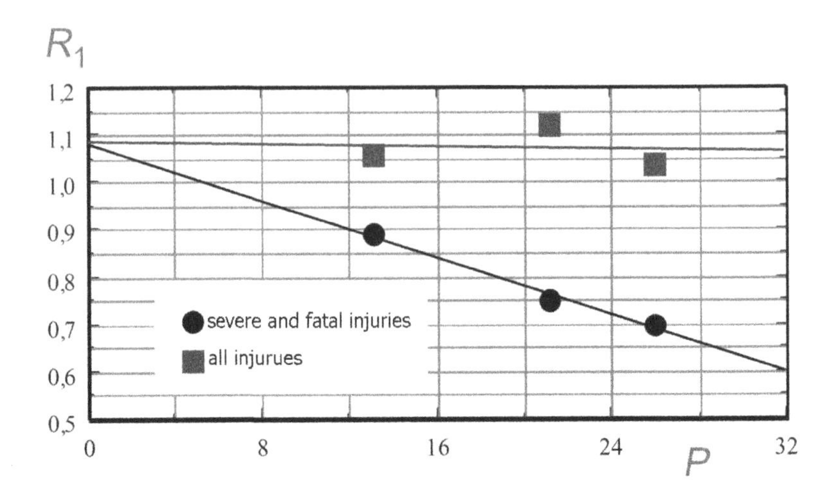

Fig. 8.26. Injury risk diagram, as function of Euro NCAP score.

Fig. 8.26 illustrates graphically the results obtained after the correction of the effect of the mass difference.

A strong correlation between Euro NCAP score and the reduction of severe and fatal injury risk can be observed, while there is no correlation between score and risk of reporting any injury.

Higher score cars reduce the injury risk by about 30% in comparison to cars without evaluation.

8.3.3 Insurance Companies Rating

One major expense that customers have to meet to operate a private vehicle is the insurance premium. This expense is compulsory in terms of civil liability insurance, and is becoming increasingly requested as far as the different risks are considered for collision, theft or fire cover.

This section will address the issue of repair costs of non-severe crashes, recognizing that economic aspects have primary importance.

At the time of repair, costs are directly born by either the vehicles owners or the insurance companies; in the end, however, repair costs are effectively born by vehicle owners anyway, being simply divided into installments over a longer time span. In particular, as concerns a total risk coverage insurance, the so-called fully-comprehensive or 'kasko' insurance in Europe, the insurance premium simply provides the owner with greater stability in terms of car repair costs.

This fact can influence significantly the appeal of a car when a purchase is being considered and, correspondingly, must be taken into account adequately during product development.

The evaluation of the premium by insurance companies is determined with reference to the total costs of reimbursing policy holders during previous years.

In many countries of the European Union the insurance premium is customized to the car to be insured; therefore the customization criteria have a clear influence on car design.

Since there is no international standard on this subject, it is appropriate to consider the example of the method applied by German insurers to calculate the insurance premium of which details can be found on the Allianz website:

$$www.allianz - autowelt.de,$$

This issue is of fundamental importance for the German market.

Other relevant information on this subject is available on the internet site:

$$www.thatcham.org,$$

where standards applied by English insurers are reported. A very interesting manual is available on the same website of how to conceive a car with low repair costs and high quality repairs.

Given the importance of this issue, it is feasible is that European Union will impose regulations in this area in the future.

As regards the Allianz rules, three different test are considered:

- front collision;
- rear collision and
- lateral collision;

these test standards have been adapted from passive safety standards.

Front collision

The car under test is driven against a non-symmetric barrier, as shown by Fig. 8.27.

Clearly the barrier height must be higher than the vehicle impact area. The barrier must be rounded with a radius R of 150 mm and the offset O must be 40% of the vehicle width B, measured without taking the side mirrors into account.

The only possible contact of the barrier with the vehicle must be in the area of increased thickness.

The vehicle test conditions are the following:

- Ignition must be on: Each passive safety device must work properly, including air-bags, belt pre-tensioners, window bags, etc., because their possible repair cost will be taken into account.
- Useful load corresponding to a 50th percentile male dummy with safety belts.

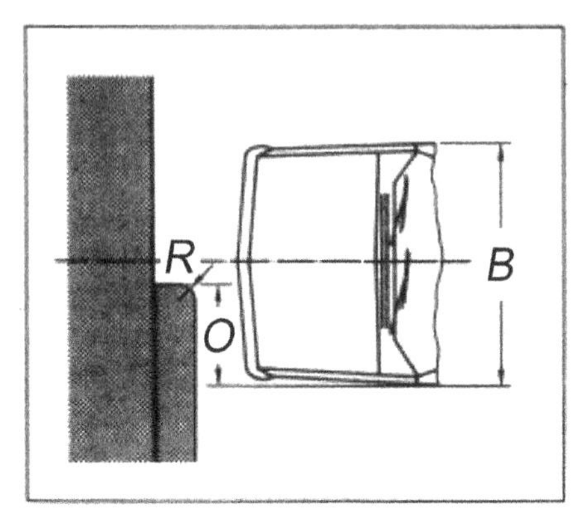

Fig. 8.27. Scheme of insurance companies test simulating a low speed front crash, against a non-symmetric barrier.

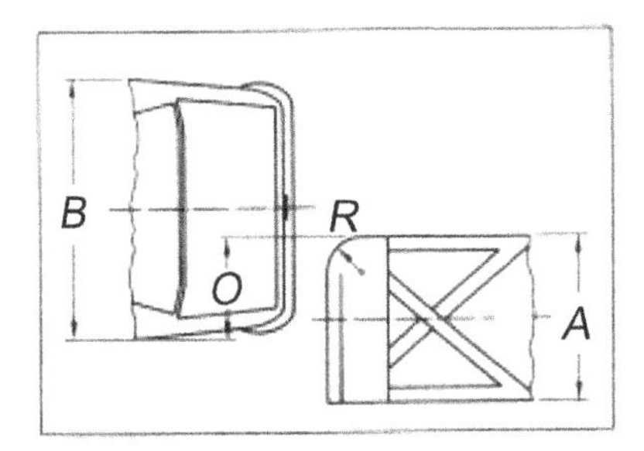

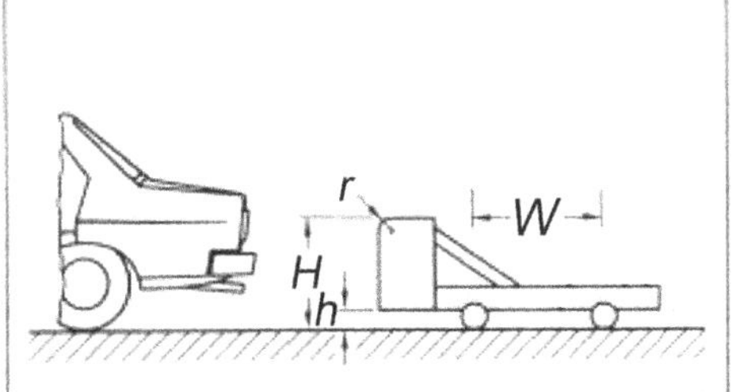

Fig. 8.28. Scheme of insurance companies test simulating a low speed rear crash, against a non-symmetric moving barrier.

- Full tank, brakes released and gearbox in neutral.
- Impact speed of 15^{-0}_{+1} km/h.

After the impact, the axle geometry and the gaps between the body components are measured to assess non visible damage.

A new barrier is under development with a inclination of 10° (top view) to enable the first contact area of the barrier to be an exterior part of the body.

Rear collision

The test is made by means of a moving barrier, as shown in Fig. 8.28.

It must comply with the following specifications:

- The offset O is the same as for the front collision.
- The overall mass of the moving barrier is 1,000 kg.

- The barrier width A is 1.2 m.
- The barrier wheelbase W must be greater than 1.5 m.
- Total barrier height H is 700 mm, while ground clearance h is 200 mm.
- The vertical barrier rounding R is 150 mm, while the horizontal rounding r is 50 mm.
- Impact speed is again 15^{-0}_{+1} km/h.

The test conditions are the same as in the front collision test; this test is made on the same vehicle following front collision.

A new test procedure is under development where the barrier is launched against the vehicle with a path inclined 10° with respect to the vehicle symmetry plane in such a way as to impact the exterior side of the body first.

Lateral collision

Currently an actual test procedure is not available; damage is considered as reference, where the following parts of the left side must be repaired:

- Door with its glasses, locks and weather strips;
- Body panels surrounding the side and the bottom of the door.

However a new procedure is under development, in which a moving barrier is made to impact the car under test as shown schematically in Fig. 8.29.

Results evaluation

The damage to the car after the three tests is assessed and the repair costs are calculated. These costs are evaluated starting from the official list prices of the car manufacturer's service network, including spare parts, standard repair time and hourly manpower cost.

The three repair costs are weighted with reference to the relative frequency of their occurrence; at this time, the weights are the following:

- Front collision 54%.
- Rear collision 30%.
- Lateral collision 16%.

The result is divided by the mean repair cost of all insured cars; in this way an average damage index is obtained.

For a given car type, also an average mean frequency index is calculated, starting from actual accidents of similar cars already insured. This index takes into account the statistics that some types of car are usually owned by customers with specific driving habits which may be more or less risky.

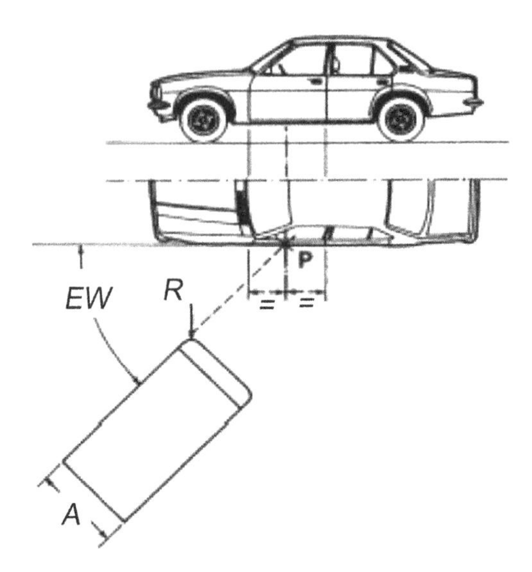

Fig. 8.29. Scheme of the insurance companies test under development, simulating a low speed side crash, against a non-symmetrically moving barrier.

The product of the two percentages above provides the average *total repair need* (TRN) of a given car that is used to calculate the insurance premium. A figure of 100 is obtained by a car perfectly in the average. For instance:

- For $TRN < 40$, the insurance premium is 320 euro.
- For $60 < TRN < 69$, the insurance premium is 580 euro.
- For $90 < TRN < 99$, the insurance premium is 840 euro.
- For $TRN > 800$, the insurance premium is 10,000 euro.

Advertising such figures that are available to the public has created an unofficial but very effective measurement of car vulnerability to small accidents and, as with Euro NCAP, this figure is considered to be an effective new product specification in the drive towards car design excellence.

The corresponding design rules could be as follows:

- Very stiff longitudinal beams; these should permit the front collision test to be undertaken without major deformation, and support an inexpensive sacrificial front crash-box positioned just behind the bumper so as to be easily replaceable.
- Engine compartment with an appropriate lay-out to enable broken parts to be replaced without need to disassemble the powertrain.

- Sufficient space between the components at the front of the car in order to avoid damage; for instance, radiators, blowers and their frames should not be damaged during insurance company crash tests.
- The same concept should apply also to headlights, that should be positioned in relatively protected spaces.
- Easy replacement of parts damaged during tests.
- Spare parts pricing policy which takes into account sales volume. For instance, a frequently damaged part (side mirror cover) should not be incorporated in expensive assemblies (side mirror internals) to avoid expense of repairs.

8.3.4 Aging Resistance

A car's endurance, or resistance to age, is a function that can be evaluated objectively by driving it for a specified distance without failure. In this context it is necessary to clarify what is meant by failure; during the life of a car, customers demand few faults and little deterioration with respect to the car when new.

In the case of the body, this includes requirements concerning:

- Ergonomy and packaging,
- Thermal comfort,
- Acoustic and vibration comfort,
- Structural integrity and
- Passive safety,

which can deteriorate only within an acceptable range of tolerance; in addition, nothing that can affect vehicle availability (i.e. ready to perform its function) can occur except through the fulfillment of scheduled maintenance.

In this sense failures include, therefore, not only the breakdown of mechanical or electric parts, but also any noises that are not detected on the vehicle when new, leakages of lubricant, aesthetic corrosion, changes in dynamic behavior, increase in fuel consumption, changes in control feedback, etc.

It is hard to forecast a-priori how a vehicle will be used, because use is conditioned by the life and driving style of the customer; in addition, the applied loads can depend on unforeseeable events.

Therefore, endurance specifications are assigned statistically, often using the parameter B_{10}, which defines the endurance achieved without any failure by 10% of the population of vehicles produced.

As a reference value for this parameter, it can be assumed that adequate for today production would be approximately:

- $B_{10} > 200{,}000$ km, for cars and commercial vehicles;
- $B_{10} > 400{,}000$ km, for buses;
- $B_{10} > 800{,}000$ km, for heavy duty trucks.

As the technology and market evolve continuously, it is possible that these values will change in the future.

Such travelling distances make it almost impossible to perform the tests needed to assess endurance experimentally, in the standard interval (3÷4 years) devoted to a new vehicle development following design and prototype manufacturing, with a sufficient level of confidence; nor can this task be delegated reliably to numerical simulation of the vehicle.

In the case of cars, a life of approx. 200,000 km implies, on average, 4,000 h of driving time assuming standard driving tasks; six months are allocated to this task, as usually occurs, then to perform two complete sets of tests on two different prototypes generations (where the second implements the corrections to the first), the test time must be shortened by at least 3÷4 times.

Phenomena that can influence the endurance of a vehicle can be classified according to the following categories:

- fatigue;
- wear;
- corrosion;
- shocks and collisions.

External fatigue loads that can stress body components arise from two different sources: the tires and the engine.

Tires apply to the chassis longitudinal, lateral and vertical forces, changing over time; the first and second act with the frequency (on average low) of vehicle acceleration, braking and cornering events along the path of the vehicle; the latter act with higher frequencies that depend on the shape and spatial density of the obstacles overcome.

The engine usually stresses the chassis both at low frequencies, determined by the schedule of maneuvers (acceleration, releases, shifts) and at higher frequencies determined by its reciprocating motion of the pistons inside the engine.

Other periodic forces may be amplified when these forces excite the natural frequencies of the different sub-systems and structures; this is particularly relevant in the case of the chassis structures and the transmission.

Most body components are prone to fatigue being made of metals with resistance which can be described according to Wöhler's model; a threshold of load amplitude (fatigue limit) exists for these materials which, if not exceeded, can be applied indefinitely without any damage.

For this category of parts, test times can be reduced by applying techniques which remove the periods of load history below the fatigue limit. This can be done precisely by analyzing the load time histories that will be applied during bench tests or by driving cars in more strenuous tests that apply more severe loading.

Using this last approach the total distance driven during the real life of a car can be condensed into about 50,000 km of heavy use.

Wear, which is the removal of material on parts in relative motion, is a function of the level of friction generated between the parts; on the body, wear applies mainly to elastomer parts (bushings, weather strips, sliding seals, etc.) and also to small mechanisms.

According to the hypothesis proposed by Theodor Reye about 140 years ago, wear depends on wasted friction energy and can therefore be accelerated by increasing loads, with attention to temperatures that can affect the mechanical properties of materials.

A wear test for components can be reliably performed on test benches, where contact conditions are made more severe according to empirical procedures.

Corrosion is caused by the chemical action of many agents (humidity, salts, other chemical compounds and aerosols) on parts exposed to the atmosphere or splashed by the wheels; since this action is not constant throughout the life of the car, the test can be accelerated by exposing entire cars or components to corrosive humidostatic rooms for a certain period.

Another method, just as effective as the first, is to drive the car through acid water pools over a certain portion of the fatigue test course.

Accelerating wear and corrosion testing remains empirical and is defined according to the past experience of the vehicle manufacturer.

Vehicle resistance to shocks and collisions must be examined through artificially reproduced events.

This applies to crash tests against barriers, required according to regulations, where chassis components must not interfere with occupants as a consequence of the collision.

There are also non-regulated shocks, where it is good practice to verify that no critical situations for occupants occur; one example of this category is the accidental collision of a wheel against the curb, as a consequence of a mistaken maneuver; it is obvious that in these cases body mounting interfaces structural integrity is not requested at any rate.

The designer must guarantee that there are no partial or hidden ruptures undetectable by the driver. Linkages and suspension mounts must feature a rupture load at least 50% higher than the collapse load, where deformations become permanent; deformations, prior to rupture, must alter suspension geometry in such a way as to be easily noticeable by drivers, in order to encourage interrupting the trip to undertake the necessary repairs.

Vehicle life is simulated by separate tests reproducing specific situations; fatigue tests are more difficult because they are determined not only by their duration but also by the conditions of vehicle use.

Each manufacturer has decided to design vehicles for the most demanding conditions, accepting high safety margins for ordinary use; loop courses have been

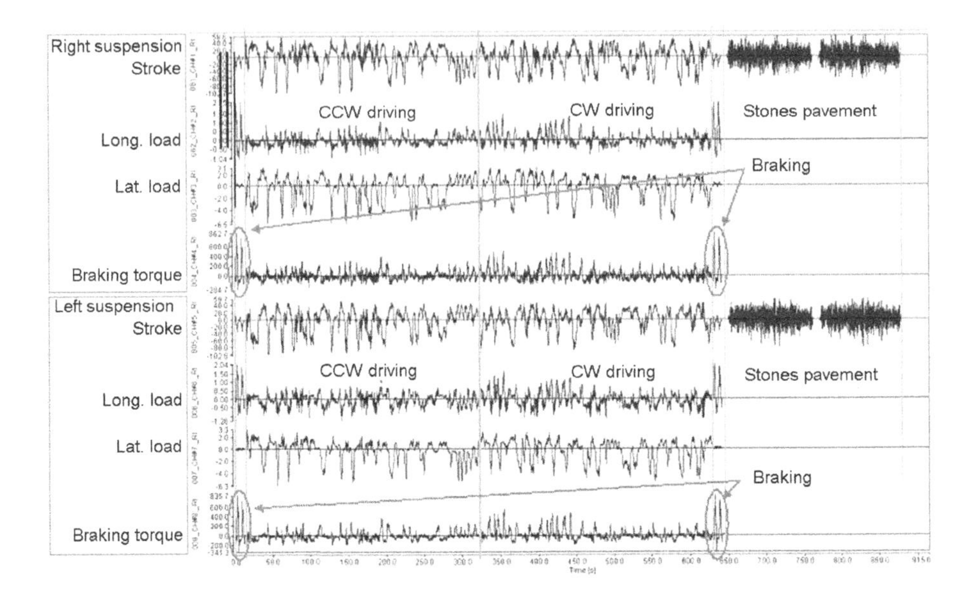

Fig. 8.30. Records of suspension strokes, showing longitudinal and lateral loads on rear suspensions of a medium size car, driven on a fatigue loop clockwise (CW) and counterclockwise (CCW).

developed that are characterized by many bends, bumpy and uneven stretches of track (artificially damaged tarmac, stone pavement, dirty road, etc.) and rail crossings; part of these courses is dedicated to water-crossing.

Such loops, if driven at high speed, can concentrate 200,000 km of real life into about 50,000 km, which can be driven in about 1,000 h; this time corresponds to about two months, assuming three driving shifts on the same car, and including test interruptions to maintain and inspect the test prototype.

Load conditions to be analyzed using mathematical models or applied during bench testing are also derived from this kind of loop.

A common test loop includes straight stretches which are long enough to allow the car to reach the highest acceleration and deceleration conditions; curves are driven at the slip limit.

The length of the loop is not relevant, since it will be repeated in both driving directions so as not to stress the vehicle in a selective way, until the total driving distance is reached; instead loop length is conditioned by the need to apply all tests suitable for simulating the most demanding driving situations. Usually such loops are between 20 and 30 km in length.

The pavement must provide a high friction coefficient in order to stress suspensions and the chassis as much as possible. Sometimes, on long straight stretches, signals can be used to request additional maneuvers (decelerations, accelerations, slaloms).

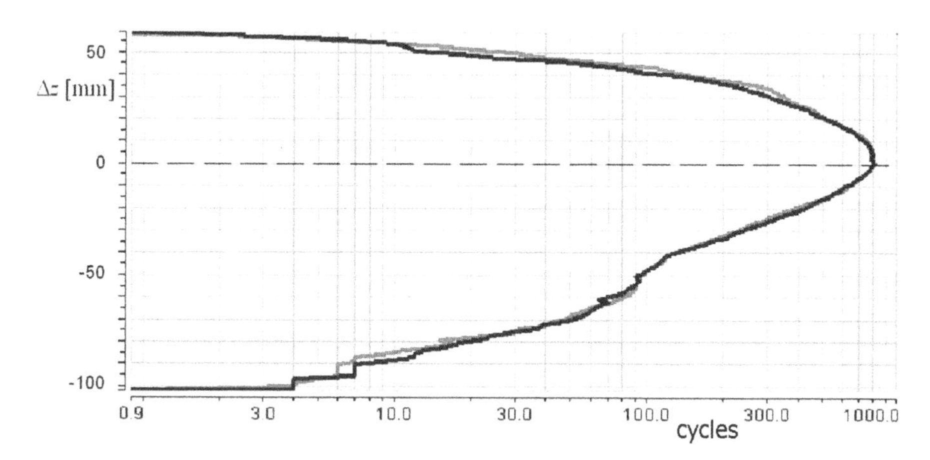

Fig. 8.31. Histogram showing the cycle count of different suspension stroke classes Δz of right and left rear suspensions, on a fatigue loop.

Fig. 8.30 shows a record of the main force components acting on a medium size car on a loop of this kind.

Instead of vertical forces, the measured suspension strokes are shown; vertical forces may be calculated from the suspension characteristics. Brake torque has also been added to separate transmission effects from those of the brakes in longitudinal forces.

These records originate from test bench load conditions following mathematical elaboration; the same conditions can be applied to finite elements analyses, which are usually integrated into multibody models in order to simulate the entire vehicle. Reorganization of the test cycles can be performed according to the rain flow method; the result is a set of load histograms, shown in Fig. 8.31, Fig. 8.32, Fig. 8.33, representing suspension stroke as well as longitudinal and transversal accelerations.

Accelerations are derived by forces with reference to the sprung mass and can be applied to different cars as well, as a first approximation.

These histograms define the so called *load blocks* which correspond to driving the entire loop, about 30 km long, once clockwise and once counterclockwise; the load block is applied about 2,000 times to simulate the entire vehicle life.

It should be noted that accelerations refer to the sprung mass; acceleration values apparently inconsistent with practical friction coefficients are not surprising, because vertical loads are increased by transfers due to lateral accelerations.

Fig. 8.34 shows a bench for fatigue tests of a car body, complete in its main chassis components; the actuators used to apply the different loads and torques are also shown. The front wheel can be stressed similarly.

Dynamic analysis of these forces assumes a particular importance if we want to determine the fatigue load for particular components under specified driving conditions.

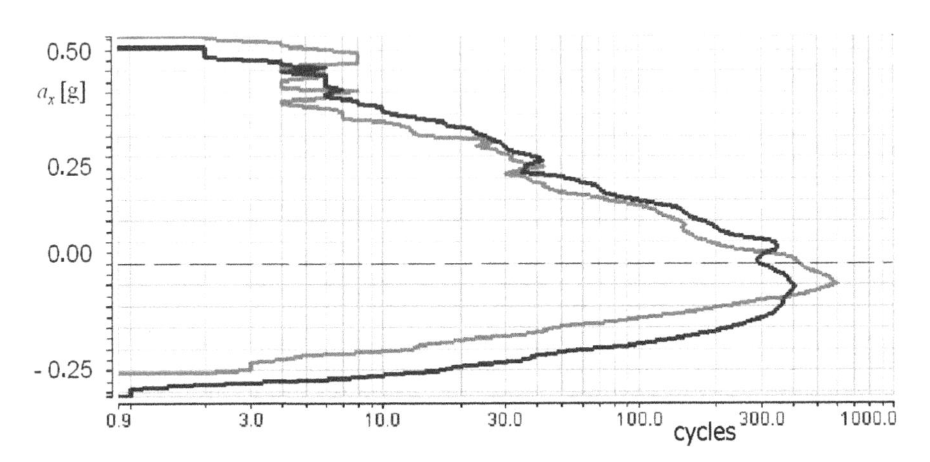

Fig. 8.32. Histogram showing the cycle count of different longitudinal acceleration classes a_x of right and left rear suspensions, on a fatigue loop.

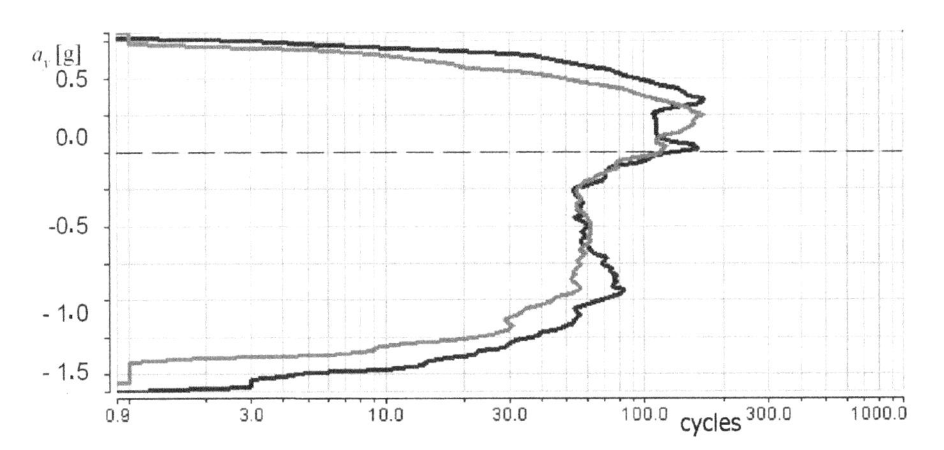

Fig. 8.33. Histogram showing the cycle count of different suspension transversal acceleration a_y of right and left rear suspensions, on a fatigue loop.

Fig. 8.34. Fatigue test bench for a car body, complete with the most important chassis components; actuators supply rear suspension forces and torques.

Usually the entire vehicle simulation applies multibody modeling techniques that are useful when the displacement levels are relevant; these models enable both the displacements and the forces exchanged in any part of the system to be calculated.

In multibody modeling, system elements are considered to be rigid bodies connected by elastic or viscoelastic couplings. In some cases, for a more precise determination of acting loads, it is necessary to also take into account the flexibility of some part, such as the car body or the twist beam of a rear suspension.

The Craig-Bampton theory enables the flexibility of specific vehicle structures to be represented by their modal synthesis which can be obtained using a finite element calculation of the modal deformations.

In practice, for a specific structure:

- The displacements of the structure, compatible with its degrees of freedom, are evaluated as the result of a unitary force applied.
- The vibration modes are calculated, in a range up to about 100 Hz.

These data are applied to the multibody model in order to determine the internal forces exchanged between the different components.

A multibody model of the vehicle system should include at least:

- the car body, as a rigid body or as a flexible body, according to the scenario under study;
- the powertrain mass and its suspension;
- the complete front and rear suspensions;
- tires, when loads are calculated from an open loop maneuver of interest.

The different parts are connected by joints with appropriate elastic and viscoelastic characteristics; the mass and inertia properties are calculated for each of them.

If we refer to the previously described scenario, the same input for the test bench of Fig. 8.34 can be adopted as input of the multibody model.

The output of this analysis should be the forces that are exchanged at the articulation points of the suspension at the car body.

These forces may be applied to finite element models used to calculate the stress histories of the component under study in order to determine the fatigue life of this element.

This calculation can be performed in two different ways, due to the fact that there is some overlap between force frequencies and natural frequencies of the component under investigation.

If there is no overlap, forces can be applied quasi-statically. A stress time history can be easily obtained via linear combination of the different effects. Instead, if there is overlap, dynamic modal techniques must be applied.

8.4 Regulations

A vehicle cannot be sold and obtain the necessary registration for driving on public roads, unless it is built according to a set of legal specifications. In Europe the agreement of these specifications with existing laws is demonstrated, by two official documents: the *certificate of homologation* and the *certificate of conformity*.

The first document proves that the vehicle is designed according to legal requirements. It is issued by the public authority in charge of this function in each country; in Italy, for instance, that authority is the Department of Transportation. The homologation certificate is issued on the basis of a technical report of the manufacturer and the completion of given tests, performed on prototypes of that vehicle.

The second document proves that any produced vehicle is identical, in terms of homologation requirements, to the tested and approved prototype; this document is issued by an appointed representative of the manufacturer, for instance the general manager of the final assembly plant. Vehicle conformity can, at any time, be verified by the public authority in charge, by inspection of samples of produced vehicles and tests for requirements of the homologation certificate.

Homologation requirements are set by government laws, which impose minimum and maximum acceptable values and the related test methods to be used for their verification; the manufacturer is free to identify the most suitable technologies to be employed for their fulfillment.

These requirements are relevant to part of the vehicle functions already introduced in previous chapters, particularly:

- outside visibility;
- minimum dynamic performance necessary to guarantee safe driving;
- occupant protection in case of collision;
- reduction of the environmental load caused by vehicle traffic, with particular reference to polluting gases, carbon dioxide, outside noise and waste produced by the disposal of older vehicles.

The above laws are issued by each national government; in the past, some laws have also been developed by international institutions to enhance the free movement and sale of vehicles in countries other than that of their origin.

The European Community has already faced, in the 1960's, the problem of harmonization of national laws, to remove any impediment to the free circulation of goods within the EC and grant the citizens of member states the availability to buy state of the art vehicles; this task has recently been taken over and completed by the European Union.

The European Union behaves like a supranational body, requiring all member states to develop laws which comply with a common standard; these supranational laws are called *Directives* and will be cited with the letter D, followed by a figure showing the year of enactment and a following number. For instance the D70/156 directive was the 156th law approved in 1970 regarding vehicle homologation; at the end of the last century and in the present one, the complete year figure is cited.

These directives are subjected to frequent updating in order to take into account new available technologies or to stimulate their application; in this case each updating may regard new statements only, but not those that have remained unchanged. Sometimes, however, new directives are issued that redefine all the issues according to new rules.

The construction of an updated situation may be quite difficult. To make directive consultation easier, they are organized in numbered groups that regard a given topic; the group number increases according to the date of issue of the first directive in the group.

In parallel with Directives, *Regulations* have been also issued, summarizing in a single document all approved test procedures relevant to given homologation functions; these documents are identified with the letter R, followed by a progressive number, unique for each title, independent of any addition or modification.

Since Directives must be established before national laws, they must be available in advance of their enforcement time; in their formulation, they provide an enforcement year for new homologations and one for new registrations.

No member state is allowed to prohibit sale, registration or circulation of any vehicle complying with the Directives in force.

In this chapter those Directives and Regulations impacting body design and its functions in the vehicle system are considered.

The following paragraphs summarize laws regarding:

- the vehicle in general,
- body shape,
- artificial external lighting,
- visibility,
- protection of occupants and pedestrians,
- thermal comfort,

referring to body components and also to those issues that will be introduced in the following chapters.

The information discussed here is an updated summary at the time of writing this book and should be considered as a general reference only; correspondingly it is suggested that anyone requiring updated guidelines look through the primary documents and check for new versions. This can be easily done by visiting the websites of the European Union dedicated to this purpose; everyone can download excerpts of the Official Gazette.

The site:

http : //ec.europa.eu/enterprise/automotive

contains a suitable research engine for looking into Directives and Regulations using keywords or their identification numbers.

8.4.1 The Vehicle in General

The D 70/156 Directive defines the homologation procedure reported in summary in the previous paragraph; according to this Directive, the manufacturer is obliged to submit an *information form*, reporting all vehicle characteristics that cannot be altered without a new homologation. The information form reports the following information regarding the issues of concern, including the manufacturer's data, the position of the Vehicle Identification Number (VIN) identifying body, chassis and engine, and the vehicle category.

The following vehicle categories are identified:

- M_1: vehicles for transporting people with fewer than eight seats, in addition to the driver's seat;
- M_2: vehicles for transporting people with more than eight seats, in addition to the driver's seat and with a maximum overall weight of 5 t[5];
- M_3: as above, but with an overall weight exceeding 5 t;
- N_1: vehicles for transportation of goods with maximum weight exceeding 1 t, but lower than 3.5 t;
- N_2: vehicles for transportation of goods with maximum weight exceeding 3.5 t, but lower than 12 t;
- N_3: vehicles for transportation of goods with a maximum weight exceeding 12 t;
- O_1: trailers with a maximum weight not exceeding 0.75 t;
- O_2: trailers with a maximum weight exceeding 0.75 t, but lower than 3.5 t;
- O_3: trailers with a maximum weight exceeding 3.5 t, but lower than 10 t;
- O_4: trailers with a maximum weight exceeding 10 t.

[5] Rules were issued before the compulsory introduction of the SI measuring system. Newer updates report masses, instead of weights.

Several pictures and a vehicle scheme showing the main dimensions are attached to the information form; the number of axles and wheels is reported, showing permanent or part time driving wheels. An outline scheme of the chassis frame, if any, should be included, indicating the material used for side beams.

Among the main dimensions, the wheelbase and interaxis (for vehicles with more than two axles) must be reported under full weight conditions; for trailers, the distance between the hook and first axle pivot must be declared; for road tractors, the saddle pivot longitudinal and elevation position must be referenced to the vehicle. All dimensions are defined by the ISO 586 standard. All tracks must also be declared.

The weight of the bare chassis frame (if any) must be declared, not including cabin, fluids, spare wheel, tools and driver; the weight breakdown on the axles must be also declared.

Also the weight of the vehicle completed with body or cabin (depending on the product sold by the manufacturer) and other items must be declared together with the weight distribution on each axle; if the vehicle is a semi-trailer, the weight on the hook must also be stated.

Finally, the maximum allowed weight has to be declared together with its breakdown on axles and hook (if any).

The transmission is described by a draft scheme, including data regarding its weight, architecture (single stage, double stage, etc.), type of control (manual or automatic), transmission ratios (gearbox and final drives) and the vehicle speed that can be obtained on any of the existing gears at an engine speed of 1,000 rpm.

Suspension schemes must also be attached, including the damping and elastic characteristics of shock absorbers and springs; allowed tire sizes must be declared.

A scheme of the steering mechanism and column must also be included; maximum design forces on the steering wheel and maximum steering angles at the wheel and the steering wheel must be declared. For these angles, the vehicle turning radii for right and left turns must be stated.

Service, emergency and parking brakes must be fully described.

Finally, the body is described, including lighting, signaling, passive and active safety devices; this topic is described further in the following sections.

A *homologation form* certifies the released homologation and reports for each of the characteristics of the information form:

- the conformity of the presented prototypes to the described items;
- the conformity of those characteristics to legal requirements;
- the positive execution of tests;
- the existence of required drawings.

The D 70/156 Directive also reports all forms to be used for information and for the certificate of conformity.

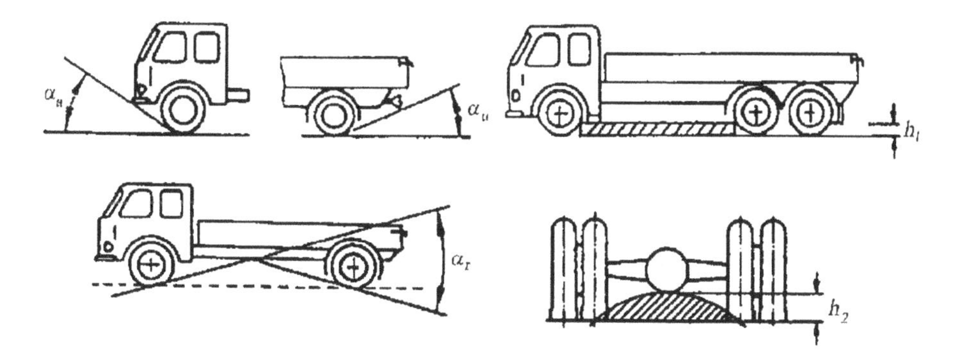

Fig. 8.35. Front and rear attack angles α_a and α_p; α_r ramp angle; ground clearance h_1 and h_2.

The D 87/403 Directive completes the previous documents with the definition of off road vehicles; these are vehicles of M_1 and N_1 categories, featuring the following characteristics:

- at least one front and one rear driving axle, one of which can be disengaged by the driver;
- at least one self-locking or locking differential;
- the grade ability of at least 30%, with no trailer;
- at least one of the following requirements:
  - angle of attack α_a of, at least, 25°;
  - angle of exit α_u of, at least, 20°;
  - ramp angle α_r of, at least, 20°;
  - ground clearance h_2, under the front axle of, at least, 180 mm;
  - ground clearance h_2, under the rear axle of, at least, 180 mm;
  - ground clearance between the mid of each axle h_1 of, at least, 200 mm.

Fig. 8.35 indicates the reported dimensions using sketches.

The front and rear attack angles measure the capacity of a vehicle to face sudden slope changes in normal forward and reverse driving without any interference between chassis and ground; the ramp angle, on the other hand, refers to a sudden slope change in both directions.

The ground clearance between axles is the maximum height of an ideal parallelogram that can be inserted between the axles and under the chassis; the ground clearance under the axle refers to the lower point between the two contact points of the wheels on the same axle.

Other geometrical prescriptions are assigned to other kinds of vehicles.

The D 91/21 Directive updates the previous ones by specifying that an M_1 motor vehicle should feature a maximum speed of at least 25 km/h; in addition, it introduces mass as a measurement instead of weight.

The same Directive establishes maximum vehicle dimensions:

- 12,000 mm of overall length;
- 2,500 mm of overall width;
- 4,000 mm of overall height.

The maximum allowed vehicle mass must be, at least, the total of vehicle curb mass plus the product of offered passenger seats multiplied by 75 kg, which is assumed to be the nominal weight of a passenger, including his hand baggage, plus the useful payload.

Mass breakdown on the axles may be calculated by positioning the passenger reference weight at the R point of each seat; sliding seats must be set at their rearmost position. Allowed baggage and payload must be uniformly distributed on the trunk or load compartment floors.

Measured vehicle mass, at prototype homologation or control of conformity, is admitted within a tolerance field of $\pm$ 5% around the declared values.

The D 92/53 Directive presents many updates of D 70/156 for the forms and homologation procedure. In this revision, specific rules are introduced regarding small volume productions and end of series productions, along with rules concerning waivers; the concept of equivalence between homologations granted by different member States is also introduced.

It is also established that each member State issuing homologation certificates must arrange statistical control plans on operating vehicles, suitable to detect possible non-compliance with the homologated prototypes; in the case of nonconformity, the issuing State must inform other States of the event and must organize the compulsory recovery plan for the existing vehicles.

All applicable Directives and Regulations are reported on the Attachment IV of this document.

Directive D 2000/53 establishes rules to control waste products from vehicles disposal by component and materials recycling.

Each year about 12 million vehicles are scrapped in the European Union; they correspond to about 0.5 % of the total production.

The rules of this Directive also address improvements to the environmental operations of companies involved in this activity.

To prevent noxious waste formation, a set of laws has been introduced to limit the use of some substances for vehicle and component manufacturers, making recycling easier and avoiding dangerous waste treatment.

Member states must adopt laws suitable for reaching the following overall targets:

- The recovery percentage (materials not sent to a landfill) of scrapped vehicles must be at least 85% of the average vehicle weight, while at least 80% of the weight must be reemployed. For vehicles produced before 1980, lower targets can be set, but not lower than 75% for recovery and 70% for recycling.

- The recovery percentage must reach 95% by 2015 and the recycling percentage 85% of the average vehicle weight.

For this purpose, components and materials must be code labelled, to enhance identification and classification for selective recovery.

Scrapped vehicle treatment must include:

- batteries and LPG vessels removal;

- removal of explosive materials, such as air bags;

- removal and separated collection of fluids, like fuels, lubricants, cooling fluids, brake oil, air conditioning fluids and others, unless they are necessary to parts reemployment;

- removal of all components containing mercury.

Other scheduled operations include:

- catalyst removal;

- selective removal of parts containing copper, aluminum and magnesium;

- tires and big plastic elements removal (bumper, dashboard, reservoirs);

- glass removal.

The economic accomplishment of these operations implies a number of additional design rules:

- banning certain materials, such as asbestos, lead, cadmium, chromium, etc.

- indelible material labelling of any component;

- designing components with a reduced number of materials;

- designing an easy disassembly;

- identifying components suitable for a second life.

In the near future, it is likely that manufacturers will be obliged to accept the burden of disassembly.

8.4.2 Body Shape

Group 4: licence plate housing

Directive D 70/222, the only in group 4 at this time, establishes that plate housings must feature a rectangular flat surface of at least 520 mm (width) by 120 mm (height) or 340 mm (width) by 240 mm (height).

The median vertical line of the licence plate must lie in the left half of the body (with reference to driving direction).

The left edge of the licence plate must be inside the plane parallel to vehicle symmetry plane and tangent to the body in its maximum width point.

The licence plate must have its vertical edge parallel to the longitudinal vehicle symmetry plane; the plate plane must be vertical with a tolerance of 5°. However, for particular body shapes, a higher inclination is allowed, i.e.:

- no more than 30° upwards, if the upper edge of the plate is not higher than 1.20 m on the ground;
- no more than 15° downward, if the upper edge of the plate is higher than 1.20 m on the ground.

The lower edge of the plate must not be lower than 0.30 m on the ground, while the upper edge must not be higher than 1.20 m. However, if this statement cannot be fulfilled, a higher height is allowed, but should be as close as possible to the limit. In any case, a height over 2.00 m is forbidden.

The licence plate must be seen from any point included between four planes: Two vertical planes through the lateral plate edges with an angle of 30° with the vehicle longitudinal symmetry plane; one upper plane through the upper plate edge with an angle of 15° upwards; one horizontal lower plane through the lower plate edge. If the upper edge of the plate is higher than 1.20 m on the ground, this last plane must be 15° inclined downwards.

All the heights above must be measured at unloaded vehicle.

Group 6: Doors and running boards

The D 70/387 Directive establishes that all vehicles must be designed to allow occupants a safe entrance and exit.

All doors must be comfortably used without any danger. Doors and locks must be designed in order to avoid annoying noise at their closure.

Locks must avoid doors unintentional opening.

Side door hinges must be in front of the door, with reference to vehicle driving direction. For wardrobe doors (hinges at the opposite sides of two neighboring door, no B pillar in between), this statement apply to one door only; in this case, the other cannot be opened, unless the first has been previously opened.

The door lock and its mechanisms must withstand:

- a longitudinal and a transversal load of 444 daN each, at partially locked condition;

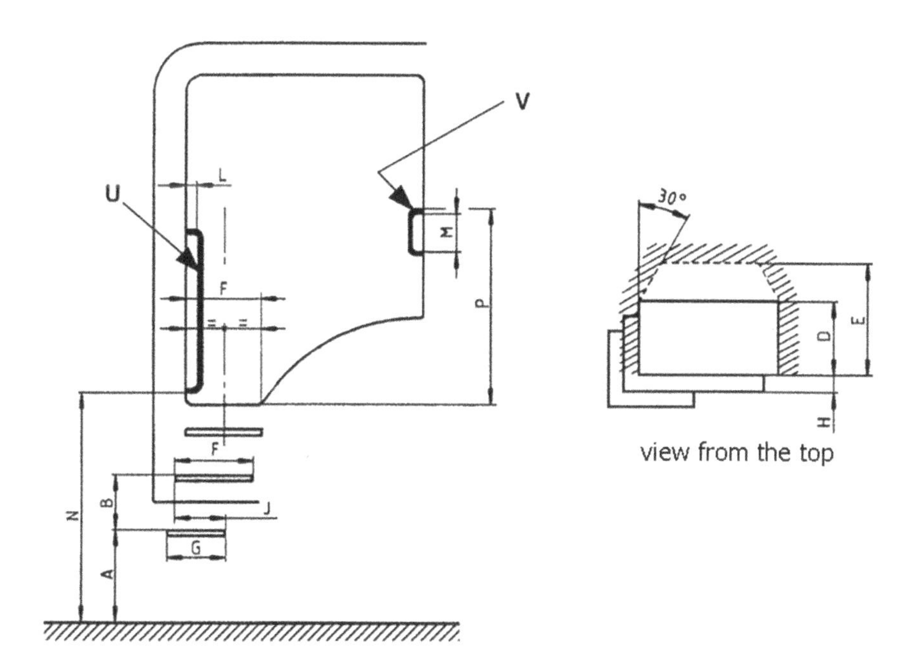

Fig. 8.36. Scheme for regulated dimensions relevant to vehicle access for M and N categories.

- a longitudinal load of 1,111 daN and a transversal load of 889 daN, at fully locked condition.

The door lock must remain closed under 30 g acceleration in both longitudinal and transversal directions.

Hinges must withstand the same forces that are applied to locks.

The same Directive shows test procedures to be used to demonstrate compliance.

The D 98/90 Directive is about the application of similar requirements to class M and N vehicles and sets requirements on their running boards and handrails.

If the vehicle floor is higher than 600 mm over the ground (only by off-road vehicles this value is increased to 700 mm), a suitable number of running boards must be provided to allow entering and exiting the vehicle.

The lower running board must not be higher than 600 mm over the ground and must be designed in order to avoid slip.

Hubs or other parts of the wheel cannot be used as running boards, unless technical reasons inhibit the application of running boards to the body.

In addition, with reference to Fig. 8.36, the following conditions must be fulfilled:

- running board depth, $D \geqslant 80$ mm;
- clearance to be penetrated by occupant's foot, $E \geqslant 150$ mm;
- running board width, $F \geqslant 300$ mm (reduced to 200 mm for off-road vehicles only);
- lowest running board width, $G \geqslant 200$ mm;
- no difference in running boards protrusions is allowed ($H = 0$);
- longitudinal overlap, $J \geqslant 200$ mm.

To make climbing of running board safer, one or more handrails must be provided. They must be positioned to be easily seizable and to not obstruct entrance.

If more than one running board is provided, the number and position of handrails must allow the climbing occupant to stand on three points, at least: two hands and one foot or one hand and two feet; handrails must be positioned to discourage climbing and descending with the face outwards.

The steering wheel may be considered as a handrail.

The height of the lowest point of a handrail over the ground, N must be below 1,850 mm (1,950 mm are allowed for off-road vehicles).

In addition, for type U handrails, the highest point of a handrail over the cabin floor, P must be below 650 mm, while for type V handrails, must be below 550 mm.

In addition the following conditions must also be fulfilled:

- handrail tube diameter, between 16 and 38 mm;
- minimum handrail length, M over 150 mm;
- dimension L over 40 mm.

Additional requirements on this subject are set by D 2001/31.

Group 16: Body projections

Prescriptions of Directive D 74/483 apply to the *outside surface*, which is that portion of the body below 2.00 m, over the ground, and over the body *base line* and that can be touched by a sphere of 100 mm diameter that is dragged over the body surface limited above.

The base line can be determined by ideally dragging a cone on the body surface, with vertical axis and an aperture of 30°; the base line is the geometrical trace of tangency points of this cone with the body surface.

While determining base line, jack rests, exhaust pipe(s) and wheels must not be taken into consideration. Near wheel openings, the body surface must be considered as ideally closed with a surface tangent to fender rims.

This so-defined outside surface must not exhibit either sharp edge elements or projections that could increase the risk or severity of injuries of to pedestrian hit or brushed by a vehicle passing-by. Also no projections are allowed that could hook to pedestrians or cyclists.

Protruding areas of this surface are not allowed with a curvature radius below 2.5 mm or made with materials over 60° Shore of hardness.

Ornaments applied to the body, protruding more than 10 mm over their attachment should tilt or disappear over a force higher than 10 daN contained in whichever direction on a plane parallel to the body surface.

Wipers articulation shafts must be protected by a cover complying with conditions above, with a surface not less than 150 mm^2.

Door handles, hinges or opening push bottoms must not protrude over 30 mm (40 mm for door handles only).

If door handles are to be turned to open the door, the handle tip must be oriented to the rear of the vehicle and be bent towards the door surface; the handle tip must move parallel to the door surface when is opened.

Every body panel rim must be rounded with no less than 2.5 mm radius and not less than one tenth of its protrusion.

Directive D 79/488 provides that door handles which not rotate parallel to door surface must be protected, in their closed position, by a rim or by insertion in a niche; their moving tip must be oriented rearwards or downwards.

The same Directive reports information forms to be used and procedures to be applied in tests.

Other minor details are reported by D 87/354, D 2006/96 and D 2007/15.

Group 45: Tall vehicles lateral protections

The Directive D 89/297, the only in this group, applies to N_2, N_3 and O_2, O_3 vehicles.

It does not apply to semi-trailer tractors and to trailers designed for long loads (as logs, pipes, etc.) where the load itself is part of the vehicle structure.

This Directive provides for the application of lateral protections, the purpose of which is to protect pedestrians or bikers from being dragged or swept away.

The *lateral protection* is essentially a continuous flat plate or one or more horizontal rails, or a combination of each.

Rails must be installed with a relative distance between 50 and 300 mm, in the case of N_2 and O_2 vehicles and of 100 mm, for N_3 and O_3 vehicles; plates or rails made of more than one piece must build up a continuous protection.

Protections must not increase the vehicle width and their surface must not be narrower than 120 mm (for each side) of the vehicle maximum overall width. Front ends must be bent inside the vehicle, while rear ends must not be narrower than 30 mm (for each side) of the overall rear tires width.

External surfaces must be smooth and essentially flat. Adjacent elements of the same protection can overlap, but the upper side must be set closer to the front or to the top of the vehicle. Empty spaces in the longitudinal direction are allowed, for no more than 25 mm in length; a rear element cannot be wider than its neighboring front element.

Bolt and rivet heads must be rounded and cannot protrude for more than 10 mm; any part of these protections must also comply with Directive D 74/483.

8.4.3 Artificial External Lighting

Group 20: Lighting and signaling devices installation

This group includes Directive D 76/756 and its updates D 80/233, D 82/244, D 83/276, D 84/8, D 89/278, D 91/663, D 97/28 and D 2007/35.

They regard devices suitable to light the road (headlights) or to emit a light signal (stop lights, blinker and hazard lights). Also license plate lights and reflector lights are considered in this category.

Some definitions are introduced by this Directive:

- Independent lights, when they have separate lighting surfaces, bulbs and casings.
- Grouped lights, when they share the same casing.
- Combined lights, when they share the same casing and bulb.
- Incorporated lights, when they have different bulbs, but share the same lighting surface and casing.

Each of these category is allowed only for specific applications.

Installation conditions on the body regard light position and visibility angles.

Visibility angles determine the minimum size of the apparent body surface on which lights can be seen by an observer (*lighting surface*). If a sheaf of straight lines is traced for the bulb center (optical centre) and the lighting surface contour, a light cone is obtained which defines a spherical segment at a certain distance from the optical center.

A horizontal plane for the optical center cuts this sphere segment according to its equator, while a vertical plane according to its meridian. The visibility angle β is measured on the equator, starting from the intersection of the light horizontal axis with the visibility surface, while the visibility angle α is measured on the meridian.

No obstacles are allowed within the area identified by visibility angles, that could impair light propagation.

High beam lights

High beam lights are compulsory on self-propelled vehicles, but are not allowed on trailers. The allowed number for this kind of light is 2 or 4.

Their lighting surface external edge cannot be positioned outside the light surface of any low beam light.

They must be installed in front of the front axle; their light must not disturb the driver directly or indirectly (through mirrors or other reflecting surface).

Light visibility must be granted within a cone of at least 5° aperture, with its axis through the bulb, in the x direction.

Devices capable of adjusting the light optical axis, in consequence of static pay load variation, must be provided. When two pairs of high beam lights are provided, one pair can automatically rotate in a horizontal plane, as function of the steering angle.

High beam lights can be switched on simultaneously or by symmetric couples. Low beam lights may be switched on contemporarily to high beam lights.

Low beam light

Low beam lights are compulsory on self-propelled vehicles, but are not allowed on trailers.

The outside edge of their lighting surface cannot be farther than 400 mm from the widest point of the vehicle body. The distance between the inside edges of their lighting surface cannot be lower than 600 mm.

The height from the ground must be within the range 500 to 1,200 mm.

They must be installed in front of the front axle; their light must not disturb the driver directly or indirectly (through mirrors or other reflecting surface).

Established visibility angles are $\beta = 15°$ upwards and 10° downwards, $\alpha = 45°$ to the outside and 10° to the inside.

The presence of reflecting or bright surfaces in the front of the car must not disturb other road users.

The inclination of the low beam axis is measured in static conditions for all load conditions to be taken into account. The beam inclination must be set between 0.5% and 2.5%, without manual adjustment, while an adjustment between 1% and 1.5% must be possible, with the driver only on board.

The initial inclination must be shown on a sticker on the vehicle.

Previous conditions must be also fulfilled by an automatic adjustment device. In this device is applied, in case of its failure, the light beam must be lowered, with reference to its position before the failure occurred.

Low beam light cannot be combined with other lights; they can be grouped or incorporated with other lights.

D 82/244 specifies how to measure beam inclination as function of actual static load.

Front fog lights

The application of front fog lights is optional on self-propelled vehicles and are not allowed on trailers. Only a couple is allowed.

The outside edge of their lighting surface cannot be farther than 400 mm from the widest point of the vehicle body.

Their height from the ground must be at least 250 mm.

No point of their lighting surface is allowed to be higher that the highest point of the low beam lighting surface.

Also in this case, their light must not disturb the driver directly or indirectly (through mirrors or other reflecting surface).

Established visibility angles are $\beta = 5°$ upwards and 5° downwards, $\alpha = 45°$ to the outside and 10° to the inside.

Fog lights cannot be combined with other lights; they can be grouped or incorporated with any other front light.

Reverse light

Reverse lights are applied to improve driver's rear visibility and to signal to other road users that the vehicle is moving backwards.

They are compulsory on self-propelled vehicles, but are not allowed on trailers; one or two are allowed in the rear of the vehicle.

Their height on the ground must be between 250 and 1,200 mm.

Established visibility angles are $\beta = 15°$ upwards and 5° downwards, $\alpha = 45°$ to the right and 45° to the left, in case of a single light, or $\alpha = 45°$ outwards and 30° inwards, in case of a couple of lights.

Reverse lights can be grouped with other rear lights; they cannot be combined nor incorporated.

Blinkers

Blinkers are applied to inform other road users about the intention of the driver to change direction; when switched on simultaneously they can be used as hazard lights.

They are compulsory for any kind of road vehicle. They are classified according to three categories (1, 2 and 5) and two installation schemes (A and B).

Scheme A applies to all self-propelled vehicles. It provides for 2 front lights (category 1), 2 rear lights (category 2) and 2 side lights (category 5).

Scheme B applies to trailers only. It provides only for 2 rear lights (category 2).

The outer edge of the lighting surface cannot be farther than 400 mm from the widest point of the body. The minimum distance between the internal edges of two adjacent lights cannot be less than 600 mm.

Front blinkers lighting surface must not be closer than 40 mm from the lighting surface of low beam lights or fog lights, if they are applied. A lower distance is allowed if the blinker light intensity is at least 400 cd on its optical axis.

Minimum height on the ground is 500 mm for category 5 lights and 350 mm for other categories; the maximum height is 1,500 mm for all categories.

If the vehicle architecture does not allow this maximum limit to be fulfilled it can be increased up to 2,300 mm for category 5 and 2,100 for categories 1 and 2.

The distance between the center of the lighting surface of a side blinker (scheme A) and a cross plane tangent to the front of the body must not be more than 1,800 mm. If the architecture of the vehicle does not allow to comply with minimum visibility angles, this distance can be increased to 2,500 mm.

Blinkers can be grouped with other lights or incorporated with parking lights; they cannot be combined.

Stop lights

They show to other road users that the driver is operating the service brake. They are compulsory on every vehicle.

The minimum distance between two stop lights must be higher than 600 mm. This distance can be reduced to 400 mm, if the overall width of the vehicle is lower than 1,300 mm.

The minimum height on the ground is 350 mm, while the maximum is 1,500 mm or 2,100 if the body shape does not allow the lower value.

Established visibility angles are $\beta = 45°$ outwards and inwards, $\alpha = 15°$ upwards and downwards. This last value is limited to 5° is the height of the light on the ground is lower than 750 mm.

Stop lights can be incorporated or grouped with position lights.

Front position lights

Front position lights are used to signal the presence of a vehicle and to mark its width, as seen from a point of view in front of the vehicle.

They are compulsory on self-propelled vehicles and trailers with an overall width over 1,600 mm.

The outside edge of the lighting surface cannot be set at a distance larger than 400 mm from a vertical plane tangent to the body side.

In the case of a trailer this dimension is reduced to 150 mm.

The minimum distance between the lighting surface of two position lights is 600 mm. The height above the ground must be over 350 mm and below 1,500 mm or 2,100 mm if the body shape does not allow the first value to be respected.

Established visibility angles are $\beta = 45°$ inwards and 80° outwards, $\alpha = 15°$ upwards and downwards. This last value is limited to 5° is the height of the light on the ground is lower than 750 mm.

They can be grouped or combined with other front lights.

Rear position lights

Rear position lights are used to signal the presence of a vehicle and to mark its width, as seen from a point of behind the vehicle. They are compulsory on every kind of vehicle.

The minimum distance between the lighting surfaces must be over 600 mm (reduced to 400 mm, if the vehicle width is lower than 1,300 mm), while the outside edge of the lighting surface must be distant no more than 400 mm from a vertical plane tangent to the side of the body.

The height above the ground must be between 350 and 1,500 mm or 2,100 mm, if the first limit cannot be respected.

Established visibility angles are $\beta = 45°$ inwards and $80°$ outwards, $\alpha = 15°$ upwards and downwards.

This last value is limited to $5°$ if the height of the light on the ground is lower than 750 mm.

Rear position lights can be combined, grouped or incorporated with other rear lights.

Rear fog lights

Rear fog lights are used to improve visibility of the vehicle as seen from a rear point of view, in the presence of fog or mist.

One fog light is compulsory on every kind of vehicle, while a second symmetric light is optional.

If only one light is applied it must be near the center of the road, depending on the driving side.

The distance between the rear fog light lighting surface and that of the stop light must be more than 100 mm.

The height above the ground must be between 250 mm and 1,000 mm.

Established visibility angles are $\beta = 25°$ inwards and outwards, $\alpha = 5°$ upwards and downwards.

They cannot be combined with other rear lights.

Parking lights

Parking lights are applied to signal a parked vehicle in a populated area. They can be used instead of position lights. They are optional on every vehicle below 6 m of length and 2 m of width and are forbidden on other vehicles.

Two parking lights must be provided in front of the vehicle and two in the rear side or one on each side of the vehicle.

The outside edge of their lighting surface must not be further than 400 mm from a vertical plane tangent to the body side.

The height above the ground should be between 350 mm and 1,500 mm or 2,100 mm if the last condition cannot be observed.

They cannot be combined with other lights.

Size lights

Size lights are positioned on the highest point of both sides of large vehicles to signal to other road users the height of the vehicle.

They are compulsory on tall vehicles with more than 2,100 mm of overall height; two of them are applied to mark the front side and two the rear side of the vehicle.

They should be applied as high and as external as possible, in symmetric positions.

Established visibility angles are $\beta = 80°$ outwards, $\alpha = 5°$ upwards and 20° downwards.

They cannot be combined, nor grouped, nor incorporated with other lights.

Non-triangular rear reflector

A reflector signals the presence of a vehicle, reflecting the light coming from oncoming vehicles.

Non-triangular rear reflectors are compulsory on every vehicle, but trailers.

The external edge of the lighting surface cannot be further then 400 mm from a vertical plane tangent to the body side, while the internal edge cannot be closer then 600 mm to its counterpart.

This distance can be reduced to 400 mm, if the body overall width is lower than 1,300 mm.

The height above the ground must be between 350 mm and 900 mm.

Established visibility angles are $\beta = 30°$ outwards and inwards, $\alpha = 15°$ upwards and downwards.

They can be grouped with other lights.

Triangular rear reflector

Triangular rear reflectors are compulsory on trailers, and forbidden on self-propelled vehicles.

The external edge of the lighting surface cannot be further then 400 mm from a vertical plane tangent to the body side, while the internal edge cannot be closer then 600 mm to its counterpart.

This distance can be reduced to 400 mm, if the trailer overall width is lower than 1,300 mm.

The height above the ground is between 350 mm and 900 mm.

Established visibility angles are $\beta = 30°$ outwards and inwards, $\alpha = 15°$ upwards and downwards.

The lower limit for α can be reduced to 5° if the height over the ground is lower than 750 mm.

They cannot be grouped with other lights.

Front non-triangular reflector

Front non-triangular reflectors are compulsory on trailers only.

The outside edge of the lighting surface cannot be further then 150 mm from a vertical plane tangent to the body side, while the inside edge of the lighting surface.

The minimum distance between the inside edges of the lighting surface must be 600 mm; it can be reduced to 400 mm if the trailer overall width is lower than 1,300 mm.

The height over the ground must be between 350 mm and 900 mm or 1,500 mm if the first condition cannot be observed.

Established visibility angles are $\beta = 30°$ outwards and inwards, $\alpha = 15°$ upwards and downwards.

The lower limit for α can be reduced to 5° if the height over the ground is lower than 750 mm.

They can be grouped with front position lights.

Lateral non-triangular reflector

Lateral non-triangular reflectors are compulsory on trailers and self propelled vehicles; on M vehicles longer than 6 m they are optional.

The number of reflectors to be applied is determined by applying the following conditions. At least one reflector must be located in the third part of the side of the vehicle, ideally divided in three equal parts; the first reflector of one side must not be further than 3 m from the front of the vehicle (for this condition the towing beam of a trailer is included in the vehicle length); two neighboring reflectors cannot be further than 3 m. The distance between the rearmost reflector and the tail of the vehicle cannot be over 1 m.

The height above the ground must be between 350 mm and 900 mm, or 1,500 mm if the body shape does not allow the first condition to be respected.

Established visibility angles are $\beta = 45°$ forwards and backwards, $\alpha = 15°$ upwards and downwards.

The lower limit for α can be reduced to 5° if the height over the ground is lower than 750 mm.

Group 21: Reflectors

Directives D 76/757, D 87/354, D 97729 and D 2006/96 report prescriptions regarding shape, dimensions and photometric performance for any kind of reflector.

Test procedures for photometric performance measurement are also available which are not commented on here for the sake of brevity.

Group 22 through 30: Lights

Many Directives report prescriptions about shape, dimensions and photometric performance for any kind of light we have examined above.

Test procedures for photometric performance measurement are also available which are not commented on here for the sake of brevity.

8.4.4 External Visibility

Group 8: Indirect vision devices

This group includes Directive D 71/127 and its modifications and updates, reported in D 79/795, D 85/205, D 86/562, D 87/354, D 88/321, D 2003/97, D 2005/27 and D 2006/96.

Indirect vision devices allow to observe parts of the road adjacent to the vehicle that are invisible to direct vision.

They could be conventional mirrors or cameras with monitors.

In the case of mirrors, the following categories are classified:

- Internal mirror: to be installed in the car passenger compartment.
- External mirror: to be installed on a point on the outside body surface.
- Watch mirror: to be installed inside or outside of the passenger compartment to allow specific fields of view, different from those of the two classes above.

Mirrors are also classified in categories:

- Category I, for internal mirrors.
- Category II and III, for main external mirrors.
- Category IV, for wide-angle watch mirrors.
- Category V, for pull-in external mirrors and
- Category VI, for additional front external mirrors.

Different category have specific fields of view we will describe later.

Category I

The reflective surface of inside mirrors must be able to include a rectangle 40 mm high and a mm wide; the following formula specifies:

$$a = \frac{150}{1 + \frac{1{,}000}{r}},$$

where r is the average of the radii of the reflective surface, to be measured according to a specification reported in the Directive.

Category II and III

The reflective surface of these external mirrors must be able to include a rectangle 40 mm high and a mm wide and a segment, parallel to the rectangle height b mm long.

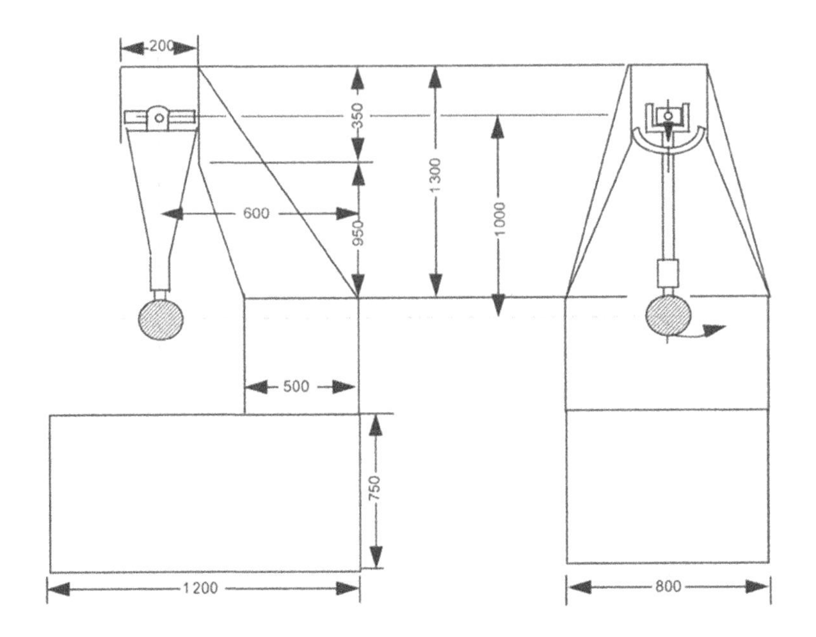

Fig. 8.37. Test machine to demonstrate mirror mounts collapsibility.

For category II:

$$a = \frac{170}{1 + \frac{1{,}000}{r}}, \qquad b = 200mm.$$

For category III:

$$a = \frac{130}{1 + \frac{1{,}000}{r}}, \qquad b = 70mm.$$

Each external mirror must be fixed to the vehicle body with collapsible mounts of which the effectiveness must be measured with the following procedure.

The test device is a pendulum that can oscillate according to two perpendicular axes, one of them perpendicular to the launch path of the pendulum. The test machine is shown in Fig. 8.37.

The pendulum bears a hammer, made with a rigid sphere of 165 $\pm$ 1 mm diameter, covered with a 5 mm rubber skin of 50 Shore A hardness.

The center of the spherical hammer is set at a distance l = 1 m $\pm$ 5 mm from the oscillation axis. The pendulum equivalent mass is m = 6.8 $\pm$ 0.05 kg.

Mirrors must be positioned in the test rig in the same position that they will actually assume in the vehicle. The test consists in letting the pendulum fall from an angle of 60° with the vertical line joining the pendulum pivot with the mirror to be hit.

Two kinds of test are established for internal mirrors.

- Test 1: the hammer must hit the mirror on the reflecting surface.
- Test 2: the hammer must hit the mirror on the edge of the mirror protection, from the side of the reflective surface and with an inclination of 45° with reference to this surface.

Again, two kinds of test are established for external mirrors.

- Test 1: the hammer must hit the mirror on its reflective surface.
- Test 2: the hammer must hit the mirror on the opposite side of its reflective surface.

In case of category II and III with a common mount with a category IV mirror, tests must be applied to the lower unit only.

The result of all these tests must be that the pendulum is able to overcome the mirror of at least 20° ± 1°, with reference to the vertical line.

In case of an internal mirror glued to the windshield, the part of the broken mount remaining on the windshield must not protrude for more than 10 mm over the glass surface.

After the test the reflective surface should be undamaged; however surface rupture is allowed if mirror fragments remain attached to their support; only very small fragments are allowed to separate from the surface in the impact point.

The following table summarizes prescriptions about the application of the mentioned mirrors categories to different kinds of vehicles. Abbreviations 'Opt.', 'Comp.', 'Proh.' stand for optional, compulsory and prohibited respectively.

Vehicle category	Cat. I	Cat. II	Cat. III	Cat. IV	Cat. V	Cat. VI
M_1	Comp.	Opt.	Comp.	Opt.	Opt.	Opt.
M_2	Opt.	Comp.	Proh.	Opt.	Opt.	Opt.
M_3	Opt.	Comp.	Proh.	Opt.	Opt.	Opt.
N_1	Comp.	Opt.	Comp.	Opt.	Opt.	Opt.
N_2 ($\leq$ 7.5 t)	Opt.	Comp.	Proh.	Opt.	Opt.	Opt.
N_2 ($>$ 7.5 t)	Opt.	Comp.	Proh.	Comp.	Comp.	Comp.
N_3	Opt.	Comp.	Proh.	Comp.	Comp.	Comp.

The driver's ocular points are defined as two points set 635 mm vertically above the driver's R point, at a distance of 32,5 mm from a vertical, longitudinal plane through the same point, each in one side. The straight line through them is perpendicular to this plane[6].

The binocular field of vision is obtained by superimposition of the right and left monocular fields of view, as shown in Fig. 8.38.

[6] The definition of the R point is given in the following chapter.

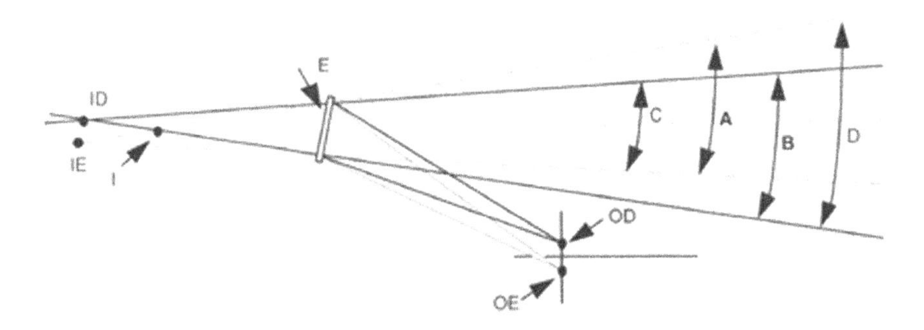

Fig. 8.38. Scheme for determining the binocular field of vision in a mirror.

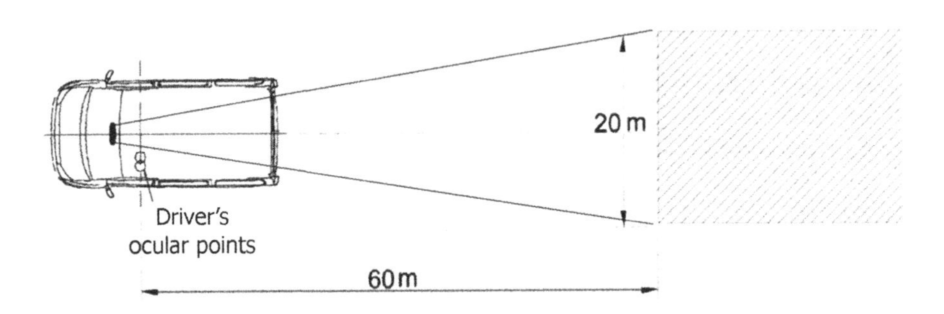

Fig. 8.39. Fields of view for Category I mirrors; the dashed area is draft on the ground.

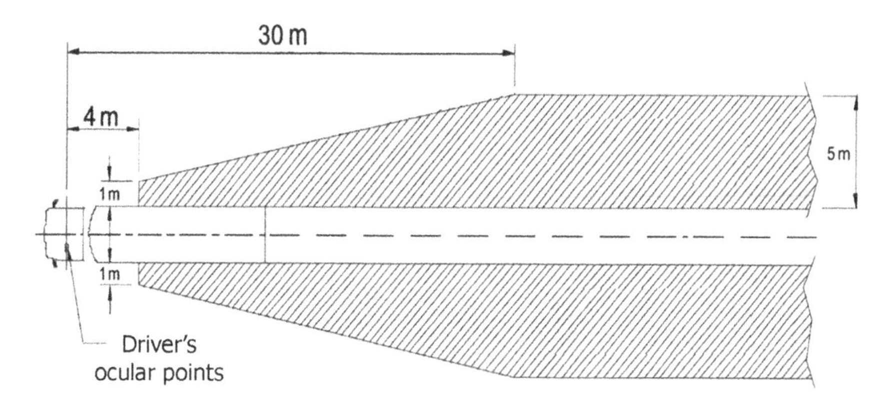

Fig. 8.40. Fields of view for Category II mirrors; the dashed area is draft on the ground.

Internal mirror (Category I)

The field of view of internal mirrors must be such that the driver can see a portion of a horizontal and flat road, aligned with the vehicle symmetry plane which is 20 m wide from 60 m behind the driver's ocular points and reaching to the horizon (see Fig. 8.39.).

External mirrors (Category II)

For external mirrors, located at each side, the field of vision must be such that the driver can see a portion of a horizontal and flat road 5 m wide, measured from a plane parallel to the vehicle symmetry plane for the most external point of the body, from 30 m behind the driver's ocular points to the horizon.

The front boundary of the field of vision starts with a width of 1 m from the plan parallel to the vehicle symmetry plane for the most external point of the body, 4 m behind the driver's ocular points (see Fig. 8.40).

External mirrors (Category III)

The field of vision of these mirrors must be such that the driver can see a portion of a horizontal and flat road 4 m wide, starting from a plane parallel to the vehicle symmetry plane for the most external point of the body, from 20 m behind the driver's ocular points till the horizon.

The front boundary of the field of vision starts with a width of 1 m from the plan parallel to the vehicle symmetry plane for the most external point of the body, 4 m behind the driver's ocular points (see Fig. 8.41).

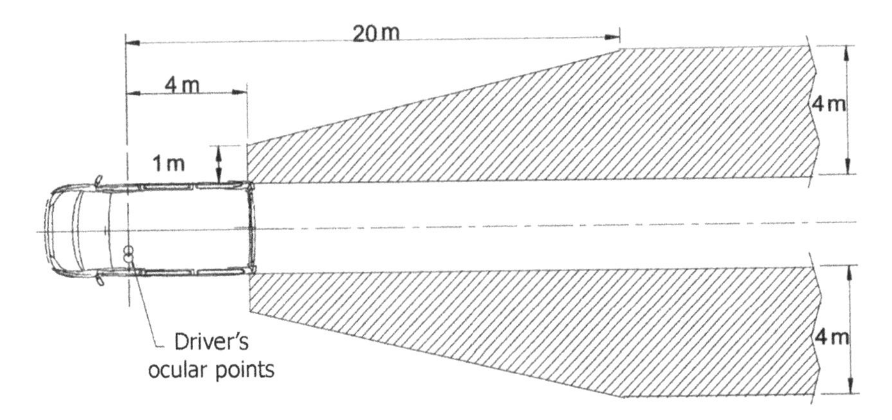

Fig. 8.41. Fields of view for Category III mirrors; the dashed area is draft on the ground.

External grand angular mirrors (Category IV)

The field of vision of these mirrors must be such that the driver can see a portion of a horizontal and flat road 15 m wide, starting from a plane parallel to the vehicle symmetry plane for the most external point of the body, from 10 m to 25 m behind the driver's ocular points.

The front boundary of the field of vision starts with a width of 4,5 m, from the plan parallel to the vehicle symmetry plane for the most external point of the body, 1,5 m behind the driver's ocular points (see Fig. 8.42).

Pull-in external mirrors (Category V)

The field of vision must be such that the driver can see a portion of a horizontal and flat road, with the following boundaries (see Fig. 8.43 a and b):

- A plane parallel to the vehicle symmetry plane for the most external point of the body.
- A plane parallel to the vehicle symmetry plane, at 2 m from the most external point of the body.
- A vertical plane, perpendicular to the vehicle symmetry plane, 1,75 m behind the driver's ocular points.
- A vertical plane, perpendicular to the vehicle symmetry plane, 1 m in front of driver's ocular points, or tangent to the front bumper, if this latter plane is closer to the driver's ocular points.

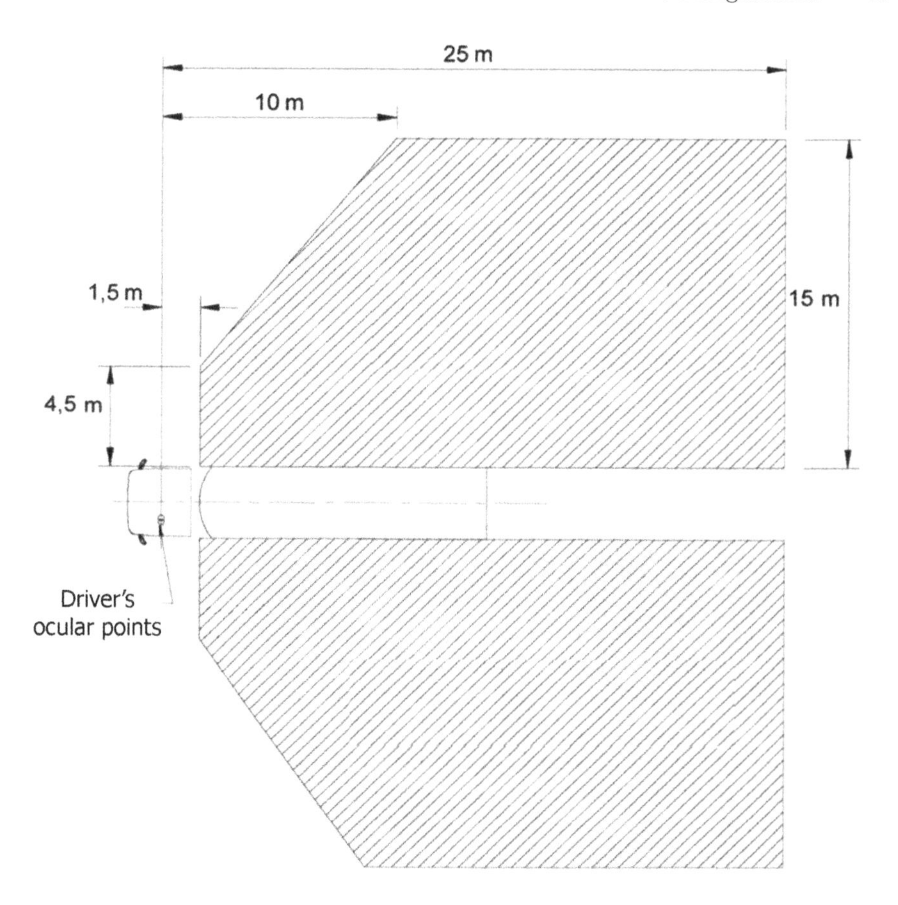

Fig. 8.42. Fields of view for Category IV mirrors; the dashed area is draft on the ground.

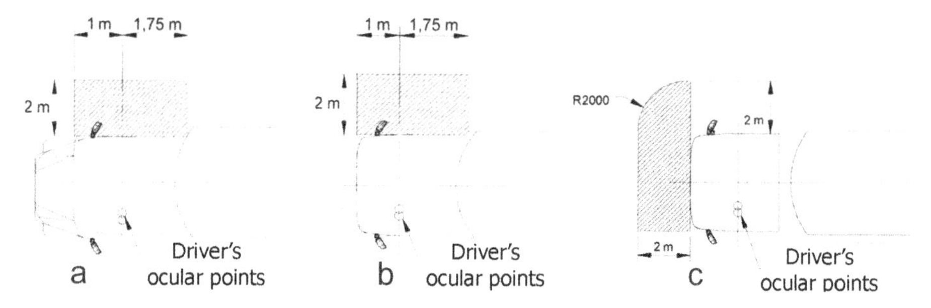

Fig. 8.43. Fields of view for Category V mirrors (a and b) and for Category VI (c); the dashed area is draft on the ground.

Additional front external mirrors (Category VI)

The field of vision of these mirrors must be such that the driver can see a portion of a horizontal flat road with the following boundaries:

- A plane perpendicular to the vehicle symmetry plane for the foremost point of the body.
- A plane parallel to the above 2 m in front of the foremost point of the body.
- A plane parallel to the vehicle symmetry plane, for the most external point of the body.
- A plane parallel to the above at 2 m from the most external point of the body.
- The front corner of this field of vision can be rounded with a 2 m radius at the driver's opposite side (see Fig. 8.43 c).

Other indirect vision devices

Other indirect vision devices can be used to improve the vision of details not reached by direct vision, in reverse drive; such devices are allowed to improve vehicle safety but they cannot substitute the established devices listed above.

These indirect vision devices can be conventional mirrors or cameras with monitors and must display at least one part of horizontal and flat road, from a vertical plane perpendicular to the vehicle symmetry plane, for the rearmost point of the body to a vertical parallel plane 2 m behind; the other boundaries are two vertical planes parallel to the vehicle symmetry plane, for the most external points of the two vehicle body sides.

If devices are applied which differ from mirrors or cameras, they must be able to detect a cylindrical object 50 cm high above the ground, with a diameter of 30 cm, set in any point of the field of vision that has been defined above.

Group 32: Front field of vision

The Directive D 77/469 and its following amendments D 81/643, D 88/366 and D 90/630 describe the conditions to be fulfilled for the driver's front field of vision for M category vehicles. Although these Directives apply to left hand driven vehicles, they are applicable to right hand driven vehicles with symmetric figures.

The requirements of these directives are completely explained in the following chapter.

Group 34: Glasses defrosting and defogging

Defrosting and defogging devices requirements are described by Directive D 78/317.

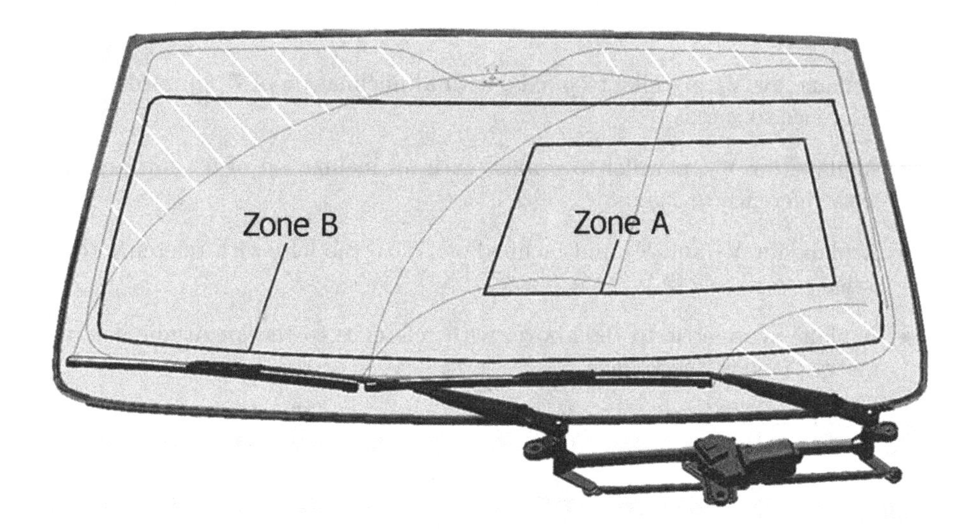

Fig. 8.44. A and B visibility zones draft on the windshield surface; the same scheme shows the minimum contour of the transparent part of the windshield and the areas covered by wipers.

By defrosting device, this Directive means a device suitable for melting frost on the windshield in order to allow direct vision. A defrosted area is an area where the frost is melted completely or partially, in such a way that external ice residuals can be completely removed using the wipers.

A defogging device is a device able to remove the layer of condensed vapor on the internal surface of the windshield in order to enable direct vision.

This Directive makes reference to the ocular points V already mentioned for indirect vision devices; on this basis, two vision zones A and B are defined.

The A zone on the windshield surface is defined by the following planes through points V, to the front of the vehicle:

- A vertical plane through V_1 and V_2 with an inclination of 13° to the left with reference to the x axis.
- A plane parallel to the y axis, through V_1, with an inclination of 3° upwards, with reference to x axis.
- A plane parallel to the y axis, through V_2, with an inclination of 1° downwards, with reference to x axis.
- A vertical plane through V_1 and V_2, with an inclination of 20° to the right, with reference to x axis.

The B visibility zone is draft on the windshield and is limited by the following lines and planes:

- A line parallel to the side edge of the transparent surface at 25 mm distance.
- A plane for V_1, parallel to y axis, with an inclination of 7° upwards, with reference to x axis.
- A plane for V_2, parallel to y axis, with an inclination of 5° downwards, with reference to x axis.
- A plane for V_1 and V_2 and inclined of 17° to the left, with reference to x axis.
- A plane symmetric to the above, with reference to the longitudinal symmetry plane of the vehicle.

The result of this geometric construction is shown in Fig. 8.44.

Any vehicle must be provided with a device able to defog and defrost the windshield in case of low ambient temperature, with the following minimum performance:

- The A zone surface must be 80% defrosted, 20 min after the start of the defrosting test; the A zone relative to the passenger must satisfy the same condition 5 minutes later.
- The B zone must be 95% defrosted, 40 min after the start of the defrosting test.
- The A zone must be 90% defogged, 10 min after the start of the defogging test.
- The B zone must be 80% defogged, 10 min after the start of the defogging test.

The defrosting test may be performed at $-8 \pm 2°$ C or $-18 \pm 3°$ C, according to the specifications of the manufacturer.

The test must be performed in a cold chamber large enough to contain the complete vehicle, and capable of maintaining the test temperature for the entire test duration and to circulate cold air. The room temperature must be stabilized for at least 24 hours before initiating the vehicle test.

After the vehicle temperature is stabilized, at least 0.044 g/cm^2 of ice are spread on the windshield with a water gun with a water pressure of at least 3.5 bar $\pm$ 0.2 bar.

During the defrosting test, the defrosting device adjustment must be set to maximum; wipers can be used during the test only if they can move and return to rest position without manual intervention.

Every 5 min the transparent surface contour must be drafted, by an observer inside the vehicle, on the internal side of the windshield.

The defogging test is made in a climate chamber similar to that cited above, in which a temperature of 3° C $\pm$ 1° C must be maintained during the test.

A vapor generator inside the vehicle produces 70 $\pm$ 5 g/h of steam for each seat present in the passenger compartment.

After the generator has been working for 5 min, one or two observers enter the vehicle and the vapor rate is reduced by the amount of seats actually occupied.

A minute later the engine is started with the test time count.

Group 35: Windshield wipers and washers

This group includes the D 78/318 Directive and its updates D 94/60 and D 2006/96.

A windshield wiper and its accessories are devices provided to clean the outside surface of the windshield.

The *wiper span* is the part of the outside wet surface of the windshield that can be reached by the wiper.

The windshield washer is a device provided to spray a detergent fluid on the outside surface of the windshield.

Each vehicle must be equipped with a wiper and a washer that work and can be switched on and off when the vehicle is in motion.

The wiper span must cover at least 80% of zone B and 98% of zone A, as defined in the previous section, in Fig. 8.44.

Wipers must have at least two working frequencies with one equal or higher than 45 strokes per minute (a stroke is a complete outward and return motion of the wiper) and the second between 10 and 55 strokes per minute; the difference between these two frequencies must be at least 15 strokes per minute.

When the wiper is switched off, the brushes must return to the rest position automatically.

Wipers must withstand a 15 s locking at the highest speed.

At 80% of the top vehicle speed or at 160 km/h (the lowest condition applies), wipers at the highest speed must comply to span requirements with the same effectiveness as at low speed.

Wiper arms can be raised to clean the windshield manually.

Wipers must operate correctly for 2 minutes at dry windshield, at $-$ 18° $\pm$ 3° C of ambient temperature.

Each vehicle must be equipped with a washer, able to withstand with no damage pressures arising from obstructed nozzles, according to the specifications of the following tests.

Test 1

The washer is completely filled with water and exposed to an ambient temperature of 20° C $\pm$ 2° C, for at least 2 h.

The nozzles are obstructed and the pump is switched on six times, for at least 3 s each.

The operation of the washer is then verified.

Test 2

The washer is filled with water and exposed to an ambient temperature of – 18° C ± 3° C, for at least 4 h.

The nozzles are obstructed by ice and the pump is switched on six times, for at least 3 s each. The device is then exposed to a temperature of 20° C ± 2° C, until the ice melts. The operation of the washer is then verified.

Test 3

The washer is filled of water and exposed to an ambient temperature of – 18° C ± 3° C, for at least 4 h.

The device is then exposed to a temperature of 20° C ± 2° C, until the ice melts.

The freezing and unfreezing cycle is repeated 6 times. The operation of the washer is then verified.

Test 4

The washer is filled of water and exposed to an ambient temperature of 80° C ± 3° C, for at least 8 h and then to an ambient temperature of 20° C ± 2° C. The operation of the washer is verified when the last temperature is stabilized.

Test 5

A layer of test mixture is applied to the outside surface of the windshield and left to dry; this mixture is a suspension of particles (chemical composition and size are specified in the Directive) in water, to simulate mud.

Washers must be filled with water and capable of cleaning at least 60% of zone A.

The available volume of washing fluid must be at least 1 l.

Occupants and pedestrians protection

Some of the regulations under this category are widely described in the chapter on passive safety such as Regulation R 94 concerning occupant protection in front collisions, and R 95 regarding occupant protection in lateral collisions; therefore they are not discussed further in this section.

Group 12: Interior surface

This group includes Directives regarding interior surfaces of passengers compartment as D 74/60 and its updates D 78/632 and D 2000/4.

According to D 2000/4, the interior surfaces to be considered are:

- Passenger compartment coverings.
- Controls.
- Roof and sunroof.

Reference zone

The potential impact areas of occupants heads is considered at first.

To do this, a spherical device of 165 mm diameter is used to simulate the head. This device is connected to an articulated support simulating the hip; the connection length must be adjusted between 736 and 840 mm.

The articulation point, representing the hip, must be set in any of the available sitting positions; it should be remembered that a bench seat may include more than one sitting position.

The following rules must be observed:

- For adjustable seats, the articulation point must be set in the H point and in a position 127 mm forward; for this second position the height is increased by that resulting from seat adjustment or by 19 mm.

- For non-adjustable seats, the articulation point must be set in the H point.

After setting this device in a vertical position, should no contact exist with passenger compartment walls, within the vertical dimensions allowed by the device adjustment, the device is inclined forwards in all obtainable positions up to 90° with reference to a vertical cross plane through H.

Potential impact points correspond to the tangential of the device head with interiors surfaces, but glasses.

The *reference zone* is defined by the envelope of the potential impact points with the exception of an area obtained by projecting the steering wheel contour horizontally, increased by 127 mm.

No dangerous unevenness or sharp edge is allowed within the reference zone.

To determine protrusion of an uneven element, a 165 mm diameter sphere is moved along the interior surface; the gradient between the displacement of the centre of the sphere, perpendicular to the interior surface, and the tangent displacement is considered: A surface is substantially flat if this gradient is lower than 1; if it exceeds 1, the protrusion is the perpendicular displacement of the sphere.

All surfaces within the reference zone must be covered by resilient material, as specified subsequently.

When the hardness of the covering material is lower than 50 A Shore, this procedure is applied after having removed these covers.

Push buttons, levers or other rigid parts protruding from 3.2 to 9.5 mm must have a minimum cross section of 2 cm^2 (measured at 2.5 mm from the top of the protruding part) and feature rounded edges with not less than 2.5 mm radius.

If their protrusion is higher they must be designed to collapse with a protrusion of less than 9.5 mm or to detach, under a force of 37.8 daN.

The resilience of the surface must comply with the following test.

A portion of the interior surface, to be tested, is set on a test bench reproducing the installation in the vehicle. The surface is hit by a pendulum whose reduced mass at the centre of the hammer is 6.8 kg, at a speed of 24.1 km/h; the shape of the hammer is the usual 165 mm diameter sphere. The maximum allowed deceleration of the pendulum is 80 g for no more than 3 ms.

Other surfaces

The other surfaces refer to those below the dashboard and in front of the H point, not including doors and pedals or behind the back rest of the front seats.

All these surfaces must comply with the specifications listed concerning protrusions.

The hand brake lever, when set on or below the dashboard, must be positioned so as not to touch, in its released position, any of the occupants during collisions; if this condition is not respected, its surface must comply with that specified for the console.

Consoles or similar elements must be designed so as not to present any protrusion or sharp edges, and must comply with one at least of the following rules:

- The side to the rear must present a rim 25 mm high with its edges rounded not less then 3.2 mm; this surface must be resilient as specified for the reference zone.
- Consoles must easily break, detach, deform or collapse under a horizontal, longitudinal force of 37.8 daN, without new protrusions appearing.

When these surfaces are layered with a material which is less hard than 50 A Shore, the specifications listed above apply after having removed those parts.

Roof

The following rules apply to the inside surface of the roof; however, they do not apply to parts that cannot be touched by the 165 mm spherical probe.

This surface above or in front of occupants heads must not present any protrusions or sharp edges.

The height of protruding parts must not be more than its width and edges must be rounded with not less than 5 mm.

Formers or ribs must not protrude for more than 19 mm and must be rounded, as stated above.

For convertibles with a textile cover, only metal arches and windshield frames are subject to the above specifications.

Sunroofs must comply with the same rules as conventional roofs.

Group 15: Seats

The Group 15 includes a number of Directives regarding seats and their anchorages to the car body; the first Directive on this subject is the D 74/408, later modified and implemented by the following D 81/577, D 96/37 and D 2006/96.

At this point it is appropriate to provide some definitions. A *seat anchorage* is the mechanical system connecting the vehicle to the car body, while the *seat adjustment device* is the mechanical device enabling the entire seat or parts of it to be adapted to the occupant sitting position and their anatomic conformation.

A seat adjustment device allows, in particular, to apply to the seat:

- a longitudinal displacement,
- a vertical displacement and
- an angular displacement.

In addition to these, another device is provided that can be called *seat displacement device*, capable of applying a longitudinal or angular displacement to the entire seat or to one of its parts in order to facilitate passenger access; this kind of device has no intermediate stable positions like the adjustment device.

A *locking device* ensures that the seat is fixed in a stable way in one of its possible positions.

Any adjustment or displacement device must include an automatic locking device.

The manual disengagement of the displacement device must be installed on the external side of the seat near the door in a position that can be reached by both occupants of this seat and the seat behind it.

In general no permanent displacement is allowed to the seat structure, its anchorage, the adjustment and displacement devices and their locking devices, as a consequence of the tests that will be described subsequently.

Permanent displacements are accepted provided they not increase the risk of injuries in the case of collision when the maximum force prescription is respected.

Locking devices must not unlock as a consequence of the test.

Displacement devices must operate properly following the tests and must be able to work at least once to allow easy egress of the occupant.

Resistance test

A longitudinal acceleration to the front of the vehicle (simulating sudden deceleration after impact) of 20 g minimum for 30 ms should be applied to the entire vehicle body or to a significant portion of it, according to a detailed test specification.

No object inside the vehicle must prevent locking devices to unlock.

Energy dissipation test

The seat is installed on a test bench using its own anchorage system, to simulate the installation on the vehicle.

If a head rest is provided, it must be installed on the seat back rest as in the vehicle; if the head rest is not installed on the seat, also the anchorage to the vehicle body must be reproduced on the bench.

If the heat rest is adjustable, it must be positioned in the most unfavorable position allowed by its adjustment devices (the position that allows presumably the highest deflections).

The head rest is hit by a pendulum simulating a human head, the usual sphere of 165 mm of diameter with a reduced mass of 6.8 kg. The pendulum must hit the head rest from the rear to the front of the seat in a longitudinal plane. The impact speed is 24.1 km/h.

These tests can be performed also using a full scale car in an impact test against a rigid barrier with an impact speed in the range between 48.3 and 53.1 km/h.

Group 10: Seat belt anchorages

Group 19 includes all regulations regarding seat belts anchorages; they are reported by the D 81/575 Directive and by its updates D 81/575, D 82/318, D 90/629, D 96/38 and D 2005/41.

Prescriptions regarding anchorage positions and working angles are not reported in this section having already been widely described in Volume I.

In any kind of vehicle, each sitting place must be equipped with seat belts with 3 point anchorage; seat belts with 2 point anchorage are permitted only in the following cases:

- Seats pointing to the rear side of the vehicle.
- Lateral seats on M_1 vehicles, if they are separated from the vehicle sides by a corridor.
- Front central seats on M_1 vehicles, if the windshield is outside the reference zone.
- Rear seats for M_3 and N vehicles if their anchorages are outside the reference zone.

The following tests are required.

Traction test

When seats are grouped in benches, all anchorages of the same group of seats are to be tested simultaneously.

The application of test loads must be as quick as possible; anchorages must withstand the load for at least 0.2 s.

For 3 point safety belts with a reel, an adaptor is applied to the belt to apply test forces from the traction apparatus.

For M_1 and N_1 vehicles, a test load of 1,350 $\pm$ 20 daN is applied by a traction device (see Fig. 8.45b) to the upper anchorages of each seat belt with a traction apparatus that reproduces the geometry of the seat belt which interacts with the upper torso.

The same test load is applied to the two lower anchorages (see Fig. 8.45a).

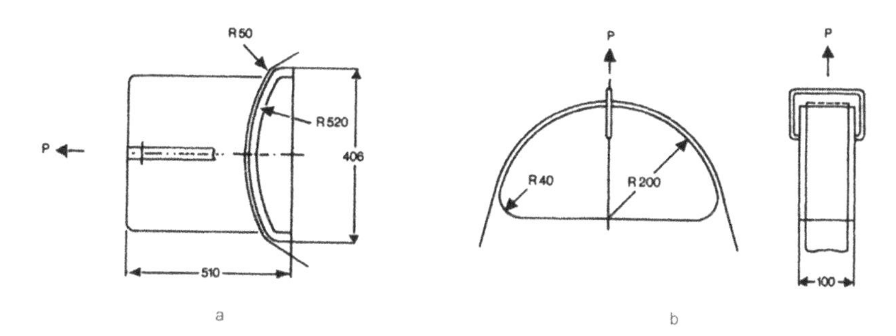

Fig. 8.45. Traction device for seat belt anchorages; a: upper anchorage for 3 points seat belts; b: anchorage for 2 points seat belts.

For M_2 and N_2 vehicles, the above test loads are reduced to 675 ± 20 daN and to 450 ± 20 daN, for M_3 and N_3 vehicles.

In the case of 2 point seat belts, a test load of 2,225 ± 20 daN is applied by a traction device, as in Fig. 8.45 a.

For M_2 and N_2 vehicles, the above test loads are reduced to a 1,110 ± 20 daN and to 740 ± 20 daN, for M_3 and N_3 vehicles.

If one of these anchorages is fixed to the seat structure, their test loads are increased by an additional force; for M_1 and N_1 vehicles, this force is 20 times the weight of the complete seat; for M_2 and N_2 vehicles, this force is reduced to 10 times the weight of the complete seat, and to 6.6 times the weight of the complete seat for M_3 and N_3 vehicles.

All anchorages must withstand these test loads; a permanent deformation or a partial rupture is allowed provided that each load is maintained for the time established.

Group 31: Restraint systems

Group 31 regards regulations concerning safety belts and restraint systems including the D 77/541 Directive and its updates D 81/575, D 82/319, D 87/354, D 90/628 and D 96/36. They are all substituted by D 2000/3, introducing further changes; more updates are introduced by D 2005/40 and D 2006/96.

According to these Directives, a safety belt is a system comprising belts, a buckle, adjusting devices and anchorages, installed in a vehicle in order to reduce the risk of injuries to occupants in the case of a collision or sudden deceleration of the vehicle; this objective is obtained by limiting the freedom of motion of the occupants' bodies.

The following kinds of safety belts can be identified.

- Sub-abdominal belt, a belt that wraps around the occupant's basin.
- Diagonal belt, a belt that winds across the occupant's chest, from one hip to the opposite shoulder.

- 3 point belt, a belt substantially made by a combination of the above types.

- 4 point belt, a belt made by a sub-abdominal element and two vertical straps.

A restraint system must at least include a seat fixed to the vehicle and a seat belt where at least one anchorage is fixed to the seat.

A child restraint system is a set of components including a combination of belts or flexible links, with a safety buckle, adjustment devices, anchorages and an additional seat or seating adaptor that can be installed on a vehicle. This system is intended, again, to limit the risk of injuries after a collision or a sudden deceleration, by limiting the freedom of motion of the occupant, in this case a child.

General issues

The rigid parts of the restraint systems, including the buckles, adjusting devices or anchorages, must not present sharp edges which could damage, wear or rip safety belts.

All parts potentially subject to corrosion must be properly protected. After corrosion tests, no alteration is allowed that could compromise system operation including any visible corrosion detectable by a qualified operator.

Buckle

The buckle must be designed to avoid any possible misuse. This implies in particular that it must be impossible that the buckle remains in a half locked position. How to unlock the buckle must be self explanatory. Any part of the buckle that may come into contact with the occupant's body must have a surface not less than 20 cm^2, with a length not less than 46 mm, on a plane at a maximum distance of 2.5 mm from the contact surface.

The buckle must open when a button is depressed or a similar device is operated. The surface to be depressed must have the following minimum dimensions, projected along the button motion trajectory:

- Recessed buttons (cannot be operated by a 25 mm diameter sphere) must have a minimum surface of 4.5 cm^2 and a minimum width of 10 mm.

- Non-recessed buttons must have a minimum surface of 2.5 cm^2 and a minimum width of 10 mm.

The button surface must be red; no other part of the buckle can feature this color.

The buckle must be locked and unlocked 5,000 times before the dynamic test; after this test the force necessary to unlock the buckle must not exceed 6 daN.

Reel

Reels allow the belt to unwind for the length necessary to automatically adapt to the occupant's body dimension when the belt is fastened.

Reels must have a locking position at least every 30 mm of belt length. The belt must return to the initial position automatically following a forward and backward motion of the occupant.

If the reel is part of a sub-abdominal belt, the winding force, measured on the free belt length between the dummy and the reel, must be between 0.2 daN and 0.7 daN. If the belt winds on a pulley or a bar, this condition must be respected by the belt, between the pulley or bar and the dummy.

If the restraint system includes a manual or automatic device to prevent the belt to completely wind on the reel, this device must not be working during the measurement of this force.

The belt must be wound and unwound on the reel 5,000 times; after this procedure a corrosion test and a dust test are performed. Successively, the belt is again wound and unwound 5,000 times. After these tests the reel must continue to work properly.

Reel with emergency lock

A reel with emergency lock is a one that during normal vehicle operation does not affect the occupant's freedom of motion. This reel includes a device to adjust the belt length to the occupant's body automatically and a locking device, to lock the belt in case of an emergency; two types are considered:

- Type 1: locking occurs due to vehicle deceleration.
- Type 2: locking is due to a combination of vehicle deceleration and belt motion or any other automatic device.

This kind of reel must:

- Lock at a deceleration value not over .45 g (M_1 vehicles) or .85 g (M_2 and M_3 vehicles).
- Not lock when the belt undergoes a deceleration below .8 g for type 1 or 1.0 g for type 2 (acceleration to be measured along the belt).
- Not lock when the lock sensor is inclined below 12° in any direction, with reference to the installation position defined by the manufacturer.
- Lock when the sensor is inclined 27° minimum for type 1 and 40° minimum for type 2, in any direction, with reference to the installation position defined by the manufacturer.

In any of the tests listed above, a belt length of not exceeding 50 mm can be unwound before the automatic device locks the reel.

If the reel is part of a sub-abdominal belt, the unwind force must not be less than .7 daN, to be measured on the belt between the reel and the occupant's body. If the reel is part of a 3 point belt, the unwind force must not be less than .2 daN or higher than 0.7 daN, when measured in the same way.

The belt must be wound and unwound for at least 40,000 times and undergo the corrosion and dust tests; a new series of 5,000 unwind and wind procedures must be completed with the system still in proper use conditions following the complete procedure.

Pre-tensioner

After the corrosion test, the pre-tensioner (and its sensor) must work properly.

An unintentional operation of this device must not cause any risk of injury to occupants.

Pyrotechnic tensioners must undergo a high temperature test without unplanned operation of the pre-tensioner or damage affecting its operation capability.

Belt

Belts must have dimensions which are appropriate to guarantee a uniform pressure on the occupant's body for the entire contact length and must not twist under any conditions. They must be capable of absorbing energy and be protected from fraying during their operation.

The minimum width of the belt is 46 mm, to be measured under a tension of 980 daN. This measurement must be made in connection with the tensile strength test.

The minimum tensile resistance allowed is 1,470 daN. The difference of tensile resistance between any two samples must be lower than 10% of tensile resistance.

Dynamic test

The belts, or the entire restraint system, are installed in the following test fixture after each component has undergone the required aging and conditioning tests, including corrosion, abrasion, high and low temperatures, light and humidity tests.

The dynamic test is aimed to verify that no part of the restraint system breaks and that no buckle, adjusting or emergency system locks during or after the test.

In addition, the dummy forward displacement for sub-abdominal belts must be lower than 200 mm and higher then 80 mm at the basin of the dummy.

For 4 point belts the above values can be reduced by 50%.

For remaining belt types, a figure between 80 mm and 200 mm must be obtained at the basin, and between 100 mm and 300 mm at the torso.

The referred measurement points are T (torso) and P (basin) in Fig. 8.46.

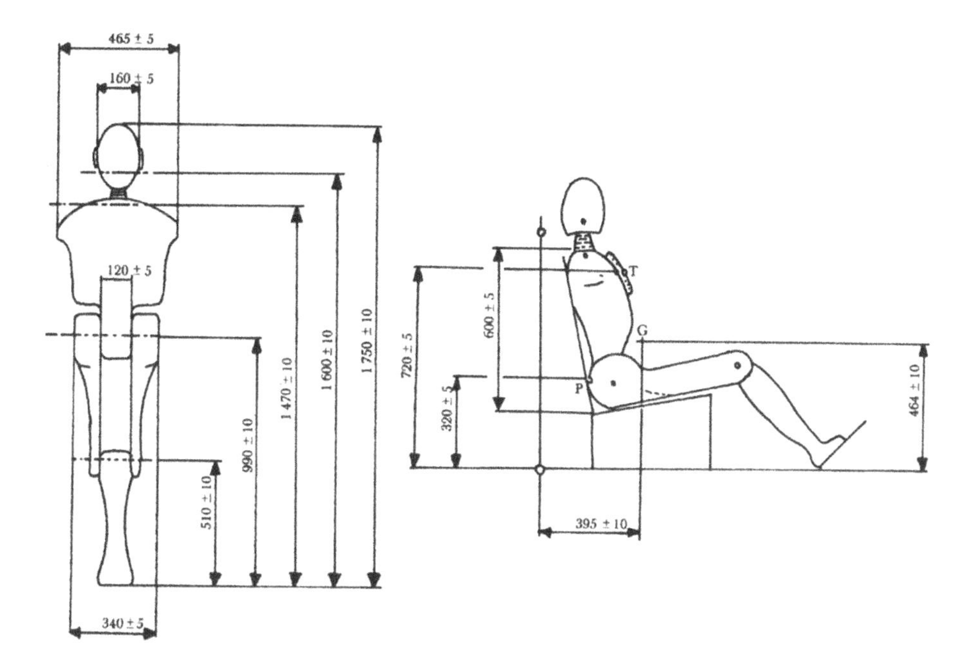

Fig. 8.46. Dummy for restraint systems tests.

The restraint system is installed on a cart featuring the seat and belt anchorages just as they are in the car, as shown in Fig. 8.47.

If a belt is part of a restraint system, it must be installed on the cart using the same portion of the structure used on the actual vehicle and this portion must be fixed on the cart.

Seats are adjusted to the dummy size, according to vehicle user manual; if no instructions are given, the seat back rest is inclined as close as possible to 25° for M_1 and N_1 vehicles, or 15° for remaining vehicle types.

The cart is pushed in such a way that its speed is 50 ± 1 km/h at the time of simulated impact, with the dummy in a stable position during the launch. The impact is simulated with a sudden deceleration applied by crushing the absorber, shown in Fig. 8.48 or an equivalent deceleration device.

Fig. 8.49 shows the band of acceptable values for the acceleration a measured in [g], as function of elapsed time t; any deceleration device, used in place of the proposed absorber, must comply with these acceptance limits.

The mass of the cart with one seat is 400 ± 20 kg. This mass is increased for restraint systems for more than one seat. Anchorages are installed according to that shown in Fig. 8.47.

The points corresponding to anchorages installation show the position where belt ends are fixed on the car, as well as the dynamometers. A, B and K

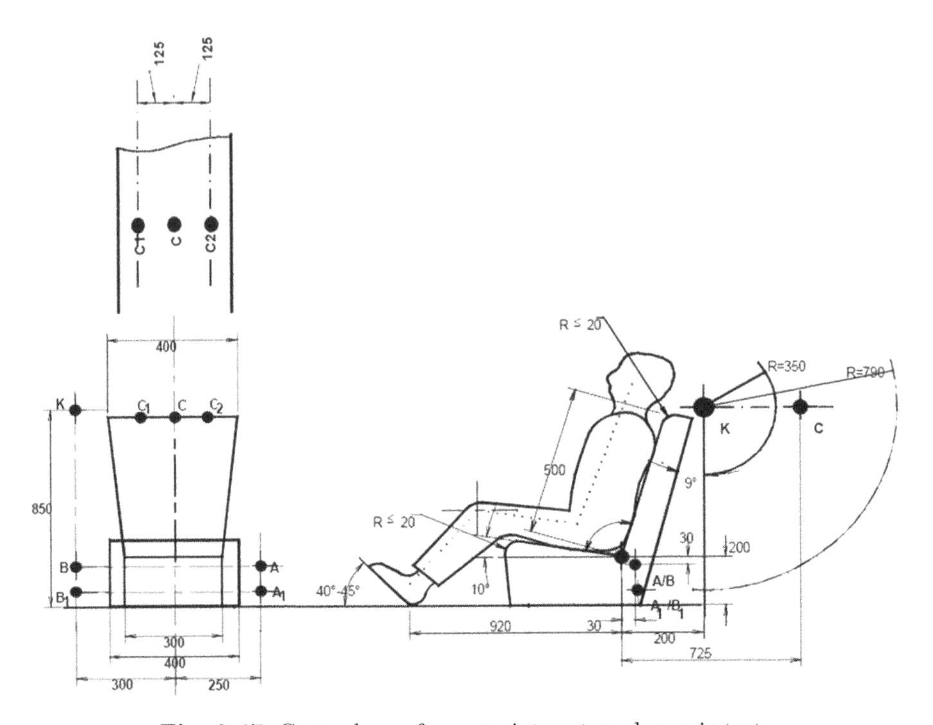

Fig. 8.47. Cart scheme for restraint system dynamic test.

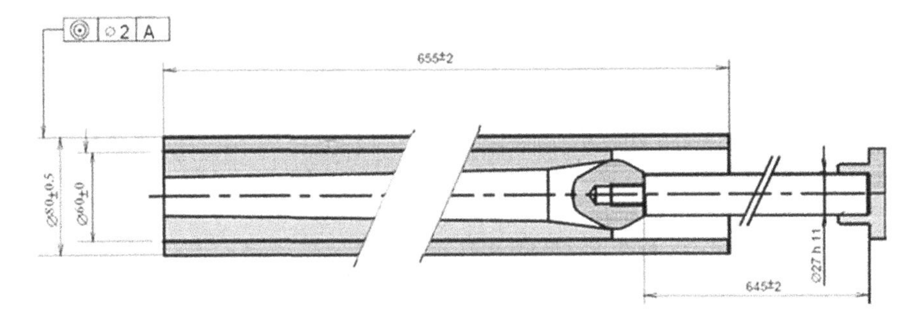

Fig. 8.48. Energy absorber used to stop the cart in the restraint system dynamic test.

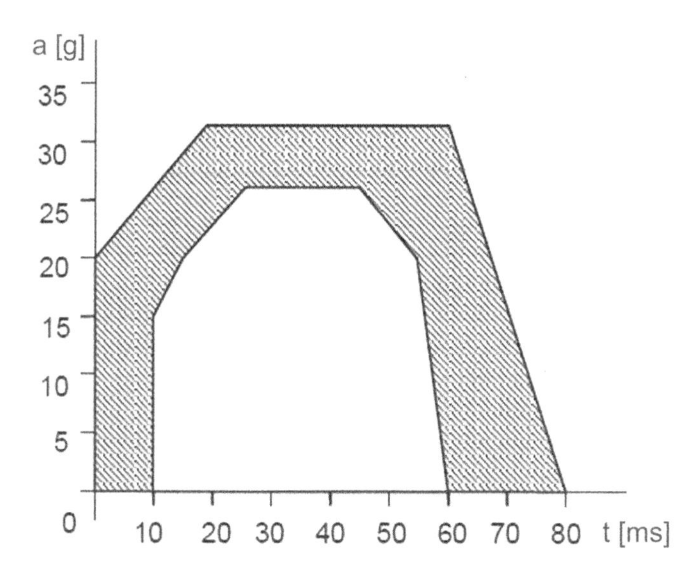

Fig. 8.49. Acceptance band for the acceleration curve a measured in [g], as function of the elapsed time t, to be applied in the restraint system dynamic test.

anchorages are chosen when the upper edge of the buckle is not further than 250 mm from the fixation point; in different cases A_1 and B_1 points are used.

The upper anchorage must not displace more than 0.2 mm longitudinally, when a load of 98 daN is applied in this direction.

No permanent deformation must result as consequence of the test on the cart and its parts which support the anchorages.

The absorber is made of:

- A steel tube as casing.
- A polyurethane tube to absorb energy.
- A polished steel nut, penetrating the polyurethane tube.
- A rod and a plate.

The main dimensions of the above parts are shown in Fig. 8.48; the Directive specifies the characteristics of the materials to be applied.

The R 44 Regulation describes the characteristics and tests to be fulfilled for child restraint systems which are not described here for the sake of brevity although it is necessary to bear in mind that child restraint systems are categorised with respect to the following classes of occupants in terms of mass:

- group 0 for children with mass lower than 10 kg;
- group 0+ for children of mass lower than 13 kg;

- group 1 for children whose mass is between 9 and 18 kg;
- group 2 for children whose mass is between 15 and 25 kg;
- group 3 for children whose mass is between 22 and 36 kg.

Group 36: Head rest

This group includes all the requirements for headrests, as described by Directive D 78/932 and its amendments of D 87/354 and D 2006/96; all this matter is also summarized by the R 25 Regulation.

The headrest is a device which has the purpose of limiting backwards motions of the occupant's head with reference to the torso, the objective being to reduce the risk of injury to the cervical rachis as consequence of an accident. This device may or may not be integrated into the seat backrest.

The application of head rests must not cause additional injury risks for the vehicle occupants, in particular dangerous protuberances or sharp edges or dangerous positions.

Any part of the head rest in the defined impact area must be able to dissipate energy, according to the following specifications.

Head rest mounts on the seat must not present rigid parts or protrusions through the upholstery, during the head rest test.

The head rest height must exceed the R point height by at least 700 mm.

If the head rest height is adjustable, the height of the device must be at least 100 mm; the lowest position of the head rest should result in a maximum gap of 50 mm with respect to the back rest. If not adjustable, a maximum gap of 50 mm is allowed.

The head rest width must offer comfortable support to any occupant; a minimum width of 85 mm at each side of the side symmetry plane is required; this device must withstand the required load without permanent deformation.

The head rest must be designed to allow a maximum displacement of the head of 102 mm, measured according to the test procedure.

Static test

The test refers to the scheme in Fig. 8.50.

The impact area is limited by two vertical planes at a distance of 70 mm from each side of the seat symmetry plane. In the other direction, the area is limited by the plane perpendicular to the reference line r, at 635 mm above the R point.

The line r is draft through the R point and the neck thorax articulation of a 50th percentile male dummy.

Using a spherical head of 165 mm of diameter, an initial force, corresponding to a torque of 37.3 daNm around R, is applied perpendicularly to the r_1 line, 65 mm below the upper end of the head rest.

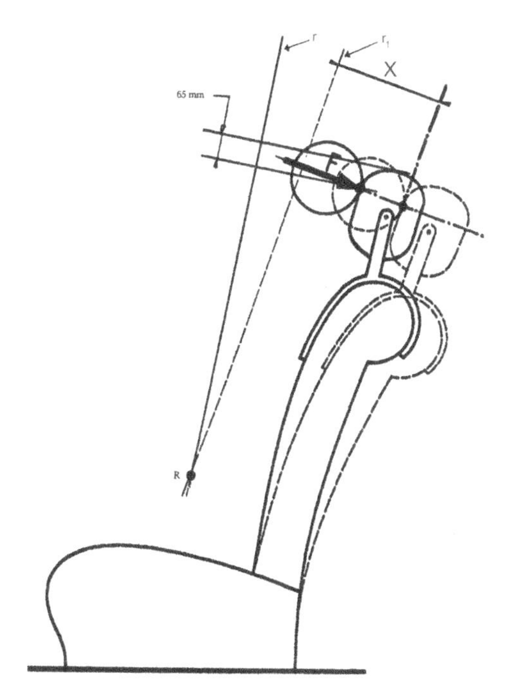

Fig. 8.50. Scheme for static load application to a head rest.

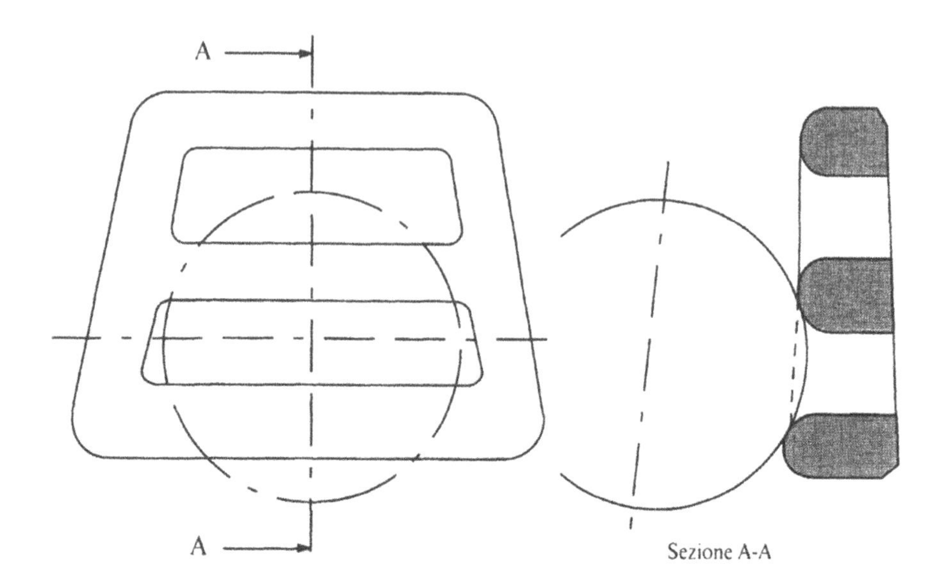

Fig. 8.51. Scheme of load application, for a windowed head rest.

The distance X in Fig. 8.50 must be lower than 102 mm.

The initial force is increased up to 89 daN in the same direction, and no rupture must occur.

If the head rest is windowed to improve rear visibility, each element of the head rest at the window contours must undergo this test in the position corresponding to maximum displacement, as shown in Fig. 8.51.

Energy absorption test

The head rest is installed on its seat which is installed on the test bench in such a way so as to not be displaced due to the impact.

The test device consists of a pendulum with a mass reduced to its percussion center of 6.8 kg. The pendulum mass is a spherical head with 165 mm of diameter, with its center in the percussion center of the pendulum.

The pendulum must impact the surface perpendicularly with a speed of 24.1 km/h.

The pendulum deceleration must not exceed 80 g for more than 3 ms.

Pedestrians protection

This regulation is not yet in force. Nevertheless, the D 2003/102 Directive reports the objectives with respect to the maximum injury severity for a pedestrian impacted by a vehicle according to assigned test procedures. D 2004/90 specifies test procedures and related test dummies.

Although these regulations were to be judged technically not applicable following a test period which ended in July 2004, they are frequently applied on a voluntary basis and have been summarized in a new regulation, released in draft for discussion, on October 3, 2007.

Here it is appropriate to examine the two original Directives that are applied by Euro NCAP for product rating.

The vehicle is in reference conditions, with two 75 kg passengers in front seats; vehicles with active or adjustable suspensions are set in the conditions suggested by the manufacturer for driving at a speed of 40 km/h.

The front area of the vehicle interested in potential impacts with pedestrians is defined according to the following procedure.

A reference line (upper line) on the bumper is the locus of contact points with a ruler, 700 mm long, and the bumper, when the ruler is kept on a longitudinal vertical plane, with an inclination of 20° rearwards and is moved in any direction, having its lower end in contact with the ground. The reference light is shown in an example, in Fig. 8.52a.

A second reference line (lower line) is obtained in the same way but with the ruler moved with an opposite (forwards) inclination: See again Fig. 8.52b.

The height of the bumper is the vertical distance between upper and lower reference lines, while the bumper edge is the contact point of the bumper with a vertical plane, inclined at 60° with reference to a vertical cross plane. See Fig. 8.52 c.

The front impact surface is the outside surface of the body, except windshield and A pillars: It includes, therefore, the hood, fenders, cowl, wiper articulations and the lower windshield frame.

A hood reference line is the locus of contact points with a ruler, 1,000 mm long, inclined at 50° rearwards in a vertical plane, when the ruler is moved with its lower end at 600 mm over the ground. In this case, see Fig. 8.52 d.

Other reference lines are draft with a ruler, 700 mm long, inclined at 45° to the inside of the vehicle, in a cross plane. See Fig. 8.52e.

Leg dummy

The leg dummy is made with two rigid segments simulating femur and tibia, covered with a foam layer and joined by a deformable joints simulating the knee.

This device has a total length of 926 ± 5 mm, and a mass of 13.4 ± 0.2 kg; other dimensions and conditions are given in the aforementioned Directive.

During the impact test, the leg dummy shown in Fig. 8.53 is launched in ballistic flight against the car, according to the scheme shown in the figure. The free flight phase should start at a distance from the body in such a way that the dummy does not rebound against the launch device before acceleration measurement. Any kind of device can be used for the launch if these conditions are satisfied.

The impact speed is 40 km/h.

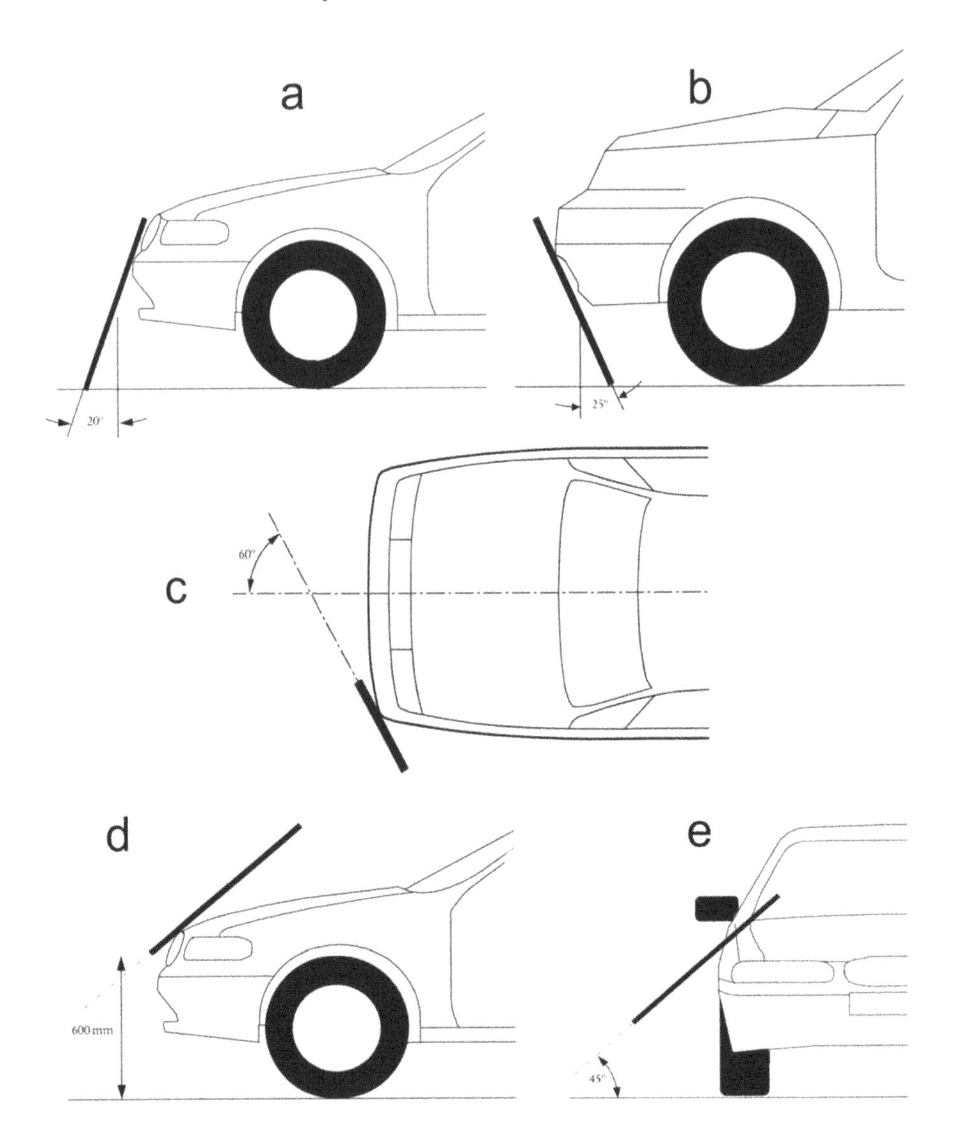

Fig. 8.52. Definition on a vehicle of the borders of potential impact areas with pedestrians.

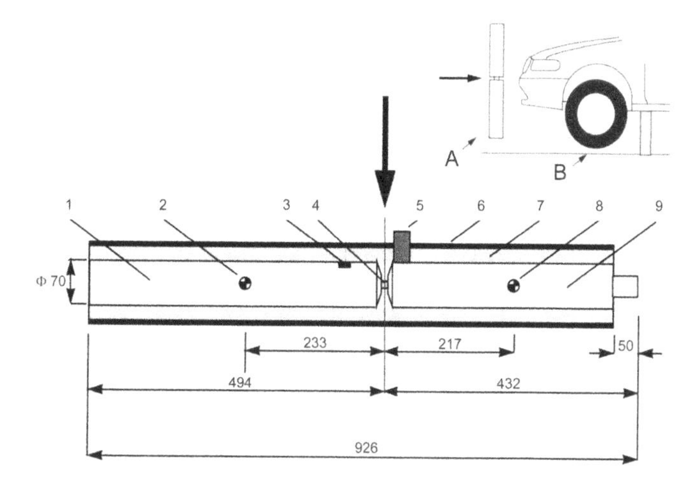

Fig. 8.53. Leg dummy. 1: tibia; 2: tibia center of weight; 3: accelerometer; 4: deformable element; 5: damper; 6: artificial skin; 7: foam material; 8: femur center of weight; 9: femur. The upper miniature represents the scheme of the impact test; A is the actual ground line, while B is the laboratory ground.

After the rebound, the maximum allowed knee deflection is 15°, the maximum knee allowed displacement is 6.0 mm and the maximum acceleration on the tibia head acceleration must be lower than 150 g.

Thigh dummy

The thigh dummy is also made of rigid material, covered by foam on the side that is exposed to collisions. It is 350 ± 5 mm long and is manufactured according to specifications in the aforementioned Directive; this device is represented in Fig. 8.54.

The thigh dummy used to simulate collisions is shown in Fig. 8.54; it is mounted on a low friction launch sled that allows the device to move only along the assigned path. The launching device can be any kind of catapult capable of obtaining an impact speed of 40 km/h.

The same dummy is used for a collision test against the front edge of the hood, as shown in the figure; the impact speed is the same.

For both tests, the sum of the instantaneous impact forces on the device must not exceed 5.0 kN and the bending moment must not exceed 300 Nm.

Head dummies

Dummies of heads of adults and children are required for tests. They are made using a rigid spherical body with an external synthetic layer fully described in the regulation proposal. Their diameter is 165 ± 1 mm, for the adult dummy and 130 ± 1 mm, for the child; their masses are 4.8 ± 0,1 kg and 2.5 ± 0.05 kg respectively.

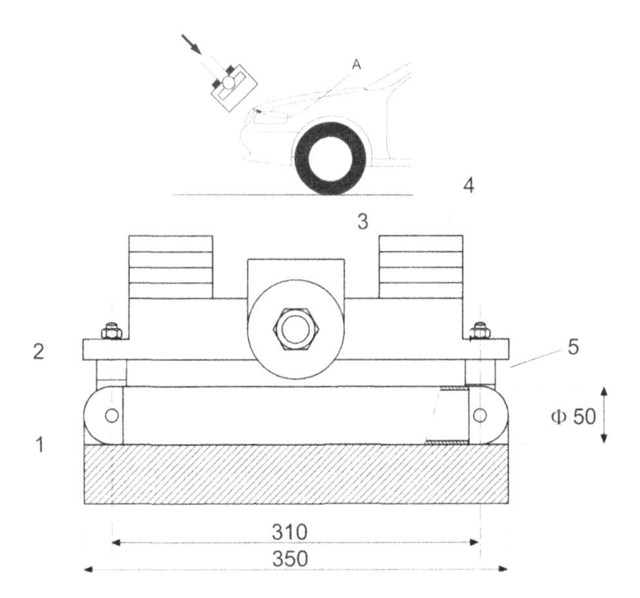

Fig. 8.54. Thigh dummy. 1: expanded foam; 2: femur dummy; 3: thrust element articulation; 4: ballast; 5: load cells. The upper figure represents the test scheme; A is the front reference line of the hood.

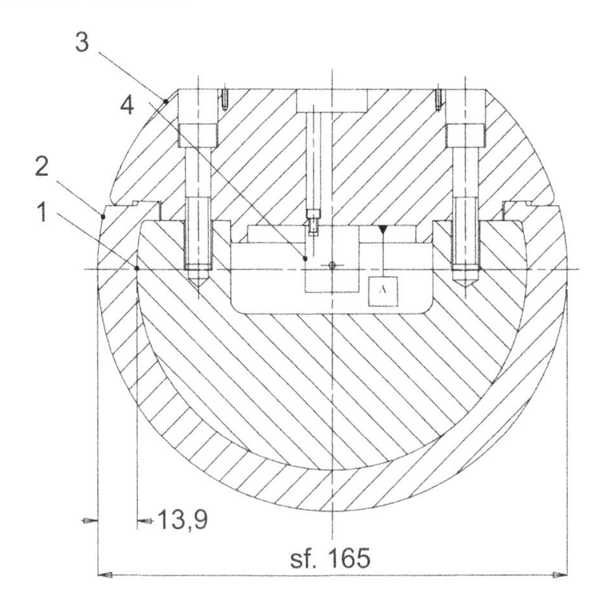

Fig. 8.55. Scheme of the adult head dummy. 1: inside element; 2: artificial skin; 3: rear plate; 4: accelerometer. The child head dummy is similar but with different size and mass.

At least 18 different impact tests must be performed, 6 for each third part of the central and lateral hood surface, in the positions that are supposed to be the most dangerous.

The head dummy, shown in Fig. 8.55 for the adult size, is in ballistic flight at the time of the impact. The phase of the free flight starts at a distance from the vehicle so as to guarantee that test results are not affected by the rebound of the dummy against the launch device.

Any launch device can be used provided an impact speed of 40 km/h can be obtained.

For any test the *HIC* index[7] must not exceed 1,000.

Only the adult head dummy is also used for an impact test against the lower rim of the windshield using the same procedure and acceptance limits.

Front protections

This section refers also to Directives D 2005/66 and D 2006/368 concerning front vehicle protections available as aftermarket supplies.

A front protection is one or more separated structure, as for instance a tubular rigid bumper or an add-on bumper, added above or below the original bumper, to protect the external surface of the vehicle body from collision damage.

Structures with mass lower than 0.5 kg, designed solely to protect lights, are not included in this definition.

These Directives extend the application of pedestrians protection rules also to such devices.

Thermal comfort

All regulations concerning the heating systems form part of Group 36, including D 78/548 and its update D 2001/56.

A heating system is any kind of device designed to increase the inside temperature of the vehicle including the load compartment, if any.

These Directives make distinctions between combustion heating systems, using a dedicated liquid or gaseous fuel, and those using heat rejected directly by the vehicle powertrain or recovery heating systems which use heat recovered from the powertrain via its cooling water, air or oil.

All vehicles in the M and N categories must be equipped with a heating system.

General rules ensure that the heated-up air entering the passenger compartment must not be more polluted than the air outside of the vehicle, and that passengers cannot come into contact with parts of the vehicle or hot air that can burn.

The additional pollution to the environment due to heating devices must be maintained within acceptable limits.

---

[7] See the definition of HIC, Head Injury Criterion, in the last chapter of this volume.

The rules regarding passenger compartment pollution are satisfied by heating systems, including a heat exchanger fed by exhaust gas or polluted air, if the all following prescriptions are applied to the heat exchanger walls washed by the heating air:

- They guarantee hermetic sealing to a pressure of at least 2 bar.
- They are made with a single piece, with no detachable parts.
- Their minimum thickness is 2 mm, for not-alloyed steels.

The same rules are satisfied by heating systems applying engine cooling air directly, with no heat exchanger if:

- The cooling air is in contact with the engine walls not including detachable parts.
- Any connections between parts are gas-tight and resistant to oil.

A test is required to demonstrate air quality.

Before commencing this test, the heating system must work for one hour at the maximum temperature in calm air (wind speed less than 2 m/s) with the windows closed. Combustion heating systems must be tested with the engine switched off. If the heating system switches off automatically before two hours have passed, the tests can be performed before the system switches off.

CO concentration in the passenger compartment is measured by sampling the air outside the vehicle, as close as possible to the air intake to the heating system and inside the vehicle, less than 1 m from the heating air exhaust.

CO concentrations must be recorded for at least 10 min. The difference between readings must never exceed 20 ppm.

A second test must be performed to assess temperatures.

Before commencing this test, the heating system must work for one hour at the maximum temperature in calm air (wind speed less than 2 m/s) with the windows closed. Combustion heating systems must be tested with the engine switched off. If the heating system switches off automatically before two hours have passed, the tests can be performed before the system switches off.

If the heated air comes from outside the vehicle, the ambient temperature must not be lower than 15°.

Surface temperatures are measured using a contact thermometer for those parts that can enter into contact with the driver during the normal on-road vehicle operation.

A maximum temperature of 70°C is allowed for bare metals and 80°C for other materials. For parts behind the seat a maximum temperature of 110°C is allowed.

Temperatures of parts that can enter into contact with passengers during the normal on-road vehicle operation must not exceed 110°C for M_1 and N vehicles, with the exception of the air vent. A maximum temperature of 70°C is allowed for bare metals and of 80°C for other materials for M_2 and M_3 vehicles.

Hot air temperature must never exceed 150°C in the center of the air vent.

Combustion heating system pollution is measured according to the following procedure.

Before commencing this test, the heating system must work for one hour at the maximum temperature in calm air (wind speed less than 2 m/s) with the windows closed. If the heating system switches off automatically before two hours have passed, the tests can be performed before the system switches off.

Emissions must not exceed the following values:

- CO $\leq$ 0,1 % in volume;
- NOx $\leq$ 200 ppm;
- HC $\leq$ 100 ppm;
- Smoke opacity $\leq$ 1°B, for gaseous fuels and $\leq$ 4°B, for liquid fuels.

Safety devices must be provided to control the heating system in the case of emergency. These must be designed in such a way that if a flame is not obtained after switching on, or this flame is extinguished during operation, the burner cannot be fed again before 4 min, or before 1 min if there is a thermoelectric flame control, or before 1 s if there is automatic flame control.

Heating systems that use water as means of heat transfer, and the related heat exchangers, must withstand a minimum pressure which is double the normal working pressure and not less than 2 bar.

If ventilators are applied, they must have a retarded switching off in the case of overheating or fuel feed interruption.

A lamp must clearly show when the system is working.

Heating systems must be fire proof in the case of overheating. This condition is fulfilled if the device is far away from potentially dangerous components and if there are thermal protection of fireproof materials with adequate ventilation.

In M_2 and M_3 vehicles combustion heating devices must be installed outside of the passenger compartment or in an hermetically sealed container.

9
Ergonomics and Packaging

Ergonomics is applied to car body design with the aim of reducing the effort required by the driver and passengers in the different conditions particularly as regards driving the vehicle. The important role of ergonomics in the safety of the vehicle justifies the presence of regulations concerning related issues in different countries: Indeed some are enforced by law, such as those concerning the assessment of direct and indirect visibility in Europe.

In the initial stages of the car body design, the first task is to define the positions of the occupants inside the vehicle which is usually chosen in order to obtain a good compromise in terms:

- optimizing postural comfort,
- allowing easy access to the commands (steering wheel, pedals, gear shift lever,...)
- reducing the effort to get in and out of the vehicle,
- guaranteeing appropriate location of the powertrain and other mechanical sub-systems without compromising the space given to the occupants inside the vehicle too severely,
- respecting standard regulations concerning direct and indirect visibility.

The aim of this chapter is to outline the fundamental factors that influence the positioning of the occupants inside the vehicle while satisfying the constraints of other functionalities and requirements.

L. Morello et al.: The Automotive Body, Vol. 2: System Design, MES, pp. 127–200.
springerlink.com

SAE and European standards and law requirement are considered as a reference. In Europe directives and rules are enforced by law. In the US, on the other hand, SAE standards are recommended practices that must be satisfied by vehicles to be sold there.

9.1 Hints on Physiology

The sensation of comfort in a vehicle is influenced by multiple factors, including the sense of safety, the state of health and psycho-physical well-being. Consequently the objective measurement of comfort is far from being a straightforward issue [3]. Discomfort feeling, by contrast, can be related to different biomechanical factors that involve the muscles, skeleton, and blood circulation system of the occupants.

A specific aspect of automotive ergonomics is that the position of the occupants is substantially fixed for long periods, while the driver must preserve the ability to act on controls and view the nearby environment for long periods of time. This implies that:

1. it should be possible to maintain the posture for long times without altering the capability to act on the commands in the case of the driver; this happens only if the angles between the main body segments are included in ranges determined by the anatomy of the articulations and of the muscular system.
2. the driver should maintain the capability to rotate his head effortlessly. This requires that the head is in vertical position and not in contact with other supports such as the head restraint.

The driver must be able to act on the pedals at all times. This is possible if the legs are able to move freely, implying that their role in bearing the load and keeping the posture is marginal. The outcomes of this are the following:

3. The seat is the main physical support and bears most of the weight of the occupant. The loads that act on the upper body are transferred by the backbone that, correspondingly, has a particularly important role in all issues related to postural comfort.
4. The legs act on the pedals with relatively small forces.
5. The driver's arms act on the commands (steering wheels and gear shift lever, for instance). When compared with the past, the widespread use of servo-systems that characterizes modern cars has lead to a considerable reduction of the effort required to act on the commands (steer, brake, and gear shift lever). Nowadays discomfort is mainly related to keeping the posture and not to the need to operate the commands.

All these factors lead to the conclusion that the backbone is the most significantly loaded area of the body and, therefore, plays a fundamental role in determining the level of discomfort. The legs and arms, on the other hand, bear relatively weaker loads and they generate a discomfort feeling only when the angles at the joints exceed the allowed ranges.

Some consideration regarding the factors that influence the discomfort level related to the main body parts, i.e. backbone, leg and arms, are useful to explain the basis of the current thinking regarding the positioning of the occupants inside a vehicle.

9.1.1 Backbone

The human body includes a skeleton, the main function of which is to provide a support to all soft parts and transmit the loads (internal and external) to guarantee equilibrium and together forming a flexible system with many degrees of freedom. In this context, the backbone is (Fig. 9.1) a very important part as it is one of the most vulnerable and severely stressed, being constituted by a series of bones, called vertebrae, separated by softer parts, the intervertebral discs, that allow relative motion. The vertebrae are grouped into four segments

- cervical,
- thoracic,
- lumbar,
- pelvic.

The intervertebral discs consist of a external cartilage body (anulus fibrosus) that surrounds a gel tissue (nucleus pulposus) mainly made of water with loose proteins in suspension. While the anulus gives compliance to articulation to allow the relative motion of the vertebrae, the nucleus acts as a shock absorber, damping the energy from the external loads. Its liquid nature also allows even distribution of the load on the vertebrae, thus minimizing stress concentrations.

Due to the large number of small and large loads and traumas, the intervertebral discs change size and their mechanical and geometrical characteristics. Such modifications can be reversed during the rest when the body is able to self regenerate. In the long term and with age the repeated loads induce irreversible modifications: The thickness and compliance of the discs tend to reduce, mainly because of dehydration of the nucleus, losing its ability to absorb shocks and enable flexible articulation. This degenerative process is usually associated with pain of various intensity. The nerves that exit the spinal cord are compressed between the vertebrae. In the long term they can be damaged leading to loss of functionality.

The motion of the backbone is controlled by dedicated muscles, contributing to the transmission of the external loads. Similarly to other muscles, the natural reaction to fatigue is the modification of the posture and a series of small movements that tend to reduce the stress concentrations spontaneously. These movements are hindered during driving by the need to maintain a nominally fixed position.

The lumbar region is the segment of the backbone that mainly influences its functionality and long term health. Because most of the weight of the head and the upper body acts on it, is the part most prone to fatigue and damage due to repeated loads. Its position influences the driver's functionality and capability to resist in a seated position for long periods of time. A very effective means of increasing the endurance of the driver inside the cockpit is therefore to provide effective support to the backbone, especially in the lumbar region. Today this is one of the main objectives of seat design.

Increasing the seat inclination relative to the vertical could provide a means of transmitting the load from the upper body to the seat directly, reducing the load on the lumbar segment. Nevertheless, a seat with a large inclination (say, more than 25 deg relative to the vertical) reduces visibility and easy access to the commands. The seat inclination is therefore a compromise between these contrasting needs.

The factors that should be checked in order to avoid over-stressing of the backbone during driving are:

- Avoid both too heavy and too little effort developed by the backbone muscles. On one side, heavy activity leads to fatigue, on the other too weak muscular activity results in a motionless position, leading the load being concentrated on the same parts of the joints and therefore over stressing and pain.
- Reduce, if possible, the compression loads acting on the lumbar segment of the backbone. Taking into account the fact that measurement of these loads is not easy, several studies have demonstrated that the contact pressure at the seat back and cushion may be a reliable indicator of the stresses acting on the backbone which can then be used as a criterion during the design phase.

Different studies of the muscular activity and the loads acting on the intervertebral discs have demonstrated [4] that both of these parameters are lower when standing than in the seated position. Additionally, the loads on the intervertebral discs are influenced by three main factors: a) vertical loads and b) curvature of the lumbar segment, c) muscular effort (Fig. 9.2)

Both vertical load and muscular effort are lower in the seated position than when standing because part of the load is supported directly by the seat. If the seat inclination is larger than 20 deg, the muscular effort decreases, reducing the load on the intervertebral discs.

For a given vertical load, the stress acting on the discs depends on the curvature. A change in the curvature modifies the stress acting in the ligaments that connect the vertebrae. Similarly to what happens in a composite structure

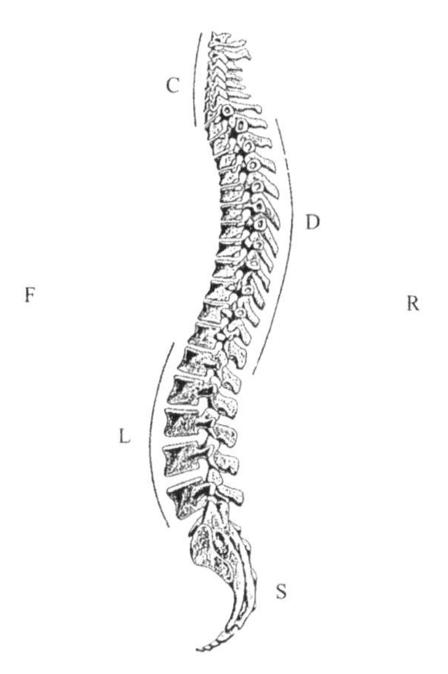

Fig. 9.1. Side view of the backbone. The frontal side is to the left (F), the rear side to the right (R). Cervical segment C); thoracic segment D); lumbar segment L); pelvic segment S).

composed of tension elements (the ligaments) and compression elements (the vertebrae), the increased tension in the ligaments increases the compression force acting on the discs. Since the tension in the ligaments is smaller when standing than in the seated position, different authors [4], [5] suggest replicating in the seated position the same curvature typical of the standing one. With reference to Fig.9.2), this is represented by configuration B, that corresponds to a considerable convexity of the lumbar part (about 50 mm) in the front, or inward, direction (lordosis). Other authors ([6]) demonstrate that a slightly smaller curvature is better suited to reduce the stresses in the seated position.

Compared to the standing position, in the seated position: 1) the thigh to trunk angle is smaller; 2) the vertical load is transmitted to the seat and not to the legs. These aspects justify the adoption, in the seated position, of a curvature of the lumbar segment smaller than in standing position. The lordosis can then reduced to about 20 mm. The suggested position is then similar to that spontaneously adopted during the rest on a side (configuration D, Fig. 9.2).

9.1.2 Joints

The different parts of the human body are connected by joints with complicated kinematic behavior which are not easily represented by a combination of hinges with fixed axis of rotation. The only exception may be the hip joint, that represents a spherical hinge with good approximation. Instead in the knee the hinge

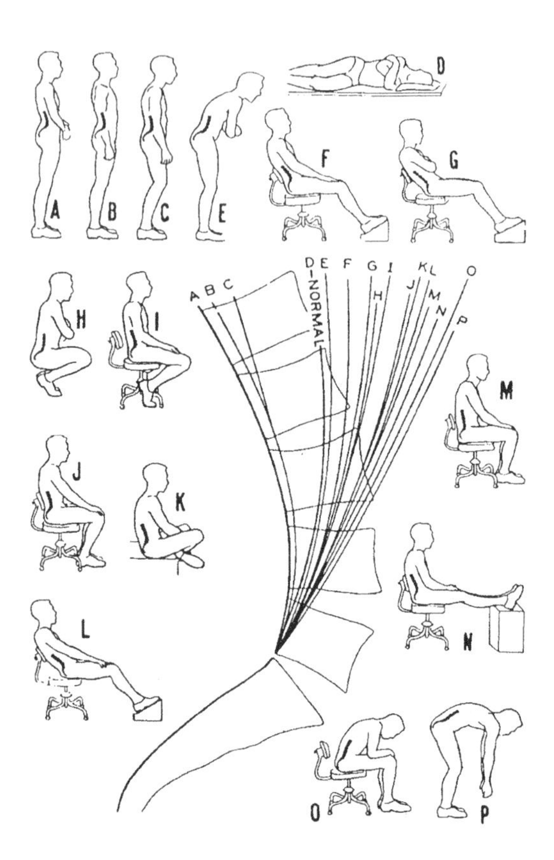

Fig. 9.2. Configurations of the lumbar segment of the backbone. When seated the configuration that minimizes the stress in the intervertebral discs is that adopted during rest on a side (configuration D).

axis is not fixed but changes depending on the angle between femur and tibia. A similar situation occurs at the ankle joint. All this must be taken into account when considering the reliable representation of the parts of the human body.

Considering the loads, the joints of the arms and of the legs are relatively small. In fact, this allows to the driver to easily act on the commands. The generalized use of servo systems for both steering and braking, on the other hand, has reduced the muscular effort needed to move the commands even in heavy vehicles. The arm and leg angles can then be set at mid span of the available range to enable the largest motion possible ([7], [8]). This allows to find accurate and swift position control of the steering wheel and of the accelerator pedal but also to apply the high level of forces required on the brake pedal during emergency braking.

Table 9.1. Angles (degrees) allowed by the main joints in the lateral plane xz of the vehicle.

ankle	α	70÷120
knee	β	40÷180
trunk-thigh	γ	60÷200
shoulder	ε	10÷45
elbow	τ	35÷230
wrist	η	135÷203

To define the posture during ergonomic analysis, a number of characteristic points (H, G, E, C, ...) are set as easily recognizable locations close to the joints which can be easily identified by touch even through clothes, as summarized in Fig. 9.3 and denoted as follows:

1. SD: line parallel to the barefoot;
2. Ch: lateral malleolus;
3. G: lateral condyle;
4. H: greater trochanter.;
5. E: acromium;
6. C: lateral epicondyle;
7. P: ulnar styloid;
8. M: 5th distal phalanx.

These points enable also the definition of the so called postural angles: i.e. the relative angles between the body segments . Fig. 9.3 indicates these postural angles: ankle (α), knee (β), trunk-thigh (γ), arm flexion (ε), elbow (τ), and wrist (η) (ref. Tab. 9.1, Grandjean [9]).

9.1.3 Effects of the Vibrations on the Comfort

The occupants of a vehicle are subject to vibrations of different amplitudes and frequency. In general, the main effect of vibrations is a reduction of the perception of comfort, or, in the case long exposure to high amplitude vibrations, the effect could be pathological on various parts of the body. Different specifications have been defined in various countries to limit the exposure to vibrations and thus preserve health, including: D2002/44, ISO 2631, British Standard 6841 .

The most important frequency range in terms of influence on the human body is that between 0.5 Hz to 80 Hz. Below 0.5 Hz the body effectively follows the vehicle in its maneuvers. Above 80 Hz, vibrations generate local effects of limited

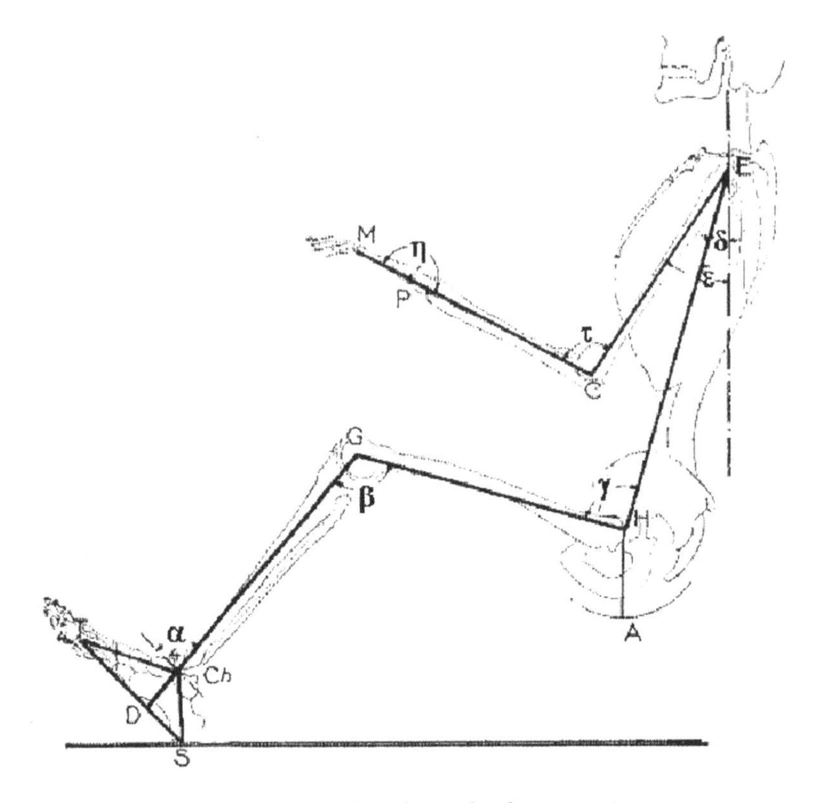

Fig. 9.3. Postural angles and reference points.

importance. In the intermediate range, with the presence also of resonances in the internal organs, even temporary or permanent damage may result if the amplitudes are too high.

Each curve shown in Fig. 9.4 illustrates, as a function of the frequency, the vibration amplitude corresponding to a given comfort level (iso-comfort curves). Curves labeled with 0, a, b, c, ..., f correspond to increasing levels of comfort. In the range 6÷20 Hz the figure shows a lower level of acceleration compared to the adjacent frequencies meaning that the sensitivity to vibration is higher. This is caused by some resonances in the internal organs that occur in that range. The increased stress and fatigue of the tissue affected by vibration justifies this increased sensitivity. Below 1 Hz the sensitivity increases again. Motions of very low frequency could excite the equilibrium organs (inside the ear), leading to nausea and motion sickness.

British Standard 6841 quantifies the amount of vibration that can be applied to a person. This is done by means of the *Vibration Dose Value* (VDV)

$$VDV = \left(\int_0^T a_w^4 dt \right)^{1/4} , \tag{9.1}$$

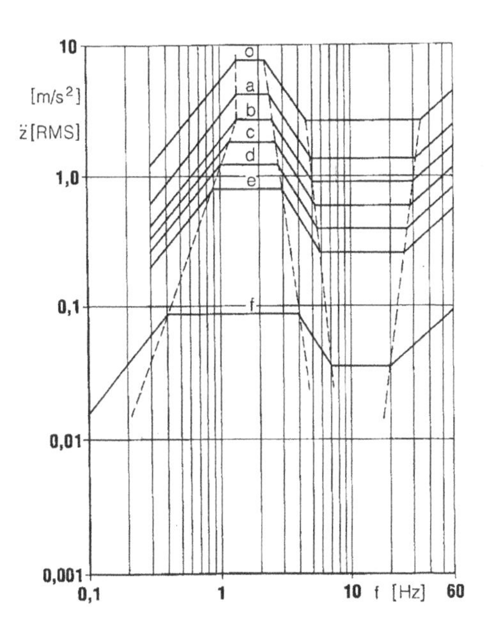

Fig. 9.4. Iso-comfort curves for vertical accelerations. The range of lower sensitivity is 1.2÷1.6 Hz. That of higher sensitivity is between 6÷20 Hz

the value of which is computed from the acceleration a_w obtained after weighting in order to take the different sensitivity to vibration into account (see, for example, Fig. 9.4).

Despite being conceived for the industrial environment, the vibration dose value (VDV) has been adopted to quantify the amount of fatigue induced by vibration to the occupants of a vehicle.

9.2 Manikins for Interior Packaging

9.2.1 Hints about Anthropometry

To define the size of an occupant, or a human body parts, it is usual to introduce the concept of percentile. A given anthropometric measure (l) correspond to a percentile x (and indicated as l_{xth}) when a fraction equal to the percentile of the population does not exceed that measure: i.e. $l < l_{xth}$. For example, the height corresponding to the $95th$ percentile is the height that is not exceeded by 95% of the population taken into account in the study.

Anthropometric data are usually given as indicated in Fig. 9.5. For each human body dimension quoted in Fig. 9.5, Tab. 9.2 and Tab. 9.3 list the anthropometric dimensions of the male and female population for some percentiles (usually $5th$,

50*th*, and 95*th*). It is important to note that each size is not related to the others and must be considered as independent from the others [10]. For example: a person with a height corresponding to the 99*th* percentile could have the shoulder breadth corresponding to the 60*th* percentile and to the arm reach corresponding to the 75*th*.

When the design process involves only one size of the human body (the stature to design the height of a door, for example), the approach related to the percentile is simple and accurate. Nevertheless, it requires some attention when the design involves a group of different body dimensions. The driving position in a vehicle, for example, does not only concern just the stature but also other dimensions such as the leg, of the upper body and of the head, which must each be taken into account at the same time. A possible solution could be to design the driver's seat considering the set of relevant dimensions all included in a large range of percentiles (for example: from 5*th* to 95*th*). One could think that this approach would satisfy a large portion of the population, leaving out a mere 5% of short people and another 5% of tall people. Since anthropometric data are not related to each other, many people (up to 50%, in this case) would consider that driving position to be uncomfortable. The experimental tests on a group of people will show that many people will be excluded by the designed cockpit. On one hand some people would consider the space for the head or the legs to be too small, whereas others would have problems in reaching the pedals or consider the seat too narrow.

Despite being in a different field, clothing is a good example of this concept. Although trousers usually come in Small, Medium or Large waist size it is possible to buy them with Short, Regular or Long leg sizes. In fact long legs do not necessarily correspond to large waist. Similarly, various shirt sleeve lengths can be chosen for a given collar size.

To avoid these risks, it is necessary to verify the design by considering a set of people with real dimensions i.e. not just a statistical analysis of a given population. This can be done experimentally or by software tools. The first case requires the construction of a mock up.

Furthermore it is important to realize that data organized in percentiles are not valid everywhere. For example: European anthropometric measures are not the same as in Far East. Additionally they change with time: Human dimensions tend to increase, phenomenon known as secular growth. Following the end of the Second World War in Europe and North America the anthropometric dimensions increased steadily although at the present this trend has nominally stabilized. Instead significant growth is now evident in developing countries, indicating the intuitive notion that the increase in dimensions is probably related to the better standard of living and nutrition.

The general conclusion is that it is important to employ anthropometric data that are up-to-date and relative to the market where the car will be used.

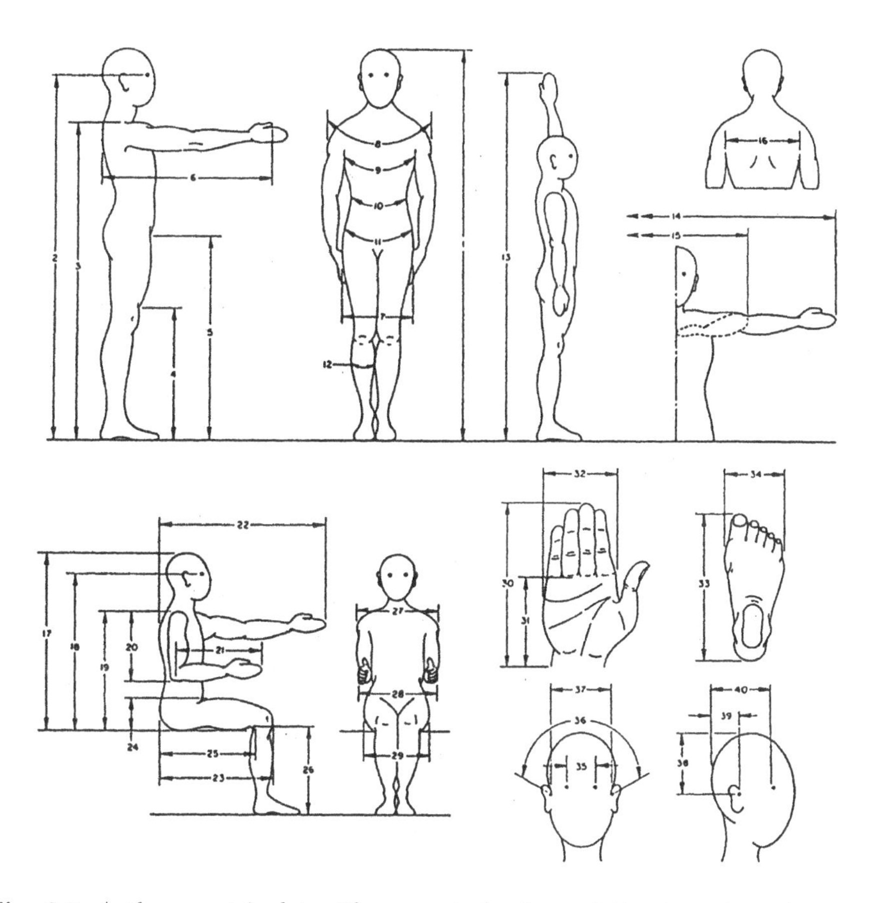

Fig. 9.5. Anthropometric data. The numerical values relative to each quote are reported in Tab. 9.2 and 9.3.

Table 9.2. Main anthropometric data as function of the percentile, male population (United States). The data are in *cm* (except when indicated explicitely).

Dimension	5*th*	50*th*	95*th*
1. Stature	161.6	173.5	184.9
2. Eye height	151.9	162.6	173.5
3. Shoulder height	134.1	143.8	152.9
4. Kneecap height	49.0	54.4	59.4
5. Crotch height	77.2	83.3	90.7
6. Functional reach	75.4	82.0	88.9
7. Hip breadth	32.3	35.3	39.1
8. Shoulder circumference	105.7	114.6	125.5
9. Chest circumference	89.2	98.3	109.7
10. Waist circumference	70.4	80.5	95.3
11. Hip circumference	87.1	95.8	106.2
12. Calf circumference	32.8	36.6	40.6
13. Overhead reach	197.6	-	227.3
14. Arm span	167.4	-	192.0
15. Arm akimbo span	79.4	85.7	92.4
16. Interscye breadth	27.4	30.5	34,0
17. Sitting height	84.3	90.7	96.5
18. Eye height	74.7	80.0	85.1
19. Shoulder height	54.1	59.2	63.8
20. Shoulder-elbow	33.4	36.3	39.1
21. Forearm-hand	44.7	48.0	51.3
22. Anterior arm reach	78.7	-	94.0
23. Buttock-knee	54.1	59.2	64.0
24. Thigh clearance	10.9	14.5	17.5
25. Buttock-popliteal	43.9	49.5	54.9
26. Popliteal height	39.4	43.9	49.0
27. Shoulder breadth	41.9	45.5	49.3
28. Elbow-to-elbow	34.8	41.9	50.5
29. Hip breadth	31.0	35.6	40.4
30. Hand length	17.5	19.1	20.3
31. Palm length	9.9	10.7	11.7
32. Hand breadth	8.1	8.9	9.4
33. Foot length	24.9	26.7	28.2
34. Foot breadth	8.9	9.7	10.4
35. Interpupillary	5.5	6.1	6.8
36. Coronal arc	33.5	35.5	37.5
37. Temporal region breadth	12.6	13.5	14.5
38. Temporal region - top of head	11.9	13.2	14.5
39. Temporal region to nape	8.5	10.2	12.4
40. Eye to nape	15.7	17.2	18.9
41. Weight	57.2 kg	75.4 kg	98.5 kg

Table 9.3. Main anthropometric data as function of the percentile, female population (United States). The data are in *cm* (except when indicated explicitely).

Dimensions	5*th*	50*th*	95*th*
1. Stature	150.0	160.0	170.4
2. Eye height	142.2	150.4	150.6
3. Shoulder height	123.0	132.9	143.4
4. Kneecap height	45.4	50.0	54.6
5. Crotch height	66.2	73.6	81.4
6. Functional reach	67.7	74.2	80.4
7. Hip breadth	31.6	34.8	36.8
8. Shoulder circumference	92.6	100.0	107.4
9. Chest circumference	78.2	88.5	103.3
10. Waist circumference	58.8	66.3	79.0
11. Hip circumference	85.4	94.4	107.3
12. Calf circumference	30.6	34.1	37.9
13. Overhead reach	185.1	199.2	213.3
14. Arm span	149.4	164.3	178.3
15. Arm akimbo span	74.2	79.5	85.1
16. Interseye breadth	31.2	35.0	39.2
17. Sitting height	78.5	84.8	90.7
18. Eye height	68.7	73.7	78.6
19. Shoulder height	53.7	57.9	62.5
20. Shoulder - elbow	30.2	33.2	36.2
21. Forearm - hand	38.9	42.4	45.7
22. Anterior arm reach	55.9	-	83.8
23. Buttock - knee	51.9	56.9	62.4
24. Thigh clearance	10.4	13.7	17.5
25. Buttock - popliteal	43.2	48.0	53.3
26. Popliteal height	35.6	39.9	44.5
27. Shoulder breadth	35.0	40.6	45.8
28. Elbow - to - elbow	31.2	38.4	49.0
29. Hip breadth	31.3	36.3	43.4
30. Hand length	16.1	17.9	20.0
31. Palm length	8.8	9.9	10.8
32. Hand breadth	6.9	7.6	8.6
33. Foot length	22.2	24.0	26.0
34. Foot breadth	8.0	8.9	10.0
35. Interpupillary	4.9	6.1	7.4
36. Coronal arc	31.7	33.9	36.3
37. Temporal region breadth	12.1	12.9	13.7
38. Temporal region - top of head	11.6	12.7	14.1
39. Temporal region to nape	8.9	10.1	11.8
40. Eye to nape	15.0	16.3	18.1
41. Weight	47.2 kg	62.2 kg	90.3 kg

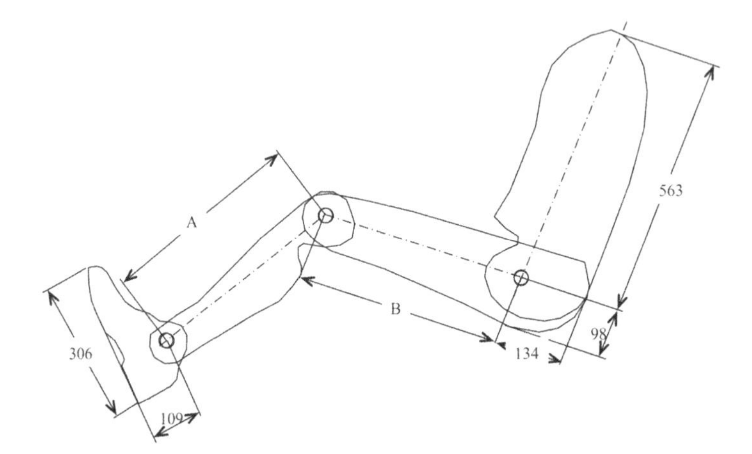

Fig. 9.6. Bi-dimensional manikin from SAE J826 recommended practice for interior packaging. Only the dimensions related to the leg (ankle-knee distance A), and knee-hip distance B) are related to the percentile, the others are fixed. The foot and the trunk are of fixed size. This choice is motivated by the need of insuring a minimum allowance to the trunk and to the foot, even if this could generate rather absurd situations when the leg size is that of small percentiles (5th or even 50th). The four elements of the manikin are hinged together. All dimensions are expressed in mm.

9.2.2 Two-Dimensional Manikins

The interior packaging of a vehicle must guarantee good ergonomic positioning of the occupants to improve their comfort. From this point of view it is important to take into account several important aspects:

- the size of the human body and all its parts change considerably from one individual to another, even amongst a homogeneous population;

- the joints between body segments allow very complex movements that are difficult to represent in a simple form.

Very simple, bi-dimensional manikins have been introduced to simplify the design process. This is especially important during the preliminary stages of the design process, when the management of an overwhelming number of parameters could lead to endless iterations to reach a satisfactory solution. The bi-dimensional manikins represent, in the lateral view, the human body segments that are more influential in the interior packaging process. This is done in a plain and basic form with a dedicated tool (the manikin) comprising a small number of rigid bodies representing the main body segments. Fig. 9.6 shows the SAE J826b .

The manikin represents the lateral profile (plane xz) of an adult male sitting at the driver's position in a vehicle with four rigid bodies that represent the foot,

Table 9.4. SAE J826 recommended practice. Dimensions depending on the percentile. A: ankle-knee distance; B: knee-hip distance

Percentile	10th	50th	90th	95th	99th
A [cm]	39.0	41.6	44.3	46.0	47.6
B [cm]	40.8	43.1	45.6	45.7	47.5

the leg, the thigh, and the torso, each articulated by fixed axis hinges. The head is not included in this manikin: Instead it will be introduced in the following section while addressing the SAE J1052 recommended practice.

The intersection of the plane of the manikin (xz) and the axis connecting the hip joints is defined as H point. The H point is considered as the reference point to locate the manikin in the vehicle and is therefore the starting point for a number of procedures. A typical example is that required by the US and EU standards concerning the direct and indirect visibility.

Another specific point is the so-called heel point (HP) at the rear corner of the heel: When acting on the accelerator pedal the HP point represents the point of contact of the foot with the floor. The importance of the HP will be pointed out in a later section since it becomes the starting point of all packaging process.

Regardless of the complexity of the knee and ankle joints, and of the almost infinite number of degrees of freedom of the backbone, the bi-dimensional manikin has only three fixed-axis hinges representing the ankle, knee and hip joints, while the trunk is a rigid body. These assumptions are only acceptable in the context of a vehicle packaging: This manikin cannot, therefore, be employed for purposes other than to represent individuals sitting in a vehicle.

Figure 9.6 shows how only the dimensions related to the leg (ankle-knee distance A), and knee-hip distance B) are related to the percentile, as reported in Tab. 9.4. The dimensions of the trunk and the foot (wearing shoe), by contrast, are fixed. This choice is motivated by the need of insuring a minimum allowance to the trunk and to the foot, even if this could generate rather absurd situations when the leg size is that of small percentiles (5th or even 50th). In fact, providing too little space around a foot wearing large shoes or boots could generate dangerous situations. For example, if the foot gets stuck in the pedals or in the lower part of the dashboard.

To represent the manikin in a even more simplified way and to quote a given posture, all parts of the bi-dimensional manikin have a reference line: Clearly for the leg the reference lines are the ankle-knee and the knee-hip lines, whereas for the foot it is the line from the heel point (HP) and tangent to the sole (ball of foot). The trunk reference line connects the hip point (H) to the shoulder point, represented by a corner on the upper part of the manikin.

The reference lines defined on the manikin allow to quote angles between them to represent a given posture. By looking at Fig. 9.3 at a first glance it should be possible to compare these angles with the postural angles measured on humans from reference points measured on their body. Nevertheless this comparison can not be done in a straightforward way since the SAE J826 manikin has fixed axis hinges that connect the body segments, while the human body has rotoidal hinges with axes that depend on the angle of the joint.

9.2.3 Head Contour

As already mentioned, the SAE J826b manikin does not include the head but just the body, the leg and the foot. The SAE J1052 recommended practice describes how to take the head into account in the packaging of a vehicle ([11]) by means of the contours reported in Fig. 9.7 and in Fig. 9.8. The contours are a means of taking into account the space required to the head, including the hair when looking in straight ahead direction. As shown in Figs. 9.7, 9.8 each contour corresponds to a given percentile, obtained by the statistical analysis of a 50% male - 50% female population. Each individual is asked to seat with a back inclination of 25° to the vertical and to look in horizontal direction straight ahead. The tangents to the contour corresponding to a given percentile, 95th for example, leave a fraction of the population equal to the percentile (95% in the case of 95th percentile) below the tangent. The head of the remaining fraction (5% in the case of 95th percentile) crosses it having a part above that plane. The head contour is therefore obtained as envelope of planes (lines in each view) that split the plane in two parts. The contour indicated with letter m in the same Figs. 9.7 and 9.8, indicates the statistically mean head profile not including hair. The much larger size of the head contours corresponds to the need to take into account a much larger number of factors that influence the position of the head in a given sitting position that are not present in the definition of the mean profile. In fact each individual selects a slightly different position on the same seat because of their different body size, posture, skeletal and muscular system, psychological attitude, just to list the main items. The mean head profile, on the other hand, is obtained as the average head size located at the same position.

According to this definition, the head contours allow to evaluate the minimum distance between the inner surfaces of the vehicle and the head of its occupants, and as such provide an important tool to evaluate the impact on the ergonomics of the design choices regarding the upper structure of the vehicle, including all its interior trimmings.

The representation shown in Fig. 9.7 refers to an occupant on a seat that can be adjusted longitudinally, as is possible with most driver's seats. Fig. 9.8, on the other side, refers to an occupant on a fixed seat, as one of the passengers.

The positioning of the head contours of a given percentile is achieved with the following steps:

Lateral view:

1. Draw a vertical line passing through the H point.
2. Draw a horizontal line, 635 mm above the H point.

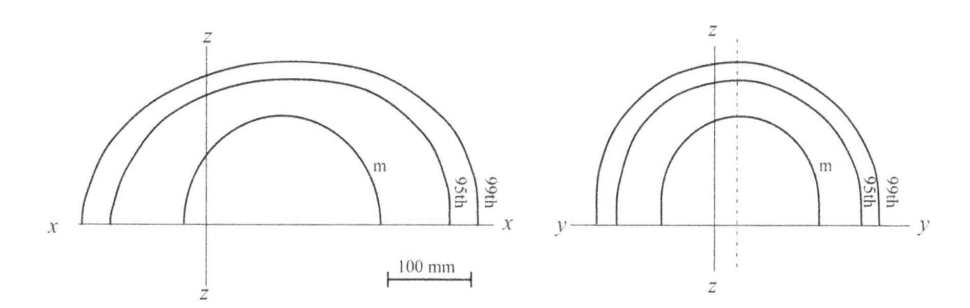

Fig. 9.7. SAE J1052: head contours for seats adjustable in horizontal position. The contours are function of the percentile. Contour m indicates the mean size of the head. The dash-dot line (-.-) indicates a vertical line passing at mid-eye distance.

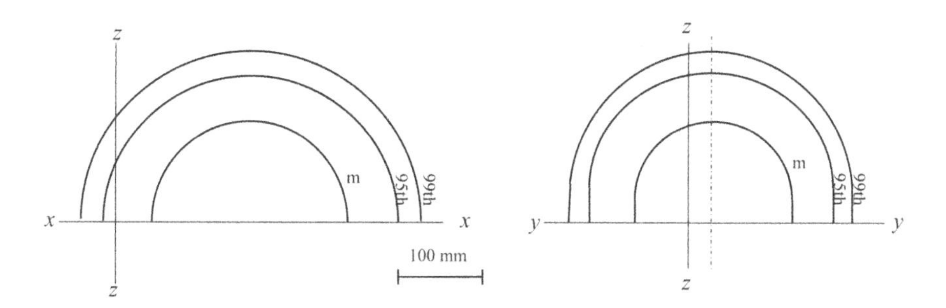

Fig. 9.8. SAE J1052: head contours for fixed seats. The contours are function of the percentile. Contour m indicates the mean size of the head. The dash-dot line (-.-) indicates a vertical line passing at mid-eye distance.

Table 9.5. Correction to the horizontal and vertical position of the SAE J1052 head contours and of the SAE J941 eyellipses as function of the inclination of the back relative to the vertical (angle δ) - Horizontally movable seats.

δ [deg]	longitudinal adjustment (x) [mm]	vertical adjustment (z) [mm]
5	-186.4	27.6
10	-137.4	25.1
15	-90.0	19.7
20	-44.2	11.3
25	0.0	0.0
30	42.6	-14.3
35	83.6	-31.5
40	123.0	-51.8

Table 9.6. Correction to the horizontal and vertical position of the SAE J1052 head contours and of the SAE J941 eyellipses as function of the inclination of the back relative to the vertical (angle δ)- Fixed seats.

δ [°]	longitudinal adjustment (x) [mm]	vertical adjustment (z) [mm]
5	-114.6	47.8
10	-51.1	45.2
15	9.6	39.0
20	67.6	29.5
25	122.7	16.5
30	175.1	0.0
35	224.6	-19.8
40	271.4	-43.1

3. Locate the head contour reference frame $xx - zz$ so that its origin coincides with the intersection of the two lines defined above, and that axes xx and zz are respectively horizontal and vertical.

4. If the inclination of the trunk $\delta = 25$ deg, (Fig. 9.3), this is the position of the head contour and no correction is necessary. Otherwise $\delta \neq 25$ deg Tabs. 9.5, and 9.6 indicate how to correct the position obtained from the previous steps as function of the seat back inclination δ.

Front view:
Locate the yy reference line at the same vertical distance from the point H as in the lateral view.

For movable seats the contour must be located so that the *mid-eye* reference line passes through the H (or R) point.

9.2.4 Three-Dimensional Manikins

The bi-dimensional manikin of SAE J826b and SAE J1052 recommended practices have been conceived as drafting tools, especially useful in the vehicle packaging phases and for the preliminary decisions concerning ergonomics. Its function is therefore limited to the very start of the design process.

As soon as the vehicle is defined as a three-dimensional object, it is necessary to introduce a type of manikin that takes the true dimensional nature of the occupants into account. Depending on the task to be accomplished, three-dimensional manikins may have different characteristics:

- three-dimensional drawing and design;
- experimental definition of the H point to verify the agreement of the various standards, for example: direct and indirect field of view, binocular obstruction.

The almost universal use of three-dimensional CAD tools has motivated the introduction of three-dimensional virtual manikins (RAMSIS, SAMMIE, JACK just to indicate three commercial codes including virtual manikins).

Three-dimensional virtual manikins enable the ergonomics effects of the design choices taken during the three dimensional design to be verified, and permit the effort required to reach commands and controls to be determined and the direct and indirect field of view and corresponding obstruction to be assessed. From the point of view of anthropometry, three-dimensional CAD tools enable manikins with dimensions consistent with the percentile databases, or to the sizes of real individuals, to be configured, allowing what has been set using percentile data to be verified in a virtual environment to ensure that it is satisfactory for the largest number of people. This allows to perform at the virtual level a number of tests that previously required the setup of expensive experimental mock-ups and measurement campaigns on a number of potential users.

The creation of target users databases, and the large number of different configurations that can be tested on them is essentially a modification of the percentile-based approach. Nowadays design can be performed directly on a number of target users with the aim of identifying a solution that satisfies most of them. This approach overcomes the most critical issue of the percentile-based approach, i.e. the implicit assumption that an individual is an assembly of body segments the dimensions of which are unrelated to each other. Even within the percentile approach, the flexibility and the parametric nature of most of these virtual tools allows to test the different percentiles apart from the "design" one. This enables a good compromise to be reached in terms of posture, visibility and accessibility also for percentiles much smaller or larger than the reference one ([10]).

Fig. 9.9 shows the three-dimensional manikin prescribed by D71/127; a similar manikin is described by SAE J826 recommended practice. Both manikins are meant as physical objects, the aim of which is to find the position of the H

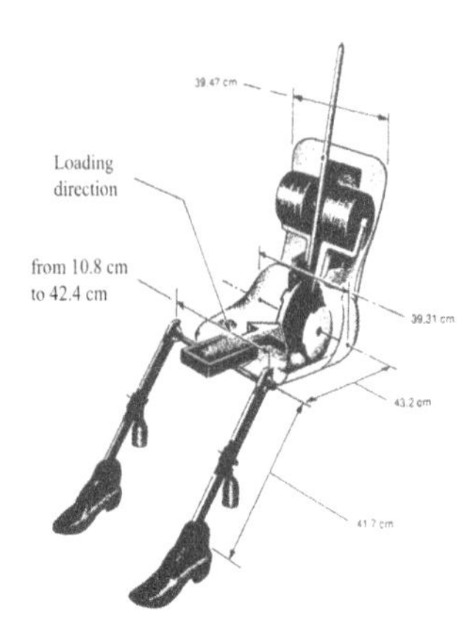

Fig. 9.9. Three-dimensional manikin as prescribed by EU standard D71/127 to find the position of the H point and torso inclination relative to the vertical (angle δ) of the driver's seat of an existing car or a prototype.

point (or SgRp in SAE terms) and the torso inclination relative to the vertical (angle δ) of the driver's seat in a real car. These data constitute the starting point of the homologation tests concerning direct and indirect visibility. Due to the critical importance for a car manufacturer of the homologation process, the same standards indicate a detailed positioning procedure of the manikin to reduce the uncertainties of this measure to the minimum.

9.2.5 SAE Quotation System

To compare the dimensions and geometrical characteristics of different vehicles, the use of a unique quotation system is very helpful to avoid sometimes complicated conversions between the many different possible choices available to the designer. In this context, the SAE J1100 ([12]) recommended practice defines how to indicate the main dimensions of a vehicle, starting from the overall dimensions down to details of relatively minor importance. Some of these dimensions are shown in Fig. 9.10 and 9.11. Some dimensions relate to the posture of the occupant, while others to the overall size and layout. SAE recommends that the measurement of the angles and characteristic dimensions is based on the use of bi-dimensional and three-dimensional manikins and not on the direct measurement on real individuals. As already mentioned, this prevents a direct comparison with respect real humans (from the characteristic points of Fig. 9.3) and similar (but not equal) dimensions measured on the manikins. For example: although apparently the same, angle L40 of Fig. 9.11 and angle δ of Fig. 9.3 are different quantities being defined in a different way. The first is measured using a well specified manikin, the second from markers located on the body that can be easily identified by touch even through clothes. To overcome this difficulty, some

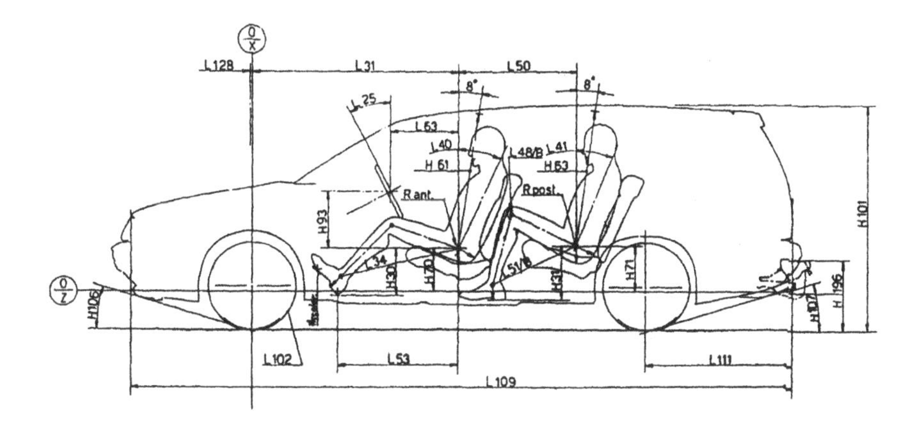

Fig. 9.10. Example of a vehicle packaging. The main dimensions are consistent with SAE J1100 recommended practice.

rules have been proposed to link measurements on humans to manikin related quantities consistent to SAE J826b and J1100 recommended practices.

9.3 Hints of Occupants Positioning

The aim of this is to outline the main factors that influence the positioning of the occupants in a vehicle considering the main constraints coming from the vehicle structure, the commands, and the need of assuring an ergonomically acceptable position. The packaging of a vehicle is a process that must fit many different subsystems within: powertrain, commands, suspensions and structure, dashboard, luggage, apart from the driver and the passengers. Since the volume is relatively small, the positioning of the occupants is, in a sense, the result of a compromise. Nevertheless the positioning process can not compromise the following needs:

- comfort,
- safety,
- functionality and accessibility to the commands,
- direct and indirect visibility.

In a vehicle as in many other tasks that a person can undertake (from office work to much more physically involving tasks such as construction), the correct posture of the backbone is an important factor in increasing the resistance to effort, load and vibrations. A correct position of the backbone is also an essential factor in avoiding long term damages to the skeletal system. More specifically, in the case of a vehicle, it is necessary to take the following factors into account:

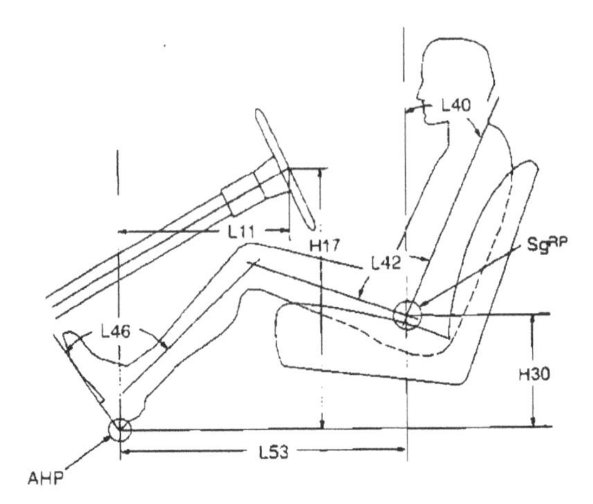

Fig. 9.11. SAE J1100 dimensions relative to driver's posture.

- The backbone is the structure that transmits the vertical load of the upper body to the legs. The transmitted force increases from the cervical vertebrae to the lumbar ones that, in comparison, are the most heavily loaded as indicated by their larger size 9.1.

- To reduce the effects of fatigue and limit the possibility of permanent damage, it is necessary to minimize the vertical accelerations.

- The driver must have an unobstructed view of the outer environment; The position must then allow the driver to maintain his head in upright position without the need of a support. The upright position is suggested also for the passenger beside the driver, even resting on the head restraint at the cost of limiting possibly limiting his view outside.

A slight inclination of the seat back is a good means of reduce the load on the lumbar vertebrae. This allows to transmit part of the upper body load directly to the seat, thus relieving the lumbar part. Nevertheless, the inclination should not be excessive so as to avoid the need for effort to maintain the head in upright position. An excessively reclined position limits the accessibility and the effort that can be applied to the commands (steering wheel, for example).

9.3.1 Basic Postures

Fig. 9.12 shows three basic postures of interest for automotive use: SE seated, RC reclined, and CR cramped: Each allows the driver to maintain the head in upright position without effort.

In principle, other positions could be adopted. In the *proned* position, for example, the driver can look forwards but is highly uncomfortable and requiring

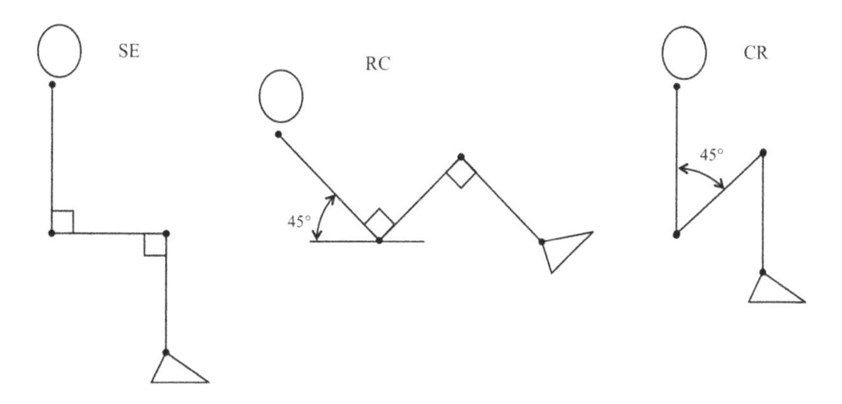

Fig. 9.12. Basic postures for automotive use. SE: seated; RC: reclined; CR: cramped.

significant effort to act on the commands, with clear concerns regarding safety: This position has been adopted only in a number of speed-record vehicles. The *upright* position, on the other hand, requires a very small footprint area but the largest vertical extension. Apart from a small number of notable exceptions (e.g. the Segway), its use is largely restricted to public vehicles for urban mobility (buses and underground, for example).

Reclined *position (RC)*

The reclined position, represented in Fig. 9.12, is characterized by a 45 deg inclination of the trunk and leg segments with respect to the vertical. The head and neck are in an upright position, although with some difficulty.

Advantages.

- Possibility to resist large vertical accelerations since most of the load is supported by the seat.
- Reduced load on the blood circulation system. The limited vertical size requires little effort by the heart to pump blood to all organs.
- Reduced vertical size. The limited cross section of the car body allows the aerodynamic performance to be improved.

Drawbacks.

- Reduced visibility: after a time, the considerable effort on the neck to maintain the upright position prevents free rotation of the head, thus introducing serious limitations on the direct field of view. The backwards visibility when maneuvering in the reverse direction (during parking, for example) is also seriously limited. This is because to look backwards the driver is compelled to lift his upper body in an almost vertical position and then to turn his neck and shoulders which can be done only if the shoulders are not in contact with the seat.

- Reduced accessibility to the commands: the stretched position of the arms requires effort and reduces the amount of torque that can be applied to the steering wheel.
- Longitudinal size: the considerable longitudinal size can be mitigated in a two row arrangement by a partial overlapping of the legs of the second row under the body of the front.

Use. This posture is seldom adopted. An almost reclined position is typical of racing cars, although with a smaller inclination of the upper body to improve visibility and the action on the commands by the driver.

Seated *position (SE)*

This basic posture has 90 deg angles between the body segments and by the vertical position of the seat back.

Advantages.

- Visibility: relative to the reclined position the neck directly supports the head in upright position. Additionally, backwards visibility during reverse maneuvers requires an easier rotation of the upper body (relative to the reclined posture), since it does not involve lifting from the seat back.
- Longitudinal size: the longitudinal size is much smaller than in the reclined position.
- Accessibility: the torque that can be produced by the driver when the steering wheel axis is almost vertical is the highest compared with other positions, the accessibility to other commands such as the gearshift and switches is better than in other postures. Also getting in and out of the vehicle is much easier in this case. Since the upper body is already upright, and the centre of gravity is at about the same level as in standing position, the effort to get in and out is low, particularly when compared to the reclined position.

Drawbacks.

- Reduced capability to resist to vertical accelerations: the seat back does not alleviate the vertical loads on the backbone. To overcome this problem, the seat can be connected to the vehicle chassis by means of a compliant suspension, enabling the dynamic loads on the lumbar segment of the backbone to be reduced, and long term damages avoided. The stiffness and damping of this suspension are designed with the objective of attenuating the dynamic excitations transmitted to the driver as much as possible.

- Vertical size: the vertical size is larger than that of the reclined position. For vehicles with a large frontal section (i.e. industrial vehicles) this may not be a limiting factor, whereas in other cases, e.g. most cars, it certainly would be.

Use. The accessibility to the commands and the visibility make the seated position most suitable for industrial vehicles. Additionally, the limited longitudinal size enables high the cargo volume, while the vertical size does not present a problem because of the large cross section of such vehicles.

Cramped *position (CR)*

Implies a small thigh-trunk angle (45°, in the figure).

Advantages. Small longitudinal, and vertical, size.

Drawbacks.

- Very uncomfortable posture which can be maintained for only a short duration.
- Reduced capability to resist to vertical accelerations: The limited curvature of the backbone induced by the position of the legs increases the loads on the lumbar segment.
- Accessibility: it is almost impossible to maneuver conventional commands such as the steering wheel and pedals, while getting in and out of the vehicle is highly uncomfortable.

Use. The adoption of this posture is almost compelled for the second row of vehicles with a small size dedicated to rear seat passengers (i.e. sub compact or sports cars).

9.3.2 Positions Adopted in Automotive Applications

The posture adopted for the front row in automotive applications is usually a compromise between the reclined and seated positions. The objective, clearly, is to combine the principal advantages of these two basic postures while limiting the drawbacks. To limit the longitudinal size, the second row has a rather more cramped posture, with smaller thigh-trunk angles, especially for compact vehicles while it does not apply for large sedans where the second row may have a posture optimized for comfort.

Different investigations performed on real vehicles and in the laboratory indicate the range of angles listed in Tab. 9.7 as being acceptable for the tasks to be performed by the driver and to minimize postural stress. The angles are defined as in Fig. 9.3.

Considering the postural angles, it is worthwhile to note that:

β In order to allow easy access to the pedals, this angle should neither be too large, in order to allow extension of the leg (β increasing), nor too small to allow bending of the leg (β decreasing). In all cases β is not a critical angle and can have wide variations about the mean value. It is suggested that the knee be in front of the mean plane of the steering wheel to avoid any interference when acting on the brake pedal. The distance from the brake pedal (not depressed) to the plane tangent to the front of the steering wheel is usually between $600 \div 660$ mm.

γ It is suggested that this angle is not smaller that 90 deg, especially for fat people. The bump on the belly of the SAE J826 bi-dimensional manikin evidences the risk of interference with the thigh.

δ A reclined position ($\delta > 25$ deg) implies a larger distance to the steering wheel and a more restricted visibility with respect to the more "seated" positions ($\delta < 25$ deg) but allows a better resistance to vertical accelerations. According to Grandjean, the recommended value for δ implies a rather vertical position of the upper body. Rebiffé, on the other hand, suggests a more reclined position in order to obtain a better weight distribution on the seat. The weight of the upper body is going more on the seat back, reducing the load on the cushion with a better general comfort. As already stated, the more reclined positions leads to increased difficulty in terms of acting on the commands, and needs a steering wheel axis in an more horizontal direction. ($30 \div 35$ deg are usual inclinations of the steering wheel axis to the horizontal; the lower bound is in the range of 20 deg). In all cases it is necessary to check that there is no risk of interference with the knees during hard braking.

α A comfortable condition for the ankle corresponds to a relatively fixed position with small variations about the mean value. The decision regarding the angle α which is rather important is usually made considering that the accelerator pedal is at mid travel position. Considering that the maximum rotation allowed by a normal ankle with no major effort is about $\Delta\alpha =$ 40 deg, the variation of α is in the range $70 \div 120$ deg. Additionally, as the travel of the accelerator pedal is about $50 \div 60$ mm, this corresponds to a rotation of the foot of ± 8 deg about point HP. If, with the accelerator pedal at mid-travel, $\alpha = 85$ deg or 90 deg, the fields of variation for the ankle are the following:

$$\alpha_1 = 85 \pm 8 \text{ deg} = 77 \div 93 \text{ deg}$$
$$\alpha_2 = 90 \pm 8 \text{ deg} = 82 \div 103 \text{ deg}$$

These angles are in the range allowed by the ankle joint ($70 \div 120$ deg, see Tab.9.1).

The angles listed in Tab. 9.7 are not unrelated to each other and to the body size. A position with an extended leg ($\gamma = 130$ deg) will be rather unusual in

Table 9.7. Angles (deg) between the main body segments suggested by different Authors. Rebiffe and Grandjean make reference to allowed ranges. Porter and Park to the mean and standard deviation (in parentheses).

	Rebiffe (1969)	Grandjean (1980)	Porter/Gyi (1998)		Park (2000)	
			men	women	men	women
γ	95÷120	100÷120	101 (6)	99 (5.2)	116 (7.6)	119 (7.6)
β	95÷135	110÷130	121 (8.1)	117 (8.6)	133 (9.9)	135 (6.6)
τ	80÷120	-	128 (20.3)	113 (17)	119 (13.7)	106 (10.9)
ε	10÷45	20÷40	50 (2.4)	40 (2.8)	-	-
α	90÷110	90÷110	93 (6.4)	92 (5.3)	102 (8.2)	99 (9.0)
δ	20÷30	20÷25	18 (3.2)	14 (3.8)	-	-
L40	-	-	18 (3.2)	14 (3.8)	-	-
L42	-	-	94 (3.8)	91 (3.9)	-	-

Table 9.8. Position of the H point as measured by Porter and Gyi [8] in the laboratory using a seating buck.

SAE dimensions	L11 [mm]	L40 [deg]	L53 [mm]	H30 [mm]	H17 [mm]
min.	322	5	577	283	580
mean	438	16	738	301.1	628
max.	602	25	889	335	689
standard dev.	48	4	67	11	24

association with a vertical trunk (small thigh-trunk angle $\gamma = 95$ deg). Taller people tend to prefer an extended leg and large thigh-trunk angle while shorter people adopt a position with a rather small knee angle and an upright position of the trunk. These two simple cases demonstrate that the choice of a posture is largely determined by the size of the body and by the relative proportion between its segments. The use of data such as those reported in Tab. 9.7 require then an experimental check to verify the design choices.

Another important factor is that the posture is also dependent on the configuration of the driver's seat. The need to read the instrument panel on the dashboard and look through the windscreen and lateral windows are factors that influence the posture. A short individual, to look the road and the instruments, will choose a posture with the trunk in a more vertical position than a tall person. Instead a tall person, for the same reasons, will choose a more reclined position of the backrest and move the seat backwards.

The approach based on the assumption that the trunk is a rigid body (SAE J826 and Fig. 9.3) neglects all variables related to the posture of the backbone and, more in detail, its lumbar segment. Some studies [3] have addressed the problem of identifying the optimal posture of the backbone by introducing an

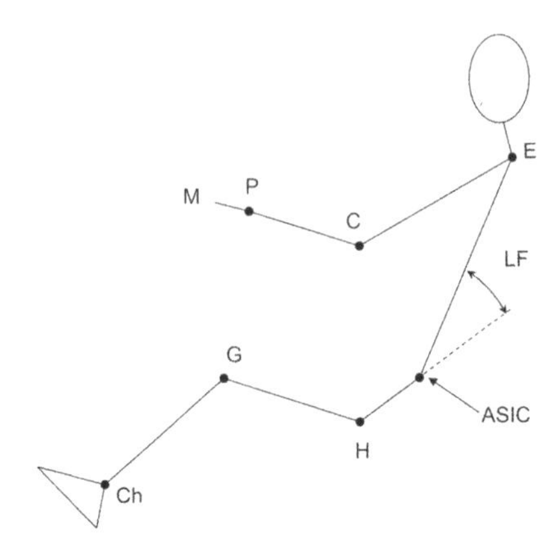

Fig. 9.13. Reference points for taking the posture of the lumbar segment into account. A reference point has been added on the anterior superior iliac crests - ASIC.

additional degree of freedom in the postural parameters. The introduction of a reference point on the iliac crests (anterior superior iliac crests - ASIC in Fig. 9.13) relates to the influence of the inclination of the hip bone relative to the lumbar segment.

Regardless of the freedom in selecting the backrest inclination, all subjects involved in the survey chose a nearly constant flexure of the lumbar segment with an angle LF=31.6 deg). As the backrest inclination influences both the trunk-thigh inclination, and the lumbar flexure, this last parameter (angle LF) is an important factor of the postural comfort. Nevertheless, the use of manikins with a rigid trunk (SAE J826B recommended practice) is justified by the fact that the angle LF varies little between different subjects. Bi-dimensional manikins with an additional joint between the hip and the backbone have been proposed recently [13] and enable the effects of the backbone curvature to be taken into account from the design stage.

9.3.3 Experimental Tools to Evaluate the Postural Comfort

Fig. 9.14 illustrates a test bench, the aim of which is to measure the posture during driving (*seating buck*). The bench is representative of the position of the driver and the main cockpit elements: pedals, seat, steering wheel, gear stick, dashboard. A screen is used to visualize the road traffic environment during driving.

Usually these kind of benches allow the main parameters of the cockpit to be tuned, i.e.: the longitudinal, vertical position of the seat, backrest inclination, steering wheel position and inclination, pedals position.

The individuals (selected from a representative sample of the population) are asked to seat and simulate driving for about 2.5 hours. During the first part of

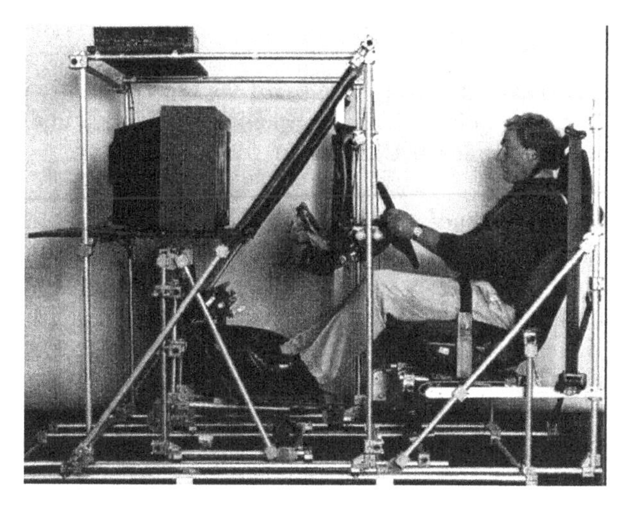

Fig. 9.14. Test bench to investigate the postural seating comfort (*seating buck*). The bench is representative of the position of the driver and the main cockpit elements: pedals, seat, steering wheel and allows to measure the angles between the body segments.

the test, each individual follows a procedure with the objective of reaching the most comfortable posture. To assess the obtained result (postural angles) this posture is kept for at least 15÷20 minutes.

During this test the postural angles are measured by means of optical markers on the characteristic points of their body (lateral condyle, greater trochanter, lateral malleolus, lateral malleolus, acromium, lateral epicondyle, ulnar styloid, 5th distal phalanx).

Also useful is the measurement of the pressure distribution on the seat cushion and backrest.

9.4 Pedals Functionality and Positioning

The consolidated trend towards larger interior volumes with respect to the overall dimensions of the car makes the positioning of the first row during the initial packaging process of great importance. This is possible only after the positioning of the pedals since the driver is always in contact with them. The positioning of the pedals is therefore the starting point in the packaging of the vehicle and has an important outcome on the definition of the entire interior space.

The pedals are attached to a part of the structure bounded by the following elements:

- Front wheel arch: The volume surrounding the front wheel that is necessary to guarantee its functionality, i.e.: steering, and vertical displacements due to the suspensions. Other important parts of the vehicle are located inside the wheel arch including the suspension, steering rods and transmission shafts (in case of front traction or four wheel drive).

- *Tunnel*: in front engine - rear wheel drive vehicle architectures, the tunnel is the structure used to integrate the transmission shaft in the floor. As described in Volume I this enables the floor level and hence the centre of gravity of the vehicle to be lowered. The tunnel is usually present also in front engine and transmission architectures: in these cases it is used to locate part of the exhaust system and the hand brake commands while increasing the resistance to longitudinal loads (in the case of a crash).

- Floor.

- Firewall: this is a structure that separates the inside of the vehicle from the engine compartment (or the front trunk in the case of rear engine architectures). The pedals are usually connected to the firewall. The steering rack may be connected to the firewall on the opposite side of the pedals, or it may be connected to some other structure (for example to a frame or subframe). In any case, the need to insure correct kinematic behavior of the steering system (Ackerman) usually requires it to be located close to the pedals.

Clearly a forward position of the pedals increases the volume inside the car. Nevertheless, the factors that limit the forward position of the pedals and the firewall are:

- Correct functionality of the steering rack requiring to locate it at a certain distance from the steering axis which is usually very close to the wheel axis. In front-transversal engine architectures, to be able to install the power-train, the steering rack is behind the engine. The forward limit to the location of the firewall may be given by the minimum longitudinal (the range is 100÷150 mm) distance between steering rack (and rods) and steering axis.

- The need to install the power-train and its architecture sets hard limits on the forward position of the firewall.

- The front part of the vehicle plays an essential role in the dissipation of the kinetic energy during a front crash. The need to limit the accelerations of the occupants requires designing a front structure with a minimum longitudinal size.

9.4.1 Wheel Arch Dimensions

The wheel arches are the volumes dedicated to the wheels. In normal cars, the wheel arches are usually inside the bodywork (covered wheels). Anyway, their large size have an important impact on the layout of the vehicle packaging. The design of the wheel arches has to be completed before proceeding with the packaging and, more specifically, with the positioning of the pedals and of the manikins.

The front wheel arches must allow all motions permitted by the suspension and steering without contact between the wheels and the fixed parts. As a rule of thumb, in 'normal use' cars, the wheel arches can be designed considering that:

- The car is in the full load conditions (load standard F for a 5 seats car);
- The maximum vertical displacement allowed by the suspension is about 70÷90 mm.
- The maximum steering angles can be evaluated considering a 100% Ackerman steering kinematics and the minimum steering radius that has to be obtained (usually in the 5÷6 m range). In a curve, the inner wheel has a larger steering angle than the outer one. For mid-size European cars, the order of magnitude of these angles is about 40 deg for the inner wheel (toe out) and 30 deg for the outer wheel. Since the vehicle must be able to make both right and left turns of the same radius, each wheel must be capable of a toe out angle of 40 deg and a toe in angle of 30 deg.

It would be possible to consider designing the front wheel arches to be compatible with maximum suspension travel and with maximum steering at the same time although this assumption could be a little too conservative. The maximum displacements of the suspension occur when the vehicle is traveling at mid to high speed, so in conditions where the steering angles are smaller than during parking. By contrast, during parking the suspension displacements are usually small but the need of maneuvering in small space requires large steering angles. The result could be to design a wheel arch that allows maximum suspension displacements just in a low to moderate range of steering angles, while the maximum steering angles are allowed for reduced suspension displacements. In Fig. 9.15 the maximum displacement of 80 mm is possible only for steering angles $-15 < \delta < 20$ deg. The maximum steering angles are compatible with only the 75% of the maximum suspension travel.

The part of the wheel that influences more the size of the wheel arch is the rounded corner between the thread and side walls. This part of the wheel determines the corner of the wheel arch closer to the front occupants (inner rearward corner). If the vehicle is front wheels driving, in addition to the tires it is necessary to take the presence of snow chains into account (if applicable). The thickness of the chains is usually in the range 12÷16 mm. Since during the use a layer of snow and dirt remains attached to them, it is necessary to consider

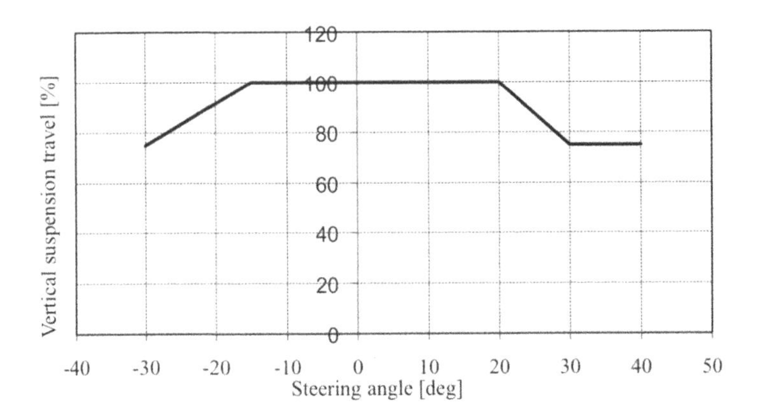

Fig. 9.15. Front wheel arch: max suspension travel as function of the steering angle that may be used to limit the size of the wheel arch.

an additional thickness of about 15 mm of allowance to avoid contact with the wheel arch surface.

The definition of the wheel arch must take into account of the thickness of the structure that separates the volume dedicated to the wheel and the inside of the vehicle and the various layers that cover it for various purposes (vibration damping, noise attenuation, trimmings). The overall thickness of all of this is in the range of (17÷35 mm).

Fig. 9.17 shows that the part of the envelope closest to the clutch pedal is the most critical one, as it is could interfere with it or with the foot. To avoid this it is necessary to consider an adequate allowance (about 70 mm from the edge of the pedal).

Rear wheel arches have a much smaller size because the main movement of the wheel is in vertical direction while the camber and steer are relatively small, even in case of four wheel steering vehicles. In all cases the space required for the suspension, snow chain (if applicable), structure and all linings and trimmings.

9.4.2 Pedals

Accelerator

This pedal is the most frequently used by the driver since it is always acted on except during braking. For this reason the posture of the driver may be optimized considering its foot acting on the accelerator at the most likely command position (mid-travel). As already mentioned, the contact point between heel and floor is usually indicated as HP (Heel Point) or AHP (Accelerator Heel Point).

The effort required to act on the accelerator should be relatively small (10÷20 N), so that the driver can easily control its position with good precision and in a displacement range compatible with the ankle joint (50÷60 mm).

To improve the control of the pedal by the driver it is important to minimize all slipping between pedal and foot (the shoe sole). Slipping, in some cases, could generate stiction with a loss of precision in actuation. If the pedal is hinged at its upper end (Fig. 9.16 a1)), it is suggested that the contact point with the foot ball lies on the line that connect the HP point to the hinge of the pedal. This alignment allows to null the relative tangential displacement between foot ball and pedal and therefore avoid the friction associated to it. Unfortunately, this alignment can occur in just one position of the pedal, so it should be realized in the most frequently used position of the accelerator (mid-travel position). In all other positions some amount of friction will be present.

An accelerator pedal hinged at the heel point (HP, Fig. 9.16 a2)) allows all problems related to friction between foot and pedal in all positions to be avoided. This solution is then usually better than the previous one in terms of performance. Nevertheless complexity is increased because the accelerator and the clutch-brake assembly become separate subsystems with all additional costs.

Brake

In contrast to what happens for the accelerator, the brake pedal requires a considerable force, at least under certain circumstances. During an emergency due to a failure to the power braking system, for example, the effort on the pedal can be up to 500 N. The expected response (braking torque) is proportional to the effort while no major requirement is given to the precise positioning. From this point of view, the brake and the accelerator pedal are then complementary. To apply such large forces, it could be necessary to act on the pedal with the foot arch with the heel not in contact with the floor. For the same reason it is important to insure that the lifted leg (the knee) does not interfere with the steering column as this could prevent the driver from braking hard enough Fig. 9.16 b). For the same reason, the foot must push just on the pedal, and any other contact with the lever or neighboring parts (the dashboard, for example) should be avoided.

Clutch

The clutch pedal requires an effort and a type of control between that required for the brake and the accelerator. The first part of the travel compensates the gaps and the compliance of the command while the second part corresponds to disengaging. The maximum effort is moderate (in the 100 N range) and the modulation of the transmitted torque requires a rather fine position control.

Not all drivers act on the pedal in same way. Some start pushing with lifted foot and, when the clutch starts disengaging, the heel goes into contact with the floor where it remains during the rest of the operation Fig. 9.16 c1), c2). Some drivers always maintain the foot in contact with the floor to have a better position control.

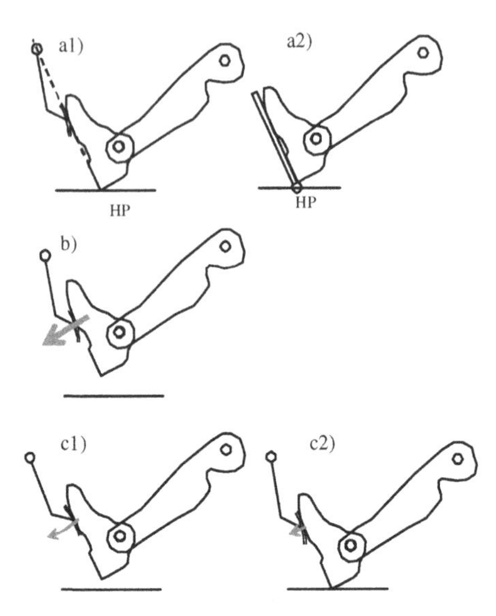

Fig. 9.16. Operation of the pedals. a1) Accelerator hinged at the upper, or at lower end a2). b) Brake: in emergency conditions the foot arch pushes on the pedal. Clutch. c1) The first part of the foot travel could be with lifted foot, c2) the disengagement is usually performed with the heel in contact with the floor to have better control.

Assembly and production issues could suggest to have a common hinge axis for all three pedals. In practice this is not possible because of the different travels that characterize them and of the different type of action. The brake and clutch pedals usually have a common hinge axis while the accelerator constitutes a separate subassembly with a different hinge. All of them are connected to the same base plate for reasons of modularity.

When not depressed, the brake pedal is usually at a higher level than the accelerator. To start braking, the right hand foot must be lifted and moved sideways. This reduces the risk of having the foot stuck between the accelerator and brake pedals with the risk of unwanted braking. (Fig. 9.17, 9.18)

9.5 Interior Packaging

9.5.1 Front Row - Driver's Position

In order to achieve a good compromise between a full exploitation of the available space inside the vehicle and a comfortable position of the occupants, the interior packaging usually starts from the definition of the area including the pedals and all neighboring components, such as firewall, tunnel, lower part of the dashboard. As the pedals are one of the main interfaces between the driver and the vehicle, the posture must be compatible with it, and its operation. Once the

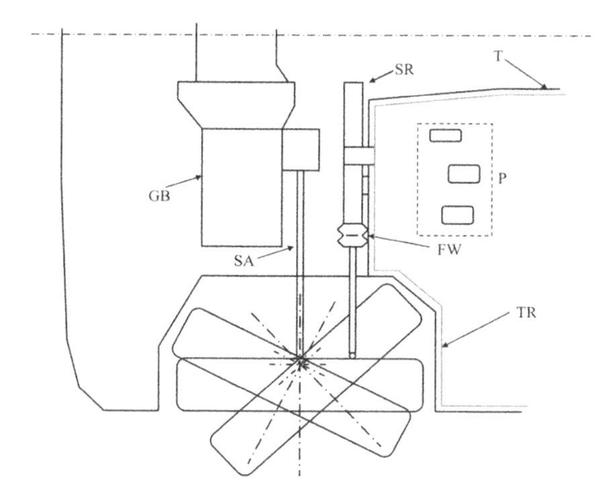

Fig. 9.17. Size of the wheel arch, steering rack (SR) pedals (P), firewall (FW), tunnel (T), trimmings (TR), transmission (GB), differential (D). Main size of the pedals group. To avoid interference, the distance between pedals should be larger than 100 mm.

anthropometric dimensions of the front and rear passengers have been decided (in terms of percentiles), the interior packaging usually starts with positioning the driver's right foot acting on the accelerator pedal (at mid-travel). The contact point of the heel with the depressed floor is indicated as AHP (accelerator heel point), it constitutes one of the milestones of the interior and body design. The rest of the driver posture can then be defined considering the angles between body segments that allow a good postural comfort (for example those reported in Tab.s 9.7, 9.8).

At a very preliminary stage, this work can be greatly simplified with the help of the SAE J826b bi-dimensional manikin. Since only the leg dimensions depend on the selected percentile, the design parameters are minimized. At this stage the head can be taken into account by means of the SAE J1052 head contours. Not necessarily the percentile of the head contour is the same as that of the manikin. The approach based on the percentile is based on the assumption that the anthropometric dimensions of each body segment is not related to the others.

Due to the rather limited range of the acceptable angles for the ankle (9.7), the choice of the AHP point and the position of the foot on the accelerator pedal largely determine the angle between the thigh and the horizontal plane and the rest of the driver's posture. A small inclination of the foot relative to the horizontal leads to an almost vertical slope of the thigh and therefore a posture close to the seated one, with a larger vertical distance between the H and HP points and smaller longitudinal space requirements (SE, Fig. 9.12).

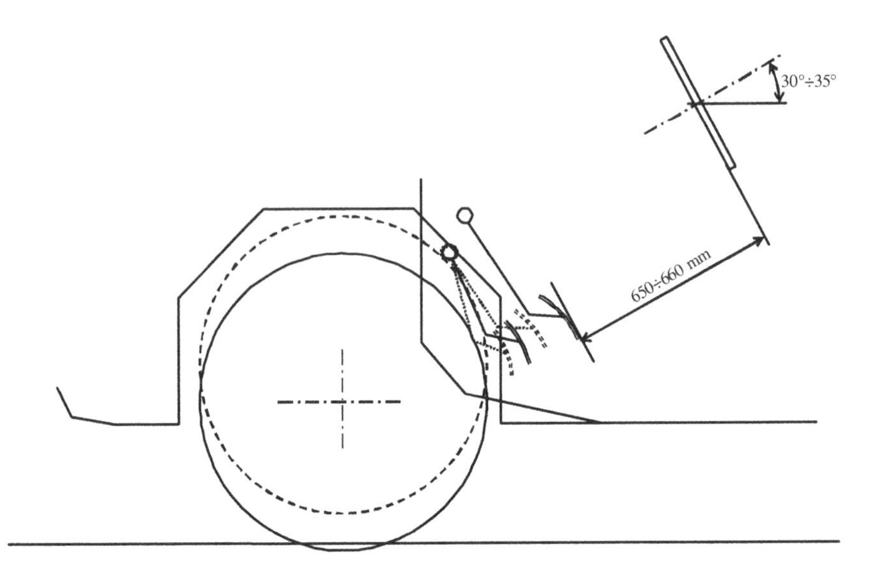

Fig. 9.18. Side view of pedals, wheel arch and steering wheel. The accelerator pedal is represented at mid-travel (solid line), full throttle and not depressed positions (dashed).

Conversely, a more upright foot leads to smaller slope of the thigh and thus a more reclined posture (RE, Fig. 9.12), with a smaller distance between H and HP points and larger longitudinal size. The current trend towards smaller vehicles for urban mobility leads to smaller longitudinal spaces for the occupants, leading to postures close to the seated one. For given percentiles this allows the longitudinal size of the seated occupants to be reduced and therefore a shorter wheelbase; or, for a given wheelbase, the available space for the occupants can be improved.

In order to accommodate people of different anthropometric sizes, the interior packaging is usually performed by considering a rather large percentile, usually not smaller than 95th. The H point associated to the rearmost normal driving or riding position, including consideration of all modes of adjustment, horizontal, vertical and tilt, is denoted the Seating Reference Point or SgRP point (SAE 1100 recommended practice), or R point. Apart from the packaging, the R point is the basis for verifying the compliance of a number of requirements, such as those related to the direct and indirect visibility.

The design, or reference, percentile may not necessarily be the largest percentile taken into account during the design process. It may be necessary to accommodate larger percentiles at the driver's seat, requiring additional seat travel in the rear direction. Even though these larger percentiles can sit in and drive the vehicle satisfactorily, their posture is usually not as good as the reference one.

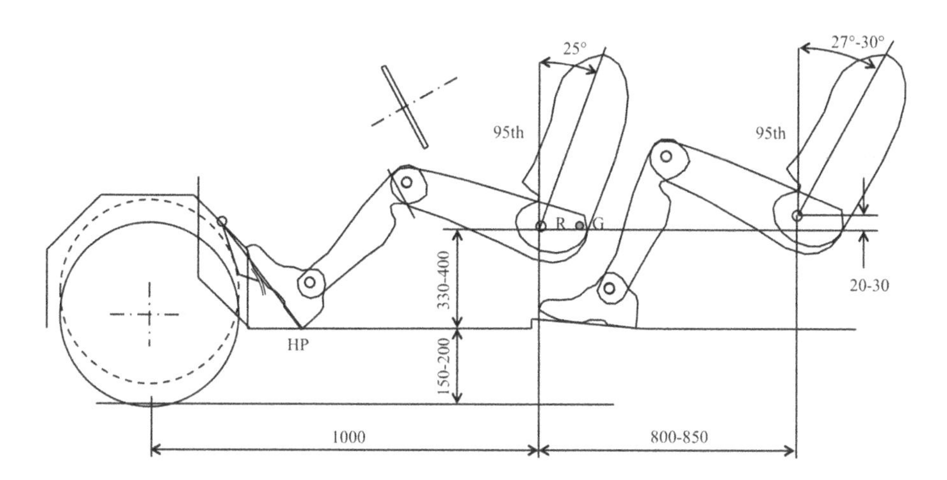

Fig. 9.19. Position of R and H point. The front and rear occupants are 95th percentile. G indicates the centre of gravity of the vehicle in full load standard.

Usually the distance between the R point and AHP point (dimension H30, SAE J1100) is about 330÷400 mm. Considering a floor to ground distance of about ≈200 mm, the ground to R point (dimension H5, SAE J1100) distance is about 530÷600 mm (Fig. 9.19), i.e. very close to the distance between the ground and the centre of gravity of the vehicle.

As illustrated in Fig. 9.19, even in the horizontal direction the H point of the front occupants lies quite close to the centre of gravity of the vehicle in the full load standard (std. F, for a five seater). The closer distance of the H point of the front occupants from the centre of gravity is one of the main reasons of the higher comfort of the front occupants during maneuvers and dynamic loads involving yaw and pitch movements.

Percentiles smaller than the reference one require careful consideration. In fact, as the reference percentile is usually rather large, the majority of people which use the vehicle will be in this category. It is then necessary to verify that the seat can be adapted to guarantee good postural comfort even to people of a rather small percentile, down to 5th percentile female, for example.

Occupant sitting aside the driver

The posture of the front passenger is usually similar to that of the driver. The absence of the pedals allows the front passenger to gain some longitudinal space for the legs that can therefore have a more stretched position. A sloping shape of the front part of the floor is very convenient for this purpose to support the foot and limit the angle of the ankle joint.

In the front view, the maximum width required by the occupants is about 600 mm at the elbow's height (Tab. 9.2, quote number 28). This width determines a relatively stringent requirement for the driver: To get the best driving performance he must be able to act on the steering wheel and other commands without interfering with any obstacle due to other passengers, the seat or the interior of the vehicle. If the driver and the front passenger are aligned in the lateral direction, the available width inside the vehicle must be at least twice that at the elbow's height of both of them (1200 mm). This size is indicated in SAE J1100 recommended practice with two dimensions measured on a vertical plane perpendicular to xz passing through SgRP:

- W3 - shoulder room - front: The minimum dimension measured laterally between trimmed surfaces 10 in (254 mm) above the SgRP.

- W5 - hip room - front: The minimum dimension measured laterally between trimmed surfaces in a region between 1 in below, 3.0 in above, and 3.0 in fore and aft of the SgRP.

9.5.2 Steering Wheel

The steering wheel must be located in a region of easy accessibility which can be determined by the reachability surfaces defined later in this chapter (Section 9.8). Its position can be evaluated to first approximation by considering the following hints:

- The distance between the lower edge of the steering wheel rim to the centerline of the brake pedal face, with the pedal in the undepressed position (L331, SAE J1100) is about 650÷660 mm;

- The angle between the vertical and the steering wheel plane (H18 SAE J1100) depends on the type of posture: In seated positions (angle δ 9.3 or L40 9.11 <15 deg) H18 ~40 deg, for $15 < \delta < 20$ deg H18 ~30÷35 deg, for more reclined positions (angle $\delta > 20$ deg) it reduces to about H18 20÷30 deg;

- The maximum diameter of the steering wheel (W9 SAE J1100) is about 340÷360 mm, with a diameter of the rim of about 30 mm;

- The y distance between the steering wheel centre and the xz plane (W7) should be at least 340÷360 mm to ensure that the driver is free to move his arms during its operation, avoiding any interference with the passenger;

- The lower corner of the steering wheel should be sufficiently above the knee to avoid interference during emergency braking and when during getting in and out of the vehicle.

9.5.3 Seat and Steering Wheel Adjustment

As indicated in the previous section, the packaging of the vehicle interior is made considering occupants of large percentiles, usually the 95th or 99th. Since the main part of the people that will use the vehicle will have much smaller anthropometric dimensions, it is necessary to verify that the designed interior is able to guarantee them good postural comfort. The 5th and 10th female percentiles manikins can be used to verify that the driver's seat and the steering wheel can be adjusted to reach acceptable postural angles (Tab. 9.7). The seat moves aft and up towards the steering wheel to accommodate shorter legs with respect to the pedals. Usually this adjustment requires a rotation of the seat in the forward direction so that the pressure on the cushion is kept almost constant, otherwise there will be an increase of the pressure in the part of the leg closer to the knee (popliteal region).

As evidenced talking about the percentile, a design based on the only anthropometric dimensions of a given percentile is not able to guarantee a good and comfortable posture to all people with dimensions falling within the percentile range. Every single individual is characterized by a practically unique combination of dimensions; therefore it is necessary to allow a number of adjustments (usually to the seat and steering wheel) that can be operated independently so that each individual can achieve the best posture. The choice of the seat back angle is left to the occupant. The design guideline is usually to limit the adjustment range to let the back inclination relative the vertical (L40 or angle δ) remain within the values that are acceptable from the postural point of view (Tab. 9.7). This is also because this angle affects the direct and indirect visibility of the driver and, therefore, is a safety critical issue.

Tall drivers tend to adjust the seat in the rearward and downward directions to reach an open posture with rather stretched legs and arms so that the steering wheel is far from the chest. Short drivers instead move it in the forward and upward directions to be able to act on the pedals and maintain good visibility, leading to a rather closed position with angled arms and the steering wheel close to the chest. The main part of the seat adjustment is performed in horizontal direction. Without dedicated vertical adjustment, the rail that allows the longitudinal movement is not horizontal but inclined in such a way that movement in the aft direction moves the seat upwards. Even though this type of coupled movement is still adopted for the least expensive vehicles, height adjustment is becoming standard even for low segment vehicles.

Although almost all road vehicles adopt adjustment systems based on the seat and the steering wheels, usually vehicles designed for the race track have a fixed seat and adjustable pedals for a number of reasons, including the lower weight required to move the pedals and the requirement that the safety belt system be connected directly to the frame and not to the seat.

9.5.4 Rear Rows

Since the passengers seated behind the first row do not need to act on primary commands as the driver does, it is not necessary for them to leave their arms free from any contact with the surroundings. The distance between one occupant and another can thus shrink. The minimum distance in lateral direction is determined by the hip size ($\approx$400 mm, dimension number 29 in Tab. 9.2).

When the available space in the lateral direction is smaller that the width at the shoulder or elbow level, to accommodate three people in a row it is necessary to arrange them with a partial overlapping of the upper body of the central passenger with that of the other two. The adoption of a seat with a central bump on the cushion and seat back compels the central passenger to seat in a slightly more forward and higher position thus obtaining the desired overlapping with his two neighbors. The bump on the seat back also provides support to his back despite the slight compromise to his postural comfort. Instead, when the seat is used by only two passengers, the central bump on the seat provides them with lateral support during cornering.

Medium to high segment sedans are usually designed to provide the rear passengers with nominally the same level of comfort level as the front passenger. Their postural angles and percentiles are chosen with similar criteria as for the front passenger. The main differences between the front and rear row involve the foot and the trunk inclination. The foot of the second row is supported directly by the floor and the adoption of a flat and horizontal surface implies open angles for the ankle outside the acceptable ranges ($\alpha_{\max} = 110$ deg, Tab. 9.7). An effective solution to this is to realize a footrest by shaping the slope of the floor at a relatively small angle and reducing the ankle extension as a consequence.

To allow improved visibility, the hip points of the rear occupants are usually designed to be at a slightly higher level than the front occupants. The vertical distance between the front and rear hip points is usually smaller than 20÷30 mm. The longitudinal distance between these points varies widely as function of the type and segment of vehicle. For medium segment sedans this distance lies in the range 850÷900 mm.

The back inclination (angle δ or L40) for rear passengers can be higher than for the front ones and it is possible to adopt values even slightly more than 25 deg. These large angles cannot be easily adopted for the driver due to the increased effort that would be required during difficult maneuvers such reversing. The larger back inclination has a positive effect on the comfort under vertical accelerations and reduces the vertical space required for the head. This has an impact on the shape of the roof that can be tapered for aerodynamic or style reasons.

In general, compact or sub-compact vehicles for urban mobility are required to be short in length. The distance between the hip point of the front and rear row is reduced accordingly and can be as low as 650÷700 mm, typically aprox. 150÷200 mm less than mid segment sedans.

Table 9.9. Main factors that influence the comfort feeling of a seat.

factor	importance [%]
subjective factors	54
vibration filtering	10
physical properties of the surface	9
pressure distribution	9
backbone support	9
thermal properties	9

The impact of the reduced longitudinal length is not as high on the front occupants as it is on the rear passengers. It is not uncommon that the front row be designed for 95th or 99th percentile, while the space left for the rear occupants is barely enough to accommodate a 5th to 10th percentile female. To reduce the longitudinal space, the posture is rather cramped, with small thigh-to-torso angles in the range $90 < \delta < 95$ deg.

One solution recently adopted in several urban concepts to overcome this problem is to partially overlap the front and rear row in the longitudinal direction. The leg of the rear passenger is then positioned at the side of the front occupants (in a similar way in which the rear passenger on a scooter or motorbike can have his legs aside the driver). This solution could lead to three seat arrangements.

9.6 Seat Characteristics

The seat is a component that influences many aspects of comfort perception. The posture, the possibility of adjustment, and the distribution of the contact pressure are just some of the aspects that have to be taken into account in the design of seats. Other important factors include the influence on the movements involved in getting in and out of the vehicle, the accessibility of commands (pedals, steering wheel, parking brake) and the instrument panel. Many more aspects of psychological origin, which are generally difficult to quantify, are also highly important including styling.

The main factors that influence the comfort of a seat are listed in Tab. 9.9. To investigate the different properties it is necessary to identify the different operating conditions and analyze each with respect to how the physical properties of the seat can influence the feeling of comfort in that condition.

Some attributes can be analyzed with a stationary vehicle, whereas others can only be assessed when the vehicle is in motion.

The aspects that should be taken into account in static conditions (when the vehicle is stationary) are:

- the support to the body to maintain the driving posture at the desired optimal configuration;
- the perceived contact stiffness on the different contact surfaces.
- the lateral support;
- the adjustment possibilities and the effectiveness of the commands that act on them;
- the tactile quality of the surface;
- the style.

Instead, when in motion, the most important factors include:

- the capability to attenuate the vibrations transmitted to the occupant;
- the absence of hindering to movement;
- the capability to maintain a good posture of the backbone and avoid excessive stresses that could be harmful in the long term;
- the distribution of the contact pressure.

9.6.1 Static Comfort

As shown by several researchers ([3], [7], [14], [15]) the pressure distribution is a reliable index that enables the vertical and lateral support given to the body to be evaluated in static conditions.

Various studies on groups of subjects demonstrated the correlation between the subjective perception of seating comfort and the pressure distribution, being very useful with regard to developing an objective criterion to evaluate the comfort level of a seat. (Some indications of this correlation are provided in the section dedicated to the functional analysis of the vehicle body.)

The seat cushion and seat back are split in to different areas, as shown in Fig. 9.21. The mean pressure in each area is evaluated as the ratio between the load acting on that area and the surface. The pressure is sometimes made non-dimensional by referring it to the maximum value or to the mean pressure on all the seat, enabling the comparison of the pressure distributions related to individuals of different weight. Another important parameter of the evaluation is the fraction of the load on the seat cushion compared to that on the seat back.

- Ischiadic pressure ratio: R_A = (mean pressure in region A)/(mean pressure on the seat cushion).
- Lateral pressure ratio: R_B = (mean pressure in B)/(mean pressure on the seat cushion).

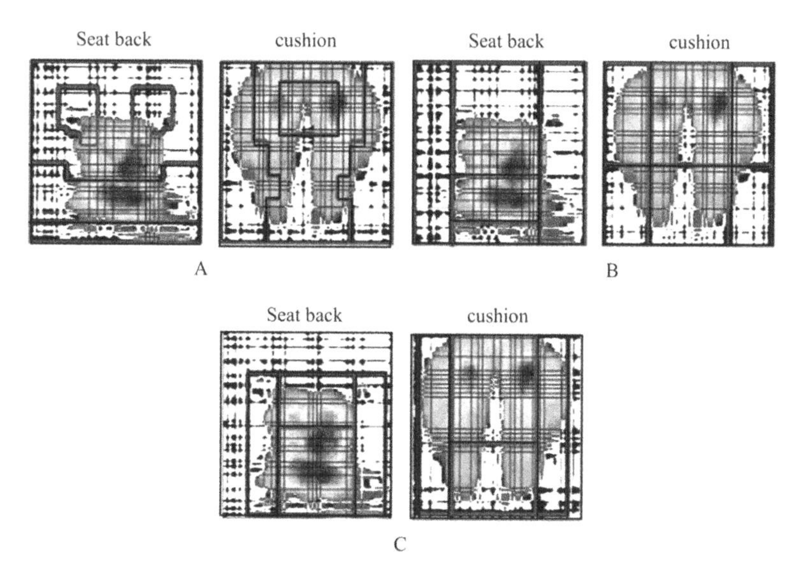

Fig. 9.20. Pressure distribution on the seat and some alternative division in regions to evaluate the comfort indexes.

- Lap pressure ratio: R_C = (mean pressure in C)/(mean pressure on the seat cushion).
- Lumbar pressure ratio: R_D = (mean pressure in D)/(mean pressure on the seat cushion).
- Symmetric pressure ratio: R_P = (mean pressure on right (or left) part of the cushion)/(mean pressure on left (or right) part of the cushion).

The study performed by Park and Kim [14] demonstrated a relationship between the seat pressure ratios and the subjective perception of the comfort. Indicating with y the comfort perception, it is possible to establish a linear relation with the pressure ratios on regions A and D of Fig. 9.21 :

$$y = aR_A + bR_D + c.$$

The value of the coefficients of the linear regression are obtained by best fitting the experimental data: $a = 56.1$; $b = 11.25$; $c = -16.16$.

All this confirms the highly important role of the lumbar part of the backbone on the postural comfort perception of the seat. In fact, the ischiadic (A region) and lumbar pressures (D region) are the most important in the definition of the subjective impression and are an index of the posture of the lumbar segment.

A more detailed description of how the pressure distribution influences the postural comfort level can be gained by means of a more complex division of

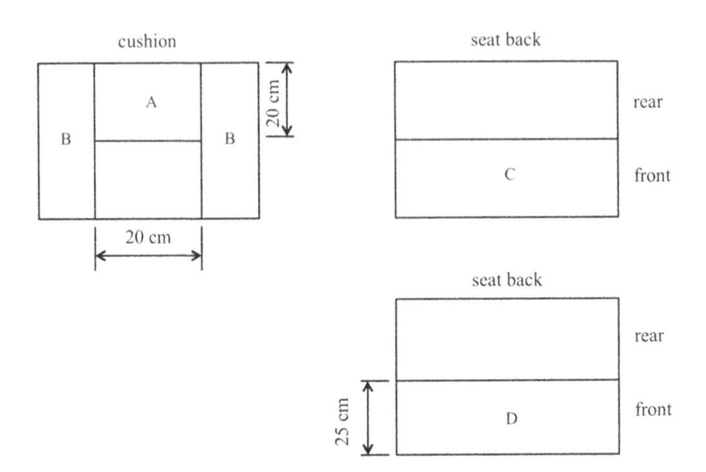

Fig. 9.21. Conventional regions on the seat cushion and seat back. Cushion: A - hip bone support region (ischiadic tuberosity). B - femurs support region. C - lap support region. Seat back: D lumbar support region.

the seat surface. Fig. 9.20 represents the pressure distribution and a possible definition of surface areas. The shape of these surfaces should take the geometric shape of the seat into account. The pressure distribution acting on these areas can be related to the subjective impression of vertical and lateral support and seat stiffness.

The analysis of the pressure distributions show that the feeling of support improves when the pressure on the ischiadic (region A) and lumbar (D) region increases at the expense of the pressure on the lateral part of the cushion (B) and on the upper part of the seat back. Similarly, increasing the fraction of weight supported by regions A and the lateral supports of the seat back and cushion (B) increases the feeling of stiffness given by the seat.

The lateral support feeling is obtained by increasing the ratio between the mean and maximum pressure in the lap region of the cushion and by increasing the mean pressure in the lateral regions of the seat back. Conversely this feeling is reduced by increasing the pressure in the lateral part of the ischiadic region and by increasing the contact surface on the upper part of the seat back.

Considering the pressure distribution on a given area, the usual indication is to obtain smooth variations and avoid concentrated pressure peaks. Pressure concentrations can reduce the blood flow. The natural reaction to this can be an increment of the muscular contraction that, in the long term can lead to local pain to the muscles and joints. A smooth pressure distribution, on the other side, reduces the muscular strain, allowing better mobility of the parts in contact with the seat, the result being that the subject continues to perform small movements. The effect is improved blood circulation in the contact region with a reduction of the discomfort in the long term. Conversely the presence of pressure peaks induces a rather fixed posture with higher muscular strain and lower capability of performing the small and spontaneous movements that allow the local stresses to be relieved resulting in a heightened feeling of discomfort.

The muscular effort required to maintain a given posture for longtime induces compensation mechanisms that attempt to alleviate the strain in the muscles and joints which usually implies a change in the curvature of the backbone, especially in the lumbar region that have the effect of modifying the loads acting on it. The subject of Fig. 9.22 a), for example, react to the discomfort by moving the upper body in the fore direction, closer to the steering wheel. This reduces the pressure in the dorsal region, improving the comfort feeling in that part. The load that is no more supported by the seat is going to the lumbar region (portion D of the seat back) or on the ischiadic portion of the cushion (portion A of the cushion). The subject could also increase the angle between the legs to gain a better support of the lower body. Instead the subject of Fig. 9.22 b) tries to stretch its upper body. The resultant load in the ischiadic region moves in the forward direction. The arms are stretched and, therefore, the subject pushes on the mid-upper part of the seat back. The curvature of the lumbar segment of the backbone increases (towards a more lordotic attitude) thus relieving the strain in the intervertebral discs. The effect is a feeling of improved comfort, albeit temporary.

The spontaneous movements that the occupants perform to compensate the effects of discomfort must be taken into account at the design stage. It is necessary to avoid that these movements induce unwanted pressure concentrations that could hider or prevent these compensation movements. At the same time it is necessary to insure that these movements do not induce high pressures in regions of the body that are not suitable to bear a load (for example the sacrum).

9.6.2 Comfort under Dynamic Loads

The vibrations transmitted to the car body when traveling should be filtered and attenuated by the seat in order to minimize potentially harmful the effects of the dynamic loads on the occupants. The seat must allow the occupants to have some degrees of freedom of movement so as to allow an adequate relieving of the stresses in the body parts while at the same time preventing movements and postures that could lead to long term damage especially at the joints and at the lumbar part of the backbone.

The evaluation of the dynamic performance of a seat can be based on the measurement of the accelerations transmitted to the occupants by the seat (cushion and seat back) when traveling on reference road surface types or proving grounds. This can be quantified by the power spectral density of the acceleration at the interface between the seat and the occupant. The disadvantage of this approach is the strict dependency on the roads selected. Additionally this measurement includes the contribution of the vehicle and that of the seat. An alternative approach to overcome these difficulties is to characterize the seat in terms of its

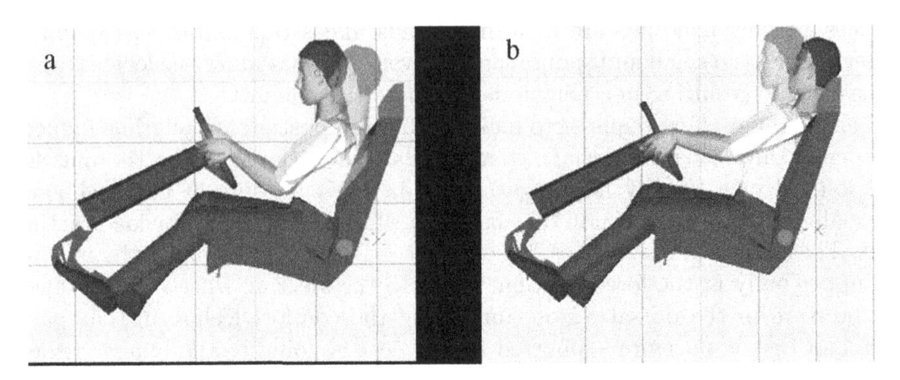

Fig. 9.22. Movements induced by fatigue. To alleviate the strain in the muscles and in the joints the driver often adopt two kinds of movements: a) movement of the upper body towards the steering wheel. This alleviates the pressure on the upper part of the seat and increases that in the lumbar and ischiadic regions; b) stretching of the upper body. This increases the rearward curvature of the lumbar segment of the backbone, alleviating the stresses in the intervertebral discs. This movement is often obtained by pushing on the steering wheel with stretched arms.

transfer functions: The seat is mounted on a shaker table that induces vibrations at its mounting brackets so that the vibration spectrum is representative of that transmitted by the vehicle chassis to the seat in real operating conditions. Measuring the acceleration on the surface of the seat (output) and on the mounting brackets to the frame (input) it is possible to define the SEAT index (Seat Effective Amplitude Transmissibility) [16]:

$$SEAT = \sqrt{\frac{\int G_{SS}(f)W_i^2(f)df}{\int G_{FF}(f)W_i^2(f)df}},$$

where:

$G_{SS}(f)$ is the power spectral density of the acceleration at the interface between the seat and the occupant;

$G_{FF}(f)$ is the power spectral density of the acceleration on the mounting brackets of the seat to the vehicle floor;

$W_i(f)$ is a weighting function that takes the sensitivity of the human body to the accelerations into account (D2002/44, ISO 2631, British Standard 6841). The domain of the integration extends over the frequency range 0.5 to 80 Hz which is relevant for the vibration comfort of the human body.

The SEAT index can be considered as the ratio between the vibrations transmitted to the seat and those that would be applied to the occupant if the seat were perfectly rigid. All this is weighted to take human sensitivity to the vibrations into account. The lower the SEAT index the better because of the lower transmitted vibrations.

9.7 Accessibility

Getting in and out of the vehicle is a fundamental issue with respect to its ergonomics. The sequence of elementary movements is influenced by several factors: the initial and final position of the subject, the geometry of the aperture, the presence of handles and constraints such as the steering wheel and A pillar, the shape of the door and its configuration in the opened position.

It is not easy to quantify such aspects with a single index; nevertheless as a general guideline, the following factors are known to improve the accessibility to a vehicle:

- Rather high position of the H point. This allows to reduce the effort to modify the distance from the ground of the centre of gravity for getting in and out. The best, from this point of view, could be to have a seat at a height requiring a simple lateral movement of the upper body. The worse is what happens in sport cars where the occupant is seated very close to the ground so that it is effectively necessary to climb in and out of it.
- Flat floor without lateral beams (sills) significantly protruding from it in the vertical and lateral directions. The sill is in fact an obstacle that must be overcome by the foot and leg.
- Flat seat cushion with low lateral humps. The lateral protuberances of the seat provide lateral support during cornering but they are also an obstacle in terms of vehicle accessibility.
- Not too intrusive dashboard so as to avoid obstructions to the path followed by the leg.
- Not too intrusive lateral aperture (door frame) so as to avoid obstructions to the path followed by the head and shoulders.

9.7.1 Getting in and Out

The sequence of single movements to get in and out of the vehicle have been analyzed by different researchers to identify indicators capable of assessing comfort related to accessibility.

The subjects can adopt rather different strategies to get in and out. Being related very little to age, gender and anthropometric dimensions, they may be driven by psychological factors. The description reported here refers to a subject getting in or out of the driver's seat (left-hand drive).

Fig. 9.23 shows the sequence of movements performed when entering the vehicle; the main steps of this sequence are as follows:

- The subject, standing by the vehicle, opens the door. The handle is usually operated by the left hand, whereas the right hand is used by a relatively small number of individuals.

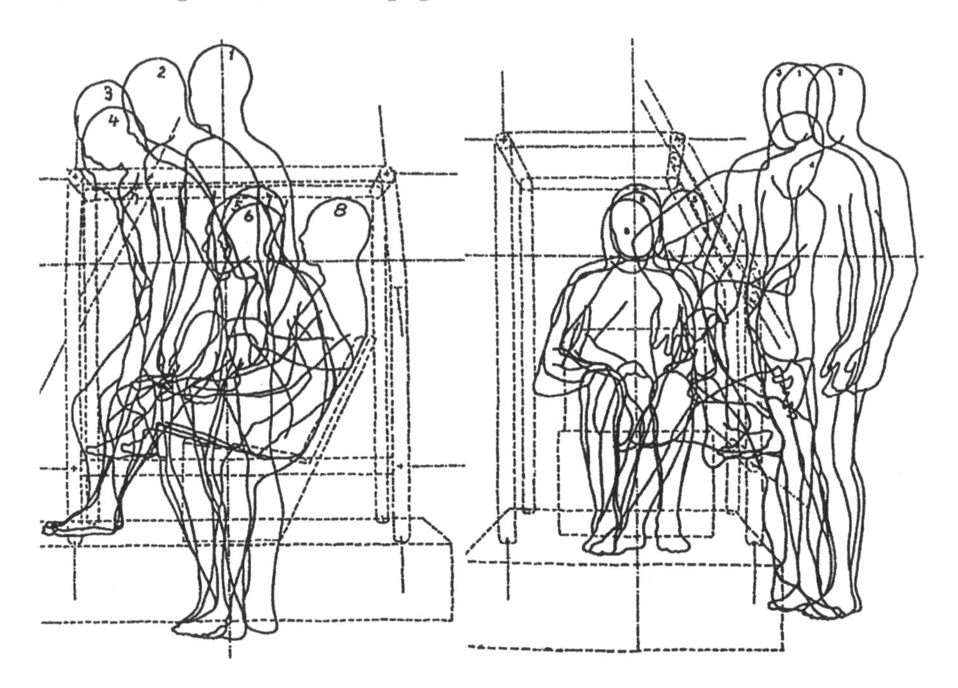

Fig. 9.23. Sequence of movements to get in the vehicle. Lateral view (left) and front view (right).

- The seated position may be reached by two alternative sequences:
  - At first the right leg is moved into the vehicle, followed by the hip and the right shoulder; the head and the left leg are finally moved in to reach the final position;
  - In some cases the seated position is reached first, with the feet still on the ground. The legs are finally moved in starting with the right. The upper body is rotated about 90 deg to reach the final position.
- Finally the door is closed.

When getting out of the vehicle the legs usually follow a sequence that is almost always the same; instead the upper body can adopt a variety of alternative strategies. The sequence can be summarized as in Fig. 9.24.

- The left leg is moved out. The rotation angle required by the leg is a function of the leg size and, therefore, of the percentile.
- The upper body is then moved by adopting one of the following alternatives:

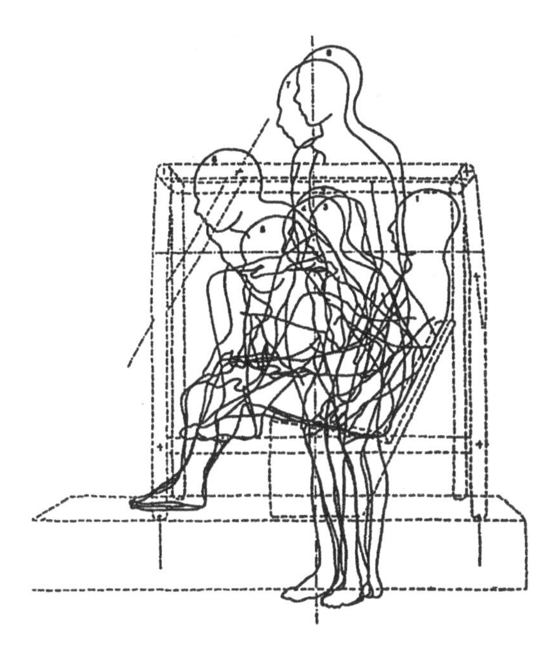

Fig. 9.24. Sequence of movements to get out of the vehicle. Side view.

- some people first rotate the upper body to the left and then start moving it out;
- other people move the upper body to the left out of the vehicle, keeping the body forward facing;

- Finally the right leg is moved out.

Getting into the vehicle, despite requiring some effort to avoid the door and its frame, is usually easier than exiting because it benefits of the contribution of the weight force acting in favor. In most cars, the center of gravity in the seated position is lower than when standing. The relatively large space that is available before getting in the vehicle is also a factor that allows an easier choice of the sequence of movements to arrive at the seated position. Conversely, many more elements influence the movements when getting out (including the steering wheel, dashboard, A-pillar and door frame, door, etc.) complicating the understanding of the aperture available. Additionally, the need to contrast the weight force complicates the exiting task. In some cases the occupant must literally climb out of the vehicle, particularly true in sport cars in which the height of the H point may be very low and the leg rather stretched.

9.7.2 Dimensional Parameters to Define the Apertures

The sequence of movements to get in and out of the vehicle should be the primary factors taken into account in the design of the apertures in order to reduce the effort required by the occupants and avoid the possibility of accidentally contacting surrounding parts.

Fig. 9.25 shows the area in car body outer surface crossed by the occupant when getting in. This area can be used to determine the size of the door and the relative position of its two pillars. For example, the slanted shape of the A-pillar in most cars necessitates it being placed in a forward position to provide enough space for the head and therefore avoid a possible contact. Similarly, the longitudinal location of the B-pillar should not constitute an obstacle to the movement of the lower body (the hip especially) when entering the vehicle.

As discussed previously, the analysis of the movements during entering and exiting introduces constraints in the definition of the structural elements nearby. Conversely the results can also be used to identify the regions where more freedom to decide the size and shape of the structure can be exerted. The area in the shade of the seat cushion, for example, is used to locate the structural node between the B-pillar and the lower sill. In the lateral view, the shape of this region extends forward, overlapping the front seat cushion. In the rearward direction it must guarantee free motion of the rear passengers legs. This allows the cross section of the B-pillar to be extended towards the front seat, providing the possibility of integrating other elements such as the winding mechanism of the seat belt.

To evaluate the doors from the ergonomic point of view, the size and shape of the aperture and the opening angle are not the only factors that should be taken into account. It is important to evaluate the possibility to reach and act on all user interfaces such as the handles, the side window commands and the side mirror commands. Figs. 9.26 and 9.27 indicate the dimensions that are relevant in the ergonomic design of the doors, including the trunk door. The reference values for the commands and handles can be defined considering the reachability surfaces, described in the following section (9.8). Other dimensions such as door size and position are depend directly on the vehicle size and segment (i.e. subcompact, compact, etc.).

9.8 Commands Reach

The driver of modern vehicles is surrounded by a large number of commands and controls ranging from the steering wheel to the radio knobs and switches. Not all of them have the same importance from the point of view of driving safety are therefore classified as being primary or secondary commands. The primary commands allow the driver to control the vehicle, and include the steering wheel, pedals, gear shift lever, hand brake, and some of the light switches: Their importance in terms of safety requires the best reach by the driver, whereas

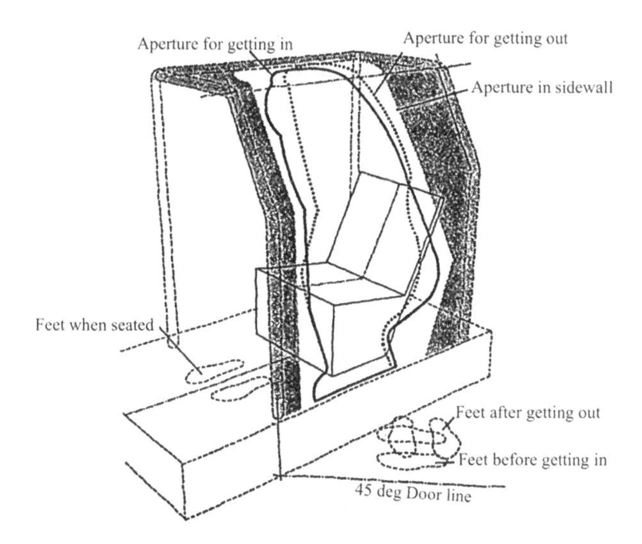

Fig. 9.25. Definition of the door aperture and frame. Area in car body outer surface crossed by the occupant when entering and exiting.

the accessibility of these commands by the other occupants is not important and should even be avoided. Instead the secondary commands activate functions that are not essential for driving the vehicle, and include the air conditioning systems controls, radio and some of the lights: The accessibility of the secondary commands has a lower priority than that of the primary commands, and so are located further from the driver in general. Most of them should be accessible also by the other occupants, e.g. the air conditioning, radio controls and the internal light switches.

The importance of the reachability issue and the need for a minimum level of standardization of the vehicle commands between different manufacturers justifies the introduction of the ISO 3958 and ISO 4040:2009 standards. ISO 3958 specifies the boundaries of passenger car hand-control locations that can be reached by hand by different proportions of male and female driver populations. ISO 4040:2009, on the other hand, specifies the location of controls in motor vehicles by subdividing the space within the reach of the driver into specific zones, to which certain controls essential to the safe operation of vehicles are assigned. It also specifies certain combinations of functions for multifunction controls and the degree to which certain indicators and warnings are to be visible.

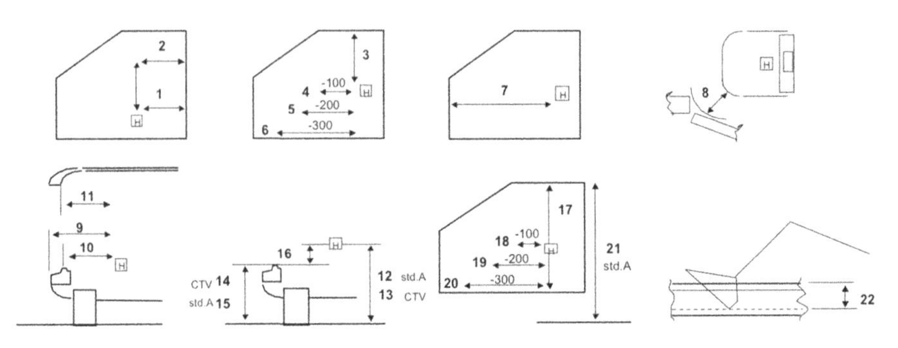

1 B-pillar to SgRp horizontal distance (0 Z H, 5 doors)
2 B-pillar to SgRp horizontal distance (500 Z H, 5 doors)
3 SgRp to roof longitudinal beam - H11 (H50-H5) (0XHR)
4 Width at -100X/H
5 Width at -200X/H
6 Width at -300X/H
7 SgRp to A pillar long distance (BX1)
8 Foot allowance (L18)
9 SgRp to outer edge of sill lateral distance EY1
10 SgRp to P1 lateral distance - AY1
11 SgRp to inner corner of sill (Ly1)
12 SgRp to ground vertical distance (H5 std.A)
13 SgRp to ground vertical distance (H5 CTP)
14 Upper sill corner (door threshold) to ground (H115 CTP)
15 Upper sill corner (door threshold) to ground (H130 S.tdA)
16 Upper sill corner (door threshold) to SgRp (H5-H115) CZ1
17 Door aperture vertical size (H41) (0 X HR)
18 Width at -100X / H
19 Width at -200 X / H
20 Widt at -300 X / H
21 Door upper threshold to ground (H50 / GZ1'A)
22 Door threshold to floor (AHP pint)

Fig. 9.26. Reference dimensions for the front doors.

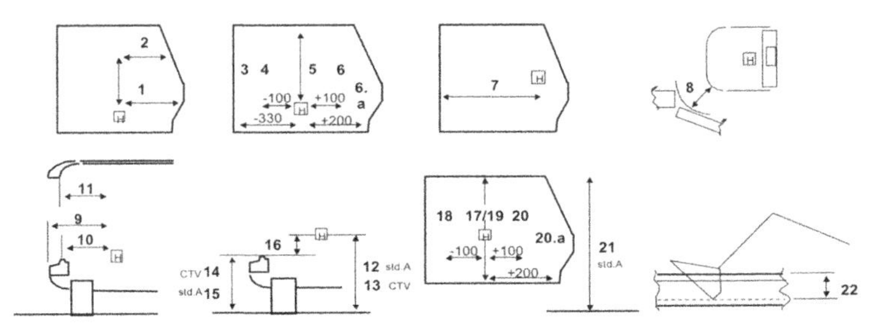

1 C-pillar to rear H point longitudinal distance (0 Z H)
2 C-pillar to rear H point longitudinal distance (700 Z H)
3 Upper threshold to SgRp vert. distance (H12) at 330 mm from H
4 " at -100 mm from SgRp
5 " same X of SgRp
6 " +100 mm from SgRp
6.a " +200 mm from SgRp
7 B pillar to H point distance (BX2)
8 Foot allowance (L19) (min distance between pillar and seat corner)
9 H point to outer edge of sill lateral distance (EY2)
10 Hpoint to P1 lateral distance (AY2)
11 H pint to inner edge of upper door threshold LY2
12 H to ground vertical distance (H10 CTP)
13 H to ground vertical distance (H10 Std.A)
14 Lower threshold to ground (H116 CTP)
15 Lower threshold to ground (H131 Std A)
16 H point to lower door threshold (CZ2=H10-H116)
17 Rear door aperture height (H42)
18 " at -100 mm from H
19 " same X of H
20 " at +100 mm from H
20.a " at +200 mm from H
21 Upper threshold to ground (H51 (GZ2' Std.A))
22 Lower threshold to rear floor

Fig. 9.27. Reference dimensions for the rear doors.

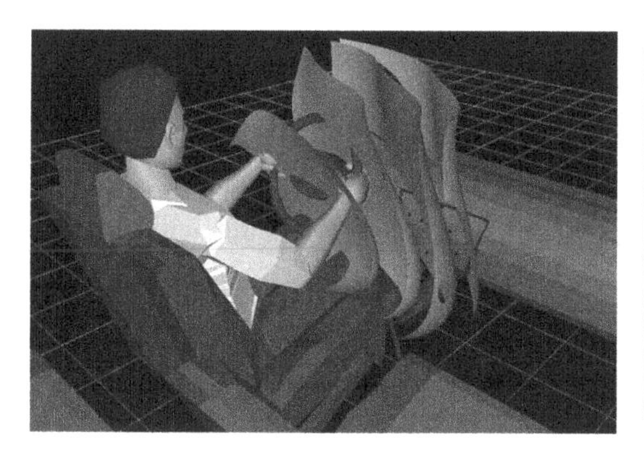
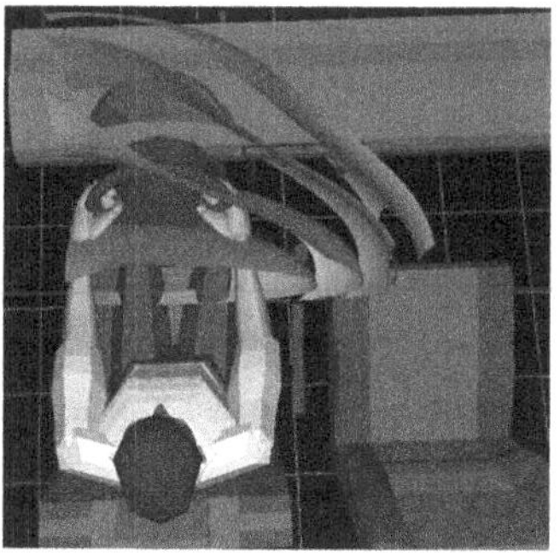

Fig. 9.28. Command reach surfaces obtained by means of a three dimensional CAD manikin. The steering wheel is located close to the surface of easier reach.

Table 9.10. Commands reach. Range of validity for UNI ISO 3958.

Back inclination (L40)	9÷33	deg
Vertical distance between H and HP point (H30)	130÷520	mm
Horizontal front seat travel (L23)	130	mm
Steering wheel diameter (W9)	330÷600	mm
Steering wheel inclination (H18)	10÷70	deg
Long. dist. btw. steering wheel center and HP point (L11)	660÷152	mm
Vert. dist. btw. steering wheel center and HP point (H17)	530÷838	mm

The command reach surfaces defined in the ISO 3958 are valid in the dimension and posture range listed in Tab. 9.10. They refer to the command of a 25 mm diameter knob with a horizontal rotation axis (Fig. 9.28). Some adjustment is required if the command requires the full hand.

The reach surfaces can be used to position the primary commands that must be reached by at least 95% of the population. In the lateral direction (y) the reach surfaces are defined from 400 mm to the left of the driver H point to 600 mm to the right, and in the vertical direction from 100 mm below to 800 mm above the same H point.

The construction of the command reach surfaces starts with the definition of the *general package factor*, G, that takes into account a number of data related to the vehicle geometry :

$$G = c_1 H_z + c_2 \delta + c_3 D + c_4 \alpha + c_5 W_x + c_6 W_z + c_7.$$

With reference to Fig.s 9.3, 9.11 the variables that influence the package factor G are:

H_z	H30	SgRp to AHP vertical distance,
δ	L40	back angle relative to the vertical,
D	W9	steering wheel diameter,
α	L25	steering wheel angle relative to the vertical,
Wx	L11	steering wheel center to AHP point longitudinal distance,
Wz	H17	steering wheel center to AHP point vertical distance,
γ	L42	thigh centerline to back angle.

Coefficients c_i have the following values:

$$\begin{array}{ll} c_1 = 0.0081 & c_2 = -0.0197 \\ c_3 = 0.0027 & c_4 = 0.00106 \\ c_5 = -0.0011 & c_6 = 0.0024 \\ c_7 = 0.0027 & c_8 = -3.0853 \end{array}$$

Apart from the analytical definition, a useful feature is that most of the CAD tools that include three-dimensional manikins implement reach surfaces to simplify the reach analysis. (Fig. 9.28).

9.9 Loading and Unloading

The aim of the ergonomic evaluation of the trunk is to identify the parameters of the aperture that ensure people of different percentiles are provided with:

- enough space for the free motion of the subject during loading and unloading of luggage or other goods;
- the reachability of the handles and grips on the trunk door or the tailgate;
- the compatibility of the loads to act on the doors and their handles.

The dimensions indicated in Fig. 9.29 can be considered as a reference to evaluate loading and unloading. The position of the trunk door and the tailgate handles can be verified by taking into account the admissible ranges for each shoulder and arm joint as listed in Tab. 9.11.

Solutions with a level cargo floor are preferred to those with structural elements in relief. Additional effort is required in the second case to lift the cargo and put it in the right place. For the same reason the surface should be flat and without protruding elements. Rear drop seat solutions are also appreciated because of the increased cargo volume. Generally a flat cargo surface is preferred.

9.10 Visibility

To understand the standard requirements concerning the visibility of a motor vehicle, it is important to recall some basic concepts regarding the functioning

Table 9.11. Allowed angles for arm and shoulder joints when opening a trunk or tailgate.

joint	min. [deg]	max. [deg]
elbow	42	98
shoulder	60	140

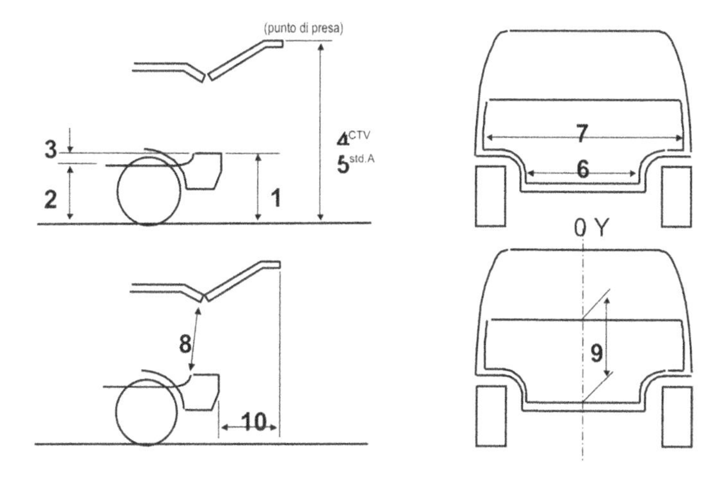

1 Trunk threshold to ground (H195)
2 Trunk floor to ground (H253)
3 Trunk floor / threshold
4 Handle to ground
5 Handle to ground for closing door Std.A
6 Minimum distance between wheel arches (W202)
7 Lower width of trunk door aperture (W207)
8 Breadth at 0 Y (L206)
9 Aperture height at 0 Y (H252)
10 Longitudinal allowance
- Effort to open trunk door
- Effort to close trunk door
- Effort on handle to open lock

Fig. 9.29. Reference dimensions for the tailgate.

of the eye. It is well known that the light enters the eye and goes through the cornea, the pupil, before being focused by the lens onto the retina. The retina is a membrane that covers the inner surface of the eyeball and is made of photoreceptor cells of the rod and cone types. The rod cells are sensitive to the low intensity light of night view, the cones allow the color view during daylight vision.

The density of the photoreceptors is not constant on the retina. The maximum density is in the so-called macula lutea or fovea centralis (fovea) and decreases with increasing distance. The result is that the sharpest visual acuity occurs in the fovea region and decreases rapidly around it.

The fovea has a very small diameter of about 0.5 mm, this means that the highest visual acuity occurs in a narrow field of view of about 2 deg about the sight line. This means that to get a precise picture of the surrounding world, the brain must move the eyes (and the head) so that the most important parts are scanned by the fovea.

The eye can move sideways of about 30 deg on both sides of the straight ahead direction. The allowed rotation on the xz plane is 45 deg up and 60 deg down. If such these movements are not enough to scan the object, the brain commands a rotation of the whole head.

The visual perception occurs with a series of rapid and unconscious eye movements that explore the object we are looking at by reading its characteristic points (looking at a person, for example, the movements are mostly concentrated on the face). A good example to appreciate the small size of the region of higher visual acuity is the process of reading a text. Even if it is possible to appreciate the size of the whole page, the eyes must move following the letters aligned in rows in order to bring them on the area of maximum acuity. The width of this region is about 10 letters.

The field of view of maximum visual acuity is the most important for the detailed view required for driving. The remaining part of the field of view is very large compared to that of maximum acuity, this part correspond to the so-called peripheral view. The peripheral view is used to perceive the presence of obstacles and signs, although the relatively coarse acuity does not allow the shape and distance to be appreciated in detail. Fig. 9.31, coming from SAE J1050a recommended practice [17], splits the field of view into a number of sections. The figure refers to the case when the head and the eyes are held fixed and the subject is looking straight ahead. As already stated, the maximum acuity occurs in a relatively small angle around the *sight line* whereas all the rest is peripheral view. The total field of view of each eye is very large; the amplitude of about 145 deg is delimited on one side by the nose, and on the other side is nearly perpendicular to the sight line. The region where the fields of view of the two eye overlap is referred to as *binocular*, amplitude of which is about 110 deg around the sight line. In the two surrounding regions only one eye is able to view at a time. The view in these two regions is denoted *monocular*. The *ambinocular* field of view is the sum of the fields of view of each eye extending to about 180 deg around the sight line.

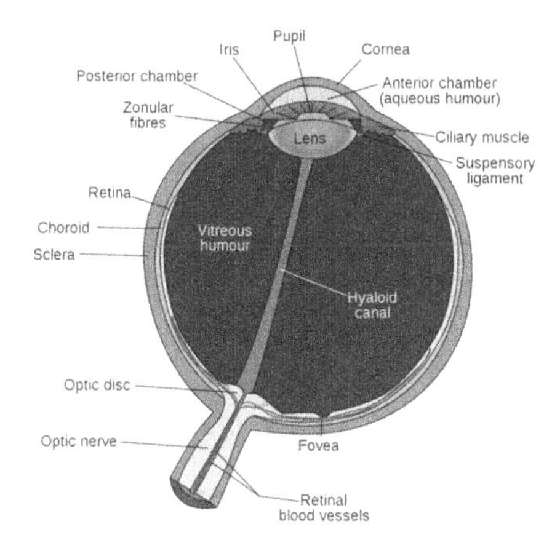

Fig. 9.30. Human eye structure. The maximum visual acuity corresponds to a 2 deg field of view. It corresponds to the very small size of the fovea.

The binocular field is characterized by a stereoscopic vision. Since only in this field it is possible to appreciate the distance that separates us from an obstacle, it is highly important for driving.

The importance of the field of maximum acuity is such that the main requirements regarding the design of the car body in terms of visibility refer to it. The aim is to ensure that this type of view extends in a sufficiently wide range to allow the driver to appreciate the surrounding environment. The standard requirements usually refer to only the line of sight, the line at the center of the small field of view with maximum acuity whereas the relatively large peripheral view is not considered.

The driver can look out of the vehicle directly forward through the windscreen or rearward with the help of the mirrors. The first field is called the *direct field of view*, whereas the second is the *indirect field of view*. The direct field of view is delimited by the two A-pillars which introduce blind regions where one or both eyes cannot see. Fig. 9.32, shows the fields of view and the relative obstructions due to the pillars. As in the case of direct view, it is possible to define mono-, bi- and ambinocular obstructions. The aim of most of the homologation standards concerning the direct field of view is to ensure that its extension is sufficiently wide, and that the binocular obstruction is not too large. The main requirements concerning these two aspects are summarized below.

9.10.1 Optical Properties of Glass Plates for Vehicle Applications

The direct view is obtained through the windscreen and the lateral windows. For safety reasons these windows should not significantly modify what is perceived by the driver in terms of image intensity, shape and position. The aim of this

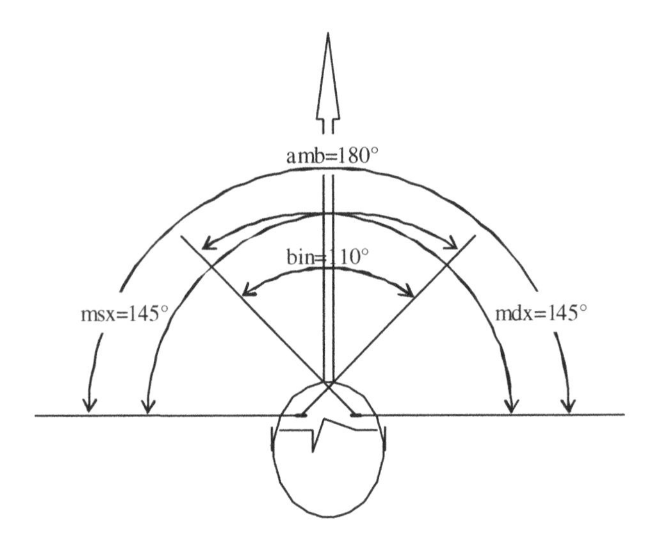

Fig. 9.31. Definition of the different fields of view in accordance to SAE J1050a recommended practice. The assumption is that both the eyes and the head are fixed in a straight ahead sight line. Apart from the objects along the *line of sight*, all the rest of the vision is of peripheral type. mdx: right eye monocular field of view; msx: left eye monocular field of view; bin: binocular field of view; amb: ambinocular field of view.

section is to provide a qualitative description of the optical defects of the glass for vehicle applications and the main design parameters that affect them.

From the regulations point of view ([18]) the main requirements are relative to the windscreen and the lateral windows, while fewer or no requirements are indicated for the other windows, including the backlight glass.

Transparency

Given a light source (s) and a receiver (r: the eye, for example), the transparency is the ratio between the intensity measured with and without the glass. For glass plates of given thickness and material, the transparency is a function of the incidence angle φ. Typically, glasses for automotive applications have a transparency that ranges from 90% for $\varphi = 0$ deg down to 80% for $\varphi \approx 60$ deg. For larger incidence, the transparency decreases rapidly to negligible values around 90 deg.

Low transparency glasses improve the visibility in intense light conditions but reduce visibility at night or in scarce light conditions. The main homologation standards (see, for example, [18]) require a minimum transparency of 75% for a sight line in the direction of travel. With white glass this leads to a maximum incidence angle of $\varphi \approx 65 \div 70$ deg; this value is less for colored glasses. The maximum allowed incidence leads to a maximum slope for the windscreen, leading to important implications on the vehicle body design.

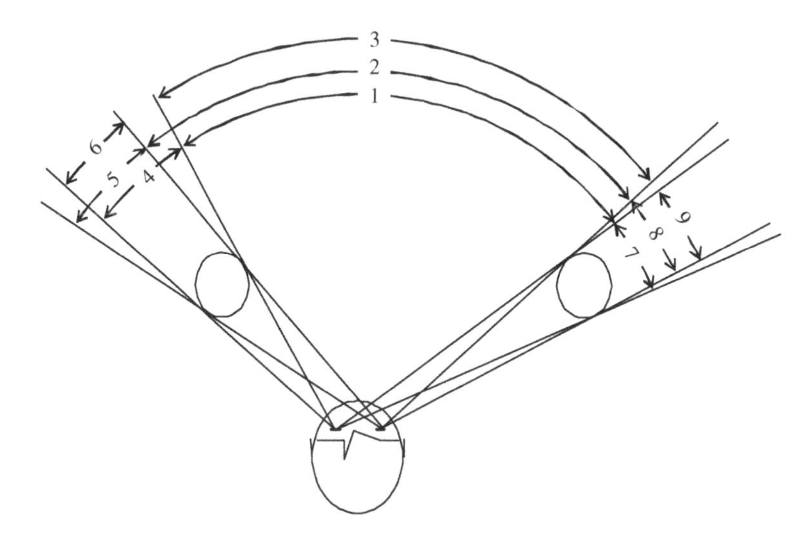

Fig. 9.32. Definition of the fields of view and of obstruction. 1: left eye monocular field of view; 2: right eye monocular field of view; 3: ambinocular field of view through the A-pillars; 4: left eye monocular obstruction (left pillar); 5: right eye monocular obstruction (left pillar); 6 binocular obstruction (left pillar); 7: left eye monocular obstruction (right pillar); 8: right eye monocular obstruction (right pillar); 9 binocular obstruction (right pillar).

Distortion

Due to the different refraction coefficients that characterize glass and air, light is slightly deviated when entering the glass surface and is deviated again when leaving it. If the glass plate were planar, homogeneous and with flat, constant thickness surfaces, the incident light ray and the transmitted one would be parallel. The glass in this case would not alter the image but it would simply shift it by a certain amount.

Because of unavoidable production tolerances, the glass surfaces will never be exactly planar and of constant thickness. Additionally, the glass itself will never be completely homogeneous. Correspondingly the incident and the transmitted light rays from a real glass plate will never be perfectly parallel. The angle δ between them is called the distortion angle (Fig. 9.34). Something similar occurs when the mid-plane of the glass is not flat but has a curvature, even if the thickness is perfectly constant.

The distortion angle can be expressed as a function of the incidence angle as

$$\delta = \delta_0 A(\varphi) + \frac{d}{r} B(\varphi) \tag{9.2}$$

where the first contribution is due to the imperfect parallelism between the two glass surfaces, while the second contribution is due to the radius of curvature r and to the thickness d of the glass plate. The construction quality of the glass is

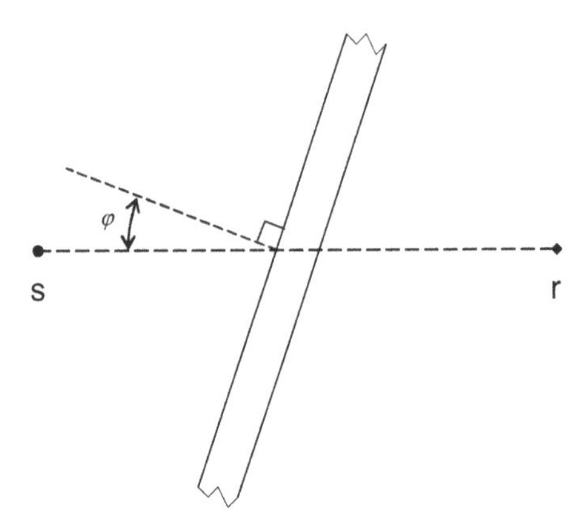

Fig. 9.33. Sketch of the transparency measurement. s: light source; r: light receiver (eye). φ incidence angle.

represented by coefficient δ_0 : The better the quality, the smaller the value of δ_0. $A(\varphi)$ and $B(\varphi)$ are positive and increasing functions of the incidence angle φ.

For a glass plate with a given thickness and quality, the distortion increases by increasing the incidence angle and reducing the radius of curvature r. The result is that a windscreen cannot be designed with a large curvature and small slope with respect to the horizontal. Most of the standards concerning direct visibility require that in each point of a given area on the windscreen the distortion is smaller than a maximum limit:

$$\delta_{\max} = 5' = (5/60)\ deg \tag{9.3}$$

Double image

A relatively small part of the incident light can be reflected inside the glass plate and emerges at some distance from the path followed by most of the light, thus producing a secondary image (secondary image). Even if this image is usually of small intensity, it tends to spoil the quality of the perceived image and therefore should be avoided.

Similarly to the distortion (Eq. 9.2), the angle between the main and the secondary image increases by increasing the incidence φ and the ratio d/r. Despite the lack of strict requirements in force today, this defect should be minimized for safety reasons.

9.10.2 Eyellipses

The reference tool for the visibility analyses is introduced by the SAE-J941 recommended practice. The two dimensional eyellipses are representative of the 90th, 95th and 99th percentile distributions of the driver eye locations. In addition to the percentile, the size of the eyellipses takes into account the longitudinal

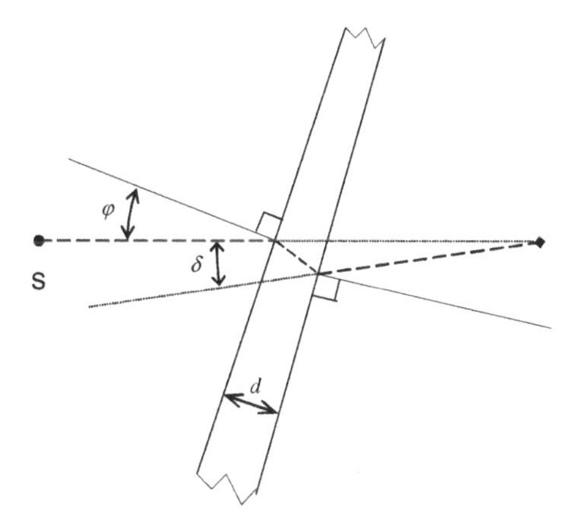

Fig. 9.34. Distortion due to the misalignment between the two surfaces of the glass plate. s light source; δ distortion angle between the incident and refracted light; φ incidence angle of the incoming light; d mean thickness of the glass plate.

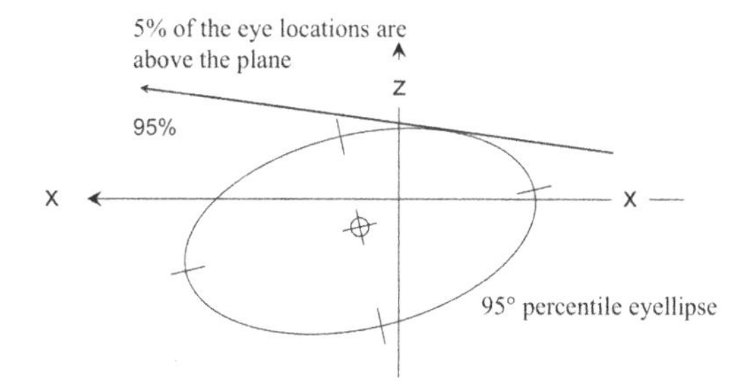

Fig. 9.35. Eyellipses from SAE J941 recommended practice. Side view.

seat travel (L23) and back inclination (L40). Different eyellipses apply for cars and trucks because of the different H-point heights (H30) and steering wheel diameters.

Fig. 9.35 shows the side view of the eyellipse corresponding to a given percentile (95th, in the figure). The contour was obtained by the statistical analysis of the eye locations of a licensed driver population with a one-to-one male-female mix.

The eyellipse of a percentile P represents the envelope of an infinite number of planes dividing the plane so that $P\%$ of the eyes are located on one side of the plane and $100 - P\%$ on the other. For example, if the plane represented by the straight line in Fig. 9.35, is tangent to the eyellipse, then 95% of the eyes will be below the line, both inside and outside the ellipse and 5% will be above it. Similarly, a straight line tangent to the lower edge of the 95th percentile eyellipse leaves 95% of the eyes above the line and 5% below it. Notice that the eyellipse

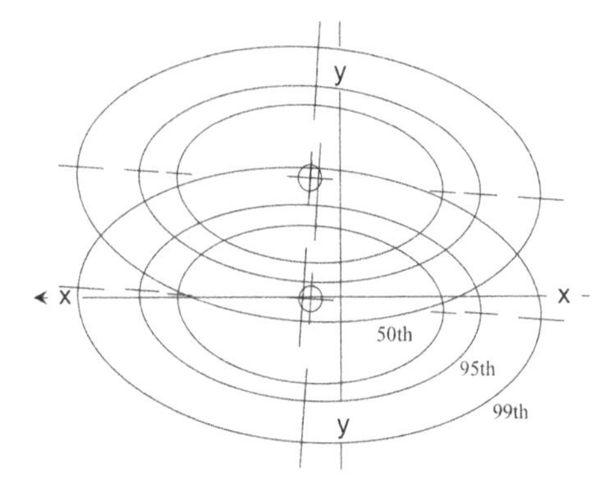

Fig. 9.36. SAE J941 eyellipses (xy plane). The xx yy reference system is part of the eyellipses, it allows to position them relative to the R (SgRp) point.

does not include 95% of the eyes. Each tangent has 95% of the eyes on one side, no matter if they are in the eyellipse or out of it.

The same definition applies for the plan view of Fig. 9.36. Since the analysis has been performed on each eye, the result is two eyellipses. The distance in the lateral direction between the points of the two eyellipses aligned on lines in y direction is about 60 mm, i.e. the mean distance between pupils (see Tab. 9.2).

The eye locations have been recorded with drivers looking straight ahead without head turning and with a back angle L40=25 deg (Fig. 9.11). The effect of different back angles is accounted for by correcting the position of the eyellipse; other viewing directions are taken into account, including head movements up to 60 deg.

Figs. 9.35 and 9.36 show that the axes of the ellipses are not aligned to the reference directions x, y, z probably due to the asymmetry of the space around the driver caused by the structure of the vehicle, namely the A-pillar and the beam at the upper border of the windscreen.

By definition, the eyellipses can be used only for sight lines compatible to the rotation of the eyes with a fixed head. Considering the previous considerations 180 concerning the motion of the eye, on the horizontal plane (xy) this rotation is about 30 deg on each side of the x direction. On the vertical plane (xz) this is 45 deg upwards and 60 deg downwards. Sight lines out of this field require a rotation of the head.

Size and construction of the eyellipses

The eyellipses can be constructed from the location of their centroid, length of the major and minor axis, and their orientation in the plan and side views. The reference frame xx, yy, zz is used as reference frame for their construction and must be considered as part of the eyellipses during the positioning procedure in the vehicle.

Table 9.12. Major axis of the SAE J941 eyellipses

L23	major axis [mm]		
[mm]	90th	9th	99th
102	109	147	216
114	122	160	239
127	135	173	241
140	147	185	254
152	155	193	262
165	160	198	267

Table 9.13. Minor axis of the SAE J941 eyellipses.

	minor axis [mm]		
	90th	95th	99th
lateral view	77	86	122
plan view	82	105	149

The major axis is reported in Tab. 9.12 as function of the percentile and of the seat travel L23 ([12]). The longer the seat travel, the longer the axis. Conversely, the minor axis is just depending on the percentile, reported in 9.13. In side view the eyellipse is tilted -6.4 deg (downward direction looking forward). In the plan view they are tilted by 5.4 deg (inward of the vehicle when looking forwards).

The location of the centroids is reported in Tab. 9.14 as function of the percentile and seat travel L23. The longer the seat travel, the more the centroids are moved forward (negative x values).

Location of the eyellipses

If the back angle (L40) is equal to 25 deg, the eyellipses (Fig. 9.37) are located with the xx and zz lines at the intersection of a vertical line passing through the R (SgRp) point and a horizontal line located 25 in (635 mm) above it.

Table 9.14. Left ($_l$ and right $_r$ eyellipse centroid of the SAE J941 eyellipses relative to the xx, yy, zz reference frame.

L23	x	z	y_r	y_l
[mm]	[mm]			
102	+1.8	-5.6	-6.4	+58.0
114	-4.6	-6.4	-5.6	+58.9
127	-10.7	-7.1	-5.1	+59.0
140	-17.0	-7.6	-4.3	+59.7
152	-20.3	-8.4	-4.1	+60.2
165	-22.9	-8.4	-4.1	+60.5

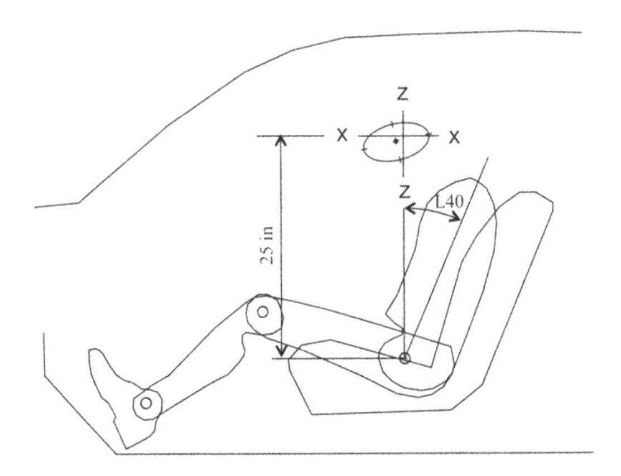

Fig. 9.37. Position in the lateral view of the eyellipses.

If the back angle is not 25 deg, the position obtained above for L40=25 deg must be modified in the side view. The X and Z displacement can be found using the following expressions:

$$X = -9.331288 + 0.404789L40 - 0.0012611L40^2,$$

$$Z = 1.067621 + 0.0156987L40 - 0.0023347L40^2,$$

where L40 is expressed in deg (deg) and the displacements X and Z in inches (in). If the back angle is smaller than 25 deg, the eyellipses are moved forward and upwards. Conversely, if L40$<$ 25 deg, the eyellipses are moved rearward and downwards.

In the plan view (the plane xy of the vehicle reference) the xx line of reference frame of the eyellipses is set at a distance from the vehicle centerline given by $0.85W7 + 0.075W3$ where $W7$ is the distance of the steering wheel axis from the centerline and $W3$ is the width inside the vehicle 10 in (254 mm) above the R point. In any case the xx reference line should be located no further inboard than $W20 + 1.1$ in.$W20$ is the distance between the R point and the vehicle centerline.

This positioning is valid for vehicles with individual driver's seat, as it is for most current cars. As a first approximation, the xx reference line passes through the R point in plan view. SAE J941 indicates the procedure to locate the eyellipses also in the case of bench seats, buses and trucks.

Critical aspects of the approach based on the eyellipses

Eyellipses have been defined by considering seats provided with longitudinal and back angle adjustments. Their construction shows the important role of the longitudinal adjustment of the size and location of the eyellipses. The driver's seat of modern cars, even of the smallest class (subcompact and compact), is provided with vertical and cushion slope adjustment. These possibilities affect the definition of the eyellipses: One approach is to use virtual manikins and repeat the procedure leading to the definition of the eyellipses and adapt them to the new conditions.

Measurement of the ambinocular field of view

Fig. 9.38 shows the procedure indicated by SAE J1050a Recommended Practice to determine the ambinocular field of view which, in the lateral view, is delimited between a line at 45 deg upwards and a line at 60 deg downwards. Viewpoints C and D are on the same plane as points F and G that are aligned in the vertical direction.

In the plan view the ambinocular field differs depending on the possibility of head turning. Fig. 9.38 shows the construction without head turning. The right limit of the field of view is a sight line that starts from the left eye, the left limit is a line that starts from the right eye. These two lines have a maximum angle of 30 deg from the straight ahead view which corresponds to the maximum angle allowed by the rotation of the eyes without head turning. The construction is as follows

- draw a line tangent to the right eyellipse at a maximum of 30 deg to the left, the tangent point is A.
- Draw a line tangent to the left eyellipse at a maximum of 30 deg to the right, the tangent point is B.

The ambinocular field of view without head turning α_p has a maximum amplitude of 60 deg and is between the sight lines passing through A and B. This angle may be less in the presence of obstructions (for example the A-pillars).

Fig. 9.39 illustrates the procedure to determine the ambinocular field of view (α_p) with head turning:

- Draw a line tangent to the right eyellipse at 30 deg to the left, the tangent point is G.
- Identify point H on the left eyellipse by drawing a line from G parallel to the yy axis. Points G and H should be at about 60 mm apart.
- Identify the hinge point J along the perpendicular to the midpoint of segment GH at a distance of 3.88 in (98.6 mm). Point J represents the intersection with the horizontal plane of the head axis of rotation. Points G' and H' represent the eye points after head turning towards the left.
- For a head rotation in the opposite direction (right), the same procedure must be repeated starting from a point G located on the left eyellipse.

The maximum allowed rotation of the head about point J is 60 deg in both the left and right directions.

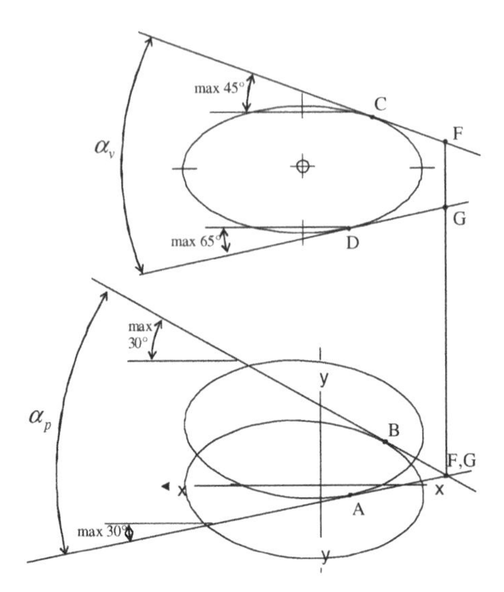

Fig. 9.38. Binocular field of view. SAE J1050a construction with eye movement only. Side view (up) and plan view (down).

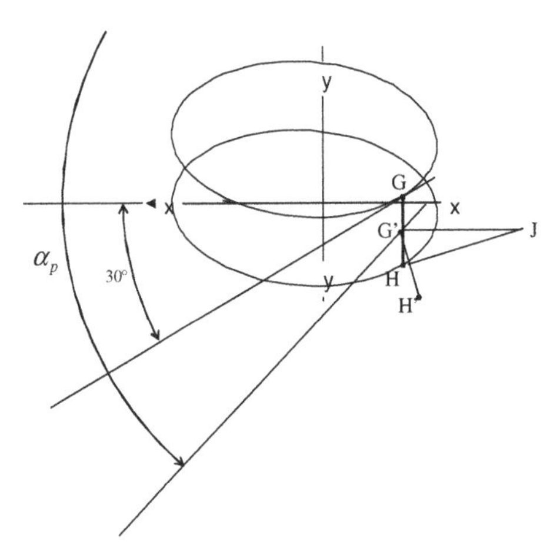

Fig. 9.39. Binocular field of view. SAE J1050a construction with eye and head movement. G, H: eye points, J: head hinge point, G', H' eye points after head movement to the left.

9.10.3 Direct Field of View and Binocular Obstruction

The essential role of the visibility in the safety of vehicle has motivated a number of standard requirements concerning a number of aspects:

- The driver should be able to view out of the vehicle through a windscreen of adequate size. In some cases, this requirement concerns a minimum aperture of the binocular field of view, whereas in others it is necessary to verify that a given number of lines of sight go through the windscreen with good enough optical properties (transparency and distortion).
- The obstruction of the non-transparent parts should not exceed given limits, defined starting from given view points. No structural part should be present in front of the eyes of the driver, except the two A-pillars. Similarly, the hood, doors and the roof should not limit too much the possibility to view out of the vehicle.
- The number of pillars in front of the driver must not be larger than two, reducing the direct field of view by creating fields of obstruction (Fig. 9.32). The requirements concerning this aspect refer to the *binocular obstruction*, that is: the field where both eyes are not able to view. This field should not exceed a given limit.

Direct field of view

The European directives concerning the direct field of view of M-class vehicles are:

- D77/649
- D88/366

Their aim is to guarantee an adequate field of view through the windscreen when the glass surfaces are clean and dry. Fig. 9.40 shows a sketch taken from the directive. The car must be in the B-load standard. The three sight lines leaving points V1 and V2 in different directions (7 deg upwards and 17 deg to the left from V1, and 5 deg downwards from V2) must intersect the windscreen in points with adequate optical properties, i.e. in these three points the transparency should be larger than 70% and optical distortion should be less than 5/60 deg. For back angles other than 25 deg, the vertical and longitudinal position of points V1 and V2 (Fig. 9.40 dimensions A, B, C, D) must be corrected as indicated in Tab. 9.15.

To avoid that parts of the car body such as the hood and the doors (belt line) reduce the available field of view too significantly, the European directive requires that any part of the vehicle, except the A-pillars, exceed the limits

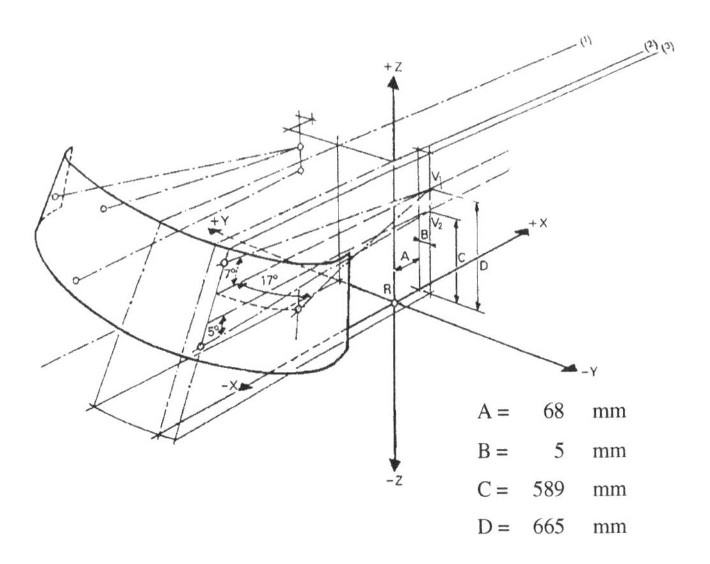

Fig. 9.40. Direct field of view following the European Directives. The longitudinal and vertical position of the visual points V1 and V2 must be modified as function of the back angle (L40).

between a horizontal plane passing through V1 and a pyramidal surface with the top at point V2 (Fig. 9.41). The pyramid has three faces at 4 deg relative to the horizontal. Part of the steering wheel can exceed the lower limit provided it stays below a plane with a 1 deg downwards slope from point V2.

The sloping and short hood used in modern cars usually does not interfere with the driver's view. In certain cases the belt line can introduce an obstruction to the lateral view, especially that on the opposite side to the driver (right door for left hand drive).

A-pillar obstruction

Fig. 9.42 shows the procedure indicated by the SAE J1050a [17] recommended practice to identify the binocular obstruction angle of the A-pillar. The reference cross section is the closer one to the left eyellipse. This section should be found by an iterative procedure considering the lateral and the longitudinal views.

If the reference cross-section falls in the direct field of view with eye movement, only the binocular obstruction is identified considering the points of view on the eyellipses. If the reference cross-section falls outside the direct field of view, the recommended practice indicates a procedure to rotate the head and eye so as to include the reference cross section within the binocular field of view:

1. Side view: find the reference cross section AA of the pillar with a horizontal plane (parallel to load standard B). This plane passes through point B as

Table 9.15. Corrections to the position of points V1 and V2 of Fig. 9.40 as function of the back angle. Positive x to the rear, positive z upwards.

L40 [deg]	Δx [mm]	Δz [mm]	L23 [deg]	Δx [mm]	Δz [mm]
5	-186	28	23	-18	5
6	-177	27	24	-9	3
7	-167	27	25	0	0
8	-157	27	26	9	-3
9	-147	26	27	17	-5
10	-137	25	28	26	-8
11	-128	24	29	34	-11
12	-118	23	30	43	-14
13	-109	22	31	51	-18
14	-99	21	32	59	-21
15	-90	20	33	67	-24
16	-81	18	34	76	-28
17	-72	17	35	84	-32
18	-62	15	36	92	-35
19	-53	13	37	100	-39
20	-44	11	38	108	-43
21	-35	9	39	115	-48
22	-26	7	40	123	-52

defined in the following. As a first approximation, this cross-section is found by cutting the pillar with a horizontal plane passing from the intersection of the eyellipse and its major axis.

2. Plan view: from point B, draw a sight line tangent to the rightmost point E of the cross section. The point B is the point of the eyellipse that is to the closest to left edge A of the cross section. Iterate the newly found point B on the side view and, if necessary, find a second approximation of the reference cross section.

3. Find the point C on the right eyellipse such that the segment BC is parallel to yy.

4. Find point D at 3.88 in (98.6 mm) on the perpendicular to segment BC through its midpoint. Point D represents the hinge point for head movements. If the cross-section is all to the right of a line at 30 deg from point B, no head movement is required. In that case the binocular obstruction is the angle between sight lines AB and EC.

5. If part of the cross section is to the left of the 30 deg sight line from point B, the head must be rotated about point D until the sight line from B' to A forms a 30 deg angle with segment B'C'.

6. The binocular obstruction angle α is between the sight lines AB' and EC'.

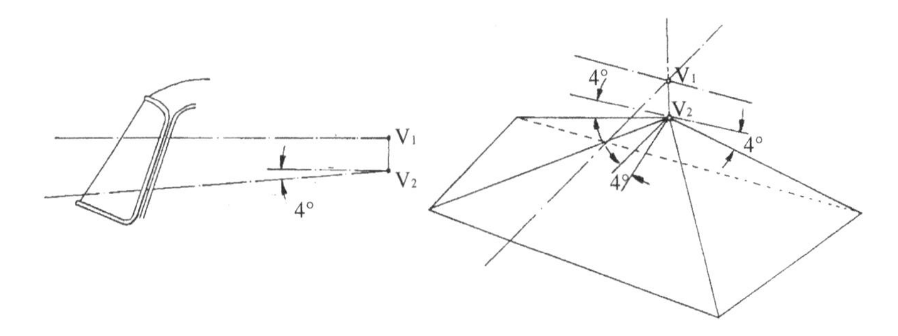

Fig. 9.41. European directive about direct field of view. Only the A-pillars can exceed the field between the horizontal plane through V1 and the pyramidal surface with top at point V2. The three planes of the pyramid are at a 4 deg downwards slope.

The binocular obstruction angle should not be larger than 6 deg. If the sight lines (AB' and EC' or AB and EC) tangent to the pillar converge (on the opposite part of the pillar relative the eyes) there is no binocular obstruction. In fact, after a certain distance from the pillar at least one eye is able to view again. Even if the driver has not a binocular view, he can appreciate the presence of obstacles, this justifies the null obstruction.

Although it would be possible to start from the eye points located to the rear of the eyellipses, this procedure considers points at the front part since this results in a higher binocular obstruction angle.

Fig. 9.44 shows the requirements of the European Directive D77/649 and D81/643, D88/366, D90/630 to determine the binocular obstruction angle of the A-pillar:

1. From point R locate points P_1, P_2, P_m given by Tab. 9.16 which represent the hinge points of the head when the driver looks out of the vehicle to the left (P_1) or to the right (P_2) on a horizontal plane.

2. If the horizontal seat travel is larger than 108 mm, the position of P_1, P_2, P_m must be corrected as indicated in Tab. 9.17.

3. If the back angle is not 25 deg, the position of P_1, P_2, P_m must be corrected as indicated in Tab. 9.15.

4. Side view: starting from point P_m, find two reference cross sections of the pillar as indicated in Fig. 9.43. The cross sections S_1 and S_2 include the structural part, all non transparent parts and the portion of the glass with optical characteristics not complying the requirements about tansparency and distortion.

5. From points P_1 and P_2 find the eye points E1, E2, E3, E4, located on the edges of two triangles with all sides of 65 mm (Fig. 9.43).

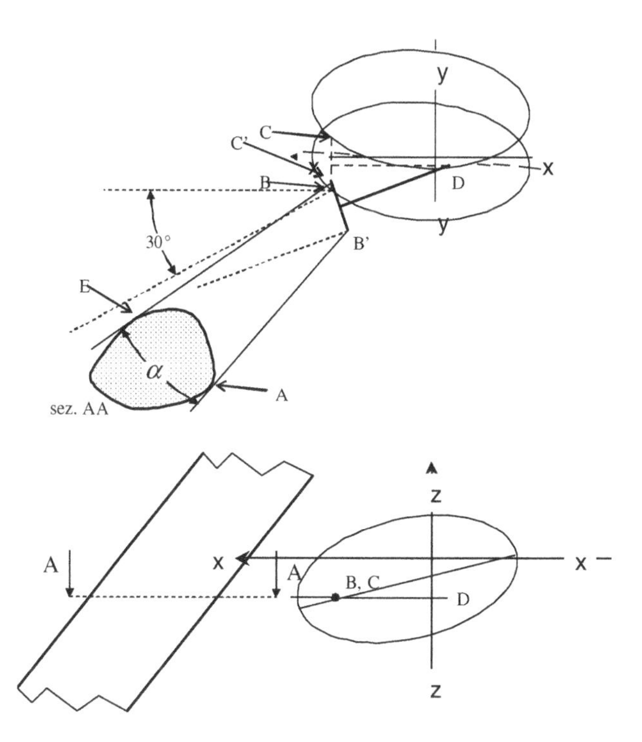

Fig. 9.42. SAE J1050a binocular obstruction of the A-pillar. The procedure leads to the definition of a field of view where both eyes cannot see (binocular obstruction). If the sight lines tangent to the pillar converge after the pillar, the binocular obstruction is null.

6. Rotate the left triangle about P1 until the sight line orthogonal to segment E1-E2 is tangent to the leftmost part of section S2.

7. The binocular obstruction is the angle between the previous sight line and that from E2 tangent to the rightmost point of section S1. It should be less than 6 deg.

Table 9.16. Coordinates of points P relative to R for a back angle of 25 deg.

	x [mm]	y [mm]	z [mm]
P_1	35	-20	627
P_2	63	47	627
P_m	43.36	0	627

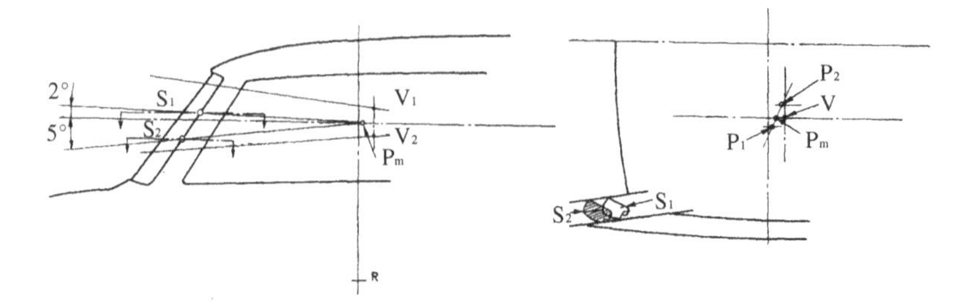

Fig. 9.43. Definition of the reference cross sections S1 and S2 of the A-pillar following the European directive.

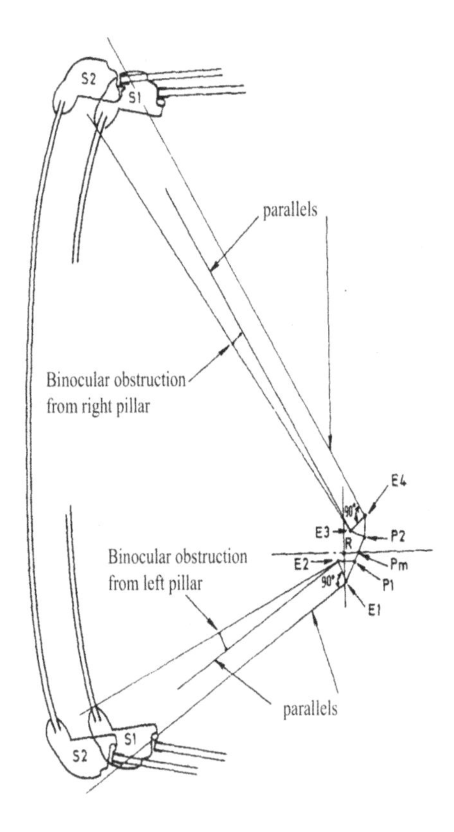

Fig. 9.44. Binocular obstruction following the European directive. The obstruction should be smaller than 6 deg. The eye points E1, E2, E3, E4 are at the corners of triangles with all sides equal to 65 mm.

Table 9.17. Corrections to the x coordinates of points P_1, P_2 for longitudinal seat travel larger than 108 mm.

horizontal seat travel L23	Δx [mm]
108 ÷ 120 mm	−13
121 ÷ 132 mm	−22
133 ÷ 145 mm	−32
145 ÷ 158 mm	−42
> 158 mm	−48

9.10.4 Indirect Visibility

The field of view of the driver is increased in the backwards direction by means of mirrors. In Europe the reference directives are D71/127, D77/649, D81/643, D88/366, D90/630.

Fig. 8.38 shows the monocular, binocular and ambinocular fields of view through a mirror. Points OD and OS represent the driver's eye points which are at a distance of 65 mm from each other, at 635 mm (25 in) on the vertical of point R. In this case, this position is not affected by the back inclination and horizontal seat travel.

The directives refer to an ambinocular view probably because when looking in a mirror it is rather difficult to appreciate the distance that from an obstacle. It is therefore accepted that part of the vision is of monocular type so that the driver is able to recognize simply the presence of the obstacle.

For class M_1 vehicles the European directive requires the following:

- Inside mirror, Fig. 8.39 (top, to the left), the driver should be able to view in a rectangle on the ground 20 m wide and extending to the infinite starting from 65 m from eye points OD and OS.
- Outside mirrors, Fig. 8.40 (top, at center), the driver should view the following rectangles on the ground extending to the infinite:
  - to the left, 2.5 m wide, starting from 10 m from the eye points.
  - to the right, 4 m wide, starting from 20 m from the eye points.

Size and location of the mirrors

The inside mirror should be of category I. The size of the reflective part must be enough to inscribe a 4 cm height and a cm wide rectangle:

$$a = 15\ \text{cm} \cdot \frac{1}{1 + \frac{1000}{r}} \tag{9.4}$$

r (expressed in mm) is the mean radius of curvature of the mirror which can be found from the minimum and maximum radii r_{min}, r_{max} at the centroid of the reflective part.

$$r = \frac{r_{min} + r_{max}}{2} > 1200 \text{ mm} \tag{9.5}$$

The inside mirror should be completely above point V1 and at a minimum distance of 350 mm from it in longitudinal direction. In order to reduce the distortion of the reflected image, it is suggested to use a planar mirror whenever possible.

The outside mirrors should be of category III. The size of the reflective part must be enough to inscribe a 4 cm height and a cm wide rectangle

$$a = 13 \text{ cm} \cdot \frac{1}{1 + \frac{1000}{r}} \tag{9.6}$$

where r (in mm) is the mean radius of curvature of the mirror.

$$r = \frac{r_{min} + r_{max}}{2} > 1200 \text{ mm} \tag{9.7}$$

Additionally, it must be possible to inscribe a 7 cm vertical segment.

In the plan view, the segment that connects the centroid of the reflective part to the center of segment OS - OD should be at less than 55 deg relative to the longitudinal direction x. Since the lower part of the outside mirror enclosure is less than 2 m from the ground, it should not increase the width of the vehicle of more than 20 cm. The width is measured without taking into account local features such as handles or other small elements on the body surface.

10
Climatic Comfort

Some knowledge regarding the physiology of the climatic comfort will be introduced in the following sections; at this stage it is useful to understand the mission of the climate control system [19][20].

Firstly thermal comfort conditions will be identified as well as system parameters capable of describing them; secondly the human body metabolic activity as function of ambient conditions will be introduced and its consequence on heat exchange will be described.

At this point it will be possible to draw a thermal balance to design the climate control system preliminary.

Afterwards the design and test criteria for the most important parts of this system will be introduced, referring to the explanation of their operation reported in Volume I.

Volume I also reports information which is relevant as concerns the design of conditioned air distribution in the passenger compartment.

10.1 Physiology Outline

Thermal comfort is defined by ISO 7730 Standard as *that mental condition that expresses satisfaction for the thermal environment where a subject is exposed.*

The complexity of describing thermal comfort with suitable words comprises also its definition.

L. Morello et al.: The Automotive Body, Vol. 2: System Design, MES, pp. 201–238.
springerlink.com

As will be seen in this section, thermal comfort depends upon a series of environmental parameters and not only on air temperature as a simplistic approach may suggest.

In fact, thermal comfort depends also on:

- air humidity,
- energy radiation,
- air motion.

In rooms where the climate is artificially controlled with a treated air circulation, also air quality, windows fogging and air stream noise are important.

In addition, the thermal comfort relates directly to the subjective perception of the human being in charge of evaluating thermal comfort.

Thermal perception could be considered as a sixth sense used by humans to perceive the surrounding environment.

10.1.1 Body Temperature Control

The human body is provided with a very efficient internal temperature control system which ensures that the internal body temperature is maintained within a narrow range around 37°C, to avoid discomfort and dangerous situations of thermal stress.

To reach this target, the thermal energy that is continuously produced by the human body (ranging form 100 W while performing sedentary activity, up to 1000 W while performing a strenuous effort) must be dissipated to maintain the temperature within the acceptable range: Thermal energy is not generated, nor dissipated uniformly over the body.

When the body internal temperature becomes too high, the body reacts in two ways. The first is the *vasodilatation* causing an increased flow of blood through the skin, increasing the thermal conduction to the external body layers. As a result the skin temperature becomes similar to that of the internal tissues.

At this point a second process is started: *perspiration*. Perspiration represents a true cooling mechanism for the body, because the energy needed to evaporate the sweat is directly subtracted from the skin.

An increase of few tenths of a degree is enough to stimulate the perspiration mechanism which is capable of multiplying the heat dissipated by the human body four-fold.

If cooling is not sufficient the body heats-up: if the internal temperature increases by more than 2°C with respect to the optimum figure of 37°C, the behavior of the subject is modified: slackness feeling is an example of this alteration.

On the contrary, if an excessive heat exchange between skin and environment takes place, the blood flow through the external layers of the skin is reduced; the skin is cooled-down, but the internal tissues are protected from over-cooling.

This is the mechanism of *vasomotor control* against the cold and relates to a narrow temperature range below the optimum.

Below this temperature the body reacts by producing thermal energy mechanically, by tremors, muscle tension or other non-voluntary activities. If this thermal energy is enough to compensate for losses, the internal temperature is maintained around the optimum value.

Behavioral control of cold includes an increase of motor activity or the usage of heavier clothing.

We should take into account while evaluating thermal comfort that hands and feet, being distant from the center of the body, are more prone to cooling.

If the measures that are put in place to hinder cooling are inadequate, the body cools-down: If the internal temperature decreases more than 2°C below the optimum value, again the normal behavior is altered for example with a feeling numbness or stiffness.

The system responsible for human body internal temperature control is very complex and not yet fully understood. Studies performed on this subject have identified two main sensors located under the skin involved in this control:

A first category of sensors are the so-called *Krause's bulbs,* located about .5 mm deep under the surface of the skin, which are sensitive to heat loss. These are spread-out across the body but are more highly concentrated where the sudoriferous glands are located, ie. on finger tips and elbows.[21]

A second category of sensors, called *Ruffini's organs,* which are sensitive to temperature increase, are located in deeper layers in the skin, particularly around the lips, nose, chin, forehead, thorax and fingers. Because of the thermal insulation provided by the skin, these sensors are slower to react to temperature changes compared with the Krause's bulbs.

Body temperature

As described in the previous section, the human body generates thermal energy continuously that must be dissipated to the environment in order to maintain the body internal temperature within acceptable limits.

The *metabolic energy* M produced by the human body should balance at least the following functions:

- personal activity M_{act};
- maintain muscular tension M_{shiv}, if required by actual body position.

The difference between metabolic energy and useful muscular work W, $M-W$ is stored in the body, causing a consequent temperature increase or dissipation to the environment through the skin and breathing activity.

The thermal heat loss to the environment may be accomplished through the following modes of exchange[1]:

- skin sensible heat;
- latent heat due to perspiration (E_{rsw});
- heat loss due to perspiration (E_d);
- heat loss due to respiration (C_{res});
- latent heat due to the water contained in breath (E_{res}).

The heat exchange of the skin includes conduction, convection and radiation; however it may be described as the sum of a convective contribution C and a radiation contribution R, taking the external clothing surface as reference.

Physiology studies of the human body have demonstrated that no cooling-down, warming-up or evaporation occur if the body is exposed to an ambient temperature within:

- 29°C÷31°C if the body is not covered by clothing;
- 23°C÷27°C if the body is clothed, in sedentary activity.

The body in this temperature interval falls in a neutral zone and no physiological control action of the body temperature takes place.

The skin temperature (t_{sk}) and the internal temperature (t_{cr}) are in neutral condition at the following values:

- $t_{sk,n} = 33{,}7°C$;
- $t_{cr,n} = 36{,}8°C$.

For engineering applications the human body is modelled as a simple cylinder ad interactions with the environment are described with elementary processes as in Fig. 10.1.

10.1.2 Thermal Comfort Condition

A subject considers the environment to be comfortable when thermally neutral, implying no sensation of being either hot or cold.

In this situation, two conditions must be met.

The first requires equilibrium between the internal temperature and the skin temperature.

The second requires that the heat produced by metabolic activity must be equal to that lost by the body during its current activity.

---

[1] The sensible heat derives from the temperature variation of a given mass due to convection, conduction and radiation or a variation of this mass, while the latent heat derives from mass state variation, as evaporation or melting.

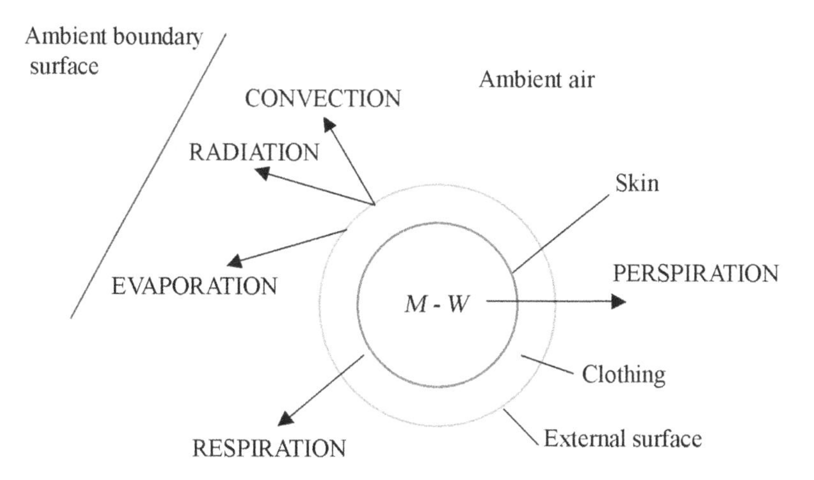

Fig. 10.1. For engineering applications the human body is modelled as a simple cylinder ad interactions with the environment are described with elementary processes.

Thermal comfort equation

The description of the balance between energies in play was proposed by Fanger[22] as way of expressing the comfort condition using physical parameters, included in the so-called thermal comfort equation.

With reference to the model in Fig. 10.1, the thermal balance equation could be the following:

$$M - W = Q_{sk} + Q_{res} = (C + R + E_{sk}) + (C_{res} + E_{res});$$

where:

- M is thermal flux generated by the human body;
- W is thermal flux exchanged with the environment;
- Q_{sk} is the thermal flux exchanged through the skin;
- Q_{res} is the thermal flux exchanged through respiration;
- C_{res} is the thermal flux due to convection, in respiration;
- E_{res} is the latent heat in respiration;
- C is the convective heat exchanged through the skin;
- R is the radiation heat exchanged through the skin;
- E_{sk} is the evaporation heat exchanged through the skin $= E_{rsw} + E_d$.

The terms of this equation often refer to the external surface of the bare body A_D, that can be inferred from the empirical Du Bois' formula:

$$A_D = 0,292 m^{0,425} h^{0,725},$$

where:

- m is the mass of the subject;
- h is the height of the subject.

The formulae of the following sections refer to the evaluation of the terms of the equation above.

Heat exchanged by the human body

A single equation can describe the combined effect of convection and radiation:

$$C + R = \frac{t_{sk} - t_0}{R_{cl} + \frac{1}{f_{cl}h}},$$

where the body surface temperature t_{sk} is given by the empirical relationship:

$$t_{sk} = 35,7 - 0,028(M - W)$$

and the operative temperature t_o (the definition of this temperature will be supplied later) is defined by:

$$t_o = \frac{h_r t_r + h_c t_a}{h_r + h_c},$$

where:

- h_c is the convective thermal exchange coefficient;
- h_r is the linear term of the radiation thermal exchange coefficient;
- t_r is the radiant mean temperature (the definition of this temperature will be supplied later);
- t_a is the ambient temperature;
- R_{cl} is the thermal resistance through clothing;
- f_{cl} is the area coefficient of clothing.

The area coefficient of clothing is given by the equation:

$$f_{cl} = A_{clo}/A_D,$$

where A_{clo} represents the external surface of the clothed subject and takes into account the thermal resistance of clothing I_{cl}. The following empirical relationships are usually adopted:

$$f_{cl} = 1,00 + 1,29 I_{cl},$$

for $I_{cl} < 0,078$, or:

$$f_{cl} = 1,05 + 0,645 I_{cl},$$

for $I_{cl} \geqslant 0,078$.

Heat exchange due to perspiration

This term (E_{sk}) includes the contributes of sweating E_{rsw} and the heat diffused by the skin E_d. The following equation is proposed:

$$E_{sk} = w\frac{p_{sk,s} - p_a}{R_{e,cl} + \frac{1}{f_{cl}h_e}},$$

where:

- p_a is the steam pressure in the ambient;
- $p_{sk,s}$ is the steam pressure at the saturation temperature t_{sk};
- $R_{e,cl}$ is the resistance to heat exchange through clothing surface;
- h_e is the thermal conductance for convection through clothing;
- w is the percentage of wet skin.

Heat exchange due to respiration

This term includes the sensible and latent heat of expired air during respiration which has a different temperature with respect to that of the ambient and contains a quantity of steam:

$$C_{res} + E_{res} = \left[0,00014M(34 - t_a) + 0,0173M(5,87 - p_a)\right]/A_D,$$

where:

- M is the metabolic heat flux;
- p_a is the steam pressure in the ambient.

Ambient parameters

From a practical point of view, it is necessary to establish which parameters are significant to rate comfort.

Once again it is necessary to stress that the human subject is not only sensitive to the ambient temperature but also to the energy wasted by its body; the parameters to be taken into account are those that influence this energy loss.

Four parameters describe the ambient from this point of view:

- the air temperature of the surrounding atmosphere;
- the mean radiant temperature;
- air velocity;
- air humidity.

In addition, two parameters describe the subject:

- the metabolic index;
- the clothing index.

The following sections will be addressed to explain the parameters above.

Air temperature

Air temperature t_a is measured around the subject but outside of its boundary layer. This temperature has influence on the heat exchanged by the human body because of air convection.

Air temperature is usually measured by means of a dry bulb thermometer and, for this reason, is sometimes called *dry bulb temperature.*

Mean radiant temperature

Mean radiant temperature t_r is the uniform temperature of a black cavity where occupants would exchange the same amount of heat as in the real non-uniform ambient.

It depends obviously on the temperature of all surfaces of the room enclosing the subject and on any other surface the human body can receive heat energy by radiation thermal exchange.

The difference between mean radiant and air temperature is positive, by definition, when surfaces enclosing the subject are warmer than his skin, is negative if the opposite condition applies.

The precise calculation of mean radiant temperature involves many assumptions on surface radiation and on multiple reflection within the enclosed space.

Assuming that all surfaces exchanging radiation energy with the human body are black, a simplified equation allowing the calculation of the mean radiant temperature can be defined:

$$T_r^4 = T_1^4 F_{p-1} + T_2^4 F_{p-2} + \ldots\ldots + T_n^4 F_{p-n},$$

where T_i are the temperatures (measured in °K) of all surfaces radiating energy to the subject, while F_{p-i} are the *view factors*, that are coefficients defining the fraction of radiating energy of a given surface pertaining the subject in question and satisfying the following condition:

$$\sum_{i=1}^{n} F_{p-i} = 1.$$

The evaluation of the view factors is the most delicate aspect of this calculation.

The direct measurement of the radiant temperature can be measured using a *globe-thermometer.* It is simply a dry bulb thermometer enclosed in sphere of treated copper (black metalline) with diameter 150 mm and possessing nominally the same absorption characteristic of the human skin.

As a demonstration of the importance of the radiating component in the thermal exchange of the human body, the actual adsorbtion coefficient of the human skin has been shown to be higher than any other substance including black metalline.

As a consequence human being are very sensitive to any variation of the mean radiant temperature.

Air speed

The air speed relative to the human body v_a plays an important role in evaluating the thermal comfort of a subject in an enclosed space with artificial climate because it influences the convective thermal exchange with the human body and is one of the control parameters of all air conditioning systems.

In hot or humid weather, any motion of the surrounding air can increase the heat loss from the human body at the same air temperature.

This fact is justified by two different mechanisms.

If the air temperature is less than the skin temperature, an increase in air speed increases convective heat exchange because a higher amount of fresh air washes the skin.

In the case that the ambient is moderately humid (between 30÷85% of relative humidity), the increased air speed will enhance also sweat evaporation by removing saturated air and replacing it with dryer air.

The air speed in question is the average speed on the boundary layer between the ambient and the human body.

Air humidity

Air humidity u_a refers to the amount of humid steam contained in a volume of air. At a given dry bulb temperature, the quantity of humid steam that can be adsorbed by the air before saturation and consequent precipitation is called *absolute humidity.*

The saturation point also called *dew point* refers instead to the maximum amount of humidity at a given air temperature.

The *relative humidity* (RH) is the ratio between absolute humidity and dew point humidity.

Relative humidity is relevant to sweat evaporation.

If RH is over 80%, most of the sweat cannot evaporate and the air surrounding the human body becomes quickly saturated. On the contrary, if RH is below 20%, mucous membranes become dry quickly with an increased risk of irritation and infection.

Air humidity has a modest influence on thermal comfort in air conditioned environments; nevertheless there are limits that should not be exceeded.

This parameter is measured by hygrometers.

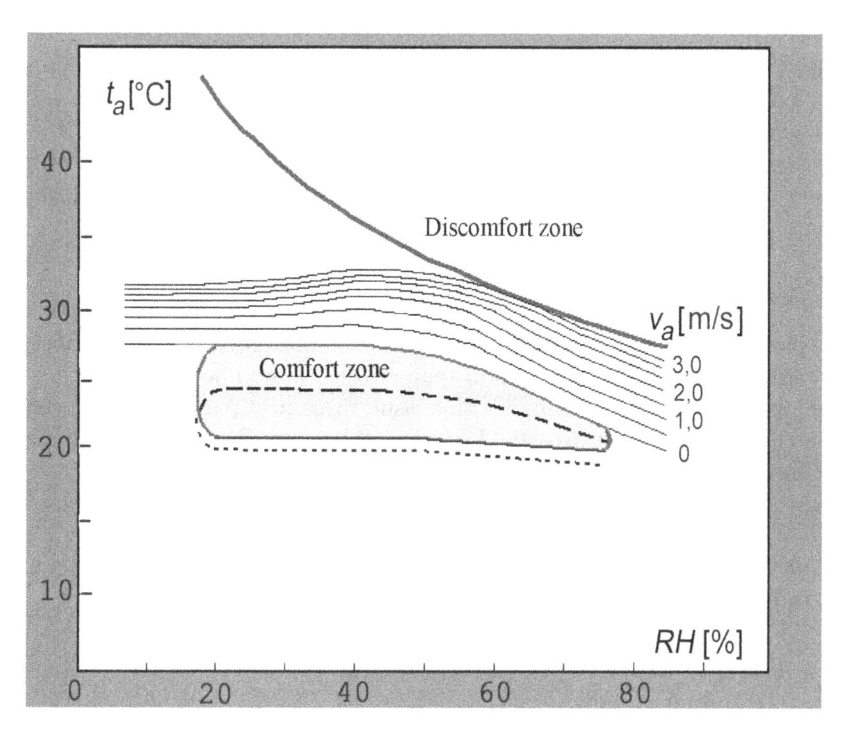

Fig. 10.2. Diagram identifying comfort and discomfort zones; the diagram is quoted with t_a, the dry bulb air temperature, RH, the relative humidity and v_a the air speed.

A way to summarize the effects of air temperature, relative humidity and air speed is reported on the diagram of Fig. 10.2.

Air temperature (dry bulb temperature) t_a is shown by the vertical axis, while RH, the relative humidity, is shown on the horizontal axis. The shadowed area in the middle of the diagram shows combinations of values of temperature and humidity that are perceived as comfortable in summer while sitting in the shade.

The dotted line surrounds an area perceived as comfortable in winter. This diagram shows that the human body can adapt to different situations of temperature.

The curves quoted with the air speed v_a show how the comfort zone in summer can be enlarged by a suitable air speed.

For example with a temperature of 23°C and relative humidity of 50%, no corrective action on the air speed is necessary; with an air temperature of 26°C and relative humidity of 70% an air speed of 1 m/s is necessary to obtain thermal comfort.

Integrated parameters

Three parameters are necessary to identify thermal comfort; in the case of significant radiation also the mean radiant temperature should be considered in place of the air temperature. To simplify this issue three new parameters are considered in the literature that are called integrated because they allow to summarize the influence of different basic parameters.

These new parameters are:

- *operative temperature* t_o, integrating the effect of air temperature and mean radiant temperature;

- *equivalent temperature* t_{eq}, integrating the effect of mean radiant temperature and air speed;

- *effective temperature* ET^*[19], integrating the effect of air temperature, mean radiant temperature and air humidity.

Effective and equivalent temperatures depend on the activities and clothing of the subject, while the operative temperature is usually independent of these parameters.

To understand how these three new parameters are defined, consider a subject leaving in the room under evaluation and moving to a new room defined as the reference: By adjusting the temperature in the reference room in order to obtain the same heat loss the subject had in the room under evaluation, the temperature in the reference room, that will be therefore an integrated temperature, can be determined.

Each of the integrated parameters must satisfy a specific condition in the reference room, as explained subsequently.

Operative temperature. The operative temperature t_o is often used as an evaluation parameter for a confined environment although is not a parameter upon which the thermal comfort of the subject depends.

It is defined as the uniform temperature of the reference room in which the subject would exchange the same amount of convective and radiant heat actually exchanged in the non uniform room under evaluation. Correspondingly it can be calculated as the average of the air temperature and mean radiant temperature, using the thermal convective conductivity between clothing and air h_c and the radiant conductivity h_r as weighting factors:

$$t_0 = \frac{h_r t_r + h_c t_a}{h_r + h_c}.$$

To overcome the difficulty of measuring thermal conductivities that depend on unknown parameters as the temperature of the outside of clothing, the operative temperature is sometimes approximated to the arithmetic average of the two temperatures:

$$t_0 = \frac{t_r + t_a}{2}.$$

In summary, to evaluate the operative temperature, the reference room must have the same air speed, the same humidity and the same radiant temperature as the room under evaluation.

Equivalent temperature. The equivalent temperature integrates the effect of convection and radiation on the amount of heat rejected by the human body.

The equivalent temperature is by definition the temperature of a reference room with the same air and wall temperatures of the room under evaluation, but with zero air speed in which the subject would exchange the same amount of heat actually exchanged in the room under evaluation.

In conclusion the equivalent temperature depends on the operating temperature with the additional effect of the air speed. With high air speed, t_{eq} is well below t_o, while for an air speed in the range around 0.1 m/s, the two temperatures are almost equal.

The reference room must have in this case the same humidity as the room under evaluation, its air temperature must be equal to the mean radiant temperature of the room under evaluation and its air speed must be zero.

Effective temperature. The effective temperature is the mostly used integrated parameters and is also applied as the evaluation parameter for overall thermal comfort of a confined environment with controlled climate.

This parameter includes the effect of temperature and humidity: two rooms with equal ET^* are perceived as being equal despite different values of temperature and humidity.

ET^* is the dry bulb temperature in a black cavity with 50% of relative humidity (the current reference room) in which the subject would exchange the same amount of heat as in the room under evaluation; the subject in the reference room must exhibit the same skin temperature t_{sk} and have the same percentage of wet skin w.

In summary, to measure the effective temperature, the reference room must have the same air speed, the air temperature must be equal to the mean radiant temperature and its relative humidity must be 50%.

Metabolism

The human body is a kind of chemical laboratory in continuous activity. Food, drink and other substances undergo a huge number of chemical reactions that together constitute the human body *metabolism*.

Metabolic processes are basically oxidation which produce thermal energy; in conclusion the potential chemical energy of food and drink (not including the fraction transformed into spare substances) and of other substances used generates thermal energy inside the human body.

This energy is the difference between the consumption of potential chemical energy and of useful work produced, and its effect will be to cause an increase in the body temperature.

This energy is also called *metabolic energy* or *energetic metabolism*.

The measurement unit for metabolism is MET, corresponding about to 58.15 W/m^2; it is the metabolism of a non working sitting subject divided by its body surface.

Since an average subject has a body surface of about 1.7 m^2, a metabolism of 1 MET corresponds to about 100 W.

The minimum value for metabolism is about 0.8 MET when the subject is sleeping and can increase up to 10 MET for heavy manual work or sport activities.

Some average values for metabolism are reported in Tab. 10.1.

Metabolism is influenced by the following factors:

- age: metabolism decreases with increasing age with an almost proportional law;
- sex: female metabolism is roughly 5% lower than male metabolism at the same age.

Although the exact value of metabolism cannot be measured directly, the following values can be assumed for reference: 44 W/m^2 for a male and 41 W/m^2 for a female subject.

Clothing

Clothing Index CLO provides a measurement of the thermal insulation due to clothing: clothing reduces the body heat loss to the environment affecting the thermal balance.

This index is a thermal resistance that is usually measured as a ratio with a reference value of 0.155 $m^2{}^\circ C/W$, representing a typical situation of clothing suitable for a sedentary office activity.

Table 10.1. Some values of metabolism measured in MET and W/m^2.

Kind of activity	[W/m^2]	Met
Bedded subject	46	.8
Seating subject	58	1.0
Standing subject	70	1.1
Office work	70	1.2
Driving a car	80	1.4
Shop working	93	1.6
School teaching	95	1.6
House keeping	100	1.7
Walking at 2 km/h	110	1.9
Laying-down bricks	125	2.2
Gardening	170	2.9
Heavy housekeeping	170	2.9
Dismantling with a pneumatic hammer	175	3.0
Walking at 5 km/h	200	3.4
Wood cutting with a motor saw	205	3.5
Ice skating at 18 km/h	360	6.2
Digging	380	6.5
Cross-country skiing 9 km/h	405	7.0
Wood cutting with an axe	500	8.6
Running at 15 km/h	550	9.5

In these conditions, the clothing index is $CLO = 1$; a bare body would have a clothing index $CLO = 0$.

The clothing index increases when the clothing is suitable for use outdoors in the winter season, while decreases in summer season.

Some values of CLO are given in Tab. 10.2.

The overall value of the clothing index CLO_t can be calculated by adding CLO_i indices pertaining to each piece of clothing, as shown in Fig. 10.3, according to the following formula:

$$CLO_t = \sum_{i=1}^{n} CLO_i.$$

This way of proceeding provides data that are usually sufficiently accurate; if more accurate data are necessary, the total heat resistance of an assigned clothing can be measured using internally heated human dummies in a climatic cell.

Also stuffing present in any piece of furniture contacting the human body, eg. car seats, exert a significant influence on heat rejection and must be taken into account.

Table 10.2. Some clothing indexes measured in *CLO* and m^2°C/W.

Type	Description	*CLO*	[m^2°C/W]
Underwear	underpants	0,03	0,005
	boxers	0,06	0,009
	vests	0,13	0,020
	undershirts	0,09	0,014
Shirts	short sleeved shirts	0,09	0,014
	long sleeved shirts	0,12	0,019
	flannel long sleeved shirts	0,30	0,047
	as above, with roundneck	0,34	0,530
Trousers	shorts	0,06	0,009
	long light	0,20	0,031
	normal	0,25	0,039
	flannel	0,28	0,043
Suits	coveralls	0,50	0,078
	gowns	0,30	0,047
Sweaters	sleeveless	0,12	0,019
	light	0,20	0,031
	light with roundneck	0,26	0,040
	heavy with roundneck	0,37	0,057
Jackets	light	0,25	0,039
	normal	0,35	0,054
Overcoats	regular overcoats	0,60	0,093
	winter jacket	0,55	0,085
	heavy winter jacket	0,70	0,109
Miscellaneous	socks	0,02	0,003
	stocking	0,10	0,016
	light sole shoes	0,02	0,003
	heavy sole shoes	0,04	0,006
	boots	0,05	0,008
Seats	wooden or metal	0,00	0,000
	padded	0,10	0,016
	padded with armrests	0,20	0,032

Fig. 10.3. The overall value of the clothing index CLO_t can be calculated by adding CLO_i indexes pertaining each piece of clothing.

10.1.3 Thermal Comfort Evaluation

The aim of this section is to define temperature and humidity conditions identifying thermal comfort. If the thermal comfort is not optimum these conditions will help us to obtain an improvement.

To evaluate thermal comfort in a confined thermally controlled space, two indices are usually applied:

- the predicted mean vote (PMV);
- the effective temperature ET^*.

Predicted mean vote

According to Fanger [22], the thermal comfort sensation is a function of the body *thermal load* L. It is defined as the difference between the thermal energy, produced through the actual activity of the subject, and the heat loss which would occur if the skin temperature and the transpired heat E_t were maintained within comfort limits for that activity.

In comfort conditions, thermal load L should be 0.

Fanger proposes to evaluate the comfort conditions by defining an index which is empirically correlated to the thermal load via the following formula:

$$PMV = [0,352^{(-0,0362M/A_D)} + 0,032]L,$$

where:

- M, is the metabolism;
- A_D, is the body surface, according to Du Bois.

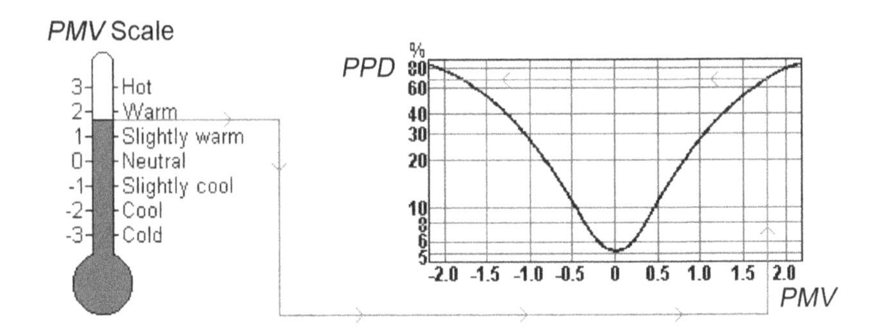

Fig. 10.4. Qualitative score scale of PMV and correlation with PPD, the percentage of people dissatisfied.

PMV is the *predicted mean vote*, a function of the activity of the subject, clothing, dry bulb temperature, radiant mean temperature and relative humidity of the environment.

PMV is the average evaluation of the environmental comfort of a significantly large group of subjects; PMV scale includes 7 scores, where three negative scores are for the feeling of cool, three positive scores are for the feeling of hot and the central value of 0 corresponds to a neutral thermal sensation

In Fig. 10.4 the scale description and the corresponding score are shown on the left-side.

Also at zero score for PMV, some individual subjects in the sample may be dissatisfied by the level of thermal comfort even though each subject is performing the same activity with the same clothing: comfort evaluation is subjective.

To evaluate the percentage of subjects who consider a given environment to be uncomfortable, a new parameter has been introduced: PPD the percentage of people that are dissatisfied:

$$PPD = 100 - 95^{-(0,03353 PMV^4 + 0,2179 PMV^2)}.$$

People assigning -3, -2 or $+2$, $+3$ on the PMV scale are classified as being dissatisfied.

On the right side of Fig. 10.4 an empirical correlation between PMV and PPD is represented.

Effective temperature

The effective temperature is the mostly applied comfort evaluation parameter and depends on the percentage of wet skin and on the permeability of clothing:

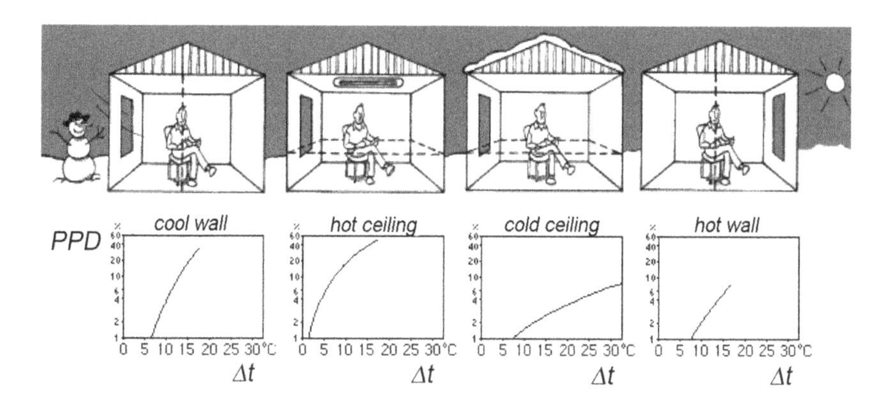

Fig. 10.5. Some example of local discomfort caused by vertical and horizontal thermal gradients.

For this reason it is more sensitive to the relative humidity than PMV and this sensitivity increases as the humidity increases.

In summer the comfort zone is characterized by ET^* between 22,8÷26°C and relative humidity between 20% and 60%.

In winter the comfort zone is characterized by ET^* between 20÷22,9°C and, again, relative humidity between 20% and 60%.

Local discomfort

Although a subject could be in a neutral comfort situation, sone parts of his body cold be exposed to particular conditions caused by unevenness in thermal parameters: This situation is called *local discomfort.*

This kind of discomfort cannot be corrected by changing average air temperature or humidity, but acting locally.

The most frequent causes for local discomfort are the following:

- too hot or cold room walls;
- too hot or cold floor;
- asymmetry in radiant thermal flux;
- air draft.

Fig. 10.5 illustrates some examples of local discomfort caused by horizontal and vertical temperature gradients; the diagrams below each situation image represent PPD as a function of the temperature difference between a reference section (dotted in the drawing) and one of the walls around the room in question. The first three situations are relative to winter heating, whereas the last corresponds to summer cooling.

Vertical temperature gradient

This effect is caused by the thermal gradient existing in closed room as a consequence of the change of air density due to temperature. This phenomenon causes a feeling of hot at the head and of cool at the feet.

Experimental results report that a difference of 3°C between head and foot is rated as discomfort by 5% of interviewed subjects. While a higher head temperature at the head that at the foot may generate dissatisfaction, the opposite situation is never rated as discomfort.

Floor temperature

A different floor temperature may affect foot comfort.

It is inappropriate to consider floor temperature because the heat loss only affects local comfort; this heat loss depends on the thermal conductivity of the floor and shoes.

It has been reported that sitting people accept a floor temperature 1°C higher than standing people.

ISO 7730 specification provides for a floor temperature in the range between 19÷29°C.

Radiation asymmetry

This kind of discomfort arises when a part of the body is subject to a heat radiation and the radiation flux is not uniform.

This situation can be described by radiant temperature, measuring the difference of this temperature on the two opposite sides of a same plane portion.

A hot ceiling with a cool window is a cause of higher discomfort than hot walls and a cool ceiling.

Air drafts

Discomfort occurs when an undesired local cooling is caused by a concentrated air stream.

It should be noticed that the human body is relatively sensitive to air motion since it stimulates thermal sensors to convey signals to the brain. Many alert signals that are irrelevant independently may together cause high annoyance.

To describe air speed fluctuations, the *turbulence intensity* τ is frequently considered; it is defined as:

$$\tau = 100\frac{\sigma_a}{v_a},$$

where σ_a is the standard deviation of air speed and v_a the average speed of the flow.

Fig. 10.6 reports some empirical correlations between PPD and air speed v_a at a given air temperature t_a with different values for the turbulence intensity τ.

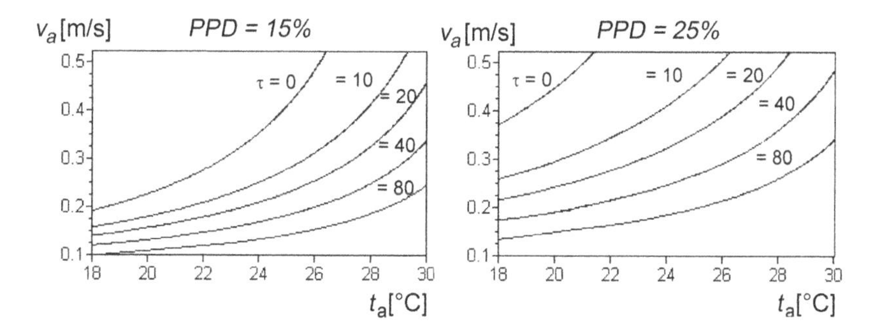

Fig. 10.6. Some empirical correlations between PPD and air speed v_a at a given air temperature t_a with different values for the turbulence intensity τ.

10.2 Passenger Compartment Energy Balance

The function of the air conditioning system is to supply thermal energy at a controlled temperature in such a way as to improve thermal comfort conditions. The first step necessary is to understand which are the parameters that play a role in the thermal balance of the passenger compartment.

The heat flows entering and exiting the passenger compartment can be summarized in a thermal balance equation where the thermal power introduced by the climate system is compared with a series of not controllable terms.

In the expression:

$$W_{imp} = W_d + W_i + W_p + W_m,$$

- W_{imp} is the power introduced by the climate system;
- W_d is the power exchanged between the passenger compartment and outside by means of convection and conduction;
- W_i is the power radiated through glasses;
- W_p is the power introduced by passengers because of metabolism;
- W_m is the power introduced by the powertrain because of the higher temperature of its parts.

Each term of this equation is described in details subsequently. Furthermore some other parameters must be taken into account including temperature and humidity.

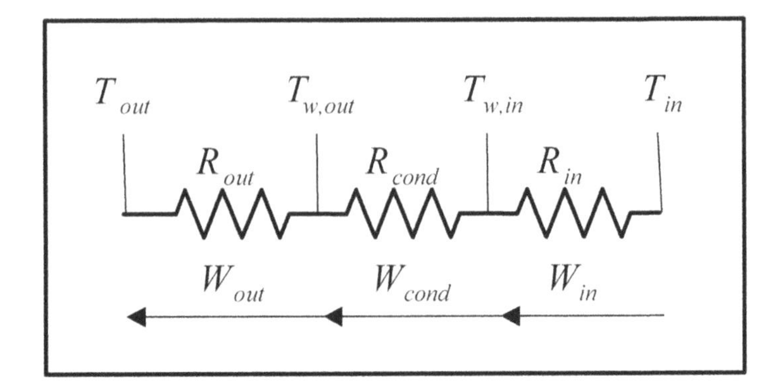

Fig. 10.7. Scheme of heat flow and temperature between the passenger compartment (at right) and the outside environment (at left).

10.2.1 Exchanged Heat

Comfort conditions require that the air temperature of the passenger compartment is different from the outside temperature.

The exchanged heat for convection and conduction is proportional to:

- the temperature difference between the passenger compartment and outside environment, Δt;
- the total thermal transmissivity between the passenger compartment and the outside environment, K;
- the surface of the walls surrounding the passenger compartment, S.

The thermal transmission coefficient K represents the power transmitted through the compartment surface divided by the temperature difference between the inside and outside. This coefficient integrates conductive and convective heat exchange.

Consider the scheme shown in Fig. 10.7 where the thermal flow is from the inside (on the right) to the outside (on left), where:

- T_{out} is the outside temperature,
- $T_{w,out}$ is the temperature of the outside surface of the vehicle,
- $T_{w,in}$ is the temperature of the inside surface of the passenger compartment and
- T_{in} is the inside temperature.

The temperatures will remain steady because of the resistances opposing the heat flow through the walls.

These resistances are the opposite of the conductivity coefficients:

- $R_{in} = \frac{1}{K_{in}}$ is the inside convective resistance,
- $R_{out} = \frac{1}{K_{out}}$ is the outside convective resistance,
- $R_{cond} = \frac{1}{K_{cond}}$ is the conductive resistance of the walls surrounding the passenger compartment,
- $R = \frac{1}{K}$ the total resistance.

In a typical case of the winter season:

$$T_{out} < T_{w,out} < T_{w,in} < T_{in},$$

and the total heat flow through the walls is:

$$W = (T_{in} - T_{out})KS$$

Since the resistances are in series, the total resistance is the sum of the different contributions:

$$R = R_{in} + R_{out} + R_{cond}$$

as the total conductivity will be given by the following equation:

$$\frac{1}{K} = \frac{1}{K_{in}} + \frac{1}{K_{out}} + \frac{1}{K_{cond}}.$$

Convective heat exchange depends on the air speed on the boundary surfaces and therefore K_{out} depends on the speed of the car, while K_{in} depends on the distribution and speed of the inside ventilation air.

In an average car with normal glass surface:

- $K = 5$ W/m^2°C,
- $S = 10$ m^2.

These reference values apply in the average and refer to state-state conditions; during transients the heat power can be 3 or 5 times higher.

This rough example refers to homogeneous walls between the passenger compartment and the outside environment: a more correct approach would include the many different contributions of smaller surfaces; in fact the thermal conductivity of the glasses and body walls are different.

Tab. 10.3 shows the influence of the different boundary surfaces of the passenger compartment on the total heat flow; the relevant contribution of glasses can be observed: Therefore the larger the glass surface, the poorer the thermal insulation of the passenger compartment.

Table 10.3. Flow breakdown through the different boundary surfaces of the passenger compartment, by winter heating.

Part name	Percentage of total
Roof	11
Glasses	18
Floor	13
Dashboard	12
Doors	8
Remaining parts	38

10.2.2 Radiated Heat

Particularly in summer, an additional amount of heat due to sun radiation is brought into the equation.

The effect of sun radiation depends on the surface and the color of the car body. Its contribution is two-fold: The oncoming radiated energy through the glasses, and the energy absorbed and again radiated by the body.

The combined effect of these contributions, which depend on the relative position of the sun and on weather conditions, can exceed 1,000 W/m^2.

10.2.3 Passengers Metabolism

The previous sections have shown how the metabolism is influenced by human activity: Taking into account the reduced amount of activity while riding or driving in a car, an average value of 100 W per passenger can be assumed for this part[23][24].

10.2.4 Powertrain Power

This contribution depends on many factors, including the car power demand and the engine size; an average value could be set around 300 W.

10.2.5 Air Conditioning System

As explained in Volume I, passenger comfort is provided by the air conditioning system which is often called the HEVAC system, an acronym which stands for heating, ventilation and air conditioning system, the main functions of the system and its subsystems.

Fig. 10.8 shows a cross section of an HEVAC system to indicate the function of each component; another scheme is shown in Volume I.

The heating subsystem includes a radiator R exchanging heat between the engine coolant and the air in the passenger compartment.

The ventilation subsystem includes the vents to introduce a certain amount of air into the passenger compartment from the outside through AE and from

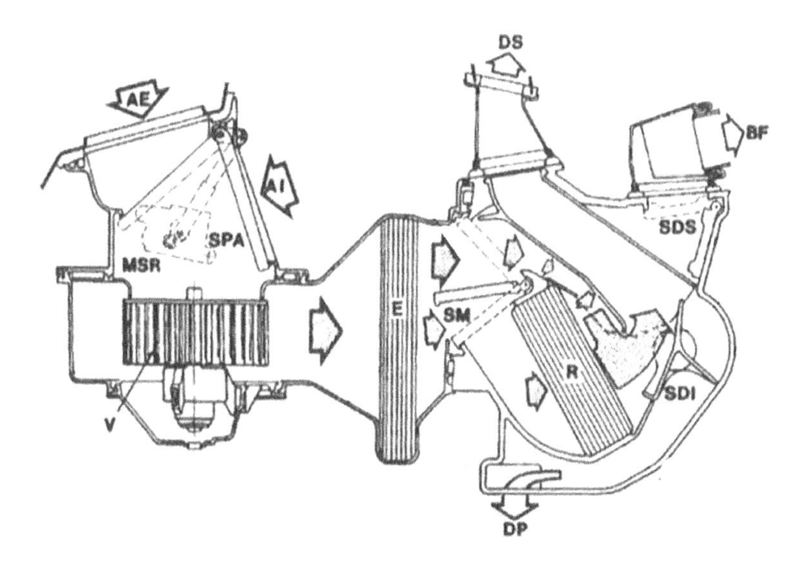

Fig. 10.8. Cross section of the HEVAC system, showing the ventilator V, the evaporator E and the heater R; the treated air can come from the outside or the inside through the vent MSR and cross the evaporator alone or also the heater, by means of the vent SM.

the inside via AI. Air circulation is activated by ventilator V. The air can be used directly as it is or its temperature and humidity can be adjusted through the evaporator E and the heater R.

The evaporator E is part of air cooling system including a condenser, a compressor and an evaporation valve, usually located in the engine compartment. Through these four components, a refrigerating fluid performs a thermodynamic cycle capable of reducing the temperature before entering the evaporator.

A set of valves, that can be controlled manually or automatically, perform a number of functions:

- a valve MSR allows the ambient or the passenger compartment air, depending on which exhibits conditions closer to the desired value, to be introduced into the HEVAC system ;
- a valve SM allows the air to cross the evaporator (to reduce its temperature and humidity) and the heater (to heat it, if necessary);
- Valves SDI, SDS and DS allow the air to be distributed to the passenger compartment in the most appropriate way and directed towards the feet or head of the occupants or towards the windshield.

Air speed can be adjusted by changing the voltage of the electric ventilator.

The thermal power introduced into the passenger compartment may be calculated as follows:

$$W_i = Q\rho c_p(T_{in} - T_{tt}),$$

where:

- Q is the air mass flow treated by the HEVAC system,
- ρ is its density,
- c_p is its specific heat,
- T_{in} is the passenger compartment air temperature,
- T_{tt} is the treated air temperature.

Car HEVAC systems are usually designed as to have the treated air crossing the evaporator in any case in order to ensure that the humidity is reduced while setting the evaporator temperature to the appropriate value.

If the air leaving the evaporator is too cold, its temperature will be increased by heating.

The heater can work in two different ways:

- by dividing the air crossing and passing-by the heater appropriately, as shown in Fig. 10.8;
- by ensuring all the air crosses the heater, adjusting the coolant flow appropriately using a valve.

10.3 Hevac System Design and Testing

This sections addresses the heating and cooling subsystems since air distribution has been already covered in Volume I.

10.3.1 Description

The scheme of a complete HEVAC system, including the engine compartment components, is shown in Fig. 10.9.

The cooling system pumps the heat from a cooler source, the passenger compartment, to a hotter environment, the exterior, by exploiting an inverse thermodynamic cycle.

The heat transfer occurs because a refrigerating fluid undergoes a thermodynamic cycle including pressure changes (via the compressor C and the expansion valve VE) and a physical state transformation (via both heat exchangers evaporator and condenser).

The thermodynamic cycle can be represented on a plane where the horizontal axis represents the enthalpy h of the working fluid and the vertical axis the

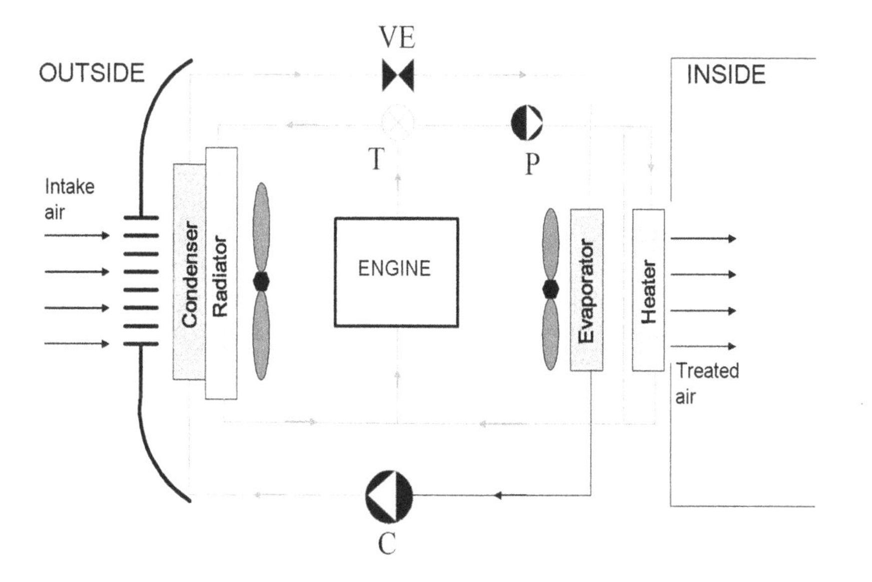

Fig. 10.9. Scheme of the HEVAC system, including engine compartment components: the cooling system comprises a condenser, the expansion valve VE and the compressor; the heating system comprises the engine, as heat generator, a heater and the circulation pump P.

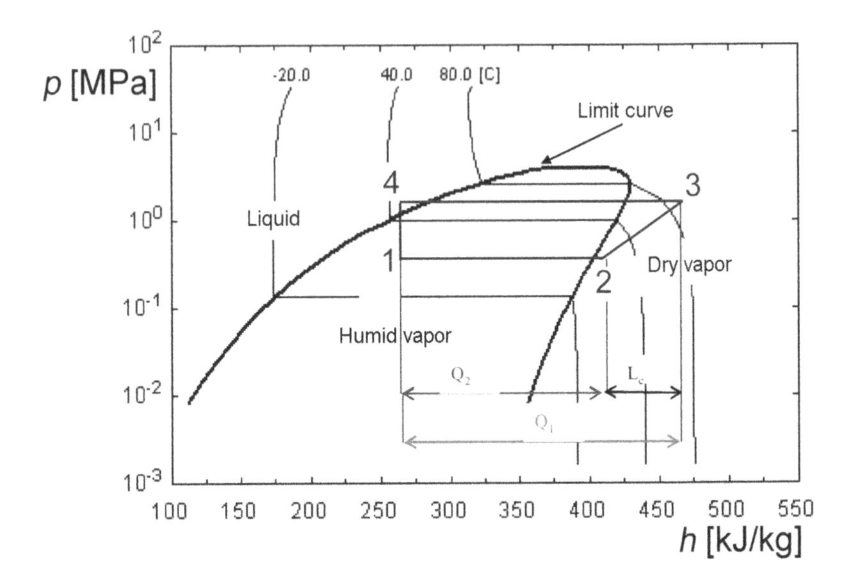

Fig. 10.10. Qualitative Mollier's diagram for an air conditioning system fluid. The limit curve identifies the points where the fluid changes its state.

pressure p. This representation is called Mollier's diagram in the case that the working fluid is not a perfect gas, but undergoes changes of state as a result of the thermodynamic cycle.

A qualitative representation of this diagram is shown in Fig. 10.10; on this diagram, the *limit curve* represents the points where the fluid changes its physical state.

The typical bell shape of this curve exhibits a maximum at the so-called *triple point* or *critical point*; it divides the plane (h, p) into three areas, where the physical states correspond to dry vapor, liquid and humid vapor (vapor with droplets of condensed liquid).

More precisely, in the area to the right of the limit curve there is dry vapor, whereas in the area inside the bell there is humid vapor. In this area a *vapor title* can be defined as the ratio between vapor mass and total mass, changing from 0 (on the left side of the bell) to 1 (on the right side of the bell).

A typical thermodynamic cycle of a refrigerating fluid is represented on this plane by the quadrangle 1-2-3-4.

The purpose of the evaporator is to extract heat from the passenger compartment and is crossed by the air entering through a set of ducts, as shown in Fig. 10.8.

The temperature of the evaporator walls must be below that of the environment to be cooled. The heat subtracted from the air is absorbed by the working fluid that changes its state from point 1 to point 2, with a curve that can be represented by a constant pressure line, if the pressure loss for crossing this component is negligible. The working fluid state changes from humid vapor (1) to dry vapor (2).

The absorbed heat must now be dissipated to the outside environment.

To make this heat exchange feasible, it is necessary to raise the fluid temperature to a value which exceeds the temperature of the air of the outside environment. The compressor provides this function, sucking the low pressure vapor coming from the evaporator (point 2) and compressing it to a higher pressure value (point 3). With this pressure increase, the temperature also increases. The compression work is provided by the vehicle engine, moving the compressor through a mechanical transmission, usually a V belt.

The 2-3 curve can be represented by means of an isoentropical curve by neglecting the thermal exchange with the environment. It should be noted that is convenient that the starting point of the compression line corresponds to a dry vapor state because fluid droplets could damage the compressor.

The working fluid, now in a gas state, enters a second heat exchanger that enables the dissipation of the heat to the external environment, according to the curve 3-4.

Again it is assumed that the pressure remains unchanged during this transformation; in a first phase the vapor is cooled down, before being condensed to the liquid state. At the condenser outlet (4) the working fluid is high pressure liquid.

This liquid enters now the expansion valve VE so that its pressure is reduced; this pressure reduction is accompanied by cooling down to reach the conditions represented in the point 1; the transformation line is isoenthalpical and the fluid returns to its initial condition.

The exchanged heats Q_1 and Q_2 and the compression work L_c are represented by the corresponding enthalpy variations on the (p, h) plane.

A common refrigerating fluid used in HEVAC systems is R134a, as mentioned in Volume I.

10.3.2 Cooling System

Here reference is made to the detailed description of the components of this system provided in Volume I.

The evaporator has the dual function of extracting the heat from the passenger compartment air and reducing its humidity. In fact the heat exchanged with the air has a sensible part:

$$M_a c_p \Delta t,$$

and a latent part, coming from water condensation:

$$M_c \lambda_c,$$

where:

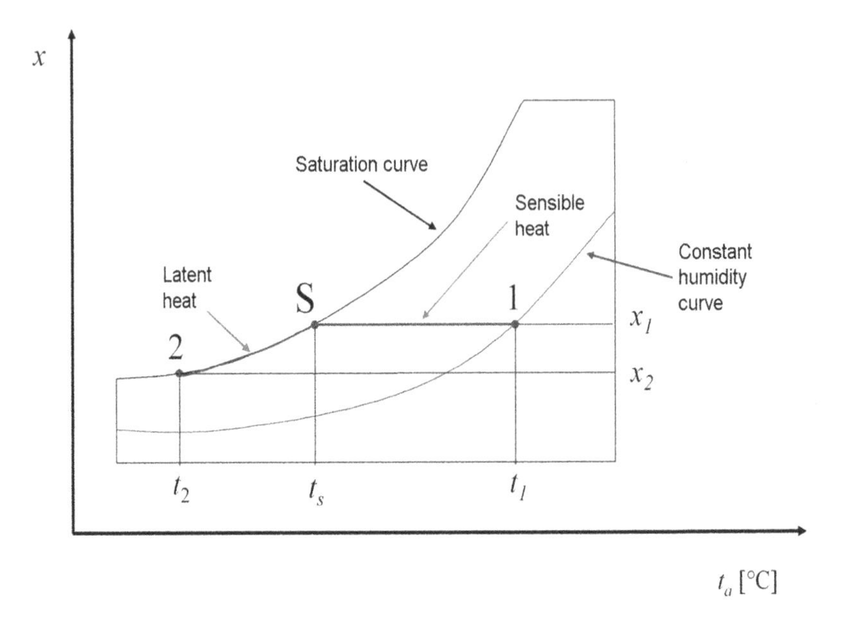

Fig. 10.11. Psychrometric diagram of humid air; the horizontal axis represents the air temperature t_a, the vertical axis the vapor title x.

- M_a is the air mass flow rate crossing the evaporator;
- c_p is the air specific heat;
- Δt is the air temperature change between the evaporator inlet and outlet;
- M_c is the working fluid mass flow rate;
- λ_c is the specific condensing heat.

The transformation of the air can be represented with a psychrometric diagram.

The qualitative diagram in Fig. 10.11 shows the vapor title x on the vertical axis and the air temperature t_a on the horizontal axis. In this case the vapor title is the ratio between water vapor mass and air mass. This diagram shows the so called *saturation curve*, the locus of points with the highest vapor content at a certain temperature; in the same way, a family of curves with constant relative humidity UR can be built up.

The transformation of the air crossing the evaporator can be identified by a curve such as 1-S-2 and includes two parts; the first part with constant humidity to the saturation point, and a second part where both temperature and humidity decrease.

The treated air has both a temperature lower than initially and a lower vapor content ($x_2 < x_1$).

Evaporators used in today's cars are cross flow heat exchangers, with laminar plates and fins. The working fluid enters from the top and flows through plates with both internal (fluid side) and external (air side) fins. Fins are necessary to increase exchanging surfaces and consequently improve exchange efficiency.

For sake of space, the evaporator is contained within a rectangle of no more than 300 mm long, with a depth of about 60 mm, usually with two rows of plates.

The total exchanged heat is typically in the range 2 to 7 kW.

10.3.3 Heating System

The scheme of the heating system is also shown in Fig. 10.8 comprising:

- the engine radiator;
- the thermostatic valve T;
- the water pump P;
- the heater and, obviously,
- the engine.

The system includes two parallel cooling circuits for the internal combustion engine generating a relevant quantity of heat. During engine warm up the thermostatic valve shuts off the circuit including the radiator in such a way that all the available heat is sent to the heater.

In this way the engine coolant heats the passenger compartment air. When the cooling fluid reaches an appropriate temperature in the range between 88÷90°C, the thermostatic valve opens and the radiator circuit is activated as the generated heat exceeds that which the heater can receive.

The heater uses the same technology and has the nominally same size as the evaporator.

The design power is typically in the range 5 to 13 kW.

10.3.4 Design Example

The diagram in Fig. 10.12 is the actual Mollier diagram for the chlorofluorocarbon R 134a; with respect to the qualitative diagram shown in Fig. 10.10, this shows other curve families including:

- constant vapor title x curves, ranging from .1 to .9, inside the limit curve, where clearly $x = 0$ on the left and $x = 1$ on the right;
- the constant temperature curves ranging from −60°C to 240°C;

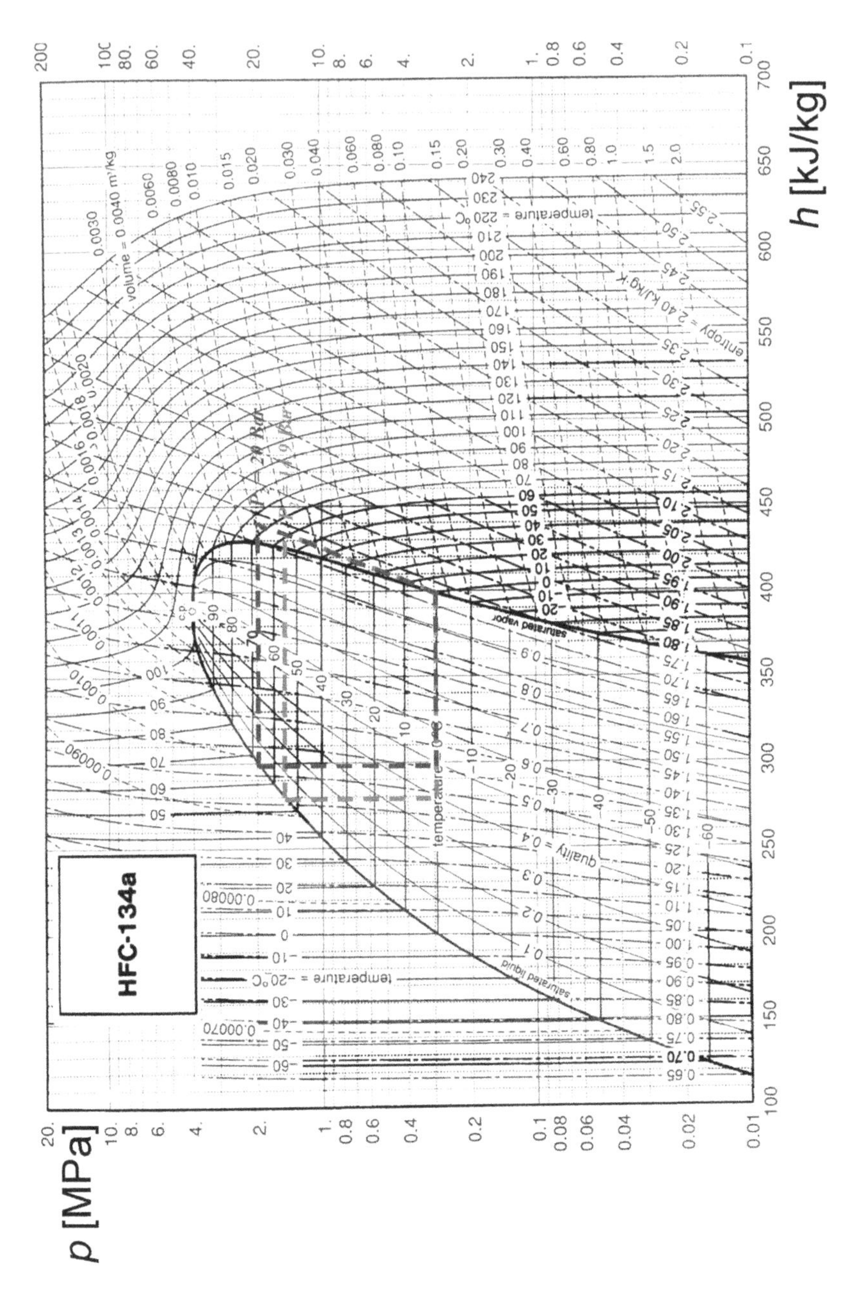

Fig. 10.12. Mollier's diagram for chlorofluorocarbon R134a; in addition to the curve introduced in Fig. 10.10, also the same title, temperature, specific volume and entropy curves are introduced. The example cycles are numbered as in the previous figure.

- the constant specific volume curves ranging from 0.003 m^3/kg to 2.0 m^3/kg;
- the constant entropy curves ranging from .65 kJ/kg°K to 2.55 kJ/kg°K.

It should be noticed that the constant temperature curves are horizontal in the area inside the limit curve because the temperature remains constant during changes of state; the length of the horizontal segment is a measure of the latent heat of evaporation.

The cycle points are numbered as in Fig. 10.10.

When designing a cooling cycle, the temperatures $t_1 = t_2$ and $t_3 = t_4$ cannot be set arbitrarily. The temperature t_1 cannot be higher than the ambient temperature from which the heat should be taken; furthermore temperature t_3 cannot be lower than the ambient temperature to which the heat should be transferred.

The exchanged heat Q_{12} can be expressed by the formula:

$$Q_{12} = KS\Delta t,$$

where:

- K is the thermal exchange coefficient;
- S is the heat exchanging surface;
- Δt is the temperature difference between air and coolant.

Clearly the smaller the value of Δt, the larger the exchanging surface and hence the cost of the heat exchanger. Cooling system design aims to contain the size of heat exchangers, both for evaporation and condensation. The objective is to enhance the heat exchanger performance without cost penalty .

The evaporation process occurs with a working fluid comprising a mixture of vapor and liquid droplets. This fact can negatively affect compressor valves operation and can dilute lubricants with consequent rapid wear of any moving part of the compressor.

Correspondingly it is necessary to compress the working fluid just in the vapor phase which requires completing the evaporation and compressing the vapor only at that time.

At this stage it is appropriate to consider a design example of a cooling cycle to provide, for example, a thermal power of 5 kW and assuming the following conditions:

- condensation temperature $t_3 = t_4 = 55°C$;
- evaporation temperature $t_1 = t_2 = 0°C$.

R134a is the working fluid.

Reading the corresponding values for pressure and enthalpy on the Mollier diagram of Fig. 10.12:

Point 2:

- $p_2 = .29$ MPa;
- $h_2 = 400$ kJ/kg;
- $s_2 = 1.7$ kJ/kg°K.

Point 3:

- $p_3 = 1.49$ MPa;

can be found at the intersection of the pressure line corresponding to p_3, at the temperature of 55 °C and the isoentropic curve with $s_3 = 1.7$ kJ/kg°K, identifying an ideal adiabatic compression through point 2.

The thermodynamic parameters of point 3 that can be achieved from the Mollier diagram are:

- $h_3 = 433$ kJ/kg;
- $t_3 = 61.5$ °C.

Point 4 is on the saturation curve at 55 °C and therefore:

- $h_4 = 281$ kJ/kg.

The enthalpy h_4 of point 3 is equal to h_1 of point 1, because the expansion 4-1 in the expansion valve is assumed to be isoenthalpic. Point 1 is characterized by:

- $t_1 = 0$°C ;
- $h_1 = 279.4$ kJ/kg.

The cycle defined in this way is represented in Fig. 10.12.

The cooling heat is given by:

$$Q_{12} = h_2 - h_1 = 399 - 279.4 = 119.6 \text{ kJ/kg}.$$

The working fluid mass rate necessary to obtain the desired power of 5 kW can be calculated using:

$$F = P/Q_{12}.$$

to obtain:

$$F = 0.042 \text{ kg/s}.$$

The compression work is:

$$W = (h_3 - h_2)F,$$

corresponding to:

$$W = 1.43 \text{ kW}.$$

The *coefficient of performance* COP is defined as the ratio between spent mechanical energy and transferred heat:

$$COP = \frac{P}{W} = \frac{h_2 - h_1}{h_3 - h_2}$$

and therefore:

$$COP = 3.5.$$

For a new value for the condensation pressure $p_3 = 20$ bar with the same cooling heat, the cycle will be modified as follows:

Point 2:

- $p_2 = 2.9$ bar;
- $h_2 = 400$ kJ/kg;
- $s_2 = 1.73$ kJ/kg°K.

Point 3:

- $p_2 = 20$ bar;
- $t_2 = 67.5$°C.

Point 3 is identified by the intersection between the line of constant pressure p_3 and the isoentropic curve $s_2 = 1.73$ kJ/kg°K, corresponding to an adiabatic compression through point 3, corresponding on the Mollier diagram to:

- $h_3 = 439.5$ kJ/kg;
- $t_3 = 75.2$°C.

Point 3 lies also on the saturation line at 67,5°C and therefore:

- $h_4 = 300$ kJ/kg.

The enthalpy h_4 of point 4 corresponds to the enthalpy h_1 of point 1, due to the nature of the expansion 4-1 through the expansion valve. In point 1:

- $t_1 = 0$°C;
- $h_1 = 300$ kJ/kg.

Hence the new value for COP decreases to:

$$COP = 2.44.$$

10.3.5 *Testing*

The mission of an HEVAC system is usually represented by two basic tests:

- the first evaluates the maximum performance of the air cooling system, ie. cooling down the passenger compartment starting from an initial condition corresponding to a long stop in the sun in a hot country;
- the second evaluates the maximum performance of the air heating system, ie. heating up the passenger compartment after a long stop in winter time in a cold country.

Traditionally hot and cold weather conditions are assumed to correspond to a southern state of the United States and to a country of Northern Europe respectively.

Both tests are usually performed in a climatic wind tunnel, where the air temperature, speed and humidity are controlled; solar radiation is simulated using a set of high power lamps.

Typically in a climatic wind tunnel, an artificial wind simulates the car speed up to 100 km/h which is lower than in conventional wind tunnels for aerodynamic research, although high air temperatures $t_a \geq 40°\text{C}$ and sun radiation of 1.000 W/m^2 can be recreated.

A second major difference between climatic and aerodynamic wind tunnels is that for climate testing the vehicle engine must be in a condition to work properly, being a significant source of heat and mechanical power; for this reason a chassis dynamometer (sets of braked rollers) is used to simulate road load.

Climatic wind tunnels are also applied to identify operational conditions in hot weather for each component working at high temperature conditions and to verify cold weather performance including engine startability and drivability.

The significant amount of energy needed to maintain the temperature at extreme values suggests the opportunity to develop specialized cold and hot climatic tunnels.

Summer mission

The car under test is monitored with thermocouples in various key positions in the passenger compartment; the climatic wind tunnel typically uses a temperature of 43°C±1 with simulated sun radiation on the car roof of about 900 W/m^2. Passenger dummies equipped with thermocouples sit in the car. Conditions are maintained until the temperature at the dummy head is in the range between 63÷65°C.

The engine is started and the air conditioning system is adjusted to maximum cool; external air valve is set to recycle. Then the car is driven according to a driving cycle with speeds between 30 and 90 km/h in calm air.

The wind tunnel must assure the temperature of 43°C±1 and a relative humidity of 30%±3% for the entire duration of the test. An acceptable test output is achieved if the following temperatures of the dummies heads are reached:

- after 30': 24°C;
- after 60': 21°C;
- after 90': 21°C;
- after 120': 25°C.

Winter mission

The car under test is monitored with thermocouples in various key positions in the passenger compartment; the climatic wind tunnel typically uses a temperature of −10°C±1 whereas the lamps simulating sun radiation are switched off.

After a time of 8 hr., usually enough to reach steady temperatures conditions, the engine is cranked and the test is started with the external air valve open. Also in this case the car is driven according to a driving cycle with speeds between 30 and 90 km/h in calm air.

Heating system adjustments are set to maximum with air directed towards the passengers' feet only.

An acceptable test output is achieved if a temperature of 12°C at the dummies heads is reached in 10'.

The main issue for rapid passenger compartment warm up in this test is usually the limited amount of thermal energy available in the cooling system after engine start in very cold weather.

For extreme climatic conditions, additional heaters must be available which can be either heat accumulators, that store a suitable amount of heat when available to accelerate engine warm up in cold conditions, or, in certain cases, other heaters which using additional burners or electric resistances.

Deicing and defogging

Additional tests evaluate the deicing and defogging of the windshield and front side windows to guarantee minimum visibility performance; these tests are also performed in climatic wind tunnels.

This test is usually made at an ambient temperature of −3 °C. When temperatures reach steady-state, a steam generator is set in the car which is able to generate 70±5 g/h of water steam for each seat in the passenger compartment; this figure corresponds to the steam physiologically produced by an average adult person.

After 5' of operation of the steam generator, two operators occupy the front seats and the amount of generated vapor is reduced accordingly.

The engine is started and the HEVAC system is set in the defogging conditions suggested.

During the 10' test, cleaned areas are contoured with a felt-tip pen each second minute, obtaining a result similar to that shown in Fig. 10.13.

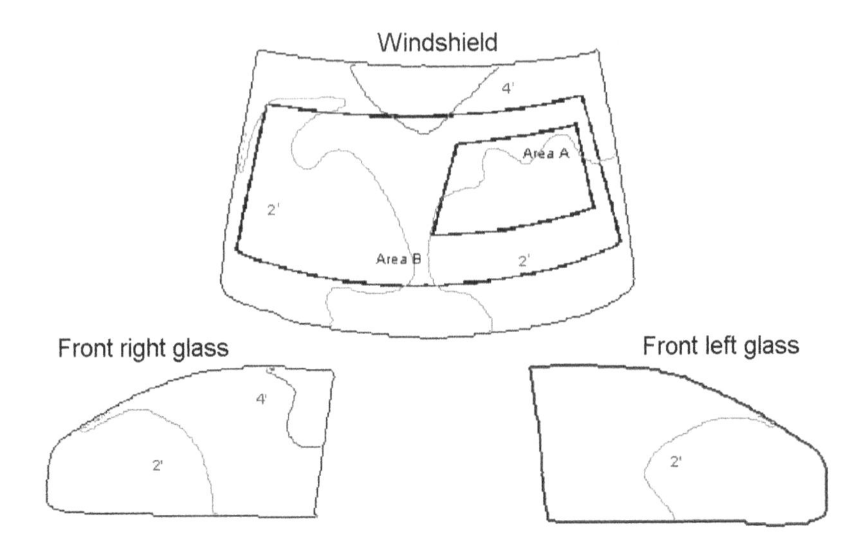

Fig. 10.13. Cleaned areas on front glasses during warm up tests; A and B areas refer to primary and secondary visibility.

The D 78/317 directive requires that 90% of the windshield (in Area A) and 80% of the windshield (Area B) are cleaned in 10'.

The deicing test is made in a similar way at a temperature of −18 °C. After stabilization, 0.044 g/cm^2 of water is sprayed on the external surface of glasses. After 9' the wipers are switched on; starting from the 10th minute cleaned up areas are contoured in the same way as before.

The same directive requires that 80% of the windshield (in Area A) is cleaned in 20' and 95% (including area B) is cleaned in 40'.

Reliability

Reliability targets apply to system components and overall performance; the system endurance must be equal to the expected car life with the condition of applying the suggested maintenance schedule, including pollen filter and V belt substitutions and exchangers periodic cleaning.

Inevitable fluid leakages should be counterbalanced by appropriate accumulator design.

11
Noise, Vibration, Harshness

When traveling, all vehicles are subject to several dynamic excitations. The induced vibration and noise has a number of effects on the vehicle and its occupants ranging from the integrity of the structure to the perception of comfort and driving performance.

High levels of vibration and noise tends to reduce comfort, the outcome being increased fatigue of the driver that in the long term has an impact on safety. The dynamic excitation induces repeated loads on the structure that reduce its life due to fatigue, hence influencing the safety and reliability. Dynamic loads acting on the tires have a negative influence on the longitudinal and side slip forces, reducing the accelerations that can be achieved during traction, braking and cornering. The result is again reduced safety.

These are just a few examples of how the noise and vibrations influence several aspects that ultimately may affect safety.

The vibrations that occur in a vehicle involve a wide range of frequencies ranging from below 1 Hz for maneuvering loads up to about 10 kHz for acoustic excitations. The effect does not only depend just on the nature and the intensity of the excitation, but also on the dynamic behavior of the structure and the acoustic cavity inside the vehicle that can amplify the effect due to structural and acoustic resonances.

Vibro-acoustic phenomena are usually classified considering the conventional frequency ranges. With reference to Fig. 11.1 the following frequency bands can be defined:

- *Ride* (0÷5 Hz) which includes the low frequency accelerations due to vehicle maneuvers and rigid body oscillations of the car body on the suspensions.

L. Morello et al.: The Automotive Body, Vol. 2: System Design, MES, pp. 239–363.
springerlink.com

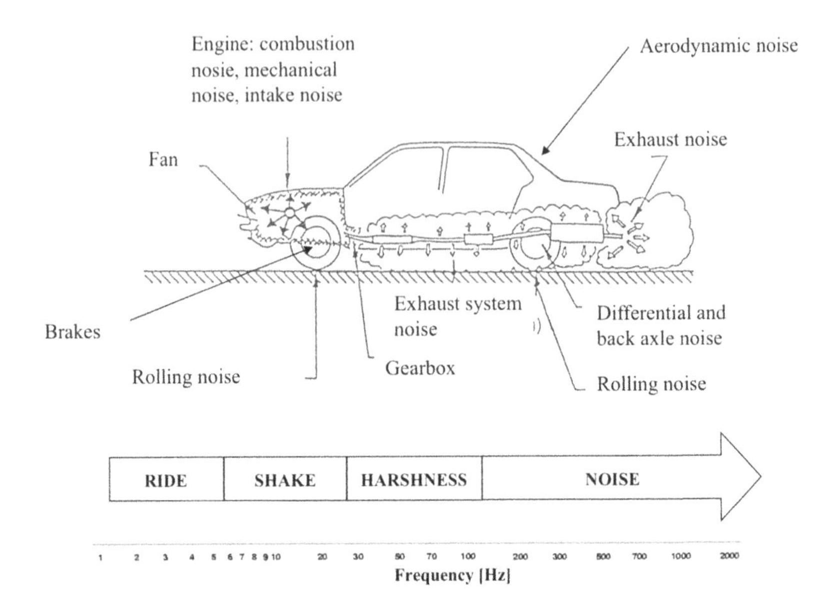

Fig. 11.1. Main sources of vibration and noise and indication of the main conventional frequency bands used to classify the vibratory phenomena.

- *Shake* (5÷25 Hz) which includes the first resonances of the main subsystems connected to the vehicle chassis such as the engine and its suspension, and the unsprung masses on the tires.

- *Harshness* (25÷100 Hz) which includes the resonances of the car body as a flexible structure. This frequency range represents a partial overlapping of the frequencies that are perceived as vibrations with the noise. High intensity acoustic vibrations included in this range are sometimes perceived by the ear as pressure variations, this is usually referred to as *boom*.

- *Noise* (>100 Hz). For frequencies higher than 100 Hz the vibration is perceived by the ears as noise. The vibration feeling is significantly attenuated and can be barely appreciated by the tactile senses. Hence the band above 100 Hz is usually referred to as noise.

The perception of noise is due to air pressure variations within the ear. The noise may enter the interior of the vehicle through apertures; alternatively exterior noise may cause vibrations of the panels that form the vehicle body producing pressure variations inside the vehicle and hence interior noise.

Depending on the mechanism at the base of the propagation, it is common to refer to two different vibration paths:

- **Structure borne:** the vehicle subsystems transmit dynamic forces to the vehicle body directly by means of the connection points and structural interfaces. This induces vibrations in the structure and in the car body panels. The structural vibrations are transmitted to the air that surrounds the occupants inside the vehicle thus causing noise. For example, the engine vibrations are transmitted by the engine suspension to the chassis; the vibration of the chassis causes the car body panels such as the firewall and the floor to vibrate, therefore inducing acoustic pressure variations inside the vehicle and hence noise.

- **Air borne:** The pressure waves that propagate outside the vehicle induce vibrations in the car body panels that, in turn, induce pressure waves inside the vehicle. The vibration of the engine block and covers, for example, induces pressure variations in the engine compartment; the vibration induced in the panels surrounding the engine is transmitted to the inside of the vehicle as noise. This example illustrates an air borne path including also a structural element; instead, in some cases the existence of holes or direct air connection between an enclosure and another allow a direct transmission path through the air. The low attenuation commonly experienced in this latter case suggests the need to minimize this kind of transmission path already in the design stage.

In all cases, the final phase of the noise transmission is in the air around and inside the ears of the occupants. Distinguishing between air borne and structure borne noise is then a matter of considering the transmission between the source and the panels that surround the interior of the vehicle.

As a first approximation, structure borne transmission can be considered to dominate below 500 Hz while the air borne becomes increasingly important above 1000 Hz.

11.1 Sensitivity to Noise

Macroscopically the human ear can be divided into three main parts (Fig. 11.3):

- outer ear – the visible ear is a flap of tissue that is also called the pinna;

- middle ear – constituted by the auditory canal, the tympanic membrane (elastic membrane that separates the external auditory canal from the tympanic cavity) and from the small bones (malleus, incus, stapes);

- inner ear – formed by three semicircular canals (devoted to the balance sense) and from the cochlea, connected to the cochlear nerve.

The outer ear collects and concentrates the pressure waves to the auditory canal. The reflections of the sound on the convolutions of the flap produce a first

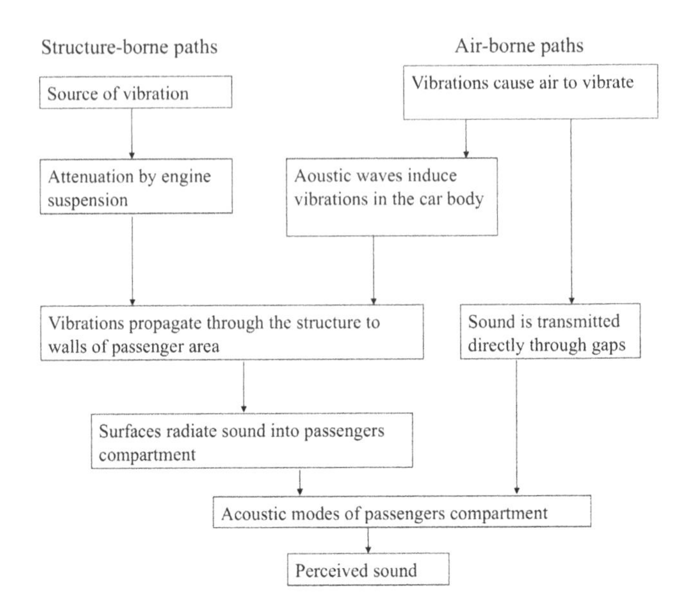

Fig. 11.2. Noise transmission paths. Structure borne: the vibrations in the structure induce vibrations in the panels surrounding the interior of the vehicle. This generates pressure waves that are felt as noise. Air borne: the pressure waves that propagate out of the vehicle may enter it through direct connections (holes) or through the panels.

processing of the sound and introducing some preferential direction of hearing. The middle ear amplifies the pressure waves and transmits them to the tympanic membrane. The vibrations of the membrane is transmitted to the chain of little bones attached to it.

The cochlea (*cochlea*) is a channel (cochleal duct) with a two and half turns spiral. Within the cochlea there are three fluid filled ducts: the tympanic canal, the vestibular canal, and the middle canal. Only the vestibular canal is connected directly to the oval window that receives the motion from the staples that acts on it as a piston.

The lower canal (tympanic canal) ends with a round window, closed by a membrane that separate it from the tympanic cavity. The tympanic and the vestibular canals are connected together at the end of the cochlea by a hole, called helicotrema. These two ducts form than one hydraulic volume filled by a liquid (perilymph, Fig. 11.4).

The central duct (cochleal duct) is filled with another liquid (endolymph). The membrane that separates the cochleal and the vestibular canals is flexible, whereas the lower one is rigid. This membrane is covered by a membrane (basilar membrane).

The organ of Corti is supported by the basilar membrane that constitutes the true sensory organ. The waves produced by the movement of the staff in

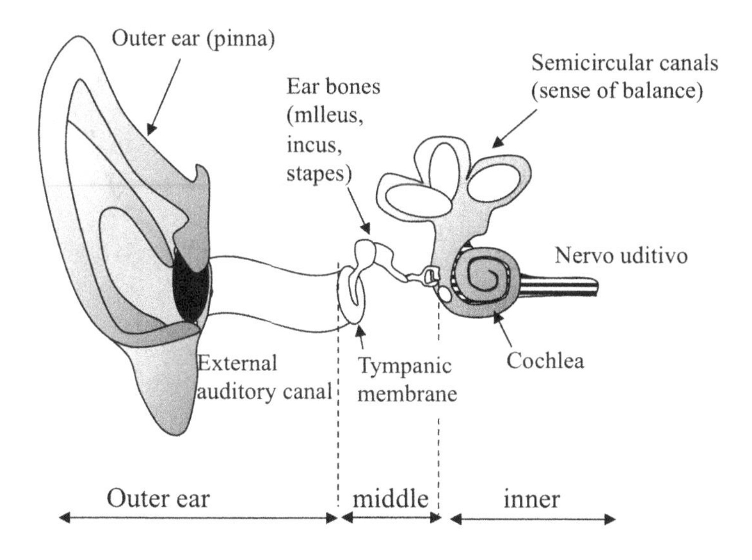

Fig. 11.3. Scheme of the anatomy of the human ear.

the perilymph deform the cochlear duct and in the organ of Corti so that the hair cells are compressed on the tectorial membrane. The resulting deformation induces electrical signals that are transmitted along the cochlear nerve and sent to the brain.

In the inner ear, the oscillations of the staff are transformed in oscillations of the perilymph. This is allowed by the compliance of the circular window at the end of the tympanic canal. The dynamic compliance of the cochlea is not constant over frequency. The staff movements induce vibrations in the perilymph. These vibrations reach a maximum amplitude at a distance from the oval window (the input of the cochlea) depending on the frequency. The higher the frequency, the smaller the distance of the peak amplitude from the oval window (and beginning of the vestibular canal).

The low frequency excitations reach the maximum amplitude close to the end of the spiral (helicotrema). For a given frequency the maximum along the cochlea of the vibration amplitude corresponds to a maximum of the sensitivity to that frequency. Fig. 11.6 shows the sensitivity to different excitation frequencies as function of the distance along the basilar membrane. The sensitivity to high frequency is maximum at short distance from the oval window, for example the peak amplitude corresponding to a 8 kHz signal is maximum at 5 mm distance. By converse, the sensitivity to low frequencies is maximum close to the helicotrema, a 300 Hz signal, for example, reaches a maximum at about 27 mm.

The basilar membrane is a continuous structure, and the vibration of one part corresponding to a given frequency propagates to the neighboring ones nominally tuned to different frequencies. The result is that a pure tone may not be perceived

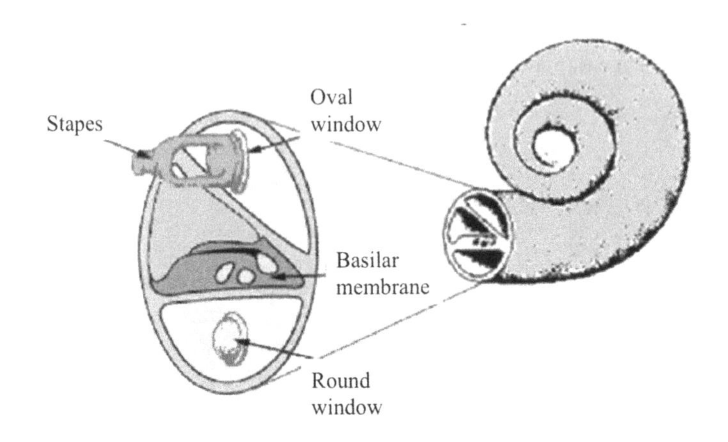

Fig. 11.4. Scheme of the cochlea.

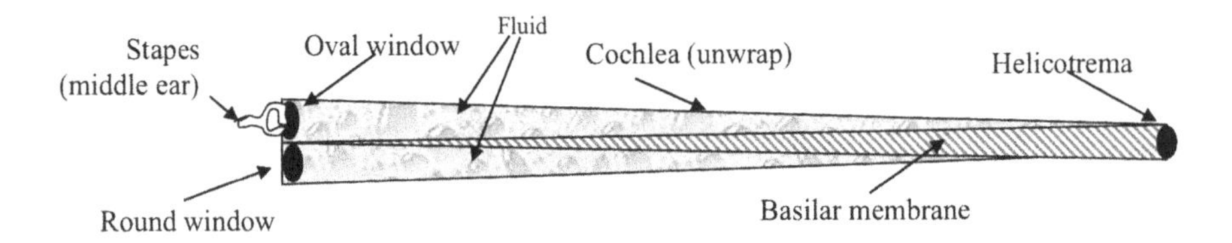

Fig. 11.5. The development of the cochlea.

if it is superimposed with one of larger intensity, even if at a slightly different frequency since the fainter sound is "masked" by the louder one, even through the frequencies are different.

Masking demonstrates a logarithmic dependence on the frequency. At frequencies below 500 Hz, a low intensity noise can be appreciated as a different tone if its frequency is relatively closer than that needed to discern two different frequencies that are both above 500 Hz.

Fig. 11.7 shows the intensity of the noise that can mask a fainter one. The masked noise correspond to the minimum of each curve (shown with a dot). Each curve indicates the intensity of the noise that is necessary to produce the same output from a given segment of the basilar membrane. The minimum of each curve correspond to the maximum sensitivity of that portion of the membrane. Due to the logarithmic scale, the frequency amplitude of each curve increases with increasing frequency.

Because of masking, at least for sinusoidal signals, the perceived tone (or height) corresponds, to the minimum of the curve associated with a given segment of the basilar membrane. Referring to Fig. 11.7, a 900 Hz signal will be perceived by the segment with maximum sensitivity at 1 kHz as a 1 kHz noise.

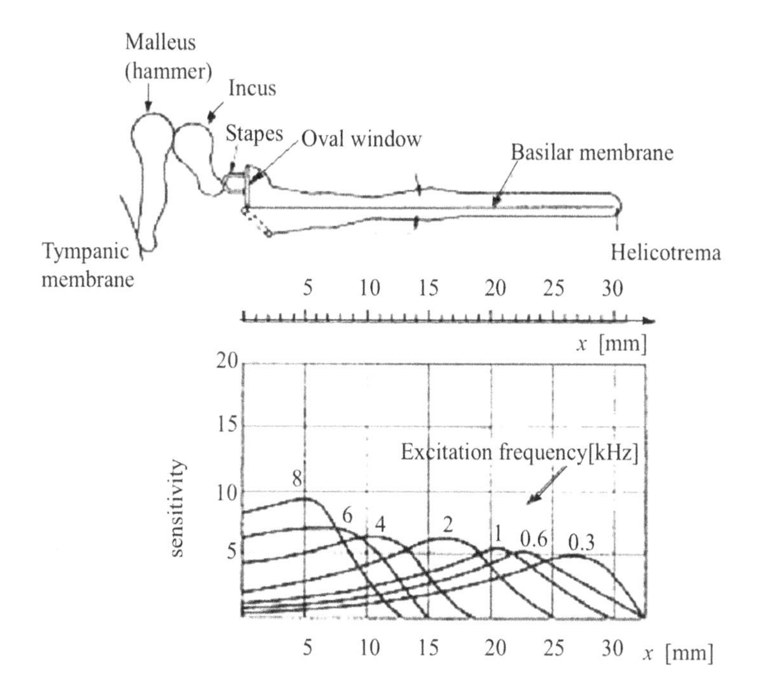

Fig. 11.6. The sensitivity of the basliar membrane as a function of the distance from the oval window. For high frequencies (8 kHz o 6 kHz) the maximum sensitivity is reached close to the oval window; for low frequencies (0,3 kHz) sensitivity is maximum at the end of the cochlea (elicotrema).

The auditory response is not only function of the frequency and amplitude content, but also of the temporal sequence of the signal essentially due to the reaction times of the functions involved in the hearing system. The subjective reaction times induce a "temporal masking" , such that a sound of finite duration masks the perception of sound events if they are separated by a too short a time interval. Similarly to frequency masking, temporal masking also exhibits a bell shaped dependence on frequency.

Intensity measures

The aim of the following section is to summarize the main physical quantities used to measure the sound intensity.

Sound intensity level

This is the power per unit surface that flows through a given area. At a given time t its value is

$$\vec{I}_a(t) = p(t)\vec{v}(t), \tag{11.1}$$

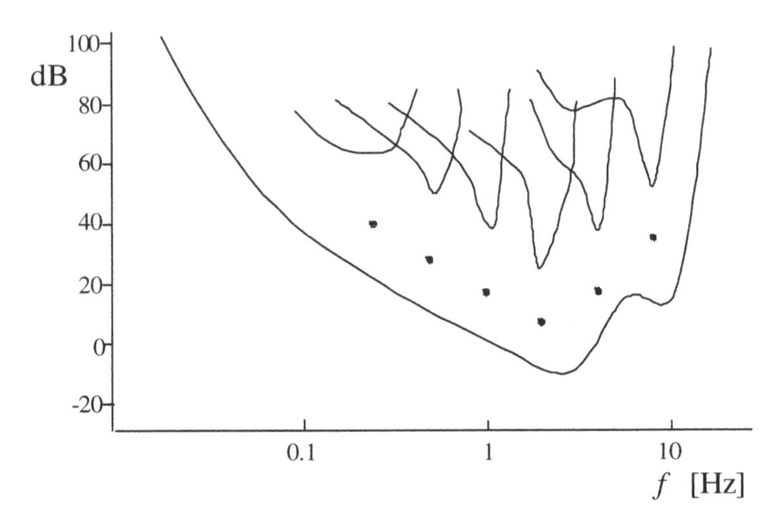

Fig. 11.7. The level of signal necessary to mask a weaker signal. The masked sound corresponds to the minimum of each curce (marked with a point).

where $p(t)$ is the pressure, $\vec{v}(t) = u\vec{i} + v\vec{j} + w\vec{k}$ is the speed vector of the fluid. Due to the Euler equation, the acceleration $\vec{a}$ of the fluid particles is proportional to the pressure gradient (∇p)

$$\vec{a} = -\frac{1}{\rho}\nabla p; \tag{11.2}$$

for a unidimensional propagation along the coordinate r, with speed u,

$$\frac{\partial u}{\partial t} = -\frac{1}{\rho}\frac{\partial p}{\partial r}. \tag{11.3}$$

To understand the physical meaning of the Euler equation, it is useful to consider a piston that moves in an cylinder opened at the end opposite to the piston. If the piston travels with constant speed, at steady state the pressure is constant in all the fluid column pushed by the piston. Conversely, if the piston accelerates, the fluid in contact with is compressed between the piston and the rest of the fluid column, that will tend because of its inertia to maintain a constant speed. The result is a pressure gradient, the resulting pressure wave propagating in the fluid column with the speed of the sound. Similarly, the accelerations perpendicular to the surface of a vibrating plate produce pressure waves that propagate around it.

The speed u of Eq. 11.3 can be computed as the integral of the acceleration $\partial u/\partial t$:

$$u = -\int \frac{1}{\rho}\frac{\partial p}{\partial r}dt \tag{11.4}$$

and the intensity in direction r is:

$$I = pu = -p \int \frac{1}{\rho} \frac{\partial p}{\partial r} dt. \tag{11.5}$$

The sound intensity can be evaluated experimentally approximating the integral and the pressure gradient in Eq. 11.5 by applying the finite difference method. If p_A and p_B are the pressures at two points at distance $r_B > r_A$ from the source

$$I = -\frac{p_A + p_B}{2} \int \frac{1}{\rho} \left(\frac{p_B - p_A}{r_B - r_A} \right) dt \approx -\frac{p_A + p_B}{2\rho\,(r_B - r_A)} \int (p_B - p_A)\, dt \tag{11.6}$$

Pressures p_A and p_B can be measured by two microphones close to each other (6÷50 mm).

Usually, the intensity is averaged over time interval T; in this case the definition becomes:

$$I_a = \frac{1}{T} \int_0^T p(t)\vec{v}(t)dt. \tag{11.7}$$

Sound Pressure Level - SPL

The sound pressure level is defined in logarithmic scale from the root mean square (RMS) of the pressure (p_{RMS}) as:

$$L_p = 10 \log_{10} \left(\frac{p_{RMS}}{p_0} \right)^2 = 20 \log_{10} \left(\frac{p_{RMS}}{p_0} \right), \tag{11.8}$$

p_0 is a reference pressure (p_0 =20 μPa).

Sound Power Level

The sound power level is defined as:

$$L_W = 10 \log_{10} \frac{W}{W_0}, \tag{11.9}$$

where W is the RMS of the power output from the source and W_0 is a reference power (W_0 =10^{-12} W).

Sound Intensity Level - Sound Power Density Level

The sound intensity level is defined as:

$$L_I = 10 \log_{10} \frac{I}{I_0}, \tag{11.10}$$

where I is the RMS value of the sound intensity of the source and I_0 is the reference intensity ($I_0 = 10^{-12}$ W/m^2).

To evaluate the variations of the quantities that measure the sound level, measured in dB, it is useful to take into account of the values reported in Tab. 11.1. The perception of doubling the intensity corresponds to a 10 dB increase. A 3 dB variation, on the other hand, is barely appreciated.

Table 11.1. Percived effects of an sound level variation.

Variation	Perceived effect
3 dB	Barely appreciated
5 dB	Clearly appreciated
10 dB	Perception of intensity doubling

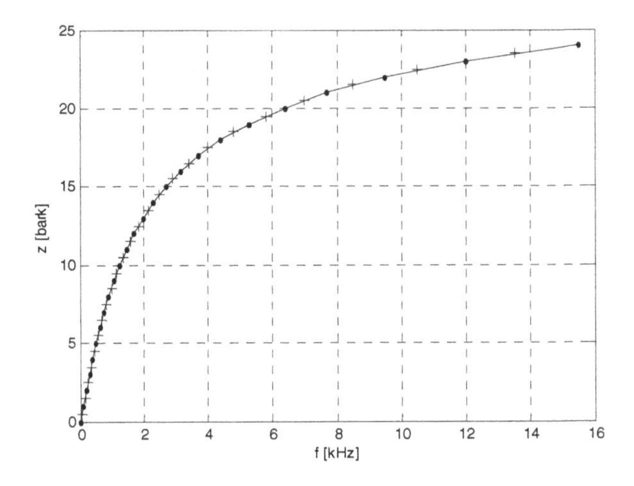

Fig. 11.8. Correspondence between the frequencies Hz and in Bark. Note that the frequency intervals in Hz corresponding to the same Bark interval increase with frequency.

Frequency perception

As mentioned already concerning frequency masking, the ear cannot discern between sounds of slightly different frequency. Correspondingly it is possible to define critical bands in which it is not possible to distinguish between two sounds of different frequency. The partition of the audible frequency range in 24 critical bands has led to the introduction of the bark as measurement unit of the frequencies perceived by the ear (from Barkhausen, the researcher that introduced the phon as the unit of perceived intensity).

Tab. 11.2 shows the correspondence between the frequencies in Hz and in Bark. Note that the frequency intervals in Hz corresponding to the same Bark interval increase with frequency.

In Tab. 11.2 the correspondence between Hz and Bark is shown numerically. The first two columns refer to the limits of each interval. For example, the interval 0÷1 Bark corresponds to 0÷100 Hz. Instead the successive columns refer to the mid-frequency of each band (f_c) and to the frequency interval (Δf_G). In a similar way to that mentioned regarding Fig. 11.8, the interval Δf_G increases with frequency.

Table 11.2. Correspondance between frequencies in Hz and in bark. The first two columns indicate the boudaries of each interval. Columns 3, 4, 5 and 8, 9, 10 indicate the central frequency of each band (f_c) and the frequency interval (f_G).

z [Bark]	f_l, f_u [Hz]	f_c [Hz]	z [Bark]	Δf_G [Hz]	z [Bark]	f_l, f_u [Hz]	f_c [Hz]	z [Bark]	Δf_G [Hz]
0	0				12	1720			
		50	0.5	100			1850	12.5	280
1	100				13	2000			
		150	1.5	100			2150	13.5	320
2	200				14	2320			
		250	2.5	100			2500	14.5	380
3	300				15	2700			
		350	3.5	100			2900	15.5	450
4	400				16	3150			
		450	4.5	110			3400	16.5	550
5	510				17	3700			
		570	5.5	120			4000	17.5	700
6	630				18	4400			
		700	6.5	140			4800	18.5	900
7	770				19	5300			
		840	7.5	150			5800	19.5	1100
8	920				20	6400			
		1000	8.5	160			7000	20.5	1300
9	1080				21	7700			
		1170	9.5	190			8500	21.5	1800
10	1270				22	9500			
		1370	10.5	210			10500	22.5	2500
11	1480				23	12000			
		1600	11.5	240			13500	23.5	3500
12	1720				24	15500			
		1850	12.5	280					

Intensity Perception

The sensitivity of the human ear is not constant over the nominal range 20 and 20.000 Hz; instead the sensitivity is relatively low at the extremes of the range (below 100 Hz and above 10 - 15 kHz) whereas it reaches a maximum in the intermediate frequency range.

The curves of Fig. 11.9 show, for each frequency, the sound intensity (or, more specifically, the sound pressure level - SPL, in this case) corresponding to the same intensity perception. These curves are usually referred to as "isophonic curves", and enable the perceived intensity to be quantified with a measurement unit called phon. At a frequency of 1 kHz, 40 phon corresponds to a sound pressure level of 40 dB.

The isophonic curves show that the maximum sensitivity of the human ear falls in the range 3 and 4 kHz, whereas the sound pressure necessary to give a certain feeling is lower than for higher or lower frequencies.

For high intensity (80÷90 phons), the curves are relatively flat compared to the low intensity curves (10÷20 phons). Considering the curve at 20 phons, for example, at 3 kHz 15 dB are enough to produce the same feeling as 75 dB at 31 Hz (75-15=60 dB difference). Conversely, considering the 90 phons curve, a 90 dB intensity at 3 kHz results in the same feeling as 110 dB at 31 Hz (110-90=30 dB difference).

The availability of a filter with a frequency response equal to the isophonic curves of Fig. 11.9 would allow an output similar to that of the human ear to be obtained and to devise an instrument capable of quantifying the perceived intensity. Taking into account that such a filter may not be straightforward to implement, simpler filters have been devised for this purpose. Fig. 11.10 shows some of these frequency weighting curves representing the frequency response of filters that enable the intensity perceived by the human ear to be quantified. The different curves of Fig. 11.10 take into account that the sensitivity is not the same for different sound intensities. Curve A is relative to intensities in the range 40 and 70 dB, curve B from 70 and 100 dB, curve C from 100 dB and higher.

The *sone* is an alternative measurement unit for perceived intensity which was proposed by S. Smith Stevens so that 1 sone is equivalent to 40 phons. The sone scale is devised in a way that doubling the sone, the perceived intensity also doubles which corresponds to increasing the sound pressure level by a factor of 10 dB (ref. to ISO 226 [2003] and DIN 4563 / ISO 532 [1975] standards, based on the research work of E. Zwicker and H. Fastl).

The suffix G (soneG) indicates that the loudness is computed from frequency groups, suffixes D or F refer to free field measurements (D and F stand for direct or free field), suffix R refer to room field or diffuse field.

The relation between the loudness in sone and phon of Fig. 11.11 shows that up to 30÷40 phon the loudness in sone is almost constant. Above 40 phon the loudness increases rapidly.

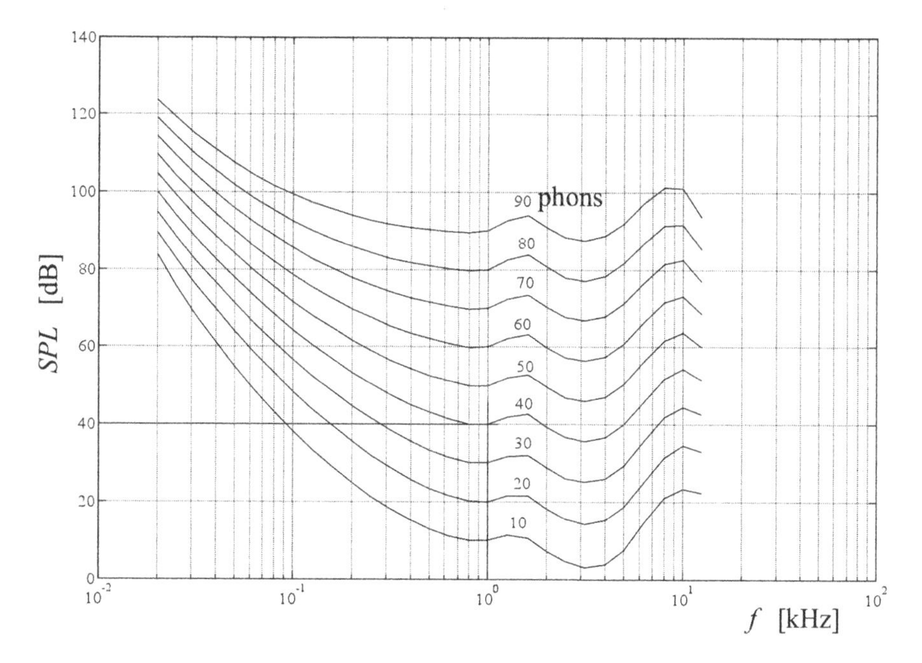

Fig. 11.9. Isophonic curves: each curve indicates the intensity measured as a function of the frequency corresponding to the same perception of intensity heard.

11.2 Sources of Noise and Vibration

The aim of the following section is to provide a short description of the main dynamic excitations that act on the car body:

- road surface,
- rolling of the tyre,
- engine and powertrain,
- brakes,
- aerodynamics.

11.2.1 Road Surface

The road profile is usually rather complex, potentially involves concentrated obstacles such as: bumps, depressions, potholes, tracks, junctions, distributed regular obstacles such as cobblestones, or random surfaces such as asphalt, concrete and dirt road. The main frequency content of these excitations is usually below 200÷300 Hz depending on vehicle speed. As opposed to other sources, for a given speed, road excitations cannot be modified since they depend on

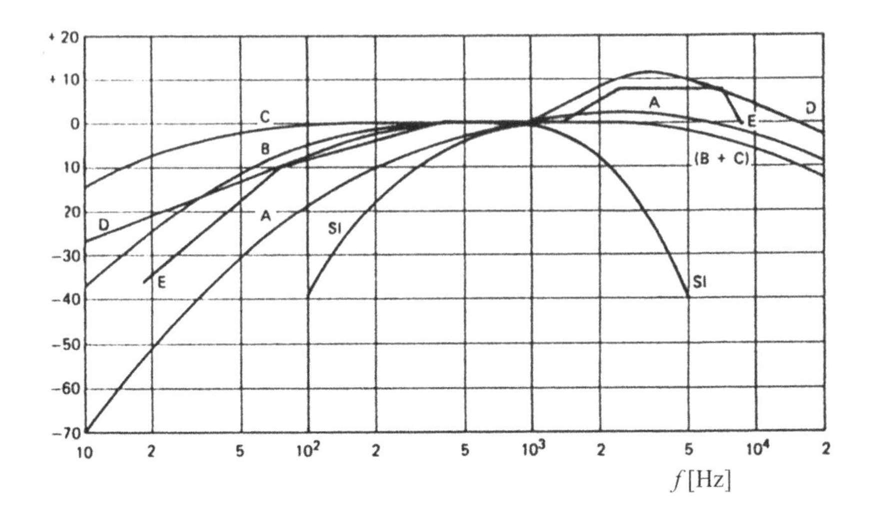

Fig. 11.10. A, B, C - type weighting curves.

the surface characteristics. To reduce the vibrations coming from the road, the main design parameters are related to the vehicle suspensions, one of the main functions of which is to filter such excitations.

For the sake of simplicity, road irregularities can be classified in the following categories:

- random profiles,
- long wave undulations,
- asperities.

Random profiles

Due to its construction, the road surface is never completely flat: Irregularities of different shape, amplitude and distribution are always present which can be characterized by measuring the vertical profile $z(x)$ as function of the distance x traveled along the road (10 km, for example). The same measurement can be repeated for different roads of the same type or on a different part of the same road. These data could be measured by installing a vertical profile sensor on a car and then recording the profile during a large number (n) of travels of the same distance on different roads of the same type. The profile sensor should be able to measure irregularities from some tenths of a mm up to a several cm. As each road profile will be different from others, the profile of each part of road can be considered to be a sample of a random population and analyzed using random signal analysis methods.

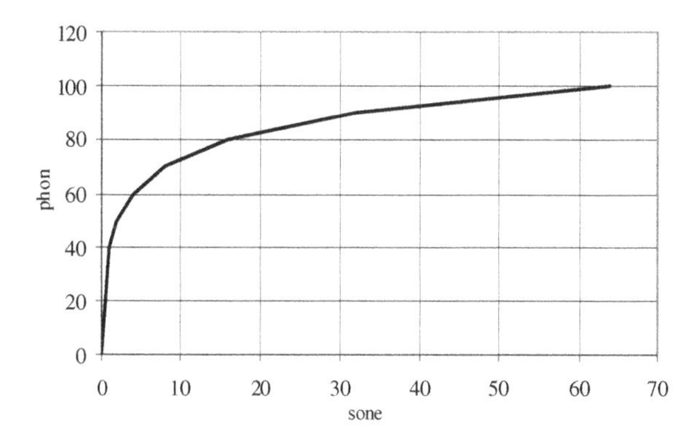

Fig. 11.11. Relationship between sound intensity measured in phon and sone.

The same kind of data collection and analysis can be repeated for different types of roads: highway, urban road, light dirt, heavy dirt. Each type can be characterized separately.

For a given road type, the i-th sample is characterized by the profile $z_i(x)$ $i = 1, \ldots, n$ as function of the traveled distance x (Fig. 11.12 shows four of these samples).

For a given distance traveled $x = x_1$ is possible to compute the main statistical indexes, such as the mean and the mean square value of the $z_i(x_1)$ data obtained from the different samples

$$E[z(x_1)] = \lim_{n \to \infty} \frac{1}{n} \sum_{i=1}^{n} z_i(x_1), \tag{11.11}$$

$$E[z^2(x_1)] = \lim_{n \to \infty} \frac{1}{n} \sum_{i=1}^{n} z_i^2(x_1). \tag{11.12}$$

The mean and the mean square value are the first and second order averages respectively. Similarly to the mean square value, it is possible to define a second order average between the values at distance x_1 and at distance $x_1 + \lambda$, where λ is a constant value. The result is the autocorrelation function:

$$R(x_1, \lambda) = E[z(x_1)z(x_1 + \lambda)] = \lim_{n \to \infty} \frac{1}{n} \sum_{i=1}^{n} z_i(x_1)z_i(x_1 + \lambda) \tag{11.13}$$

Looking at Eq. 11.12, and 11.13 the mean square value can be obtained from the autocorrelation function when $\lambda = 0$

$$R(x_1, \lambda = 0) = E[z^2(x_1)]. \tag{11.14}$$

In a similar way to the mean, the mean square value and the autocorrelation function, it is possible to define higher order averages. In general, all these averages are functions of the coordinate x.

If all averages are not function of the coordinate x, the random process (the profile) is called *stationary*. In this case the explicit dependence on coordinate x can be omitted from the expression of the averages, which therefore can be indicated with reference to just the process variable (z): $E[z]$, $E[z^2]$, $R(\lambda)$.

The averages defined in this way (Eq.s 11.11, 11.12, 11.13) consider, for a given coordinate x, the values corresponding to different profiles belonging to the same population formed by a large number of profiles. Thus they are considered to be *ensemble averages*.

Conversely, it is also possible to take just one profile $z_i(x)$ into account, and analyze it as a function of the coordinate x:

$$\langle z_i \rangle = \lim_{m\to\infty} \frac{1}{m} \sum_{j=1}^{m} z_i(x_j), \tag{11.15}$$

$$\langle z_i^2 \rangle = \lim_{m\to\infty} \frac{1}{m} \sum_{j=1}^{m} z_i^2(x_j). \tag{11.16}$$

Similarly, the autocorrelation function becomes:

$$\phi(\lambda) = \langle z_i(x) z_i(x+\lambda) \rangle = \lim_{m\to\infty} \frac{1}{m} \sum_{j=1}^{m} z_i(x_j) z_i(x_j+\lambda). \tag{11.17}$$

If the coordinate x can be considered to be a continuous variable over a travel distance L, the values can be added with an integral instead of a sum of discrete values:

$$\langle z_i \rangle = \lim_{L\to\infty} \frac{1}{L} \int_{-L/2}^{L/2} z_i(x) dx, \tag{11.18}$$

$$\langle z_i^2 \rangle = \lim_{L\to\infty} \frac{1}{L} \int_{-L/2}^{L/2} z_i^2(x) dx, \tag{11.19}$$

$$\phi(\lambda) = \langle z_i(x) z_i(x+\lambda) \rangle = \lim_{L\to\infty} \frac{1}{L} \int_{-L/2}^{L/2} z_i(x) z_i(x+\lambda) dx. \tag{11.20}$$

A stationary process is said to be *ergodic* if all ensemble averages (those made considering for a given value of x different profiles belonging to the set of different profiles z_i $i = 1, \dots, n$) are the same as the corresponding averages along each profile (z_i):

$$E[z] = \langle z_i \rangle, \tag{11.21}$$

$$E[z^2] = \langle z_i^2 \rangle, \tag{11.22}$$

$$R(\lambda) = \phi(\lambda). \tag{11.23}$$

Returning to the example of the driver that travels the same distance along different roads, if each road is of the same type, in a similar state of maintenance

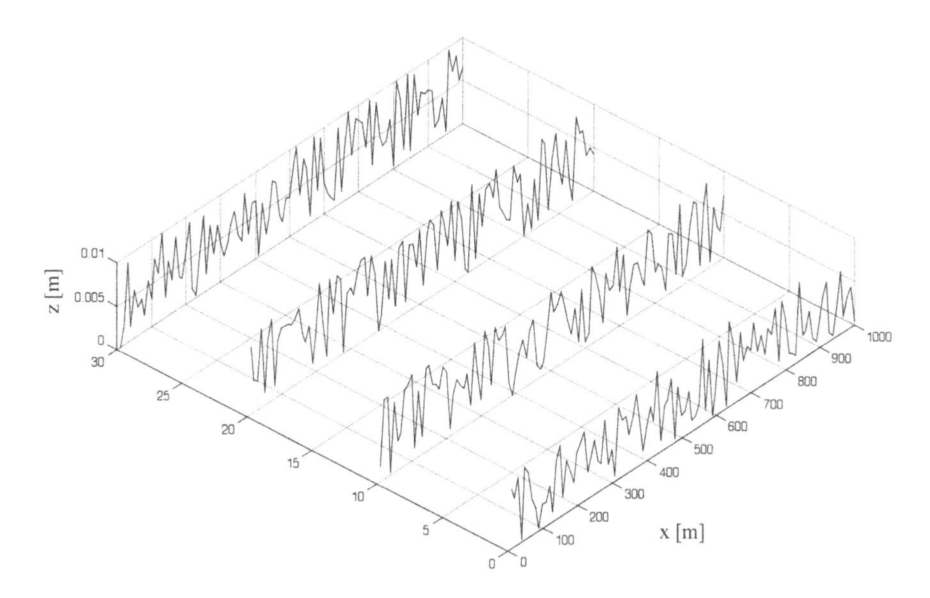

Fig. 11.12. Four different road profiles $z(x)$ as function of the traveled distance x on roads of the same type.

and with a surface of the same type, it is reasonable to assume that the averages along a single path with varying x coordinate are the same as those on different paths (z_i) at a given distance x, i.e. the random profile is ergodic.

The autocorrelation function of an ergodic profile is be obtained by averaging the product of the profile z at two points, one leading the other by a fixed distance λ (in [m])

$$R(\lambda) = \lim_{L \to \infty} \frac{1}{L} \int_{-L/2}^{L/2} z_i(x) z_i(x + \lambda) dx. \tag{11.24}$$

Instead of the distance λ, the autocorrelation function can be expressed as function of the number of cycles per unit distance ν by means of its Fourier transform:

$$R(\lambda) = \int_{-\infty}^{\infty} S(\nu) e^{i\nu\lambda} d\nu. \tag{11.25}$$

Function $S(\nu)$ is the power spectral density of the profile. If the signal z were a function of the time t, the autocorrelation function would be related to the time shift τ (in [s]) between two points sliding along the data record, and the power spectral density to a time domain frequency (in [rad/s] or in [Hz]). From the dimensional point of view, the product $\nu\lambda$ in the complex exponential of Eq. 11.25 must be adimensional or, bearing in mind that $e^{i\nu\lambda} = \cos(\nu\lambda) + i \sin(\nu\lambda)$, should

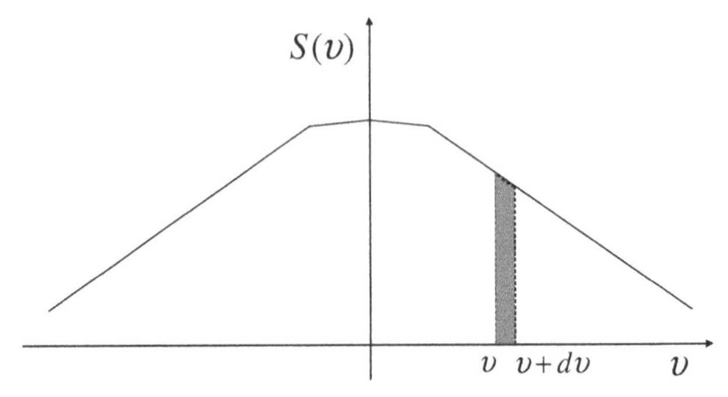

Fig. 11.13. Power spectral density as function of the spatial frequency. The area under the graph is the root mean square of the profile.

be expressed in [rad]. As λ is a distance (in [m]), ν is an angle per unit distance ([rad/m]), representing the inverse of a wavelength ([m/rad] or [m/cycle]).

Apart from the factor 2π, the power spectral density can be obtained as the direct Fourier's transform of the autocorrelation function

$$S(\nu) = \frac{1}{2\pi}\int_{-\infty}^{\infty} R(\lambda)e^{-i\nu\lambda}d\lambda. \tag{11.26}$$

It can be demonstrated that $S(\nu)$ is an even function of the spatial frequency ν, in the interval $[-\infty, \infty]$.

Considering 11.24, for $\lambda = 0$ the autocorrelation function is equal to the mean square value ($R(\lambda = 0) = E[z^2]$). Substituting in Eq. 11.25,

$$E[z^2] = R(0) = \int_{-\infty}^{\infty} S(\nu)d\nu. \tag{11.27}$$

The result is that the mean square value is the area under the diagram of the power spectral density $S(\nu)$ as function of the spatial frequency ν (see, for example, Fig. 11.13). In other words, $S(\nu)d\nu$ is the infinitesimal contribution of the frequency interval $[\nu, \nu + d\nu]$ to the mean square value.

As the total force F acting on a beam due to a distributed load is the integral of the load per unit length $p(x)$, similarly the power spectral density can be thought of as the density of the mean square value $S(\nu)$ per unit frequency ν. From the dimensional point of view, the mean square value $E[z^2]$ is a squared length ([m^2], in SI units), the spatial frequency ν is an angle per unit length ([rad/m]), and the power spectral density $S(\nu)$ is therefore a squared length per unit spatial frequency ([m^2/(rad/m)]=[m^3/rad]).

The definition of the power spectral density of Eq. 11.26 implies that the spatial frequency has positive and negative values. From both the experimental and physical point of view, the spatial frequency can have only positive values. It is then usual to introduce an experimental power density $G_{\text{exp}}(\nu_{\text{exp}})$ [m^2/(cycle/m)]. $G_{\text{exp}}(\nu_{\text{exp}})$ and $S(\nu)$ [m^2/(rad/m)] are related to each other,

$$G_{\text{exp}}(\nu_{\text{exp}}) = 4\pi S(\nu). \tag{11.28}$$

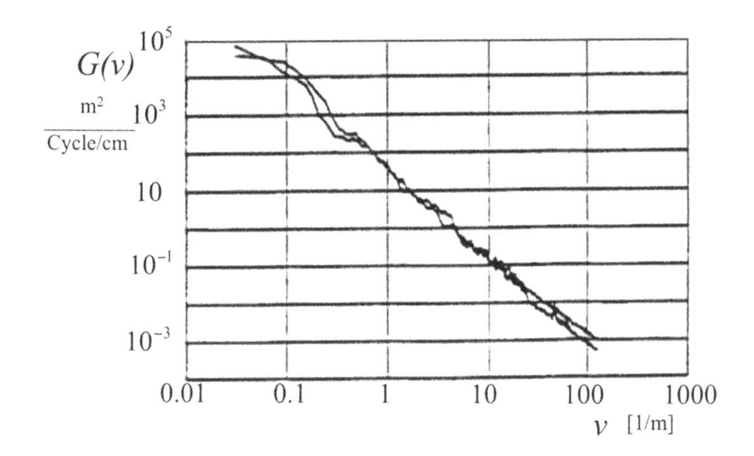

Fig. 11.14. Power spectral density typical of a highway.

The term 4π is in part (2π) due to the conversion of the spatial frequency from [rad/m] to [cycle/m], and in part (the remaining factor of 2) to the definition of $G(\nu_{\rm exp})$ only for positive values of $\nu_{\rm exp}$

$$E[z^2] = R(0) = \int_0^\infty G_{\rm exp}(\nu_{\rm exp}) d\nu_{\rm exp}. \tag{11.29}$$

To simplify expressions, the subscript $_{\rm exp}$ will be dropped in the following to indicate the experimental power spectral density and spatial frequency, instead denoted simply as $G(\nu)$ (with ν in [cycles/m] and $G(\nu)$ in [m^2/(cycle/m)].

Fig. 11.14 shows the power spectral density measured on a highway in double logarithmic scale. The diagram shows that the smaller the spatial frequency, the larger the contribution to the mean square value. Apart from the overall decreasing shape, the diagram indicates that no frequency range prevails over others. This behavior is quite general and the similar shape of the power spectral density also applies for other road surfaces with gravel and dirt, despite significantly different amplitudes in the various cases.

Instead roads paved with cobblestones exhibit different characteristics: In this case a frequency contribution is exhibited corresponding to the average number of cobblestones per unit distance plus its higher and lower order harmonics, dominates over a background similar to that of Fig. 11.14.

The power spectral density of Fig. 11.14 can be approximated by a decreasing linear function in a double logarithmic plane, i.e. a decreasing exponential:

$$G(\nu) = C_0 \nu^{-N}; \tag{11.30}$$

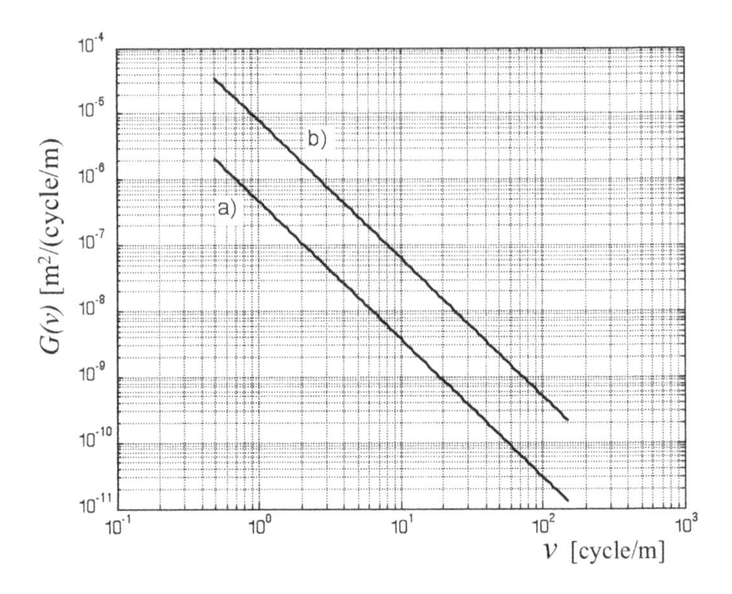

Fig. 11.15. Power spectral density of two different type of road finish. The upper line is related to a gravel road, the lower to a hyghway.

coefficient C_0 is related to the type of road finish, while N is almost the same for all types:

$$C_0 = 4.8\ 10^{-7} \qquad \text{highway,}$$
$$C_0 = 8.1\ 10^{-6} \qquad \text{gravel,}$$
$$N = 2.1.$$

Fig. 11.15 shows the power spectral density as obtained from Eq. 11.30. The upper line is typical of a gravel or dirt road, the lower to a highway.

It is known that three of the low frequency vibration modes of the vehicle correspond to the rigid body motion of the body on its suspensions. The lower frequency mode is characterized by the up and down vertical motion, the second by the pitching of the body around a transversal axis, the third by roll around a longitudinal axis. These modes can be excited by the road profile in different ways depending on the phase of the vertical displacement exciting the front and rear axles.

If the wheelbase (l) includes an even number of half waves of the road profile, the excitation of the front and rear axle is in phase, and there will be a maximum excitation of the up and down vibration mode, while the excitation of the pitching mode will be relatively low. Conversely, if the wheelbase includes an odd number of half waves, the front and rear axles are excited out of phase, and the maximum excitation of the pitching mode and minimum excitation of the up and down mode will occur. This is shown in Fig. 11.16.

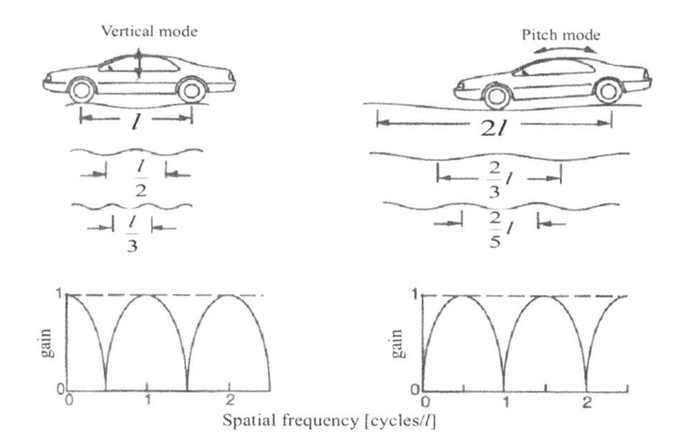

Fig. 11.16. The up and down rigid body mode of the car body is excited when the wheelbase includes an even number of half-wavelengths. The pitching mode, by converse, is excited when the wheelbase includes an odd number of half-wavelengths.

Road unevenness does not only affect the longitudinal direction but also the transversal direction, such that the right wheels move on a profile with similar but not identical characteristics (in terms of power spectral density, for example) as the left wheels. If the left and right wheels are excited out of phase there will be an excitation of the roll mode. The experimental measurements show that the road profile in longitudinal and transversal directions are not unrelated with each other. Irregularities that develop in the longitudinal direction with a wavelength comparable to the wheelbase also develop in the lateral direction and affect all its track. In other words, the longitudinal and transversal wavelengths are similar and therefore long (compared to the wheelbase) wavelengths do not affect the roll motions. Conversely for short wavelengths the right and left profiles are little related with each other and may affect the roll motion.

The road profile has been characterized as function of the longitudinal distance x from an arbitrary reference point leading to a power spectral density expressed as function of the spatial frequency. Dimensionally speaking the spatial frequency is expressed in [cycles/m] or [rad/m], and the power spectral density of the profile in $[\mathrm{m}^2/(\mathrm{cycles/m})]$.

This is because the road profile is a topographic variable and can be expressed directly as a function of geometrical coordinates on the road surface. However this enables the characterization of the road while revealing very little about the way the vertical dynamics of the vehicle would be excited by the road profile since the response on the vehicle is function of the *time* rather than the *spatial* dependence of the road irregularities. The link between these two dependencies is the speed V.

It is then necessary to describe the road profile as function of the time frequency ([Hz=cycles/m]) instead of the spatial frequency ([cycles/m]). Because of this, the power spectral density must be expressed in [m^2/(Hz)] instead of [m^2/(cycles/m)].

If the vehicle travels with speed V, the contact patches of the wheels will move with frequency:

$$f = V\nu. \tag{11.31}$$

Eq. 11.31 can be substituted in Eq. 11.30:

$$G(f) = C_0 \left(\frac{f}{V}\right)^{-N}; \tag{11.32}$$

the above substitution does not modify the power spectral density, that still represents the density of the mean square value as function of the spatial frequency

$$E[z^2] = \int_0^\infty G(f) d\nu, \tag{11.33}$$

taking Eq.11.31 $d\nu = df/V$

$$E[z^2] = \int_0^\infty \frac{G(f)}{V} df = \int_0^\infty G^*(f) df$$

the power spectral density G^* is now consistent with the time frequency,

$$G^*(f) = \frac{G(f)}{V} = C_0 V^{N-1} f^{-N}. \tag{11.34}$$

Because coefficient $N > 2$, the power spectral density G^* increases with the speed V. Fig. 11.17, shows the same power spectral density of Fig. 11.15 as function of the time frequency. The lower curve is relative to a speed of 50 km/h, the upper one to 150 km/h. For a given frequency the excitation increases with increasing the speed.

Long wavelength humps and depressions

A long undulation is a relatively smooth altimetric variation with a longitudinal dimension that is much larger than the contact patch of the tires, caused occasionally by a depression of the road surface. The dynamic forces caused by this type of irregularity imply essentially vertical forces. In a similar way to the random profile, the frequency range involved by this excitation is function of the vehicle speed.

Fig. 11.18 shows the amplitude of the frequency spectrum of an undulation for different traveling speeds. The figure shows that the larger the speed, the larger the range of frequencies involved by the excitation, but the smaller the contributions at low frequency. Intuitively, for larger speeds, the time required to negotiate the obstacle reduces and, correspondingly, the excitation frequency

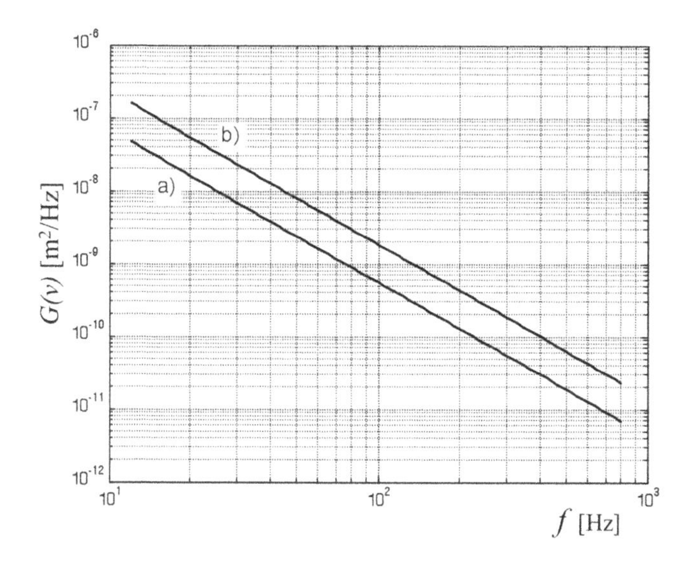

Fig. 11.17. Power spectral density of the road profile as function of the time frequency. a) 20 km/h, b) 180 km/h.

increases. At 120 km/h a 6 m long, 30 mm deep depression involves frequencies up to 10 Hz. As a rule of thumb, if the vehicle moves with speed V and the hump or depression has a length L, the time needed to negotiate the obstacle is L/V. The inverse of this time is approximately half the maximum excitation frequency f_{max}

$$f_{\text{max}} = 2\frac{V}{L}. \tag{11.35}$$

Sharp obstacles

Asperities are sharp variations of the road profile such as potholes, manhole covers, rails,.... with wavelengths which are similar to or smaller than the longitudinal size of the contact patch of each tire.

The forces that the tire produces when negotiating obstacles of this type include significant longitudinal and vertical components. Additionally, if the obstacle is oblique relative to the traveling direction, lateral components arise. Fig. 11.19 shows the variation of the vertical and longitudinal forces acting on the wheel hub as function of the traveled distance when crossing a cleat with rectangular cross section.

The test is made by rolling the tire on a large diameter steel drum at the very low speed of 2 m/s. A steel beam of rectangular cross section is mounted on the rolling surface of the drum parallel to the axis. The distance between the wheel and drum axes is kept constant so as to provide an adequate preload

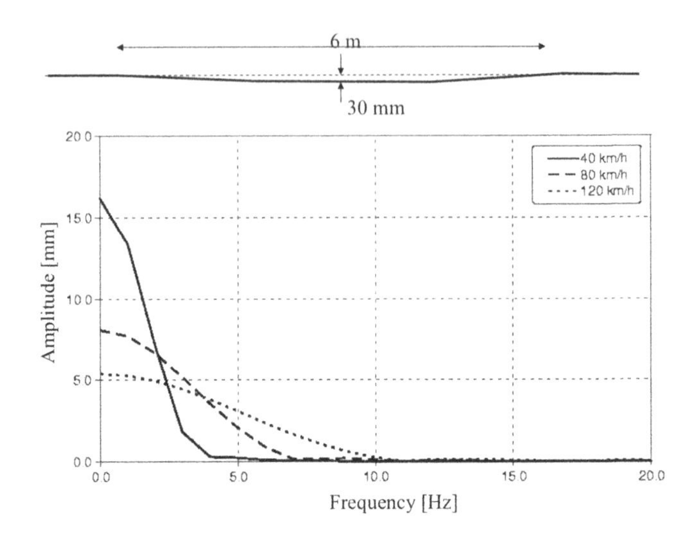

Fig. 11.18. amplitude of the frequency spectrum. Increasing the speed the contributions shift at high frequency.

between the two. The different curves refer to different preloads, showing that the longitudinal components of the force are of the same order of magnitude as the vertical ones. At the beginning of the contact, the longitudinal forces oppose the relative motion, whereas at the end their direction reverses. The figure shows that the longitudinal force changes its direction when the vertical forces reaches the maximum, a characteristic which is typical at low speeds. For higher speeds (10 m/s, for example), the maximum vertical and longitudinal forces develop at the same time, during the first phase of the contact. The resultant is in the upwards–backwards direction, as shown in Fig. 11.20. The angle between the vertical and the resultant force R is approximately 30°. The maximum longitudinal and vertical forces are then related by angle θ

$$\Delta F_{x\,\max} = \Delta F_{z\,\max} \tan\theta, \qquad \theta \approx 30^\circ. \tag{11.36}$$

The simultaneous presence of longitudinal and vertical forces that develop when crossing a sharp obstacle, together with their impulsive nature, requires the presence of shock absorbing also in the longitudinal direction. This is usually obtained by means of elastomeric mounts that connect the suspension arms to the frame and that work in parallel to the vertical motion of the suspension.

In contrast to what happens with long wavelength obstacles such as humps and depressions, the deformations of the tire that occur when crossing an obstacle are quite relevant and can be evaluated by means of the so-called equivalent (or transferred) obstacle, i.e. by the displacement in vertical and horizontal direction

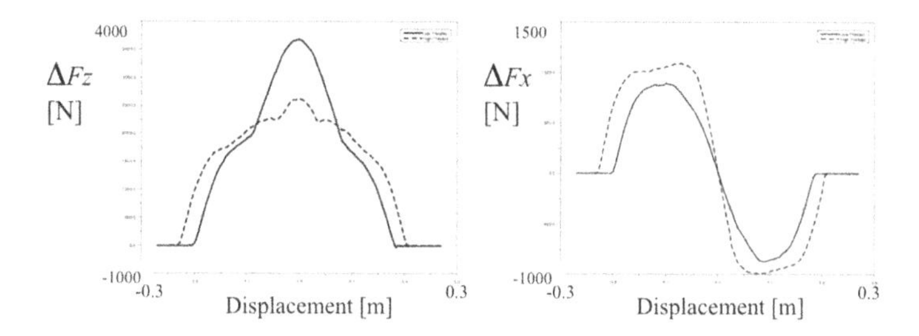

Fig. 11.19. Vertical and longitudinal force variation when crossing an obstacle with rectangular cross section (25 mm height, 100 mm long, 2 m/s traveling speed).

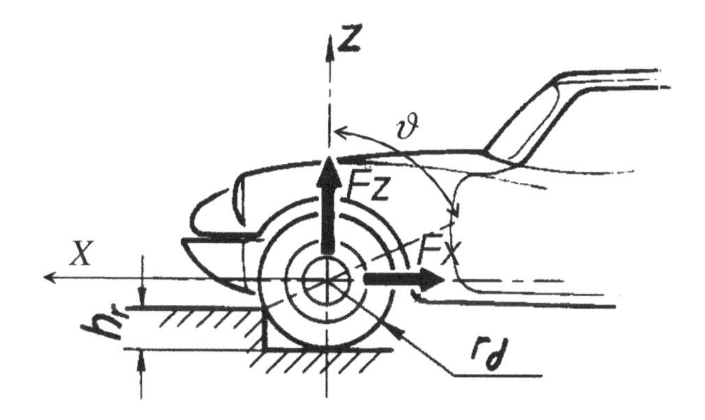

Fig. 11.20. Diagram of the maximum forces acting on the tire when crossing an obstacle at high speed. θ is the angle between the maximum variation of the vertical and the corresponding longitudinal force.

obtained as the ratio between the measured forces developed by the tire and its radial and longitudinal stiffness:

$$\begin{aligned} \delta x &= \frac{\Delta F_x}{K_x} \\ \delta z &= \frac{\Delta F_z}{K_r} \end{aligned} . \tag{11.37}$$

where ΔF_x and ΔF_z are the longitudinal and vertical forces, K_x and K_z are longitudinal and radial stiffnesses.

Fig. 11.21 shows the radial deformation of the tire obtained from the forces measured during the experiment of Fig. 11.19. The solid line represents the tire deformation (δz) obtained from Eq. 11.37, whereas the dashed line is the profile of the obstacle. It is worth noting that this approach enables just the overall deformations of the tire to be taken into account. All the dynamic effects of the tire involved in the transient during and after the impulse are not considered.

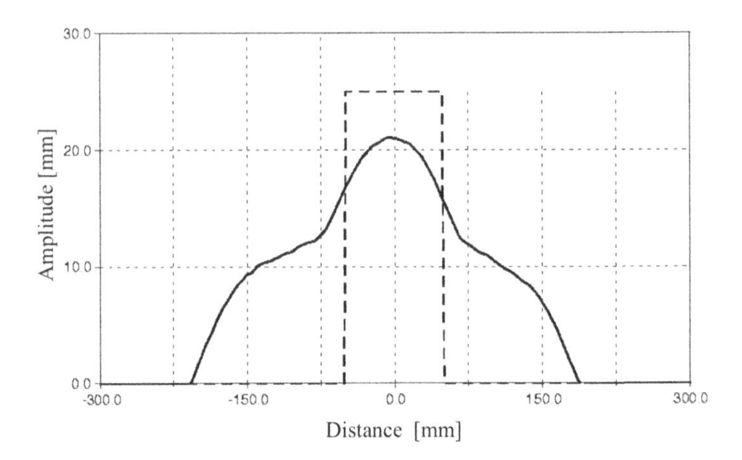

Fig. 11.21. Radial deformation of the tire obtained from the forces measured during the experiment of Fig. 11.19. Tire deformation (solid line) and profile of the obstacle (dashed) (height 25 mm, length 100 mm, 2 m/s speed).

Similarly to long wavelength obstacles, the dynamic excitation due to sharp obstacles is related to the vehicle speed. Fig. 11.22 shows the frequency spectrum of the equivalent obstacle of Eq. 11.37 at different speeds. As the size of the cleat (25 mm x 100 mm) is smaller than the depression of Fig. 11.18 (30 mm x 6 m), the time involved in crossing the cleat is shorter, at same speed, the excitation frequencies involved in crossing the cleat are larger.

11.2.2 Wheel

The wheel subsystem (tire, rim, disc brake rotor, bearings) gives rise to various types of dynamic excitations due to rotation. The frequency of such excitations is related to the rotating speed Ω.

Mass unbalance

Regardless of the nominal symmetry of the wheel, the center of gravity of the rotating part is never on the axis of rotation. The distance between the centre of gravity and axis of rotation is called eccentricity (ϵ).

The inertia forces due to rotation of the mass m of the wheel with eccentricity ϵ rotate with the same speed of the wheel on a plane perpendicular to its the rotation axis. The components of the unbalance force in an inertial frame x, z are:

$$F_x = m\epsilon\Omega^2 \cos \Omega t, \qquad F_z = m\epsilon\Omega^2 \sin \Omega t; \tag{11.38}$$

the product $m\epsilon$ is the *static unbalance*.

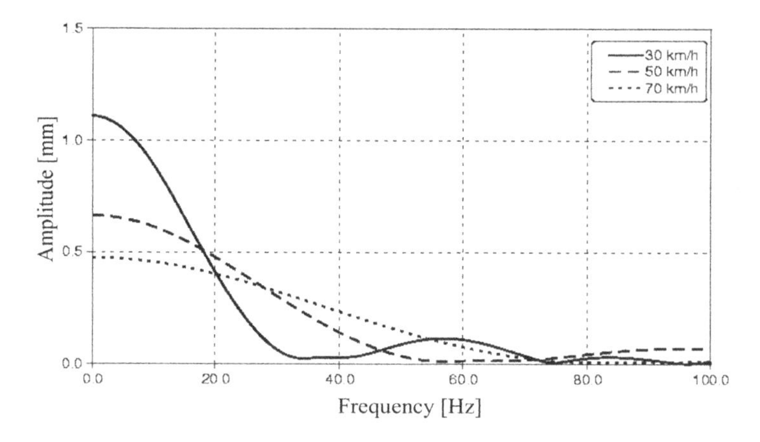

Fig. 11.22. Frequency spectrum of the equivalent obstacle.

In a similar way to the centre of gravity, the main axis of inertia of the rotating part is never aligned with the axis of rotation: The angle (χ) between these two axes is the dynamic unbalance, and the result is a torque about the x and z axes:

$$M_x = -\chi(J_t - J_p)\Omega^2 \sin \Omega t, \qquad M_z = \chi(J_t - J_p)\Omega^2 \cos \Omega t, \tag{11.39}$$

where J_p is the polar moment of inertia about the wheel axis and and J_t is the moment of inertia about an axis that lies on a plane perpendicular to the wheel axis and passing through the center of gravity of the wheel.

The product $\chi(J_t - J_p)$ is the *dynamic unbalance*. Also in this case the excitation frequency is equal to the angular speed of rotation of the wheel.

The static and dynamic unbalances can induce vibrations that can involve the steering wheel and, hence, can be perceived by the driver directly. If the excitation frequency is the same as one of the natural frequencies of the suspension or the steering system, the amplitude of vibration can be amplified considerably by resonances, negatively affecting the level of comfort.

The aim of the balancing operation is to reduce the dynamic forces acting on the wheel. Even if this does not modify the natural frequencies of the wheel and suspension, the effect is to reduce the amplitude of the vibrations, especially those of the steering wheel. Additional advantages are also a reduction of the tire tread wear, a reduction of the stresses on the bearings, suspension, and steering system, with an improvement of the fatigue life and reliability.

As on any other rotor, the aim of the balancing operation performed on the wheels is to reduce the static and dynamic unbalance. This is done by adding (or eliminating) small balancing masses (m_e) to a radius R_e (balancing radius), usually close to the outer diameter of the rim. If the unbalance added by the balancing masses is of the same in magnitude and of opposite phase than the original one,

$$m_e R_e = m\epsilon, \tag{11.40}$$

the static unbalance is nulled. In other words: the effect of the balancing masses is to reduce (or eliminate) the distance between the center of gravity and the rotation axis. In practice the balancing operation will never be perfect, a certain amount of residual unbalance will always remain, depending on the accuracy of the instrumentation and of the procedure.

To reduce the static unbalance of automotive wheels, it is possible to either add just one mass on the mean plane of the wheel, or add two masses (of half the size) at the same angular location but on opposite sides of the rim (on the rim flanges, for example) to compensate the static unbalance ($m\epsilon$) without affecting the principal inertia axis of the wheel (angle χ), while avoiding introducing a dynamic unbalance.

Conversely, to reduce the dynamic unbalance, it is necessary to add two balancing masses still on opposite sides of the rim, but on opposite diametral locations (180 deg) to reduce the dynamic unbalance (χ) without affecting the static unbalance.

Excitations due to tire shape and stiffness irregularities

Because of the unavoidable tolerances in the shape and size of the tire and rim, the surface of the wheel will never be completely axisymmetric. The unavoidable errors with respect to an ideal toroidal surface can be expressed as a series of harmonics of the angular coordinate about the wheel axis. Fig. 11.23 shows the first four contributions of such series. The first harmonic induces an eccentricity a), the second an oval shape b), the higher order harmonics induce lobed shapes such as c) and d).

Fig. 11.23 e) on the other hand, shows the basic mechanism that leads from the geometrical errors to the vertical and longitudinal forces. For simplicity, the figure shows just the first harmonic (eccentricity). As the radial stiffness of the tire is much higher that of the suspension, the rolling radius is not constant even when traveling on a flat surface. The wheel axis is then subject to vertical displacements that induce vertical forces on the suspension. Starting from relatively low vehicle speeds, the angular speed of the wheel is higher than the first natural frequency of the corner (Fig. 11.24) but lower than that of the unsprung mass.

$$\sqrt{\frac{K_s}{M_s}} < \Omega < \sqrt{\frac{K_p}{M_{ns}}}; \tag{11.41}$$

This implies that the displacements of the sprung mass induced by the irregularities of the wheel are negligible compared to those of the unsprung mass. The vertical displacements of the wheel induce a movement of the suspension and, therefore, transmits forces to the sprung mass.

In the frequency range between the natural frequencies of the sprung and the unsprung masses (Eq. 11.41), the amplitude of the radial excitation coming from

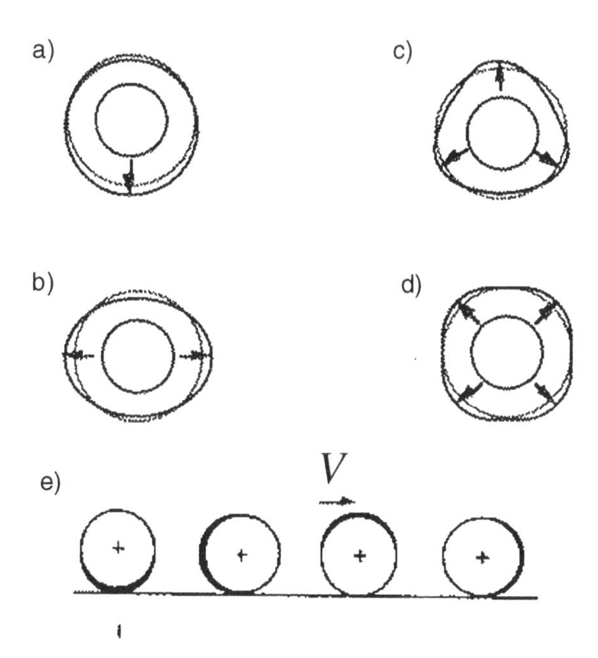

Fig. 11.23. Deviations from cylindrical shape of a tire. First harmonic induces an eccentricity (a), second harmonic to oval shape (b), third and following to lobed shapes (c,d). Explanation of the longitudinal and vertical forces due to shape errors (e).

the geometrical imperfections is almost independent of the angular speed of the wheel. In this context, the amplitude of the excitation decreases with increasing order of the harmonic.

The internal resonances of the tire can greatly amplify the effect of the excitation. The frequency analysis of Fig. 11.25 shows the harmonic contributions to the vertical force as function of the speed. Where the n-th harmonic of the angular speed (f_n=$n/2$) coincides with one of the natural frequencies of the tire (f_p), a resonant peak occurs on that harmonic

$$n\frac{\Omega}{2\pi} = n\frac{V}{2\pi R_0} = f_p. \tag{11.42}$$

Considering, for the tire a first resonance of f_p =70 Hz (since it typically occurs in the range 70 $\div$ 90 Hz), and a rolling radius $R_0 = 0.3$ m, the speed of the vehicle corresponding to the resonance of the n-th harmonic is

$$V_{nth} = \frac{2\pi R_0}{n} f_p$$

the resonance corresponding to the 3rd harmonic is at 160 km/h, that corresponding to the 4th is at 120 km/h, and the 5th harmonic has a resonance at 95 km/h, as shown in Fig. 11.25.

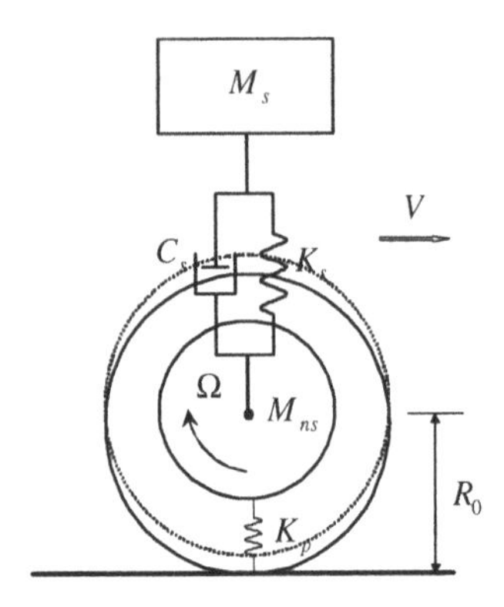

Fig. 11.24. Quarter car model - effect of the geometrical irregularities (eccentricity). M_s sprung mass; M_{ns} unsprung mass, K_s, C_s equivalent stiffness and damping; K_p radial stiffness of the tire.

In addition to the vertical forces, the geometrical irregularities induce also longitudinal forces which can be explained by taking into account the relation between the vehicle speed V and angular speed of the wheels Ω, through the rolling radius R_0

$$V = \Omega R_0.$$

As the moment of inertia of the wheels is negligible compared to that of the vehicle, the speed of the vehicle is constant over the time interval involved in the tire irregularities. The small variation of the rolling radius R_0 due to the geometrical irregularities induces an angular acceleration with a torque of:

$$T = I_p \dot{\Omega}; \tag{11.43}$$

where I_p is the polar moment of inertia of the wheel and all parts that rotate with it as a rigid body (hub, disc brake rotor,...). This torque corresponds to a variation of the longitudinal force ΔF_x. Considering that the ground applies the force ΔF_x on the contact patch, at a distance R_c (loaded radius) from the wheel axis:

$$\Delta F_x = \frac{T}{R_c}. \tag{11.44}$$

Because of the production tolerances, also the radial stiffness is not constant but varies instead with the angular position of the wheel. The variation of the radial stiffness induces a variation of the rolling and of the loaded radii. The outcomes are similar to those induced by the geometrical imperfections, i.e. vertical and longitudinal forces with frequency multiple of the wheel angular speed.

Fig. 11.26 shows the spectrum of the vertical acceleration of the driver's seat mount at a speed of 120 km/h on an highway. Indexes 1,...,4 indicate the peaks corresponding to the first four harmonics of the angular speed of the wheel. Their effect is dominant for frequencies above 20 Hz. At lower frequencies, instead, the

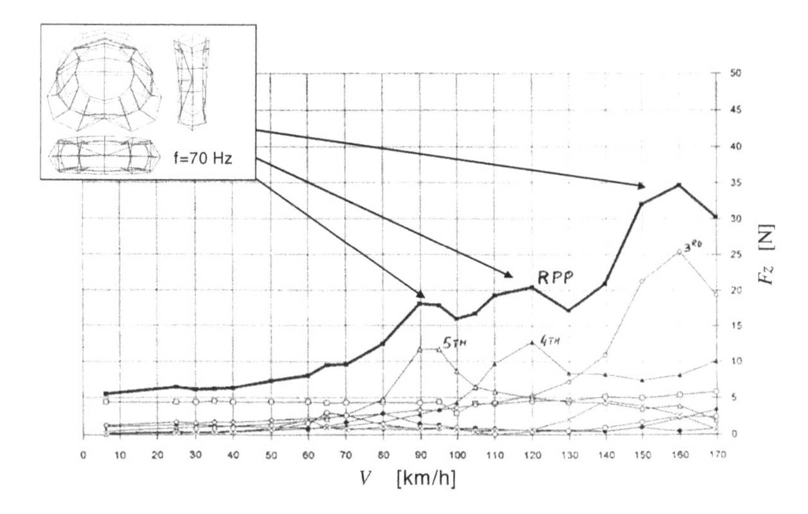

Fig. 11.25. Harmonic components of the tire vertical force F_z. The peaks correspond to the excitation of the first resonant frequency (90 Hz) by different harmonics.

larger contribution is due to the engine suspension vibrations (usually at about 15 Hz) and to the suspensions (1÷2 Hz). When the frequency of one of the harmonic excitations due to the wheels is the same as the natural frequency of the vehicle body or one of its subsystems (for example the steering line, one of the vehicle body modes, the local modes of the panels, the acoustic cavity modes,...) the response is amplified and induces a resonance.

Tire tread

The harmonic excitation due to the geometrical and stiffness irregularities of the tire is related to its structure and production tolerances, it is also present in the case of sleek tires. Another relevant source of dynamic excitations is the tread-road interaction, that involves mechanical and fluid dynamic effects. The road roughness, and the impact between the tread elements and the road surface, induce vibrations in the tire carcass and in the tread. Other vibrations are caused by the friction forces exchanged through the contact patch. Because of the tire mechanics, these forces increase from the leading edge of the contact patch to the trailing one. Qualitatively speaking, each part of the tread entering the contact is not deformed in the tangential direction. Because of the different speed between the ground and the carcass, the tangential deformation increases and reaches its maximum close to the trailing edge of the contact patch.

This effect is amplified at high temperature. In this case the tread pattern elements stick to the road surface (stick snap) due to the adhesion forces, and the accumulated tangential deformation is released suddenly when the contact ends causing tangential vibrations.

As in the previous cases, the excitation frequency related to the contact of the tread pattern is related to the angular speed of the wheel. At 120 km/h

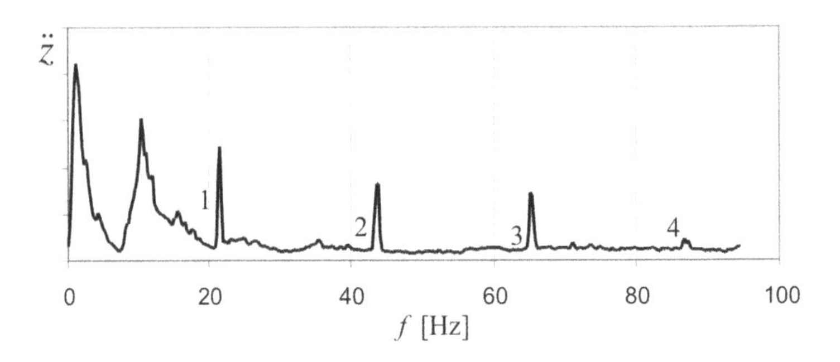

Fig. 11.26. Frequency spectrum on a seat rail at 120 km/h. In evidence the harmonics of the wheels angular speed.

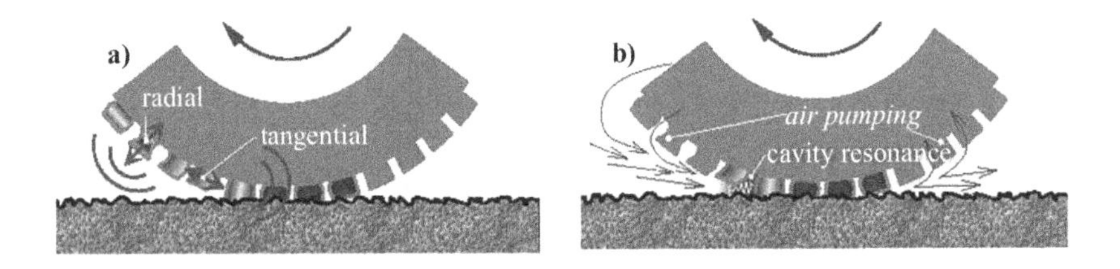

Fig. 11.27. Dynamic excitations due to the tread. a) radial and tangential resonance of the tread elements; b) cavity resonance of the air trapped in the tread pattern.

the fundamental excitation of a tread pattern with 60 elements of the same size is equal to about 1.2 kHz. The adoption of a tread pattern with a size that varies randomly along the tire circumference allows to reduce a pure tone excitation frequency when traveling. The effect of the random variation of the tread pattern is to make the impact of following elements of the pattern not periodic. The periodicity is thus that of the wheel rotation, with reduced high frequency vibrations that fall in the acoustic range.

The vibrations induced by the tread pattern are transmitted to the tire that vibrates with its own dynamics and resonances, related to the carcass and air cavity within. These vibrations are transmitted to the rest of the vehicle through the structure or to the air surrounding the wheel.

Fig. 11.28 shows the result of a noise intensity measured at the driver's left ear position. The peaks are due to the resonance of the tire structure (A), and to the internal air cavity (B). The frequency of these peaks is constant; the third peak (C) is proportional to the vehicle speed since it is caused by the excitation due to the tread pattern.

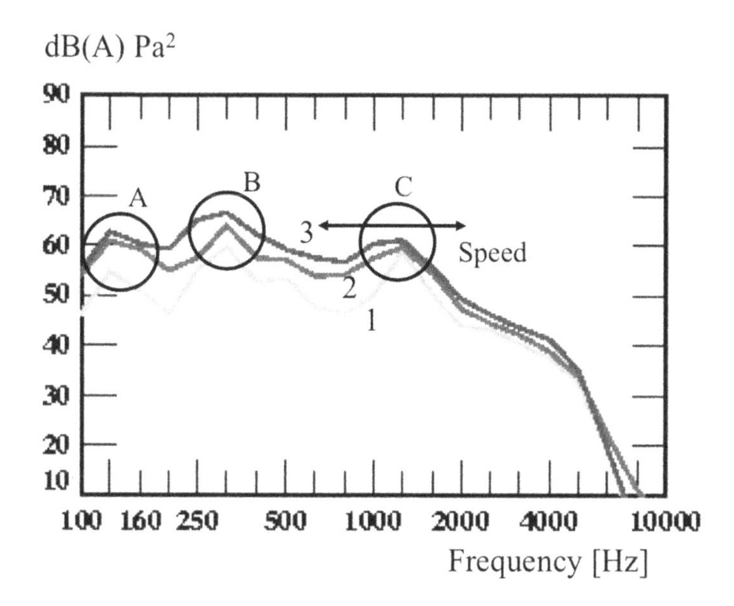

Fig. 11.28. Noise intensity at the left driver's ear. The measurement is performed on a drum test bench at 120 km/h. A) structural resonance of the tire, B) internal air cavity resonance; C) tread contact excitation. 1) left and right smooth drum surface; 3) left and right rough drum surface; 2) left drum with rough surface, right drum with rough surface.

Aerodynamic phenomena caused by the wheel close to the contact patch constitute another relevant source of noise. The so-called air pumping relates to the reduction of the volume produced by the tire deformations at the leading edge of the contact patch followed by an expansion at the trailing edge. The variations of the volume produce an air flow in the channels between the tread pattern grooves and the ground. The complexity of the grooves induce vibrations in the air with an amplification of the vibrations through an organ pipe mechanism.

11.2.3 Engine

The internal combustion engine produces vibrations due to the movement of its mechanical parts and to the thermodynamic cycle. The crankshaft and the alternating masses of the pistons and connecting rods generate centrifugal and alternate inertia forces that are transmitted to the engine block.

The inertia forces can be computed to a first approximation by considering the masses of the piston (m_p), and the crankshaft (m_a), while the contribution of the connecting rod (m_b) can be split in two parts, added to the piston ($m_b/3$) and to the crankshaft ($2/3\ m_b$).

The alternating and rotating masses are then:

$$m_o = \frac{m_b}{3} + m_p, \\ m_r = \frac{2}{3} m_b + m_a. \tag{11.45}$$

Under the assumption that the angular speed ω of the crankshaft is constant, the axial force on the piston (F_o) and the radial force (F_r) on the crank can be computed as

$$F_o = m_o r\omega^2 \left(\cos\theta + \lambda \cos 2\theta + \dots\right), \\ F_r = m_r r\omega^2. \tag{11.46}$$

where $\lambda = r/l$ is the ratio between the radius of the crank pin (r) and the length (l) of the connecting rod. Taking the crank angle θ into account (Fig. 11.29), the resulting inertia forces on the axis of the crankshaft are:

$$F_y = m_r r\omega^2 \sin\theta, \\ F_z = -r\omega^2 \left[m_r \cos\theta + m_0 \left(\cos\theta + A\cos 2\theta + \dots\right)\right], \tag{11.47}$$

where $A = \lambda + \lambda^3/3 + 15/128\lambda^5 + \dots$ The forces acting on the axis of the crankshaft include 1st, 2nd, 4th, and higher orders of the angular frequency ω of the crankshaft. Usually the first two of harmonics are larger than the following ones.

In multi cylinder engines, the inertia forces of one cylinder can be compensated by the other cylinders by appropriate phasing and geometrical layout of the engine. With four in-line cylinders , for example, the first order forces can be compensated whereas the second orders add together. Countershafts (two in a four cylinders) with unbalance masses that rotate at twice the speed of the crankshaft can be added within the engine block to compensate this effect.

Even though the inertia forces F_y of Eq. 11.47 in multi cylinder engines can balance each other, their distribution along the crankshaft axis can generate torques of the first and second order that act in directions perpendicular to the crankshaft axis. These torques are function of the number of cylinders and the architecture of the engine (with six in-line cylinders, for example, the torques tend to balance each other).

The second source of vibrations of the engine is the pressure variation involved in the thermodynamic cycle and, especially, during the combustion. Fig. 11.30 shows the cylinder pressure during a cycle and the torque produced on the crankshaft.

The torque can be expanded in its harmonic contributions; as in a four stroke engine the thermodynamic cycle repeats every two rotations of the crankshaft, the fundamental harmonic is the order 0.5. Higher order terms are also involved due to the sharp changes of pressure during the combustion. This means that the higher the number of cylinders, the more the torque produced by each cylinder can be distributed along the rotation of the crankshaft to reduce torque fluctuations. The resulting variations of the engine speed can be attenuated by increasing the moment of inertia of the flywheel and/or installing a torsional spring and

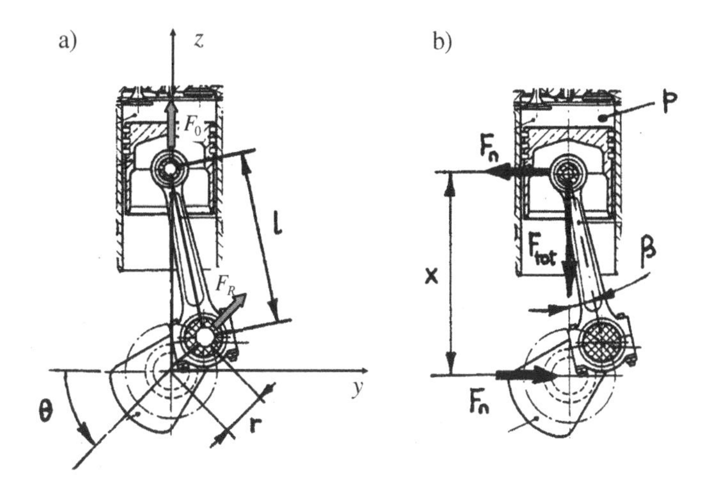

Fig. 11.29. Crank and piston geometry and forces. a) inertia forces, b) pressure forces.

damper between the flywheel and the input of the transmission. Other means such as double mass flywheels can be adopted for the same purpose, i.e. to reduce the dynamic excitations transmitted to the rest of the power train.

In a two cylinder engine one combustion occurs per revolution of the crankshaft; in four cylinder engines two combustions per revolution; in eight cylinders, four combustions per revolution, and so on. This means that for a four cylinder engine with rotational speeds between 900 and 6000 rpm, the fundamental excitation frequency is in the range 30 and 200 Hz.

The 0.5 order is not the lowest one, the unevenness between following combustions imply orders lower than the (0.25, for example).

As the inertia forces are proportional to the acceleration and, therefore, to the square of the engine speed, the excitation due to the pressure forces prevails at low speeds whilst the inertia forces dominate at high speed.

Minor effects due to other moving elements such as the camshaft and the valve-train add to the those produced by the crankshaft and pistons. Fig. 11.31 illustrates the frequency analysis of the acceleration measured on the block of an eight cylinders engine for increasing speeds from idle (800 rpm) to 2100 rpm. For each different engine speed (rpm), the diagram represents the amplitude of the frequency components (Hz). This diagram is usually called "waterfall" or "cascade" diagram. Due to the three dimensional nature of the diagram, the amplitude is reported with a color coding (grey scale in this case). As expected for eight cylinders, the fourth order is the fundamental component.

Engine vibrations produce dynamic strains and forces in the elements (engine suspension) that connect it to the vehicle structure. Vibrations in the vehicle body are excited that are amplified by the resonances of the various structural parts. The two vertical bands at about 300 to 400 Hz in Fig. 11.31, are due to

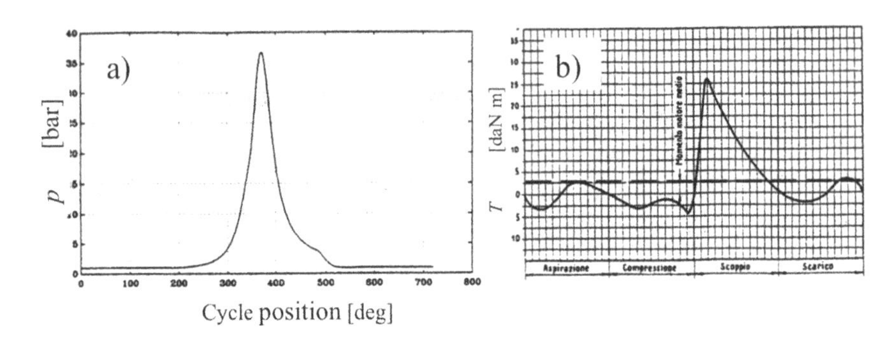

Fig. 11.30. a) Pressure as function of the crankshaft angle; b) torque on the crankshaft.

the resonance of one of these modes (in this case the engine block) excited by the various engine orders.

The vibrations of the engine block surfaces also induce pressure waves in the engine compartment that propagate as air borne noise, transmitted to the interior of the car through the vibration of the structural panels (the firewall, for example).

Fig. 11.32 represents the contribution of the different parts of an engine to the total emitted acoustic power.

In addition to the combustion forces, the intake and exhaust flows constitute another source of dynamic excitation. The pressure variations in the turbulent gas flow, that induce vibrations in the intake and exhaust pipes that radiate then noise in the engine compartment. Some of the vibrations are also transmitted to the engine block that propagates it and radiates other acoustic noise.

Fig. 11.33 shows the frequency analysis close to the output of the exhaust pipe during a ramp up. The part related to the engine orders is due to the combustion and to the periodic motion of the gas flow through the exhaust valves. A considerable part of the spectrum is constituted by wide band and uniform excitation due to the turbulence and other aeroacoustic phenomena.

Apart from the noise radiated by its surface, the low frequency vibrations of the exhaust pipe are also transmitted directly to the vehicle body by structure borne paths through the suspension elements used to connect it to the vehicle underbody.

The fuel injection system, alternator, fan, starter, compressor of the air conditioning system, and the power steering system are some of the various auxiliaries powered by the engine. The different mechanisms involved in their operation (mechanical, fluid dynamics, electromechanical) correspond to other sources of noise and vibration, that are transmitted by air borne paths or by the structure. The pumps also transmit noise through pressure waves in the pipes. Usually the accessories are powered by the belt, chain or other type of drives with a transmission ratio typical of each auxiliary which also influences the presence

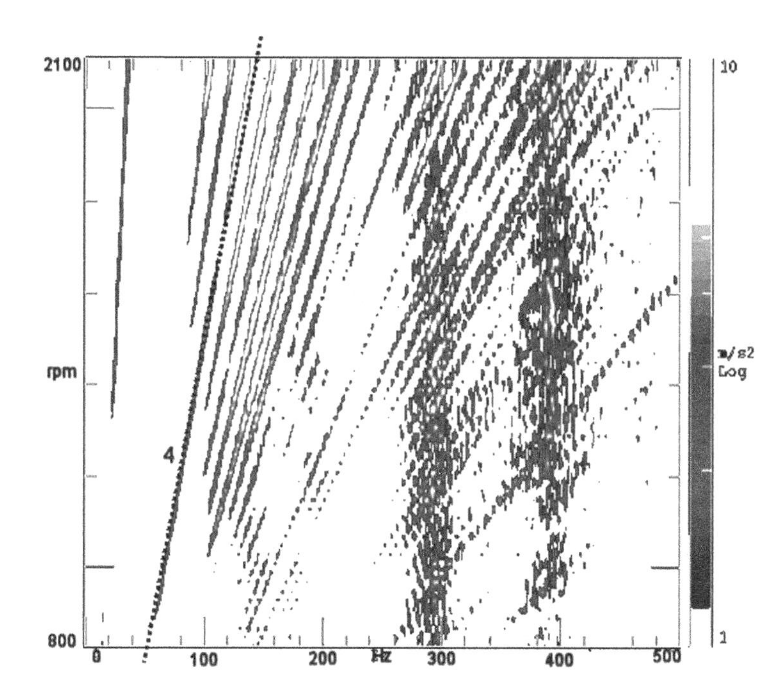

Fig. 11.31. Frequency spectrum of the acceleration on the engine block. 8 cylinders engine. For each engine speed (rpm), the amplitude of the harmonics is reported with a color coding (grey-scale) as function of the frequency. The dotted line evidences the fourth harmonic.

of principal harmonic components of the noise and vibration as function of the engine speed.

Chain and timing belt drives are the two main solutions used to drive the camshafts, both of which produce noise during operation.

Roller chains produce a noise related to the impact of each roller against the toothed sprocket and the slipping against other parts such as tensioners and guiding elements.

Timing belts produce noise because of the contact of each tooth against the pulley. The high internal damping of the rubber that constitutes the belt reduces all harmonic components except the fundamental to negligible levels. A possible source of noise in this case is the pumping of the air to and from the volume that could be trapped between pulley and belt.

At high temperatures, as the surface of the belt becomes stickier, the belt drive may emit wide band noise due to the stick and slip between the belt and the pulley. This is especially relevant for non synchronous belts used in auxiliaries transmissions. Apart from the contact against the pulleys, belt transmissions can produce noise because of the lateral vibrations of the segments traveling from one pulley to the next. These vibrations can be excited at resonance by the engine orders and the geometrical defects of the pulleys. Apart from the noise

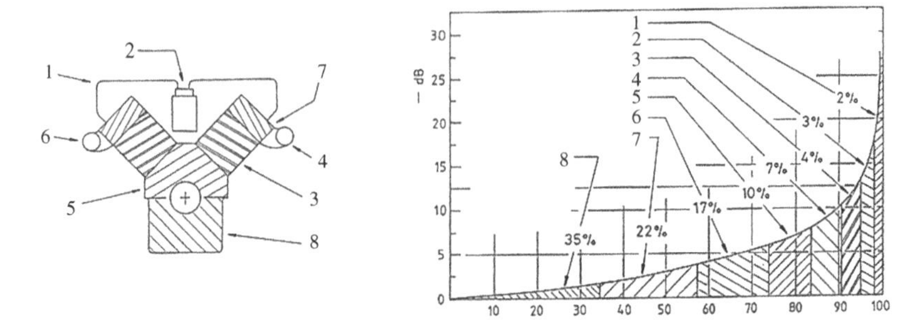

Fig. 11.32. Cumulative contributions of the total acoustic power emitted by the engine. 1) high pressure pipes, 2) fuel pump, 3) engine head, 4) exhaust, 5) engine block, 6) intake, 7) head covers, 8) oil sump.

and vibration issue, such vibrations can induce fatigue in the drive that affects its reliability.

A completely different excitation is produced by the power train because of sudden variations of the throttle command. This happens when the driver pushes on or releases very fast the accelerator. Similarly, the sudden engagement/disengagement of the clutch produce impulsive torque excitations to the power train. The result is a torsional vibration of the transmission. The torsional vibrations are transformed by the wheels in longitudinal vibrations of the whole vehicle that could involve frequencies up to about 10 Hz (Fig. 11.34).

11.2.4 Transmission

The gearbox is a source of noise and vibration due to the meshing between the gears. The most important sources of gear noise are rattle and whine.

In most of the gearboxes for automotive applications all gear sets are constantly meshing together; at any specific moment, while the engaged set is transmitting power, the others run idle. Because of the torque fluctuations, the shafts are subject to torsional vibrations, this induces idle gears to bounce against each other in the circumferential gap in between. Even though small, this gap is necessary for the functionality of the gear set to allow lubrication of the teeth during contact and avoid them seizing.

The rattle can be amplified by the torsional resonances of the transmission (usually below 100 Hz). The acceleration measured on the gearbox during rattling is shown in Fig. 11.35. The impulses may be not always repeat periodically, and can change with time and depending on the nature of the excitation coming from the other elements of the transmission.

The gear whine, is the noise produced at the harmonics of the tooth passing frequency. It is generated by the transmission error at the loaded gear meshes

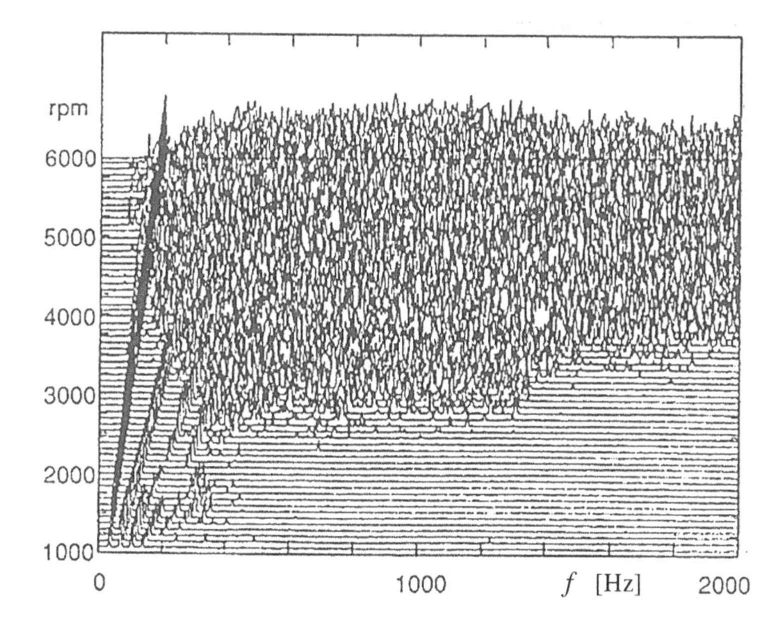

Fig. 11.33. Waterfall diagram of the acoustic noise at the output port of the exhaust pipe. The engine speeds is between idle and 6000 rpm.

and can become unacceptable if the excitation is magnified by internal gearbox resonances.

Although the involute profile used in cutting the teeth enables, in principle, a constant transmission ratio to be obtained, in practice the transmission obtained during a mesh is never constant. Manufacturing tolerances lead to errors in the ideal profiles. Elastic compliance is another source of errors: the deformation modifies the teeth profile when transmitting a torque, this induces a transmission error even in ideally cut gears. The effect of the compliance could be the impact between the top of a teeth and the root of the meshing one. To avoid impacts highly loaded gears are cut with profile corrections that reduce the thickness close to the tip. Although effective this correction is not able to eliminate the problem, especially in gears that can transmit torques over a wide range, as in the automotive case.

Fig. 11.36 shows the transmission error of 28 teeth gear that runs at 500 RPM. The transmission error is mainly at the mesh frequency and its higher order harmonics.

Even in the ideal case of a constant input speed, the transmission errors lead to a variable output speed. The resulting angular accelerations produce a fluctuation of the transmitted torque. On one side, the effect is a torsional vibration. On the other side, the torque ripple leads to a variation of the force acting on the bearings. These dynamic forces induce vibrations that propagate in the gearbox

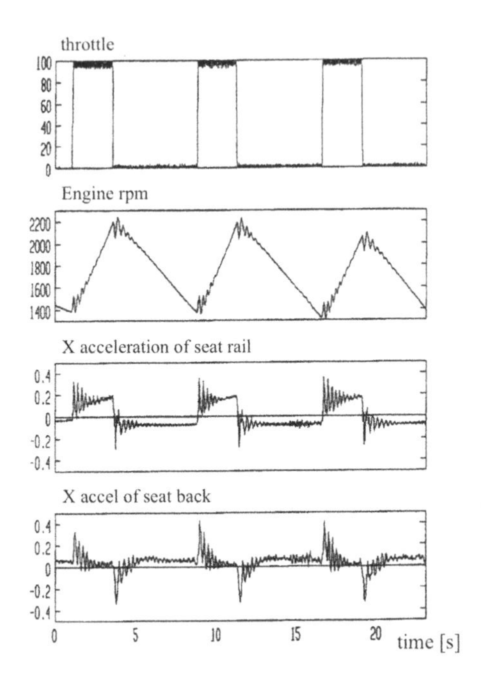

Fig. 11.34. Transients due to fast throttle up and down.

housing and, finally, lead to acoustic noise that propagates within the engine compartment.

Other sources of excitation related to the engaged gear (in contrast to the idle ones) are the friction forces between the teeth, the air and oil flow around them.

Both for rattle and whine, the dynamic excitations are transmitted at first by the shafts, then by the bearings and gearbox, before being finally transmitted to the air. The entire airborne transmission path is less important as the gearbox walls are excited very little by the acoustic pressure inside it.

The rotation of the other elements of the transmission represent another source of vibrations. In particular the prop shaft connects the transmission to the rear differential (in some cases the rear differential is integrated with the gearbox while the engine and clutch are at the front). When the differential is integrated in a rigid axle suspension, the two ends of the shaft are subject to relative displacements of the same amount as the suspension (typically more that 150 mm for cars). In the case of independent suspensions, the differential is connected to the frame and transmits the power to the wheels through semi-shafts. Also in this case the prop shaft is rather long and must tolerate smaller but still significant displacements between the two ends.

Flexible joints and supports are then installed along the prop shaft to separate it in spans of smaller length and tolerate the relative displacements. The

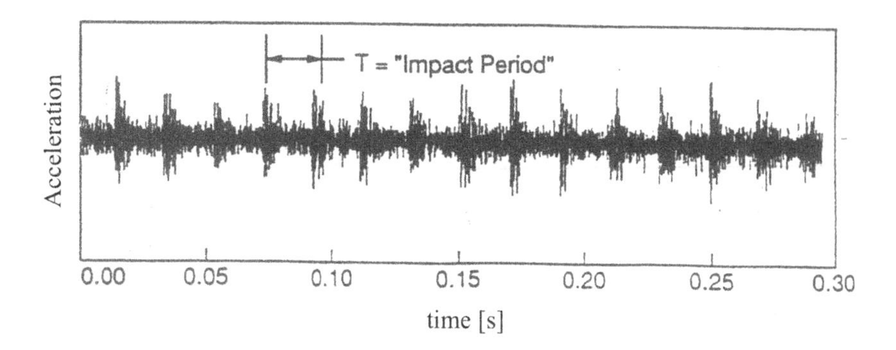

Fig. 11.35. Acceleration measured on the gearbox during rattling of a gear set. The impulses correspond to the bouncing of the gears against each other in the (small) circumferential gap between them.

unbalance distributions affecting each span, the bow of their axis, and the non axisymmetric characteristics of the supports induce dynamic excitations when the shaft is rotating. These excitations are transmitted to the vehicle structure via the supports.

The unbalance excitation is amplified when the angular speed of the shaft is the same as one of its natural frequencies (see [25]).

Universal joints are sometimes installed along prop shafts to compensate for the relative displacement and misalignment between the two ends. Considering a single universal joint, the transmission ratio between the input and output speeds (ω_i and ω_o) is not constant

$$\frac{\omega_o}{\omega_i} = \frac{\cos\beta}{1 - \sin^2\beta\cos^2\phi_i}, \tag{11.48}$$

where β is the angle between the input and output shafts and ϕ_i is the rotation of the input shaft. Because of the variation of the transmission ratio also the transmitted torque is not constant. Under the assumption that the transmission does not dissipate power it must be

$$M_o\omega_o = M_i\omega_i; \tag{11.49}$$

the output torque M_o is then

$$M_o = M_i\frac{1 - \sin^2\beta\cos^2\phi_i}{\cos\beta}. \tag{11.50}$$

If the input torque M_i is constant, the output one has a ripple between the two values

$$M_{o\,\max} = \frac{M_i}{\cos\beta}, \qquad M_{o\,\min} = M_i\cos\beta. \tag{11.51}$$

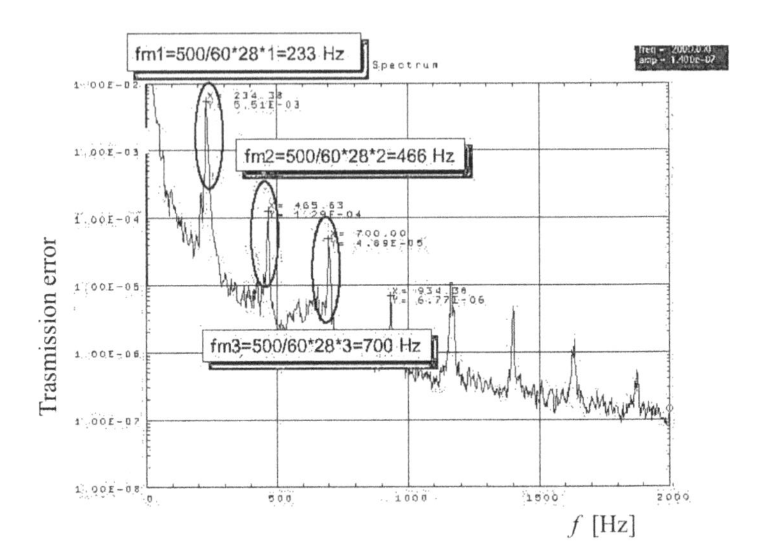

Fig. 11.36. Frequency spectrum of the transmission error of a 28 teeth gear running at 500 rpm. The peaks correspond to the fundamental meshing frequency and its harmonics.

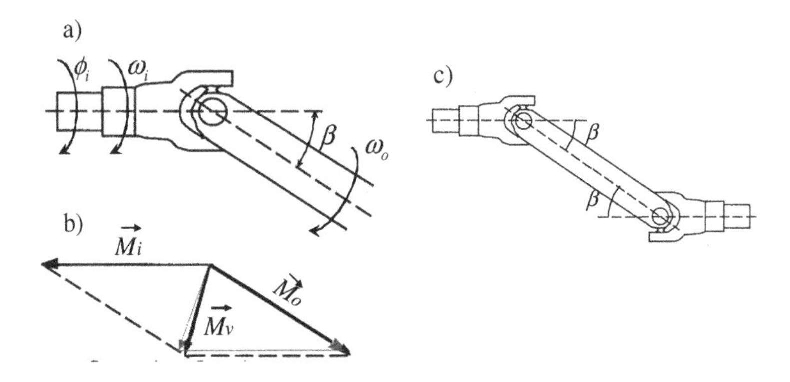

Fig. 11.37. a) Universal joint. ϕ_i input speed angle. β input to output shaft misalignment; $\vec{M}_i$, $\vec{M}_o$ input and output shaft torques, $\vec{M}_v$ constraint torque due to supports. b) torque diagram corresponding to case a). c). Shaft with two universal joints that compensate the transmission error.

Fig. 11.37 b) shows the free body diagram of the universal joint of Fig. 11.37 a). M_v is the reaction torque of the bearing support. As the output torque M_o is variable, also the bearing reaction torque M_v is variable and can potentially induce vibrations of the support and the neighboring structure. Additionally, the angular accelerations of the output shaft require inertia torque components:

$$\dot{\omega}_o = \omega_i^2 \frac{\sin^2 \beta \cos \beta \sin 2\phi_i}{\left(1 - \sin^2 \beta \cos^2 \phi_i\right)^2}; \tag{11.52}$$

due to the not negligible inertia of the output shaft.

The term $\sin 2\phi_i$ at the numerator of Eq. 11.52 indicates that the fundamental harmonic of the inertia torque acting on the output shaft of Fig. 11.37 is twice the angular speed of the input shaft.

From the purely kinematic point of view, a shaft including two universal joints with the same orientation and same misalignment β, allow a constant transmission ratio between the first and the last shaft (Fig. 11.37 c) to be obtained. From the qualitative point of view, the speed variations that are induced in the intermediate shaft by the first joint are compensated by the second.

In any case, the intermediate shaft is subject to angular accelerations that induce variable inertia torques, adding variable components to the supports in addition to those caused by the variable torque (M_v) transmitted to the intermediate shaft.

The variability of the loads acting on the prop shaft support bearings can cause high levels of vibration in the vehicle structure. Therefore elastomeric elements are usually integrated to filter such vibrations.

11.2.5 Brakes

The noise and vibration phenomena that may be involved during braking are many, as the terms used to indicate them.

At low frequency (0÷10 Hz) the ABS system generates pressure waves in the hydraulic circuit that are transmitted to the driver through the brake pedal. The variations of the braking torque determine a variation of the tire brake force that leads to longitudinal vibrations of the suspensions and the vehicle.

The vibrations at frequencies between 30÷40 Hz and 100 Hz are indicated as judder or shudder and can be perceived on the steering wheel, the brake pedal and the floor: Such vibrations are due to a ripple of the braking torque about a mean value that can be caused by a number of factors, for example: a not flat surface of the discs, thermo-elastic instabilities in the brake, uneven deformations of the brake during its installation, etc... The typical feature of the judder is the dependency of its frequency from the angular speed of the wheel, confirming that the primary cause is related to the unevenness of the friction surface.

When the temperature of the brake is high, judder can be caused by the welding on the disc of some material of the pads, altering the surface properties and causing a torque ripple. Similarly, an irregular surface of the disc can cause hot

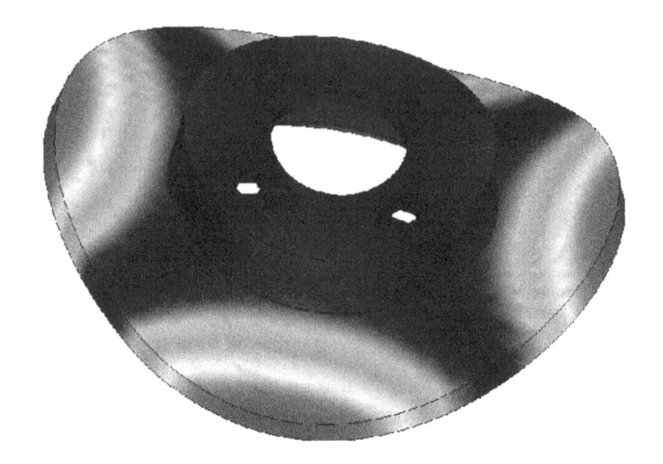

Fig. 11.38. Mode shape of a disk brake with two, orthogonal, nodal lines.

spots on its surface that can modify the local frictional behavior and, therefore, a torque ripple.

Similar vibrations can occur also in drum brakes, the difference being that the cause in this case could be the imperfect cylindrical shape of the drum.

Squeal is associated with vibrations in the range between 100 and 1.000 Hz caused by the dynamic coupling between the disc and the caliper due to the friction forces produced by the pads. The vibrations are amplified by the lowly damped resonances of the disc and the caliper, and transmitted to other elements of the suspension that can result in further amplification. Fig. 11.38 shows one of the modal shapes of a disc rotor in the range of frequencies typical of the squeal. Experimental investigations show that the squeal is more likely to occur for light braking (less than 10 bar brake pressure) and during the last phase of deceleration (at speeds lower than 50 km/h).

11.2.6 Aerodynamics

The aerodynamic field around the vehicle is characterized by turbulence and complex vortex structures. The reduction of the noise and vibration generated by other subsystems of the vehicle (engine and tires, for example) has led to an increased importance of the aerodynamic noise over the past twenty years .

Fig. 11.39 shows some of the most characteristic vortex structures around the vehicle body. The variation of the pressure due to the non stationary field produces noise and vibration that propagates in the vehicle and that therefore can be perceived by the occupants.

The most relevant aeroacoustic sources are:

- local separation of the flow, as in the wake of the A pillar, mirrors, cargo racks and boxes;

- separation in the wake of the vehicle;
- turbulence in the boundary and in the separation regions.

Turbulence, with vortices of different size, results in wide-band noise. The wakes and separation regions can be characterized by stationary vortices and periodic vortex shedding that can result in tonal excitation with a dominant frequency. The typical example is the vortex shedding in the wake of a long cylindrical body in a flow perpendicular to its axis. The structure of the wake is characterized by vortexes that separate from the cylinder on opposite sides of the cylinder section. This wake is usually referred to as von Karman wake after the well-known German researcher (Fig. 11.40).

The frequency of the vortex shedding is dominated in a wide Reynolds range ($150 < Re < 4 \cdot 10^4$) by the Strouhal number S, that links the frequency f to the diameter D and the flow speed V

$$\frac{fD}{V} = S; \tag{11.53}$$

For $\text{Re} > 10^3$ the Strouhal number is almost constant:

$$S = 0.21 \qquad \text{Re} > 10^3. \tag{11.54}$$

Considering, for example, a beam (such as that of a luggage rack) with a diameter $D = 20$ mm: If the flow has a speed of $V = 30$ m/s, the frequency of the vortices is:

$$f = \frac{VS}{D} = \frac{30 \text{ m/s } 0.21}{20 \; 10^{-3} \text{ m}} = 300 \text{ Hz}. \tag{11.55}$$

The separation of each vortex induces a force perpendicular to the flow and the beam axis. This excites the bending vibrations of the beam perpendicular to the flow. These vibrations can be greatly amplified by the resonances of the beam.

Fig. 11.41 shows the spectrum of the aerodynamic noise at the driver's left ear location during aeroacoustic wind tunnel tests.

The spectrum exhibits three main frequency bands:

The first, up to about 400 Hz, is dominated by the so-called body shape noise. This is due to the aerodynamic structures with a size similar that of the vehicle, for example the stationary vortex in the wake of the A pillar (Fig. 11.39).

The frequency band above 400 Hz, is dominated by the infiltration noise. Sometimes this noise is generated at some point around the vehicle (the mirrors, for example) and enters through the seals or other gaps. In other cases, this noise is produced by the resonance of the cavity that hosts the seal. These cavities can behave as Helmoltz resonators excited by the flow.

Above 2000 Hz the cause is mainly the nonstationary flow around small components such as mirrors, aerials, grills, racks. Also the underbody provides an important contribution due to its rather complicated shape.

The sound pressure levels measured in the vehicle when traveling on the road are compared in Fig. 11.42 to those measured in the aeroacoustic wind tunnel.

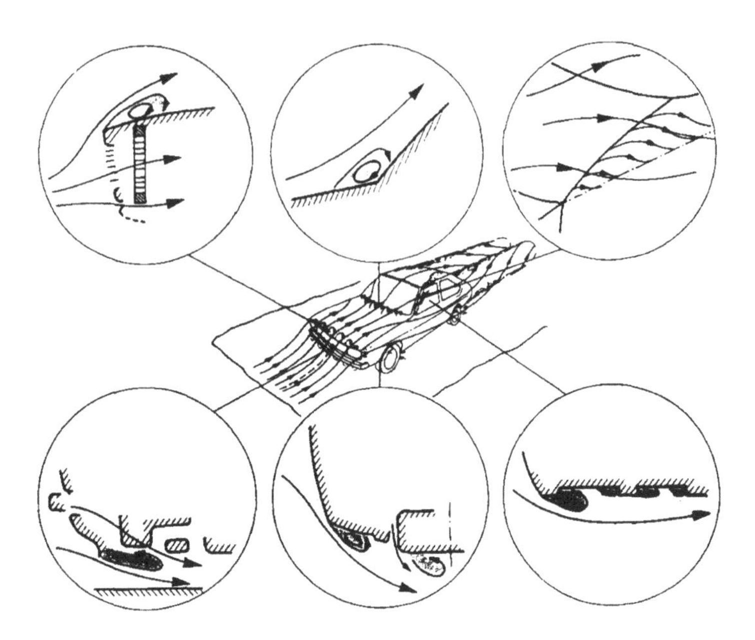

Fig. 11.39. Aerodynamic vortexes around the vehicle body.

The diagram shows that for increasing speeds the contribution of the aerodynamic noise increases more rapidly than the other contributions, until it becomes dominant at speeds typical of traveling on a highway.

11.3 Dynamic Behavior of the Body and Modal Analysis

11.3.1 Dynamic Equations

Under the assumption of the small displacements, the velocity vector of a generic point of the structure can be expressed in an inertial reference frame as:

$$v = a\dot{q} + bq + c, \tag{11.56}$$

where $q = (q_1, q_2, \ldots, q_n)^T$ is the column matrix including all Lagrangian coordinates of the system. Such coordinates enable the deformed configuration of the structure to be determined as function of time t. If the displacements are negligible compared to the size of the structure, matrices a, b, c are a function of the position of the point (x, y, z) but not of time t nor the displacements q. From the dimensional point of view, c is a speed column matrix, the elements of matrix b are the inverse of a time and the elements of a are numbers.

In the case of a continuous structure, the deformed shape is described by an infinite number of Lagrangian coordinates q denoting the displacement of each

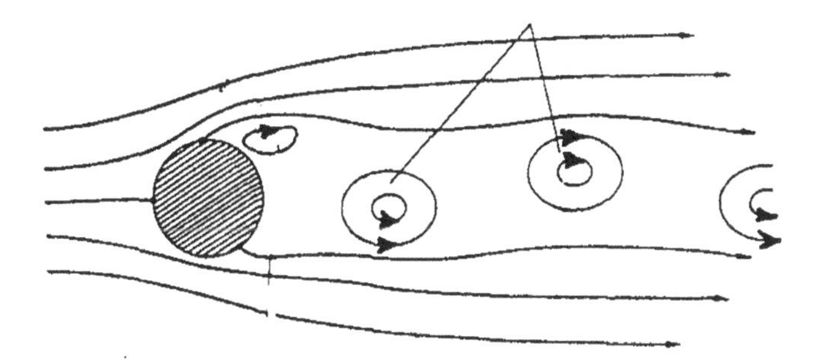

Fig. 11.40. Vortex shedding in the wake of a long cylindrical body in a flow perpendicular to its axis. The structure of the wake is characterized by vortexes that separate from the cylinder on opposite sides of the cylinder section. This wake is usually referred to as von Karman wake. The vortex that is going to leave the cylinder generates a downward force.

point of the continuous material. In this case the Lagrangian coordinate is a continuous function of the position and time $q(x, y, z, t)$.

Considering the Lagrangian coordinates as a continuous function $q(x, y, z, t)$ of the position and time would lead to a description of the dynamic behavior in terms of partial differential equations. The assumption at the base of the finite elements method is the ability to discretize the structure: i.e. to approximate Eq. 11.56 with a finite number of displacements. In this case, a single element represents a limited portion of the structure the displacements and speeds of which are approximated by a finite number of Lagrangian coordinates. These coordinates, called element degrees of freedom, represent the displacements of special points of the element called nodes. Considering a single element, matrices a, b, and c can be determined from the so-called shape functions by taking the kinematics of the element into account.

The expression of the speed v allows to obtain the kinetic energy, that includes the contribution of each elemental volume of mass dm.

$$\begin{aligned} T &= \frac{1}{2}\int_{vol} v^2 dm = \\ &= \frac{1}{2}\int_{vol} \left(\dot{q}^T a^T a\dot{q} + 2\dot{q}^T a^T bq + \right. \\ &\quad \left. + q^T b^T bq + 2\dot{q}^T ac + 2q^T bc + c^T c\right) dm \end{aligned} \tag{11.57}$$

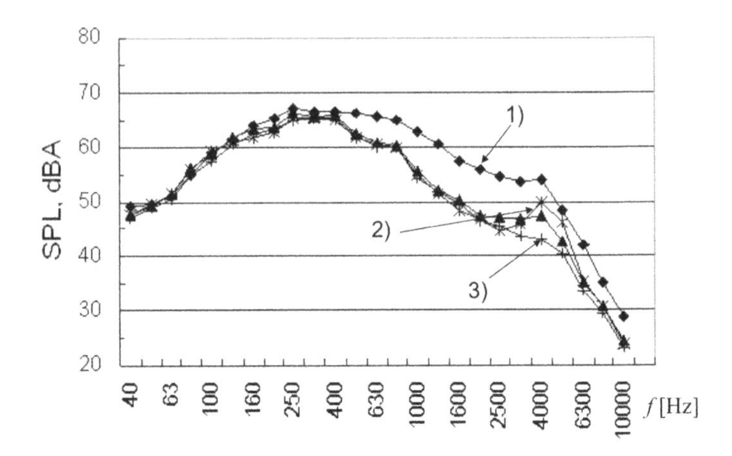

Fig. 11.41. Spectrum of the aerodynamic noise at the driver's left ear location during aeroacoustic wind tunnel tests. 1) standard vehicle; 2) taped doors; 3) no mirrors no wipers.

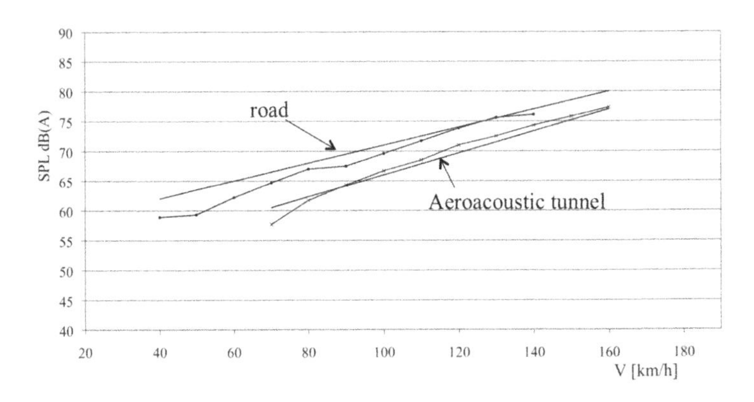

Fig. 11.42. SPL measured on road and wind tunnel tests. The aerodynamic noise increases with speed.

The last three terms of the kinetic energy are constant or linear functions of the displacement q and velocity $\dot{q}$ which do not contribute to the Lagrange equation. Therefore the kinetic energy includes three contributions:

$$T = T_2 + T_1 + T_0; \tag{11.58}$$

in which the first term (T_2) includes the contribution of the time derivative of the degrees of the Lagrangian coordinates:

$$T_2 = \frac{1}{2}\dot{q}(t)^T \left(\int_{vol} a^T a \; dm \right) \dot{q}(t) = \frac{1}{2}\dot{q}(t)^T M \dot{q}(t), \tag{11.59}$$

M is the mass matrix which is real, symmetric and of order $n \times n$. If all the degrees of freedom of the model have an associated mass, it is also definite positive. Formally M can be both the mass matrix of an element or of the structure. The only difference is its size and the number of degrees of freedom represented.

Term T_1 in Eq. 11.58 includes the mixed contribution of the displacement and velocity:

$$T_1 = \dot{q}(t)^T \left(\int_{vol} a^T b \; dm \right) q(t) = \dot{q}(t)^T N \; q(t). \tag{11.60}$$

Finally, T_0 involves the displacements alone:

$$T_0 = \frac{1}{2}q(t)^T \left(\int_{vol} b^T b \; dm \right) q(t) = \frac{1}{2}q(t)^T K_i \; q(t). \tag{11.61}$$

Despite being related to the kinetic energy, dimensionally matrix K_i is a stiffness matrix relating energy (T_0) to the square of the displacement. By definition (Eq. 11.61) matrix K_i is symmetric.

Considering the contribution of the potential energy V, the Lagrangian function results in:

$$L = T_2 + T_1 + T_0 - V, \tag{11.62}$$

where the potential energy V is a function of just the generalized displacements q. In the linear case, it can be expressed in terms of the stiffness matrix K_e

$$V = \frac{1}{2}q(t)^T K_e q(t). \tag{11.63}$$

Combining all quadratic terms of displacement q , it is possible to define a new potential term U:

$$U = V - T_0 = \frac{1}{2}q(t)^T K q(t), \tag{11.64}$$

where the overall stiffness K includes an elastic and an inertia contribution

$$K = K_e - K_i. \tag{11.65}$$

The Lagrangian function is then

$$L = T_2 + T_1 - U. \tag{11.66}$$

The equations of the motion can be obtained from the Lagrangian equations

$$\frac{d}{dt}\left(\frac{\partial L}{\partial \dot{q}}\right) - \frac{\partial L}{\partial q} + \frac{\partial \mathcal{R}}{\partial \dot{q}} = Q, \tag{11.67}$$

Q includes the generalized forces acting on the degrees of freedom of the structure. The Rayleigh dissipation function $\mathcal{R}$ takes the viscous dissipative forces into account

$$\mathcal{R} = \frac{1}{2}\dot{q}(t)C\dot{q}(t) + \dot{q}(t)Hq(t), \tag{11.68}$$

C is the viscous damping matrix (symmetric) and H is the circulatory matrix (anti-symmetric).

The Lagrange equations enable the equation of the motion to be obtained in the second order form [26]

$$M\ddot{q}(t) + (C + G)\dot{q}(t) + (K + H)q(t) = Q(t) \tag{11.69}$$

The gyroscopic matrix G is anti-symmetric, its origin corresponding to the contribution T_1 to the kinetic energy (Eq. 11.60)

$$G = N - N^T = -G^T. \tag{11.70}$$

In principle the generalized force vector $Q(t)$ may include an external (generalized) force for each degree of freedom of the structure. Usually only a relatively limited number of degrees of freedom are subject to external forces making it possible to rewrite $Q(t)$ as an input selection matrix T_{in} and a full vector F including the forces (or torques, or pressures) acting on the structure:

$$Q = T_{in}F \tag{11.71}$$

T_{in} is a matrix that applies, with a appropriate gain, the physical forces acting on the structure to the corresponding degrees of freedom of the model.

The matrices of Eq. 11.69 are usually computed by the finite element codes starting from the geometry, the material properties, the constraints and the forces acting on the structure. As the geometry could be very complicated (a vehicle structure is a clear example) the number of elements necessary for its discretization could be very large. The number of elements is then largely dependent on the geometric complexity and only in part on the desired accuracy.

The equation of the motion (Eq. 11.69) can be expressed in terms of a state vector x and input vector u:

$$x(t) = \begin{bmatrix} q(t) \\ \dot{q}(t) \end{bmatrix}; \qquad u = F;$$

by determining the acceleration from Eq. 11.69 and adding the identity $\dot{q} = \dot{q}$:

$$\begin{aligned} \dot{q}(t) &= \dot{q}(t) \\ \ddot{q}(t) &= -M^{-1}(K+H)q(t) - M^{-1}(C+G)\dot{q}(t) + M^{-1}T_{in} \end{aligned} \tag{11.72}$$

in matrix form

$$\dot{x}(t) = Ax(t) + Bu(t), \tag{11.73}$$

the state matrix A and input matrix B are:

$$A = \begin{bmatrix} 0 & I \\ -M^{-1}(K+H) & -M^{-1}(C+G) \end{bmatrix}; \qquad B = \begin{bmatrix} 0 \\ M^{-1}T_{in} \end{bmatrix}.$$

The system response is the superposition of the free response to the initial conditions and the (forced) response due to external forces $Q(t)$.

In addition to the state equation, it is possible to add an output equation to obtain the output variables as function of the state which can be the displacements, velocities or accelerations at the locations of special interest for the analysis, e.g. the seat attachment points or points on the steering column. In some cases the outputs are the displacements, speeds or accelerations of a sensor used for monitoring or for a closed loop feedback system. In a linear system the outputs can be expressed as a combination of the state and input variables,

$$y(t) = Cx(t) + Du(t), \tag{11.74}$$

If the output is a combination of the displacements matrix, C does not have null components corresponding to the first half of the state vector

$$C_q = \begin{bmatrix} T_{oq} \; 0 \end{bmatrix}; \quad D_q = 0. \tag{11.75}$$

Similarly, if the output is a combination of the speeds:

$$C_{\dot{q}} = \begin{bmatrix} 0 \; T_{o\dot{q}} \end{bmatrix}; \quad D_{\dot{q}} = 0. \tag{11.76}$$

In some cases the output is function of the acceleration, for example when an accelerometer is mounted on the structure to measure the levels of vibration.

As the acceleration is not included among the states, it could seem appropriate to obtain it via the time derivative of the speed. However, although this is possible, in practice this operation is not numerically robust because the time derivative of a response affected by noise (even numerical noise) amplifies the effect of the noise significantly. An alternative to this approach consists in considering that the acceleration is already provided by the second of Eq. 11.72 which can be included in the output equation:

$$y = T_{o\ddot{q}}\ddot{q}, \tag{11.77}$$

in matrix form:

$$y = T_{o\ddot{q}} \begin{bmatrix} -M^{-1}(K+H) & -M^{-1}(C+G) \end{bmatrix} \begin{bmatrix} q(t) \\ \dot{q}(t) \end{bmatrix} + \begin{bmatrix} M^{-1}T_{in} \end{bmatrix} u, \tag{11.78}$$

The $C_{o\ddot{q}}$ and $D_{o\ddot{q}}$ matrices that enable the output acceleration to be obtained are then:

$$C_{\ddot{q}} = T_{o\ddot{q}} \left[-M^{-1}(K+H) \; -M^{-1}(C+G) \right]; \quad D_{\ddot{q}} = \left[M^{-1} T_{in} \right]; \tag{11.79}$$

the direct input-output link ($D_{o\ddot{q}}$ matrix) takes into account the fact that the acceleration is linked to the external forces by Newton's law.

11.3.2 Free Response

In general the response that follows the given initial conditions is obtained solving the homogeneous equation Eq. 11.73:

$$\dot{x}(t) = Ax(t); \tag{11.80}$$

the displacement and the speed at a given time $t = t_0$ are given by:

$$x(t_0) = \begin{bmatrix} q_0 \\ \dot{q}_0 \end{bmatrix}.$$

The solution of Eq. 11.80 is expressed in exponential form:

$$x(t) = x_0 e^{\lambda t}. \tag{11.81}$$

In general vector x_0 and exponent λ are complex numbers which can be found by substituting Eq. 11.81 in Eq. 11.80 which leads to an eigenvalue problem:

$$\lambda x_0 = A x_0. \tag{11.82}$$

If the system is not damped, not gyroscopic and with no circulatory forces ($C = G = H = 0$) it is possible to solve directly the homogeneous equation in the second order form (Eq. 11.69) without passing through the state space:

$$M\ddot{q} + Kq = 0; \tag{11.83}$$

The solution is again of the exponential type:

$$q = \phi e^{st}. \tag{11.84}$$

In general the displacement ϕ and the exponent s are complex quantities that can be found by substituting Eq. 11.84 in Eq. 11.83:

$$\left(Ms^2 + K \right) \phi = 0. \tag{11.85}$$

This is an eigenvalue problem. An alternative form of this equation can be obtained if matrix M is positive definite; in this case it is possible to determine its inverse M^{-1}.

Left multiplying both terms of equation 11.85 by M^{-1} and letting $\lambda = -s^2$,

$$(M^{-1}K - \lambda I)\phi = 0, \qquad (11.86)$$

This is an eigenproblem which is formally the same type of Eq. 11.82. As the first form of this eigenvalue problem (Eq. 11.85) does not require the computation of the inverse of mass M, it is usually preferred in numerical computations.

The solution with $\phi = 0$ is not taken into account as it indicates simply that a system with null initial conditions remains still. The solution ϕ is different from zero only if the matrix that it multiplies is singular, i.e.:

$$\det(Ms^2 + K) = 0. \qquad (11.87)$$

If matrix M is positive definite, the properties of the eigenvalues $\lambda = -s^2$ result from the stiffness matrix K. If also K is positive definite, then all eigenvalues λ are real and positive ($\lambda_i > 0$, $i = 1, \ldots, n$). If K is only positive semi-definite, some of the eigenvalues could be null ($\lambda_i \geq 0$, $i = 1, \ldots, n$). This last case is typical, for example, when some of the degrees of freedom are not constrained.

If all n eigenvalues are positive, the values of s correspond to n complex conjugate pairs with each pair lying on the imaginary axis of the complex plane,

$$s_i = \pm j\omega_i = \pm j\sqrt{\lambda_i} \qquad i = 1, \ldots, n; \qquad j = \sqrt{-1}. \qquad (11.88)$$

Roots ω_i are the natural frequencies of the undamped and not gyroscopic system. The eigenvectors ϕ_i corresponding to each eigenvalue λ_i are the so-called mode shapes. It is important to observe that the natural frequencies and the mode shapes are characteristics of the structure and of its mass and stiffness distribution, but not of the external forces applied to it. In fact they have been obtained from the homogeneous equation.

Returning to the homogeneous Eq. 11.84, the transient following the initial condition is a linear combination of all its possible solutions:

$$q(t) = \sum_{i=1}^{n} \phi_i (a_i e^{j\omega_i t} + \bar{a}_i e^{-j\omega_i t}); \qquad (11.89)$$

coefficients a_i and $\bar{a}_i$ can be found from the initial position $q(t_0)$ and speed $\dot{q}(t_0)$. As a_i and $\bar{a}_i$ are complex conjugate pairs, the displacement $q(t)$ is real; this is necessary because $q(t)$ represents a set of physical displacements and therefore can not have complex values.

Eq. 11.89 indicates that the transient response is the superposition of the n mode shapes ϕ_i each with the corresponding frequency. If, for example, the structure at time $t = 0$ starts with zero speed from a deformed shape that coincides with the first mode shape:

$$q(0) = \phi_1; \qquad \dot{q}(0) = 0, \qquad (11.90)$$

the result is:

$$a_1 = \bar{a}_1 = 1; \tag{11.91}$$

$$a_i = \bar{a}_i = 0 \quad i = 2, \ldots, n; \tag{11.92}$$

during the following transient the structure oscillates with the first natural frequency (ω_1) and a deformed shape equal to the first mode,

$$q(t) = \phi_1 \cos(\omega_1 t). \tag{11.93}$$

This example can be repeated for initial conditions which correspond to other mode shapes.

11.3.3 Modal Coordinates Transformation

A very important property that relates the mass and stiffness matrices to the eigenvalues and eigenvectors is the so-called M and K orthogonality; i.e.

$$\phi_i^T M \, \phi_j \begin{cases} = 0 \\ \neq 0 \end{cases}; \qquad \phi_i^T K \, \phi_j \begin{cases} = 0 \\ \neq 0 \end{cases}; \qquad \text{per} \begin{cases} i \neq j \\ i = j \end{cases}. \tag{11.94}$$

which can be rewritten in matrix form

$$\phi^T M \, \phi = m; \qquad \phi^T K \, \phi = k \tag{11.95}$$

the so-called modal matrices m and k are diagonal. This property enables the decoupling of the equations of motion of the undamped and nongyroscopic system based on the use of the mode shapes as a means for representing the displacement vector $q(t)$. Instead of using the canonical base, vector $q(t)$ is represented as a linear combination of the n mode shapes ϕ_i. The coefficients $\xi_i(t)$ of such combinations are the coordinates in the modal base, usually referred to as modal coordinates or modal degrees of freedom

$$\begin{aligned} q(t) &= \phi_1 \xi_1(t) +, \ldots, + \phi_n \xi_n(t) \\ &= [\phi_1, \phi_2, \ldots, \phi_n] \, [\xi_1(t), \xi_2(t), \ldots, \xi_n(t)]^T \\ &= \phi \xi(t). \end{aligned} \tag{11.96}$$

The modal coordinates represent the contribution of each mode shape to a given deformed shape of the structure.

The modal transformation of Eq. 11.96 can be substituted in the dynamic equation of the undamped system,

$$M \phi \ddot{\xi}(t) + K \phi \xi(t) = T_{in} F(t), \tag{11.97}$$

By left multiplying all terms by ϕ^T , the mass and stiffness matrices M and K are transformed in the modal mass and stiffness of Eq. 11.95:

$$m \ddot{\xi}(t) + k \xi(t) = t_{in} F(t), \tag{11.98}$$

where the input selection matrix in modal coordinates t_{in} is

$$t_{in} = \phi^T T_{in}.$$

As m and k are diagonal, the dynamic equations in modal coordinates (Eq. 11.98) are decoupled. The modal coordinates enable the dynamic behavior of the undamped and nongyroscopic structure to be described as a set of n spring-mass systems not interacting with each other,

$$\begin{aligned} m_1\ddot{\xi}_1 + k_1\xi_1 &= t_{in11}F_1 + \cdots + t_{in1m}F_m, \\ \ldots &= \ldots \\ m_n\ddot{\xi}_n + k_n\xi_n &= t_{in\ n1}F_1 + \cdots + t_{in\ nm}F_m. \end{aligned} \tag{11.99}$$

Since a coordinate transformation, like the modal one, does not alter the eigenvalues of a matrix, the natural frequencies of the system in modal coordinates are the same as in physical coordinates. In other terms, the natural frequencies are a property of the structure and not of the coordinate system used to describe its motion:

$$\omega_{0i} = \sqrt{\frac{k_i}{m_i}}. \tag{11.100}$$

The eigenproblem enables the direction of each eigenvector in an n-dimensional space to be determined, while its length is not determined. This is because the eigenvectors of a matrix are obtained by solving a system of linear equations with null determinant, they can then be determined except for a scale factor. Taking this into account, it is always possible to normalize each eigenvector so that all modal masses have unit values. In this case Eq.s 11.98 become:

$$\ddot{\xi} + \omega_0^2\xi = t_{in}F; \tag{11.101}$$

as the natural frequencies must still be the same regardless of the normalization, the modal stiffness matrix has the square of the natural frequencies along the diagonal

$$\omega_0^2 = \begin{bmatrix} \omega_{01}^2 & \cdots & 0 \\ \vdots & \ddots & \vdots \\ 0 & \cdots & \omega_{0n}^2 \end{bmatrix}. \tag{11.102}$$

Returning to the equations of the undamped system in modal coordinates (Eq.s 11.98, 11.101), matrix t_{in} indicates the extent to which each of the external forces is able to excite the various modes: correspondingly its elements are indicated as modal participation factors. A null modal participation factor $t_{ij} = 0$ indicates that the external force F_j is not able to excite the vibrations of the i-th mode, the typical case being when the external force is applied to node of the mode shape, i.e. to a point that does not move when the structure vibrates with that mode shape. In this case, since the force acts on a point that does not move, it cannot transfer energy to the structure.

The modal transformation based on the eigenvectors of the undamped and non-gyroscopic system, is one (Eq. 11.96) of the infinite number of alternative choices possible to represent the n-dimensional space of vector $q(t)$. Although this base has been defined from the free response of the undamped and non gyroscopic system, it can be adopted also in the presence of damping and gyroscopic forces:

$$m\ddot{\xi}(t) + (c+g)\dot{\xi}(t) + (k+h)\xi(t) = t_{in}F(t); \tag{11.103}$$

the modal mass and stiffness matrices m and k, are the same diagonal matrices of Eq. 11.95. Instead matrices c, g, and h are not diagonal. In fact, to construct the mode shapes adopted for the modal transformation, M and K are orthogonal; at the same time, however, C, G and H are not orthogonal. The result is that the system in the modal coordinates of Eq. 11.96 is coupled by the gyroscopic, damping and circulatory matrices.

In many cases gyroscopic and circulatory effects can be neglected to good approximation. Furthermore, mechanical structures usually also have relatively low damping. In this case, the effect of the damping can be taken into account in the modal equations by means of a diagonal viscous damping matrix. By doing so, the off-diagonal elements of the modal damping matrix are neglected implicitly. The rationale behind this is that, due to the low damping, its effect is relevant only in the frequency range close to the resonances, where the mass and damping forces compensate each other and the external forces are balanced only by damping. The diagonal damping allows to preserve the decoupling also in resonance conditions where off diagonal terms would introduce some transfer of energy from one mode to the others.

Similarly to the one degree of freedom system, the viscous damping coefficient of the i-th mode can be expressed as function of a the damping factor $\zeta_i = c_i/c_{i\ cr} = c_i/(2m_i\omega_{0i})$. If the mode shapes are normalized to obtain an identity mass matrix,

$$\ddot{\xi} + 2\zeta\omega_0\dot{\xi} + \omega_0^2\xi = t_{in}F \tag{11.104}$$

where the matrix of the modal damping factors is diagonal,

$$\zeta = \begin{bmatrix} \zeta_1 & \dots & 0 \\ \vdots & \ddots & \vdots \\ 0 & \dots & \zeta_n \end{bmatrix}. \tag{11.105}$$

The finite elements codes used for the structural analysis start from the data of the geometry, the material properties, the constraint and the forces. One of the main milestones during the processing following the data input is the computation of the conservative elements of the dynamic equations, i.e. of the mass (always), stiffness (always), gyroscopic (seldom) and circulatory (seldom) matrices. The dissipative phenomena, on the other hand, are influenced by a number of factors that are not always directly related to the geometry, material properties and constraints. The interface between parts, micro-slip conditions, and on environmental factors such as temperature and age play an important role in the definition of the dissipative behavior.

The result is that the reliability of the structural damping that can be obtained by with finite element modeling may be quite low. A common practice is to take the damping into account as modal damping following the modal transformation. The modal damping factors can be identified from experimental measurements performed on the complete structure or on its components. If the structure is not available because still at the design stage it is possible to refer to historical data on similar structures.

11.3.4 Mode Shapes of a Car Body

As all continuous structures, the mode shapes of a car body are infinite. The complexity of the structure makes the resonances very densely packed along the frequency axis; a body in white, for example can have from 150 to 250 modes below 200 Hz. A completely assembled body, on the other hand, has even a more modes in the same frequency range also because the mass added by the interiors, doors, glasses, engine, transmission and other mechanical parts tends to reduce many of the natural frequencies.

The car body modes can be sorted in global and local depending on the amount of structure involve in the motion. A local mode involves only a part of the structure, where most of its elastic energy is concentrated. The deformed shape of a global modes involves the entire structure and usually occur at frequencies lower than 50 Hz; the associated elastic energy is distributed in a rather uniform way. By referring to their shape, car body modes can be classified as torsional, bending on xz plane and bending on xy plane (waving). Furthermore, they can be classified by considering whether or not the motion involves the front or the rear part of the body.

Figs. 11.43 and 11.44 illustrate the deformed shapes of the global modes of lower frequency for two different types of body in white frames (without interiors, the powertrain, suspensions, doors). The installation of the interiors, floor covering doors and all trimmings modifies the mass quite substantially, without altering the stiffness; the global natural frequencies will then decrease without significant change to the mode shapes.

The global modes are very important for several reasons. The first aspect relates to the resonances that can amplify considerably the effect of the dynamic excitation from the different sources, e.g. the engine and the wheels. The large vibration amplitudes have a negative effect on the comfort.

The second aspect relates to the potential coupling between the global modes of the chassis with that of large subsystems such as suspensions and subframes that are attached to the rest of the structure with compliant and dissipative interfaces. The adoption of elastomeric mounts is common in this application to dissipate the energy associated with the relative vibration motion.

Another important aspect related to the mode shapes is that of the squeaks and rattles which are acoustic emissions that cause the perception of very poor

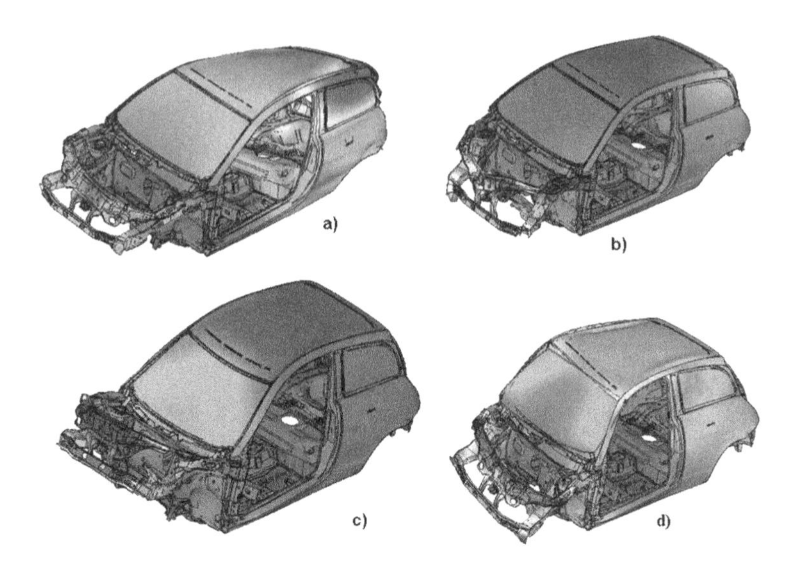

Fig. 11.43. Mode shapes of a 3 doors body in white structure. a) global torsion; b) lateral bending (waving); c) front bending; d) front torsion.

quality by the occupants. Typically squeaks and rattles occur when the vehicle is moving on uneven road surfaces and are due to the mutual displacement between connected parts. This motion, and the related acoustic emission, can be significantly amplified if the structure is excited at one of the global resonances. A typical example is the relative displacement between the dashboard and the attachment points with the body frame. Even if the displacement may be at a (low) frequency well below the acoustic range (20 Hz - 20 kHz), the generated noise has a relatively high frequency (squeak) caused by the stick and slip between the surfaces in relative motion.

Local modes involve portions of the structure, e.g. at low frequencies, large panels such as the floor, the roof and the windscreen. Low order local modes are characterized by deformed shapes with a small number of semi-waves. At high frequencies local modes involve smaller parts or a larger number of waves.

Significant effort is often devoted at the design stage to the dampen the mode shapes of the body. Global modes are difficult to damp because they involve a rather distributed strain energy density and their damping sometimes requires the application of relatively large forces at points of large modal amplitude. To this end, an effective means may be provided by using dynamic dampers. Local modes are easier to damp because their energy is more localized: correspondingly most effort is often addressed to the local modes that involve those panels in contact with interior of the vehicle, a common solution being to add a damping treatment as a layer of viscoelastic material attached to the panels that may be involved in the vibration. Despite being cost effective, the added mass is not negligible. Other solutions involve a layer of viscoelastic material sandwiched

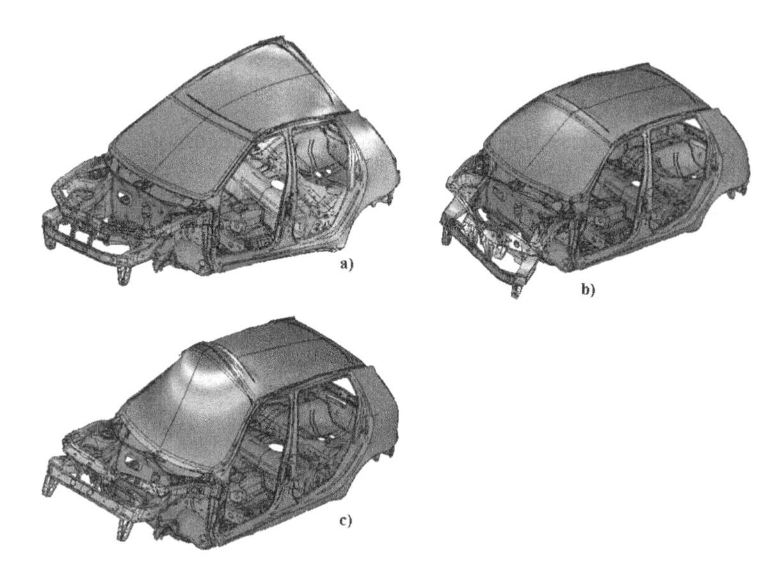

Fig. 11.44. Mode shapes of a 5 doors body in white structure. a) global torsion; b) lateral bending (waving); c) front bending; d) front torsion.

between the structure and a relatively stiff constraining layer (usually a patch of metal plate). These solutions are more costly but are much more effective in terms of the damping added and weight.

11.3.5 Forced Response

The forced response can be obtained via the time domain integration of the second order dynamic equations or in the state space form of Eq. 11.73. Assuming the system can be approximated by linear equations, it is possible to describe the input-output behavior by means of transfer functions. The application of the Laplace transform to the state equation leads to:

$$sx(s) = Ax(s) + Bu(s), \tag{11.106}$$

where s is the Laplace variable. Collecting the Laplace transform of the state vector $x(s)$

$$x(s) = (sI - A)^{-1}Bu(s), \tag{11.107}$$

and taking the output equation (Eq. 11.74) into account, it is possible to obtain the output $y(s)$ as function of the input $u(s)$,

$$y(s) = H(s)u(s), \tag{11.108}$$

where the transfer function matrix $H(s)$ is

$$H(s) = C(sI - A)^{-1}B + D. \tag{11.109}$$

It is well known from the dynamic systems analysis that the poles of the transfer functions expressed in linear state space form (Eq. 11.73), are the eigenvalues of matrix A. This can be shown considering that the analytic expression of the inverse of a generic matrix has its determinant as the common denominator. All components of the transfer function matrix $H(s)$ have then the same denominator, equal to $\det(sI - A)$. As the poles are the values of s that reduce to zero the denominator, they correspond to the eigenvalues of matrix A. Taking Eq. 11.82 into account, they are the same as the natural frequencies of the undamped system.

If the dynamic equations of the system are in modal coordinates, the Laplace transform of the dynamic equation is:

$$\left(Is^2 + 2\zeta\omega_0 s + \omega_0^2\right)\xi(s) = t_{in}F(s), \tag{11.110}$$

exploiting the modal decoupling, the transfer function between the input force and output the i-th modal displacement

$$\xi_i(s) = \frac{1}{s^2 + 2\zeta_i\omega_{0i}s + \omega_{0i}^2}\left(t_{in\ i1}F_1(s) + \cdots + t_{in\ im}F_m(s)\right); \tag{11.111}$$

Assuming, for simplicity, that the structure is subject to only one external force F:

$$\xi_i(s) = \frac{t_{in\ i}F(s)}{s^2 + 2\zeta_i\omega_{0i}s + \omega_{0i}^2}, \tag{11.112}$$

the previous expression indicates that the response of the i-th modal degree of freedom is the same as a one degree of freedom mechanical system with undamped natural frequency ω_{0i} and a damping factor ζ_i. If the system output y is a displacement, it can be obtained from Eq. 11.96 as:

$$y(s) = T_{out}q(s) = T_{out}\phi\xi(s) = t_{out}\xi(s), \tag{11.113}$$

where the output matrix in modal form t_{out}:

$$t_{out} = T_{out}\phi; \tag{11.114}$$

the substitution of Eq. 11.112 in Eq. 11.113 allows to find the input-output transfer function in modal form:

$$y(s) = H(s)F(s). \tag{11.115}$$

Due to the modal decoupling, the transfer function $H(s)$ is the superposition of each modal contribution. Using the system dynamics language, each of them is referred to as a residual:

$$H(s) = \sum_{i=1}^{n} \frac{\alpha_i}{s^2 + 2\zeta_i\omega_{0i}s + \omega_{0i}^2}; \qquad \alpha_i = t_{out\ i}t_{in\ i}. \tag{11.116}$$

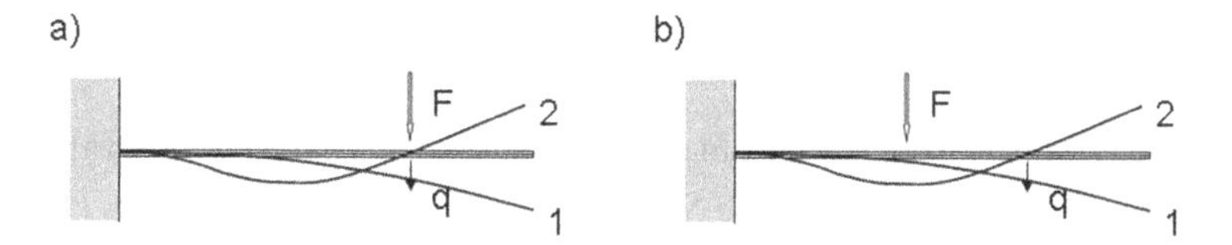

Fig. 11.45. Effect of the position of the sensor and actuator. a) colocated sensor and actuator; b) non colocated sensor and actuator. In both cases the sensor is not able to measure the contribution due to the second mode shape.

Coefficients α_i are the modal residuals representing the contribution of each mode to the output.

For example, if $\alpha_i = 0$ the i–th mode does not contribute to the output y. This does not mean that the mode is not present in the structure, but that either it can not be excited by the force F ($t_{in\ i} = 0$) or that it cannot be monitored $t_{out\ i} = 0$.

The simple clamped beam of Fig. 11.45 a) is excited and monitored by sensors at the same point and along the same direction (colocated sensor and actuator). Their position is such that the force is able to excite the first mode shape, and the sensor to measure the corresponding displacements. Conversely, they can neither actuate nor monitor the second mode. In this case $\alpha_1 \neq 0$; and $\alpha_2 = 0$; $\alpha_3 \neq 0, \ldots$

The force F of Fig. 11.45 b), on the other side can act on the first and the second mode, so that $t_{in\ 1} \neq 0$, $t_{in\ 2} \neq 0$; however the sensor is still not able to monitor the second mode.

This property can be exploited in the design of a car body to minimize the effect of some modes of vibration. If the attachment point of a source of excitation (the engine, for example) is located close to a node of a certain mode shape, the excitation of this mode shape will by minimal (i.e. null as a limit case). All this can be repeated for relevant output points such as the steering wheel support, the attachment of the pedals to the firewall or the seat attachment points. Although this is valid in the relatively narrow frequency range where the target mode shape is dominant, the same principle can not be extended to cover a wide number of mode shapes.

11.3.6 Response to a Random Excitation

If a linear system is excited with a random excitation, its response is also a random signal although with different frequency characteristics. If $G_F(\omega)$ is the power spectral density of the excitation, the power spectral density $G_u(\omega)$ of the output is [27]:

$$G_y(\omega) = |H(j\omega)|^2 G_u(\omega), \tag{11.117}$$

where $|H(j\omega)|^2$ is the square of the absolute value frequency response function, i.e. the transfer function $H(s)$ computed for a pure imaginary Laplace variable $s = j\omega$.

The mean square value of the output can be obtained as the integral of the power spectral density over the entire frequency range

$$E[y^2] = \int_0^\infty G_y(\omega)d\omega. \tag{11.118}$$

The transfer function of a lightly damped mechanical structure is characterized by sharp peaks corresponding to the resonances. A wide band excitation such as that coming from the road profile (Fig. 11.17), generates a response characterized by narrow peaks at the resonant frequencies. The smaller the damping, the sharper the peaks; the higher the modal factors α_i (Eq. 11.116), the more evident the resonance peak will be on the response. The effect is that the response is dominated by a relatively small number of frequencies with an amplitude that changes randomly with time.

11.3.7 Viscous and Structural Damping

Dissipation has been modeled in the previous equations as viscous damping that generates forces proportional to the deformation speed. With reference to Fig. 11.46, the force developed by a spring in parallel with a viscous damper is:

$$F(t) = kq(t) + c\dot{q}(t). \tag{11.119}$$

If the displacement is harmonic with angular frequency ω, it can be represented as the real part of a complex exponential,

$$q(t) = q_0 e^{j\omega t}; \tag{11.120}$$

By substituting in Eq. 11.119, the restoring force generated by the spring-damper assembly is then:

$$F(t) = (k + j\omega c)\, q_0 e^{j\omega t} = (k + j\omega c)\, q(t). \tag{11.121}$$

In the time domain, the force is a real quantity and so only the real part should be considered in the previous equation. This force is represented on the force-displacement characteristic as an ellipse with an enclosed area that increases with the angular frequency ω (Fig. 11.46 a)). In the limit case of $\omega = 0$ the damping forces become negligible and the ellipse becomes a line with slope equal to the spring stiffness k. The second part of Eq. 11.121 ($F(t) = (k + j\omega c)\, q(t)$) relates the force to the displacement by means of a complex number whose dimensions are those of a stiffness,

$$k^* = k + j\omega c. \tag{11.122}$$

It is worth pointing out that the complex stiffness of Eq. 11.122 is a mathematical tool strictly related to the assumption that the displacement is a harmonic

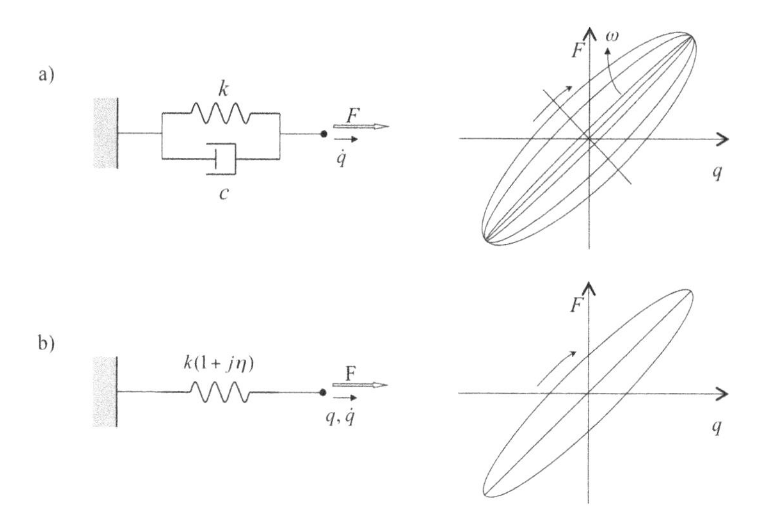

Fig. 11.46. Effect of a viscous and an hysteretic damping on the characteristic of a spring. a) viscous damping: the amplitude of the hysteresis is proportional to the excitation frequency; b) hysteretic damping modeled as a complex stiffness: the amplitude of the hysteresis cycle is constant.

function of the time. It is therefore not applicable when the displacement is not a harmonic function.

If the materials used for mechanical construction are tested with cyclic loading and unloading with constant amplitude and frequency, the force to displacement characteristics exhibit hysteresis cycles. If the material is characterized with a stress-strain curve, the area enclosed by the hysteresis cycle represents the energy dissipated in a cycle per unit volume. The amplitude of the hysteresis depends a lot on the type of material [25] [28].

On one side, the metallic materials commonly adopted for the mechanical construction (such as a car body, for example) exhibit very low levels of dissipation; conversely viscoelastic materials have wide hysteresis cycles. The large amount of energy dissipated per unit cycle justifies their use for damping structural vibrations.

The experimental data show that the hysteresis cycles of most metallic materials are relatively independent of the frequency (Fig. 11.46 b)) meaning that the energy dissipated per unit volume is just proportional to the number of cycles and not to the cycling speed. The dissipated power is then proportional to the frequency.

The easier way to model an hysteretic material with little or no dependence on the frequency can be obtained from Eq. 11.121. Removing the frequency from the imaginary term of the complex stiffness:

$$F(t) = (k_1 + jk_2)q_0 e^{j\omega t} = (k_1 + jk_2)q(t). \tag{11.123}$$

The complex stiffness is then

$$k^* = k_1 + jk_2, \tag{11.124}$$

that can also be expressed as

$$k^* = k(1 + j\eta); \qquad k = k_1; \tag{11.125}$$

where $\eta = k_2/k_1$ is the so-called loss factor.

What said about the complex stiffness is also valid for the stress-strain characteristic (σ-ε) by the introduction of a complex elastic modulus:

$$E^* = E(1 + j\eta), \tag{11.126}$$
$$G^* = G(1 + j\eta). \tag{11.127}$$

If the motion is harmonic with angular frequency ω, equations 11.122 and 11.125 indicate that, for a given real part of the stiffness k, the viscous damping equivalent to the structural damping is:

$$c_{eq} = \frac{\eta k}{\omega}. \tag{11.128}$$

The viscous damping of Eq. 11.128 is equivalent to the structural damping only at that the specific angular frequency ω used to compute the equivalency. A different equivalent damping must be used for a different frequency. It is not possible therefore to compute an equivalent damping at a certain frequency and then to use it for a broadband excitation.

In any case, the adoption of a complex stiffness to model the dissipative behavior of a structure is acceptable only if the displacements are harmonic with constant amplitude. If the motion is not harmonic there is no reason to use the complex stiffness (or structural damping) approach. It is relatively common to adopt the structural damping approach to compute the frequency response function which can be justified because the objective in this case is to determine the response at each frequency of excitation in a given range. The assumption is that the system reaches steady state conditions for each single frequency of excitation.

The structural damping approach is valid also for multi degrees of freedom systems. The equation of the motion in this case will have a complex stiffness matrix

$$M\ddot{q}(t) + (K_1 + jK_2)\,q(t) = T_{in}F(t); \tag{11.129}$$

also in this case the above dynamic equation is meaningful only if the motion is harmonic, i.e. when the force is $F(t) = F_0 e^{j\omega t}$. The mode shapes of the undamped system $M\ddot{q}(t) + K_1 q(t) = 0$, can be used to decouple the mass and the real part of the stiffness matrix. In general the modal transformation will not be able to decouple the imaginary part, except in the case that it is proportional to the first $K_2 = \eta K_1$.

$$M\ddot{q}(t) + K_1 (1 + j\eta) q(t) = T_{in}F(t). \tag{11.130}$$

The modal transformation decouples the equations of the motion of the system with structural damping and transforms it into a set of n spring-mass systems each with the same loss factor η

$$m\ddot{\xi} + k(1 + j\eta)\xi = t_{in}F. \tag{11.131}$$

The assumption of harmonic motion makes the structural damping model suitable for frequency response computations. Instead its use is rather critical for time domain computation. In this case the assumption of harmonic motion may not be valid; additionally, the complex stiffness generate complex forces that have no physical significance. It is common in this case to substitute the structural damping with the equivalent viscous damping model. If the structure is lightly damped, the effect of the damping is only relevant at the resonant frequencies. This is why a good approach is to compute each equivalent modal damping at the corresponding natural frequency,

$$c_{eq\ i} = \frac{\eta k_i}{\omega_{0\ i}} = \frac{\eta k_i}{\sqrt{k_i/m_i}} = \eta\sqrt{k_i m_i}, \tag{11.132}$$

so the equivalent modal damping factor is then:

$$\zeta_{eq\ i} = \frac{c_{eq\ i}}{c_{crit\ i}} = \frac{\eta\sqrt{k_i m_i}}{2\sqrt{k_i m_i}} = \frac{\eta}{2}. \tag{11.133}$$

The viscous damping computed using this approach is equivalent to the structural damping only at the resonant frequencies but not elsewhere. The error is negligible if the damping (either structural or viscous) is small, so that it plays a role only at resonance.

11.3.8 Model Reduction

The geometrical complexity of real structures requires a large number of finite elements for their discretization (meshing). In general, the larger the number of elements of the mesh, the more the finite element model is close to the geometry of the CAD drawing. General considerations regarding the geometrical complexity are particularly true for a car body structure comprising a large number of parts connected in different ways.

The large number of elements is then driven by the geometrical complexity, leading to finite element models with a large number of degrees of freedom. The availability at affordable cost of large computational power nowadays means that the numerical analysis of finite element problems with thousands or hundred thousands degrees of freedom is standard.

In dynamic analysis, the large number of degrees of freedom leads to a model characterized by natural frequencies covering a wide spectrum. In the case of a car body, natural frequencies range from some tens Hz for the global modes to several kHz for the local modes of small geometrical elements that concern relatively small and stiff portions of the structure.

Regardless of the number of degrees of freedom of the model, the dynamic analysis usually focuses on the modes, or on the forced response, at relatively low frequencies.

Similarly the finite element model of a vehicle subsystem such as the body or a suspension subframe could be integrated as a subsystem in a multibody software to determine the response of the vehicle to the maneuvers. The need in this case is to preserve, for each subsystem, the static, the inertial characteristics and the most important internal dynamics as they appear at the interfaces with the other subsystems. For example, a torsion beam axle is connected to the wheels, the body frame and the springs and the shock absorbers. In order to connect it to the vehicle model, the finite element model of the axle must be able to describe the interaction with the neighboring subsystems in the frequency range of interest for the vehicle dynamics.

It is then important to reduce the number of degrees of freedom without altering the accuracy of the model over the frequency range of interest.

The reduction techniques can be classified in two families depending on the type of degrees of freedom, physical displacements or modal ones, adopted for the model. The Guyan reduction is based on the reduction of the nodal displacements (physical coordinates), the Craig-Bampton one on the modal coordinates.

Guyan reduction

The equation of the motion and the output equation of an undamped structure in nodal coordinates are:

$$\begin{aligned} M\ddot{q} + Kq &= T_{in}F, \\ y &= T_{oq}q. \end{aligned} \tag{11.134}$$

For simplicity in the above equations the output is related to just the displacements and not to the nodal velocities or accelerations. The nodal degrees of freedom q can be arranged in two sets:

$$q = \begin{pmatrix} q_m & q_s \end{pmatrix}^T . \tag{11.135}$$

q_m includes the so-called master degrees of freedom that should be present in the reduced model, whereas q_s includes all other nodal displacements (subscript $_s$ stands for slave degrees of freedom) that should be eliminated. The choice of the degrees of freedom to include in the master set is arbitrary to a large extent, requiring experience and a certain degree of a priori knowledge of the behavior of the structure. The criterion is that of including in the master set the degrees of freedom associated with:

- the highest masses and stiffnesses.

- the point of application of external forces and torques
- the interface points with other substructures
- the degrees of freedom needed to describe the geometry of the deformed shape.

This rule may be simple to apply if the structure is lumped in nature, i.e. if it is constituted by relatively rigid bodies connected to each other via flexible elements. It is not simple at all in case of distributed structures, as is the case of a car body.

The partition between master and slave degrees of freedom can be extended to all the matrices of the model (mass, stiffness, input and output matrices). According to the previous discussion, if no external forces are applied to the slave set and no outputs are obtained from it, the dynamic and output equations become:

$$\begin{bmatrix} M_{mm} & M_{ms} \\ M_{sm} & M_{ss} \end{bmatrix} \begin{pmatrix} \ddot{q}_m \\ \ddot{q}_s \end{pmatrix} + \begin{bmatrix} K_{mm} & K_{ms} \\ K_{sm} & K_{ss} \end{bmatrix} \begin{pmatrix} q_m \\ q_s \end{pmatrix} = \begin{bmatrix} T_{in\ m} \\ 0 \end{bmatrix} F, \tag{11.136}$$

$$y = \begin{bmatrix} T_{oq\ m} & 0 \end{bmatrix} \begin{pmatrix} q_m \\ q_s \end{pmatrix}. \tag{11.137}$$

The assumption at the base of Guyan reduction is that the slave degrees of freedom can be obtained from the master ones with a static link, obtained from the second group of equations 11.136 in which the inertia forces are neglected:

$$\begin{bmatrix} M_{mm} & M_{ms} \\ 0 & 0 \end{bmatrix} \begin{pmatrix} \ddot{q}_m \\ \ddot{q}_s \end{pmatrix} + \begin{bmatrix} K_{mm} & K_{ms} \\ K_{sm} & K_{ss} \end{bmatrix} \begin{pmatrix} q_m \\ q_s \end{pmatrix} = \begin{bmatrix} T_{in\ m} \\ 0 \end{bmatrix} F, \tag{11.138}$$

the absence of dynamic terms in the group of equations corresponding to the slave set enables the slave degrees of freedom q_s to be determined as function of the master set:

$$q_s = -K_{ss}^{-1} K_{sm} q_m = \phi_c q_m. \tag{11.139}$$

The columns of matrix $\phi_c = -K_{ss}^{-1} K_{sm}$ are the displacements of the slave nodes corresponding to unit displacements of each of the master degrees of freedom (subscript $_c$ indicates constrained).

The relation between master and slave degrees of freedom of Eq. 11.139 is not approximated only in static conditions, as the inertia forces are null.

From another point of view, Eq. 11.139 represents a sort of kinematic link between q_s and q_m. Extending the validity of this expression to the dynamic case is the same as considering that also in dynamic conditions the slave displacements can be obtained from the master ones with the same static deformed shapes. The Guyan reduction is based on the approximation of all degrees of freedom of the structure by means of the master ones:

$$q = \begin{pmatrix} q_m \\ q_s \end{pmatrix} \simeq \begin{pmatrix} I \\ \phi_c \end{pmatrix} q_m = \phi_G q_m. \tag{11.140}$$

The mass and stiffness matrices of the reduced model can be obtained following a Lagrangian approach. Extending Eq. 11.140 also to the velocity,

$$\dot{q} \simeq \phi_G \dot{q}_m; \tag{11.141}$$

the kinetic and elastic potential energy of the structure are then approximated as function of the master degrees of freedom

$$T = \frac{1}{2}\dot{q}^T M \dot{q} \simeq \frac{1}{2}\dot{q}_m^T \phi_G^T M \phi_G \dot{q}_m, \tag{11.142}$$

$$U = \frac{1}{2}q^T K q \simeq \frac{1}{2}q_m^T \phi_G^T K \phi_G q_m. \tag{11.143}$$

The mass M_m and stiffness K_m matrices of the reduced model are then

$$M_m = \phi_G^T M \phi_G, \tag{11.144}$$

$$K_m = \phi_G^T K \phi_G. \tag{11.145}$$

The generalized force vector acting on the reduced model can be obtained by considering that the virtual work made by the generalized forces $T_{in}F$, for a virtual displacement δq, can be approximated by taking Eq. 11.140 into account:

$$\delta L = F^T T_{in}^T \delta q \simeq F^T T_{in}^T \phi_G \delta q_m = F^T T_{in\ m}^T \delta q_m \tag{11.146}$$

the input matrix of the reduced model $T_{in\ m}$ is then:

$$T_{in\ m} = \phi_G^T T_{in}. \tag{11.147}$$

Similarly for the output equation:

$$y = T_{oq} q \simeq T_{oq} \phi_G q_m = T_{oq\ m} q_m, \tag{11.148}$$

the output matrix of the reduced model $T_{oq\ m}$ is

$$T_{oq\ m} = T_{oq} \phi_g. \tag{11.149}$$

Finally, the dynamic equations of the model obtained from the Guyan reduction are:

$$\begin{array}{c} M_m \ddot{q}_m + K_m q_m = T_{in\ m} F \\ y = T_{oq\ m} q_m \end{array}. \tag{11.150}$$

In principle the Guyan reduction permits a good approximation of dynamic behavior of the complete system; however its accuracy is related to the selection of the master degrees of freedom and their number.

This choice is not easily automatized, it requires some experience or at least trial and error. In the case of models including lumped masses and stiffnesses the choice is relatively straightforward. Nevertheless, reducing an already small model may be a rather questionable choice. The reduction becomes more necessary, albeit more complicated, when the model includes thousands of degrees of freedom.

Modal reduction

The aim of the modal reduction techniques is again to reduce the number of degrees of freedom. The difference is that they act on the dynamic equations expressed in modal (rather than nodal) coordinates. Taking Eq. 11.101 into account, and assuming again that the output is related to the displacements only, the dynamic equations of the undamped system in modal coordinates is

$$\begin{aligned} \ddot{\xi} + \omega_0^2 \xi &= t_{in} F, \\ y &= t_{oq} \xi. \end{aligned} \tag{11.151}$$

Similarly to that performed for Guyan reduction, the modal coordinates can be partitioned in a master set (ξ_m) and a slave set (ξ_s), the latter being removed from the reduced model:

$$\xi = \left(\xi_m \; \xi_s \right)^T . \tag{11.152}$$

The selection regarding what to include in the master and slave sets can be done considering the frequency range of interest in the analysis. This can be estimated considering the frequency content of the excitation or the type of response that must be analyzed. If, for example, the excitation is in the 10÷200 Hz, the reduced model should include the modes that have their resonance in that frequency range. The response of the modes with higher natural frequencies (say 400 Hz) will be static to a good approximation. In a similar way to Guyan reduction, the inertial contribution of the high frequency modes can be neglected. Their natural frequency, in fact, is larger than that of the excitation:

$$\begin{bmatrix} I & 0 \\ 0 & 0 \end{bmatrix} \begin{pmatrix} \ddot{\xi}_m \\ \ddot{\xi}_s \end{pmatrix} + \begin{bmatrix} \omega_{0m}^2 & 0 \\ 0 & \omega_{0s}^2 \end{bmatrix} \begin{pmatrix} \xi_m \\ \xi_s \end{pmatrix} = \begin{bmatrix} t_{in\ m} \\ t_{in\ s} \end{bmatrix} F. \tag{11.153}$$

The response of the slave modes can then be approximated as,

$$\xi_s \simeq \left(\omega_{0s}^2 \right)^{-1} t_{in\ s} F. \tag{11.154}$$

An alternative way to compute the static response of the slave modes is to consider that the displacement in physical coordinates is given by a contribution of the master and a slave set:

$$q = \phi_m \xi_m + \phi_s \xi_s = q_m + q_s$$

The contribution due to the slave can then be obtained from the total displacement q, and the master q_m

$$q_s = q - q_m = q - \phi_m \xi_m$$

In static conditions, the total displacement is obtained from the external loads and the stiffness matrix

$$q = K^{-1} T_{in} F,$$

Similarly, the static displacement of the master modes can be obtained from Eq. 11.153 by neglecting the inertia forces,

$$\xi_m = \left(\omega_{0s}^2\right)^{-1} t_{in\ m} F = \left(\omega_{0s}^2\right)^{-1} \phi_m^T T_{in} F$$

enabling the static contribution due to the slave modes to be determined from the total static displacement and the static displacement of the master modes,

$$q_s = K^{-1} T_{in} F - \phi_m \left(\omega_{0s}^2\right)^{-1} \phi_m^T T_{in} F = \tag{11.155}$$

$$= \left(K^{-1} - \phi_m \left(\omega_{0s}^2\right)^{-1} \phi_m^T\right) T_{in} F = K_s^{-1} T_{in} F \tag{11.156}$$

The stiffness matrix K_s relates the external forces to the static displacements due to the slave modes (q_s).

The decoupling between the modal coordinates allows the response of the master degrees of freedom to be computed without considering the response of the slave modes. The dynamic equation of the reduced system is the same as that which governs the master modes:

$$\ddot{\xi}_m + \omega_{0m}^2 \xi_m = t_{in\ m} F. \tag{11.157}$$

The effect of the damping can be added in terms of viscous modal damping:

$$\ddot{\xi}_m + 2\zeta_m \omega_{0m} \dot{\xi}_m + \omega_{0m}^2 \xi_m = t_{in\ m} F. \tag{11.158}$$

Similarly as with the inertia terms, the effect of the damping on the slave modes is neglected. Eq. 11.154 is then valid also for the damped system. The output matrix t_{oq} of Eq. 11.151 is usually full. The output variables y, then, include the contribution of both master and slave modes

$$y = t_{oqm} \xi_m + t_{oqs} \xi_s \simeq t_{oqm} \xi_m + t_{oqs} \xi. \tag{11.159}$$

The slave modes response is approximated with Eq. 11.154 as a static one. The result is

$$y \simeq t_{oqm} \xi_m + t_{oqs} \left(\omega_{0s}^2\right)^{-1} t_{in\ s} F. \tag{11.160}$$

The output equation of the reduced model includes direct coupling between the input forces and the output variables. This term is the outcome of the static response of the slave modes. An alternative approach to find the output is to involve the static displacement due to the slave modes of Eq. 11.155

$$y = t_{oqm} \xi_m + t_{oqs} \xi_s = t_{oqm} \xi_m + T_o q_s$$
$$= t_{oqm} \xi_m + T_o K_s^{-1} T_{in} F$$

The advantage of this last expression is that it requires the computation of only the master mode shapes and natural frequencies. The alternative expression of Eq. 11.160 requires the computation of all natural frequencies and mode shapes.

This task could be quite time consuming in the case of matrices with large dimensions.

In general, the modally reduced model can be used to compute the response of the system in the time or in the frequency domain. To gain some insight on the system properties of the reduced order model, it is worth computing the response by means of the Laplace transform. The transfer function between the i–th master modal coordinate can be obtained from Eq. 11.158

$$\frac{\xi_{m\ i}(s)}{F(s)} = \frac{t_{in\ m\ i}}{s^2 + 2\zeta_{m\ i}\omega_{0m\ i}s + \omega_{0m\ i}^2} \qquad i = 1, \ldots, n_{master}. \tag{11.161}$$

Substituting in the output equation 11.160 it is possible to determine the transfer function between the input force F and the output y

$$\frac{y}{F} \simeq \sum_{i=1}^{n_{master}} \frac{t_{oq\ m\ i}\, t_{in\ m\ i}}{s^2 + 2\zeta_{m\ i}\omega_{0m\ i}s + \omega_{0m\ i}^2} + t_{oqs}\left(\omega_{0s}^2\right)^{-1} t_{in\ s}. \tag{11.162}$$

The direct input-output coupling is represented in this case by the static gain $t_{oqs}\left(\omega_{0s}^2\right)^{-1} t_{in\ s}$.

One of the main advantages of the modal reduction over Guyan reduction is that the selection of the master modes can be performed by considering the natural frequencies of the modes that must be included in the master set. The choice can then be performed computing only the modes that are in the frequency range of interest. The higher order modes contributes to the static mode.

Component mode synthesis - Craig Bampton reduction

Usually complex structures are assemblies of parts connected through mechanical interfaces. For example, the door of a vehicle is connected to the frame by two hinges, a lock mechanism and a number of seals. The engine and power train is connected to the frame by means of a (small) number of connection points. The modeling of such kind of structures can exploit the discrete nature of the interfaces. The idea is to model each part of the structure as subsystem that interacts with the neighboring parts through a discrete number of mechanical interfaces. In principle this is similar to what happens at a much smaller scale during the assembly of each finite element to discretize a structure. Each element is a portion of the structure that interacts with the neighboring ones by means of interfaces that are the nodes. This is why each subsystem is usually indicated as a superelement or substructure. Each superelement is characterized by internal nodes (subscript $_i$) and boundary nodes (subscript $_b$). The characteristic matrices of each superelement are built individually by means of finite element codes, they are then assembled considering equilibrium and equal displacement conditions at the interfaces. The result is a model that represents the whole structure.

Fig. 11.47 shows a structure including parts A and B. Each part is modeled as a superelement by means of the finite element method. The internal nodes ($n_{i1}, \ldots,$ n_{im}, are associated the degrees of freedom of vector q_i) interact only with nodes

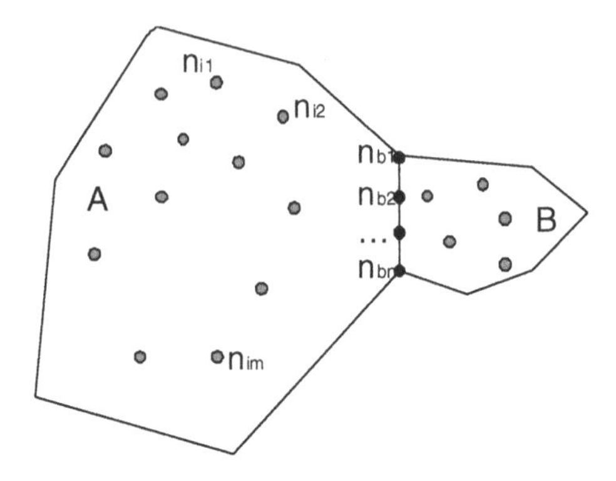

Fig. 11.47. Substructuring of a complex structure in two substructures A and B. Each substructure is discretized in finite elements. The internal nodes are indicated with subscript $_i$. The interface inodes are in common to both substructures (boundary nodes, subscript $_b$) $n_{b1}, \ldots, n_{bn}$.

of the same superelement. The boundary nodes ($n_{b1}, \ldots$, n_{bn}, are associated to the degrees of freedom of vector q_b) are in common to superelements A and B.

The above considerations regarding superelements do not necessarily imply any reduction technique; they only involve a way to organize a finite element model which takes the layout of the structure into account. The aim of the component mode synthesis (introduced by Craig and Bampton of NASA in the early '70s) is to reduce the number of degrees of freedom of a superelement preserving the possibility to assemble it to the other superelements. This is obtained by preserving the boundary degrees of freedom also in the reduced model, so that it is possible to exploit them during the assembly of the superelements.

The basic consideration is that the displacements of a superelement are given by two contributions:

- deformed shape (q_{st}) due to given displacements of the boundary nodes;
- combination of the mode shapes of the superelement with constrained boundary nodes (q_R).

The first contribution (static deformed shape q_{st}) is the same obtained by applying the Guyan reduction with the interface nodes as the only master degrees of freedom ($q_m = q_b$). All internal nodes are considered as slave ($q_s = q_i$). Taking Eq. 11.140 into account, this contribution is:

$$q_{st} = \begin{pmatrix} q_b \\ q_i = q_{i\ st} \end{pmatrix} \simeq \begin{pmatrix} I \\ \phi_c \end{pmatrix} q_b. \tag{11.163}$$

This contribution takes into account that, in static conditions, the displacements of the boundary nodes induces a deformation in all the structure, nevertheless it neglects that the inner part of the superelement is characterized by mode shapes that cannot be represented only by the boundary displacements and the associated deformed shapes.

The second contribution (q_R) adds to this a combination of the mode shapes with fixed boundary nodes. The interface nodes are fixed as if the rest of the structure (the other superelements) were able to let the displacements at the interface nodes.

$$q_R = \begin{pmatrix} q_b = 0 \\ q_i = \phi_R \xi_R \end{pmatrix} \simeq \begin{pmatrix} 0 \\ \phi_R \end{pmatrix} \xi_R. \tag{11.164}$$

Adding the two contribution:

$$q = q_{st} + q_R; \tag{11.165}$$

in matrix form:

$$q = \begin{bmatrix} I & 0 \\ \phi_c & \phi_R \end{bmatrix} \begin{Bmatrix} q_b \\ \xi_R \end{Bmatrix}. \tag{11.166}$$

The displacement vector of the model in physical coordinates (q) is approximated with a smaller number of reduced order coordinates (ξ_{CB}, subscript $_{CB}$ stands for Craig-Bampton):

$$q = \phi_{CB} \xi_{CB}, \tag{11.167}$$

where:

$$\xi_{CB} = \{q_b, \xi_R\}^T, \tag{11.168}$$

$$\phi_{CB} = \begin{bmatrix} I & 0 \\ \phi_c & \phi_R \end{bmatrix}. \tag{11.169}$$

In a similar way to Guyan reduction, the mass and stiffness matrices of the superelement can be rewritten as function of the reduced order coordinate vector ξ_{CB}, this can be done with a Lagrangian approach from the expressions of the kinetic and potential energy. Following the same steps of Eq.s 11.142, 11.143 and taking Eq. 11.167 into account

$$\begin{aligned} m_{CB} &= \phi_{CB}^T M \phi_{CB}, \\ k_{CB} &= \phi_{CB}^T K \phi_{CB}. \end{aligned} \tag{11.170}$$

Similarly, the input and output matrices are:

$$\begin{aligned} t_{in\ CB} &= \phi_{CB}^T T_{in}, \\ t_{oq\ CB} &= T_{oq} \phi_{CB}. \end{aligned} \tag{11.171}$$

The lowercase letters have been adopted for the reduced order matrices to indicate that part of the coordinates of the reduced superelement are modal.

To understand the structure of the reduced order matrices (m_{CB}, k_{CB}), the mass and stiffness matrices can be partitioned according to the boundary (subscript $_b$) and internal (subscript $_i$) degrees of freedom:

$$M = \begin{bmatrix} M_{bb} & M_{bi} \\ M_{ib} & M_{ii} \end{bmatrix}; \qquad K = \begin{bmatrix} M_{bb} & M_{bi} \\ M_{ib} & M_{ii} \end{bmatrix}. \tag{11.172}$$

Substituting the partitioned matrices M and K (Eq. 11.172) in Eq. 11.170 and taking into account of the structure of the coordinate transformation matrix ϕ_{CB} of Eq. 11.169, the reduced matrices of the superelement can be partitioned according to the boundary and modal degrees of freedom (subscript $_R$)

$$m_{CB} = \begin{bmatrix} m_{bb} & m_{bR} \\ m_{Rb} & m_{RR} \end{bmatrix}, \qquad k_{CB} = \begin{bmatrix} k_{bb} & k_{bR} \\ k_{Rb} & k_{RR} \end{bmatrix}, \tag{11.173}$$

where:

$$\begin{matrix} m_{bb} = M_{bb} + M_{bi}\phi_c + \phi_c^T M_{ib} + \phi_c^T M_{ii}\phi_c \\ m_{bR} = m_{Rb}^T = M_{bi}\phi_R + \phi_c^T M_{ii}\phi_R \\ m_{RR} = \phi_R^T M_{ii}\phi_R \end{matrix} ; \begin{matrix} k_{bb} = K_{bb} + K_{bi}\phi_c + \phi_c^T K_{ib} + \phi_c^T K_{ii}\phi_c \\ k_{bR} = k_{Rb}^T = K_{bi}\phi_R + \phi_c^T K_{ii}\phi_R \\ k_{RR} = \phi_R^T K_{ii}\phi_R \end{matrix} . \tag{11.174}$$

As eigenvectors ϕ_R refer to a superelement with constrained boundary nodes, they make diagonal the part of the mass and stiffness matrices corresponding to the internal degrees of freedom. Sub matrices (m_{RR} and k_{RR}) are then diagonal. The static deformed shapes (ϕ_c), on the other hand, do not benefit of any form of K and M orthogonality. Therefore, the corresponding sub-matrices maintain a full structure (m_{bb}, k_{bb}). Similarly, also the off diagonal sub-matrices are (m_{bR}, k_{bR}) are not diagonal.

The dynamic and output equations that can be obtained for each superelement by means of the Craig-Bampton reduction are, finally

$$\begin{matrix} m_{CB}\ddot{\xi}_{CB} + k_{CB}\xi_{CB} = t_{in\ CB}F \\ y = t_{oq\ CB}\xi_{CB} \end{matrix} . \tag{11.175}$$

The reduction of the internal degrees of freedom do no affect the boundary nodes (Fig. 11.47) the superelements can then be assembled with the same considerations underlying the assembly of each element of a finite element model. The degrees of freedom of the assembled structure is:

$$\xi_{AB} = \{q_b, \xi_{R\ A}, \xi_{R\ B}\}^T , \tag{11.176}$$

where $\xi_{R\ A}$, $\xi_{R\ B}$ are the modal coordinates that describe the internal modes of superelements A and B. The mass matrix of the assembled structure (m_{AB}) is then:

$$m_{AB} = \begin{bmatrix} m_{bb\ A} + m_{bb\ B} & m_{bR\ A} & m_{bR\ B} \\ m_{Rb\ A} & m_{RR\ A} & 0 \\ m_{Rb\ B} & 0 & m_{RR\ B} \end{bmatrix} . \tag{11.177}$$

The sub-matrices corresponding to the boundary degrees of freedom are added together, those corresponding to the internal degrees of freedom remain decoupled. The assembled matrices have then a diagonal part (that of the internal degrees of freedom in modal coordinates $m_{RR\ A}$ and $m_{RR\ B}$) and a full one (boundary and to the mixed degrees of freedom).

The Craig-Bampton reduction finds wide application in most finite element codes since it draws on some of the advantages of both Guyan and modal reduction. From the first it preserves the possibility of maintaining some physical degrees of freedom (boundary nodes) used to assemble different superelements; from the second it allows to chose what modes to include in the reduced model considering their frequency.

The superelement approach is very convenient also from the point of view of the organization of the modeling work. The models of the different substructures can be prepared as separate tasks, after the definition of the mutual interfaces. They will be assembled in a second phase. An additional benefit is that the model of each substructure can be validated by means of experiments before it is assembled to the other superelements. The advantage is a better possibility to identify the unknown parameters and to gain more insight in behavior of each substructure.

11.3.9 Cavity Modes

Together with the dynamic behavior of the body structure, another important aspect which influences the noise and vibration perceived by the occupants is the dynamic behavior of the air cavity inside the vehicle. Similarly to the surrounding structure, the cavity is characterized by vibration modes and resonances that can amplify the vibrations induced by the excitation.

Despite being characterized by resonances, the air cavity has some peculiarities relative to a conventional structure. The most evident one is that the air (as a gas) changes its pressure because of a volume change. Conversely, a solid reacts to the mechanical stress in one direction with a strain, that is a variation of its length. In the case of a uniaxial stress (σ) in a homogeneous isotropic and elastic material (such as a specimen for uniaxial characterization tests) induces a strain (ϵ) in the direction of the stress and a strain of opposite sign in the orthogonal directions. The differences between a solid and a gas are also evidenced by the different kind of variables used to characterize them. The pressure and the volume are scalar quantities, whereas the strain and stress are tensors.

Regardless of the differences between a gas and an elastic solid, an air cavity can be considered as a volume with a distributed mass and stiffness. As such the cavity is characterized by natural frequencies and mode shapes. Its response to an excitation is the superposition of different modal contributions. The configuration variable in this case is not the displacement vector but the pressure or the variation of the volume, that are scalar quantities.

The natural frequencies of a cavity with a simple geometry (such as a cylinder) can be approximated considering that the corresponding period of oscillation is equal to the time required to a sound wave to travel from one end to the next and coming back. If the length of the cavity is l the natural frequencies of the cavity are

$$f_i = i\frac{c}{2l}, \qquad i = 1, 2, 3, \ldots, \tag{11.178}$$

where c is the speed of sound in the cavity and i is an index representing the different harmonics,

$$c = \sqrt{\frac{\gamma p}{\rho}} = \sqrt{\frac{\gamma RT}{M;}} \tag{11.179}$$

γ is the adiabatic constant of the gas, p the pressure, ρ the mass density, R the constant of the gas, T the absolute temperature and M the molar mass. In the case of the interior of a vehicle the fluid is the air and the temperature is about 20 °C, so that:

γ =1.4,
M=0.029 kg/mole (mean molar mass for the air),
T=293 °K,
this leads to:

$$c = 344 \text{ m/s.} \tag{11.180}$$

Considering a prismatic cavity with sides a, b and l, the three first order natural frequencies are:

$$\begin{aligned} f_b &= \tfrac{c}{2b}, \\ f_h &= \tfrac{c}{2h}, \\ f_l &= \tfrac{c}{2l}. \end{aligned} \tag{11.181}$$

Along the same directions, in addition to the first resonances, higher order resonances occur. Increasing the frequency, the complexity of the mode shape increases; even in prismatic cavities, more than one direction is involved at the same time.

The cavity modes are characterized by regions where the pressure remains constant in an analogous way to the nodal regions (or points) of the mode shapes of a vibrating structure (i.e. the points where the mode shape has zero displacement). If the ears of an occupant are located close to the nodal region of a cavity, very little (or, at the limit, none) of this resonance is perceived, with a positive effect on its acoustic comfort.

Similarly, a source of vibration (for example a vibrating panel) located close to one of these nodes is not able to excite the corresponding cavity mode, and so the air borne noise at that frequency will be low. However the opposite is also true: if the location of the sources or the ears is close to point of maximum pressure variation, the noise intensity is amplified by the cavity resonance, resulting in low acoustic comfort.

Even if the inner volume of a vehicle is far from being a simple geometry, it is possible to determine a good estimate of the first resonances from its main dimensions, as if it were a prismatic cavity. For example a sub-compact vehicle (B segment) may have an interior with longitudinal size of 2.75 m and a width of 1.36 m. The natural frequencies estimated with Eq. 11.181 are of about 62.5 Hz and 126.5 Hz, that a quite close to the values of 65.7Hz and 132.4 obtained from a finite element model of the real volume.

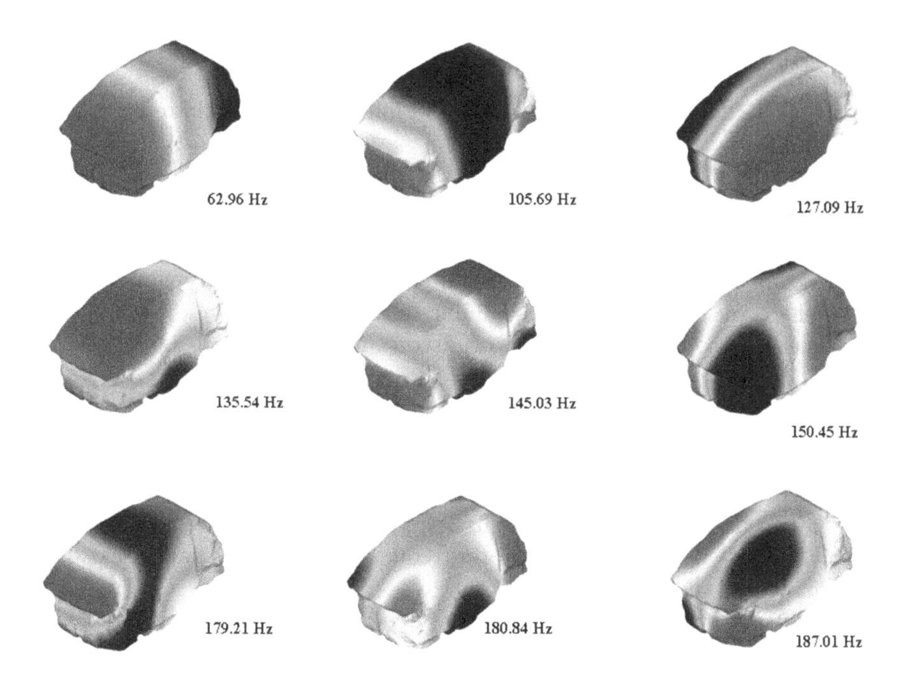

Fig. 11.48. Cavity mode shapes of a 2 volumes vehicle.

Table 11.3. Frequencies of the first cavity resonances for different classes (or segments) of vehicles.

cavity mode	sub-compact	compact	large
1st longitudinal	60÷70 Hz	55÷65 Hz	45÷55 Hz
2nd longitudinal	115÷130 Hz	100÷115 Hz	85÷100 Hz
1st lateral	130÷145 Hz	120÷135 Hz	110÷130 Hz

As the main part of the first natural frequencies of the cavity is given by the overall dimensions, vehicles of the same interior dimensions are characterized by similar natural frequencies. Vehicles of the same class or segment (A, B,.... or sub-compact, compact,...) have acoustic resonances quite close to each other, the structural resonances, on the other hand may be substantially different. Tab. 11.3 lists the cavity resonances for different vehicle classes.

Another factor that has a considerable influence on the dynamics of the interior cavity of a vehicle is the trunk. This additional cavity can interact with the interior in different ways depending on the type of connection. In some cases the two volumes are separated by a panel (as it is in most three volume cars), so that the interaction occurs due to the vibrations of the structure. In other cases the trunk volume can be connected directly to the main interior with just a relatively light cloth to separate them (e.g. as happens most station wagon cars). Sometimes the main connection is through the rear seat back that may have a certain level of acoustic transparency especially if the rear frame does not include a metallic panel.

In this context it is important to take the following aspects into account:

- the noise transmitted to the trunk can return to the main interior volume,
- the dynamics of the two cavities can interact with each other and change the behavior of the main cavity considerably. This could lead to additional resonances or it could even be exploited to realize sort of dynamic dampers of the acoustic cavity (this is similar to the tuned dynamic dampers used sometimes to reduce the effect of a structural resonance).

Unfortunately the possibility of acting at the design level on the acoustic mode shapes is relatively low. In fact it is not possible to modify the size of the interior or the characteristic of the air. The interior shape and layout, the configuration of the seats, the dashboard and the trimmings are determined mainly by other functional needs (ergonomics, accessibility, reachability), and style. It is very difficult therefore to modify the natural frequencies or mode shapes because of a need related to the interior acoustics.

Although the possibility to modify the acoustic resonances is rather small, the knowledge of the dynamics of the cavity is very important for the design of the surrounding structure or the elements along the transmission paths. The knowledge of the acoustic behavior enable the design of the dynamics of the structure in order to minimize the excitation of the most important acoustic resonances. For example, the firewall should be designed so that its structural resonances are not the same as the cavity; similarly, the engine suspension attachment points should be located close to nodal regions of the main structural modes of vibration.

11.3.10 Radiation from the Panels

The previous section has illustrated that the cavity and the structural modes can interact with each other. The vibrations of the panels induce accelerations in the air and, therefore pressure waves that propagate inside the cavity.

The basic mechanism that leads to the pressure waves is demonstrated by the Euler's equation 11.3, a plate that vibrates with speed u in direction r produces a pressure gradient along direction r proportional to the acceleration of the plate.

Another important parameter is the radiation efficiency. Close to the surface, the accelerations always lead to pressure variations, although this does not mean that the pressure wave can effectively propagate in the surrounding volume. Consider, for example, a small spherical surface with the diameter that grows and shrinks periodically. Because of the accelerations of the surface, the pressure in the surrounding air changes. The pressure wave propagates symmetrically in all directions. This is the case of a unipolar source characterized by a large radiation efficiency. If the sphere does not change its diameter, but it oscillates in one direction with constant diameter (bipole), the pressure that increases on one side corresponds to an equal pressure reduction on the opposite side. The air can move by following the pressure gradient and trying to reduce it. The propagation of the pressure wave is less effective than in the previous case and the radiation efficiency is smaller.

A vibrating panel can have regions that move in one direction and regions that move in the opposite direction at the same time. Similarly to the case of the oscillating sphere, the pressure that grows in some parts can be compensated by the reduction in other regions. The radiation efficiency in this case is influenced by the size and amplitude of the parts that move in the opposite directions and the frequency (the smaller the frequency, the more the air at higher pressure can move to fill in the pressure gap).

The radiation efficiency (σ) is the ratio between the acoustic power radiated from a source of area S and the power radiated by a piston of the same area that vibrates with a frequency large enough to make the pressure wavelength negligible compared to the size of the piston:

$$\sigma = \frac{P}{\rho c \bar{v}^2}, \tag{11.182}$$

where P is the acoustic power radiated from a given surface of area S, ρ is the mass density of the fluid, c the speed of sound, $\bar{v}^2$ the mean square speed value of the radiating surface. Two panels that vibrate with the same mean square speed value do not necessarily generate the same level of noise because of different radiation efficiencies.

For flat or nearly flat surfaces, it is possible to evaluate the sound power by following a simplified approach. The surface can be discretized in small parts each can be considered as spherical point source. The sound pressure generated by each source is:

$$p = i\omega\rho\frac{\tilde{Q}}{2\pi r}e^{-j\frac{\omega}{c}r} \tag{11.183}$$

where $\tilde{Q}$ is the volume speed of the source, r the distance from the source, and ω the angular frequency. Considering the contribution of an infinitesimal element of area dS moving with speed $v(S)$

$$d\tilde{Q}(S) = v(S)dS. \tag{11.184}$$

Adding the contributions of all surface elements dS,

$$p = \frac{i\omega\rho}{2\pi}\int_S \frac{v(S)e^{-j\frac{\omega}{c}r}}{r}dS. \tag{11.185}$$

This allows to estimate the sound pressure radiated from a flat surface that vibrates with a speed distribution $v(S)$. Each mode shape has a different speed distribution, leading to different sound pressure levels, even for the same levels of the mean square speed.

11.4 Engine Suspension

The power train, including engine with all its accessories, including the clutch, gearbox and differential, is a rather heavy and rigid subsystem connected to the frame by a dedicated suspension, i.e. by compliant and dissipative elements the main functions of which are:

- To connect the power train to the chassis transmitting the forces and torques involved in the dynamic equilibrium including the vertical and longitudinal loads when crossing a sharp obstacle such as a cleat or sharp hole. Significant torque is transmitted to the wheels (engine, gearbox and differential in one package) or to the differential (engine and gearbox transmitting the torque to a differential by means of a drive-shaft).
- To filter the dynamic excitations produced by the engine to reduce the noise transmitted to the vehicle via structural paths. A rigid link between the engine and the chassis is sometimes adopted, especially on race cars, but involves high vibration and noise levels that are difficult to accept for normal cars.

In order to reduce the vibrations transmitted by the power train to the rest of the vehicle, it is possible to exploit the dynamic attenuation that follows a resonance of a mechanical oscillator. Considering only the vertical displacements of the engine, its dynamic behavior can be described (at least as first approximation) by the simplified model of Fig. 11.49. The engine is represented by a rigid mass M connected to the supporting structure by the engine suspension, represented by the parallel of the spring of stiffness K and a viscous damper of damping coefficient C. If the displacement of the vehicle structure due to the engine vibrations is negligible, the dynamics of the system reduce to that of a single degree of freedom mechanical oscillator. The displacement q in vertical direction and the force F_{out} transmitted to the structure are governed by Newton's equation and the mechanical characteristic of the spring-damper assembly

$$M\ddot{q} + C\dot{q} + Kq = F_{in}, \tag{11.186}$$

$$F_{out} = C\dot{q} + Kq. \tag{11.187}$$

The transfer function (transmissibility function T) between the input force F_{in} acting on the engine and that F_{out} transmitted to the frame is:

$$T = \frac{F_{out}}{F_{in}} = \frac{Cs + K}{Ms^2 + Cs + K}, \tag{11.188}$$

where s is the Laplace variable.

The numerator and denominator of the same transfer function can be made adimensional dividing each by the mass M:

$$T = \frac{2\zeta s^* + 1}{s^{*2} + 2\zeta s^* + 1}, \tag{11.189}$$

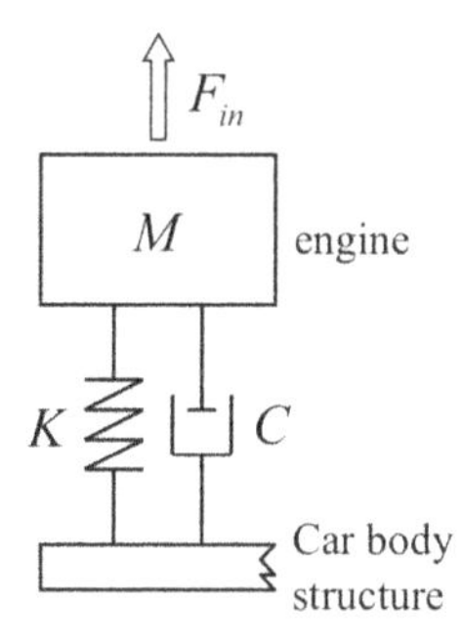

Fig. 11.49. One degree of freedom model of the engine suspension. K and C are the stiffness and damping coefficient of the suspension in vertical direction, M is the mass of the powertrain.

where $\zeta = C/(2\sqrt{KM})$ is the damping factor and $s^* = s/\omega_n$ is the ratio between the Laplace variable and the undamped natural frequency $\omega_n = \sqrt{K/M}$.

Fig. 11.50 shows the amplitude of the transmissibility as function of the non dimensional frequency $s^* = i\omega^* = i\omega/\omega_n$. The different curves correspond to different values of the damping factor in the interval $0.001 \div 0.4$.

An internal combustion engine generates dynamic excitations due to the inertia of the alternating parts and to the combustion forces. The characteristics of these forces is function of the number of the cylinders and the engine architecture (6 in line, 6 V,...). In a four stroke engine, the frequency of the combustion forces is half the number of cylinders times the engine speed. For example, in a four cylinders engine the fundamental frequency of excitation is twice the engine speed. The typical values of the excitation frequency in this case are between 20 and 200 Hz, corresponding to engine speeds between 600 and 6.000 rpm. In an eight cylinder engine, the fundamental harmonic of the combustion forces is between 40 and 400 Hz for the same engine speeds. With the engine running at idle, it is necessary to avoid that the low excitation frequencies induce shaking of the vehicle frame. Conversely at high rpm, it is necessary to reduce the acoustic excitations transmitted to the interior which could be amplified by the acoustic resonances (boom, Tab. 11.3).

All transmissibility functions of Fig. 11.50 have a common unity amplitude ($T = 1$) at $\omega^* = \sqrt{2}$, whereas at higher excitation frequencies the transmitted force is attenuated. If $\omega_{exc\ \min}$ is the minimum excitation frequency, the forces transmitted to the structure are attenuated if the resonant frequency of the suspension is smaller than

$$\omega_n < \frac{\omega_{exc\ \min}}{\sqrt{2}}. \tag{11.190}$$

For a given excitation frequency range, the lower the resonant frequency of the suspension, the higher the attenuation level. As the mass of the power train is given, the choice of the resonant frequency requires the design of the suspension

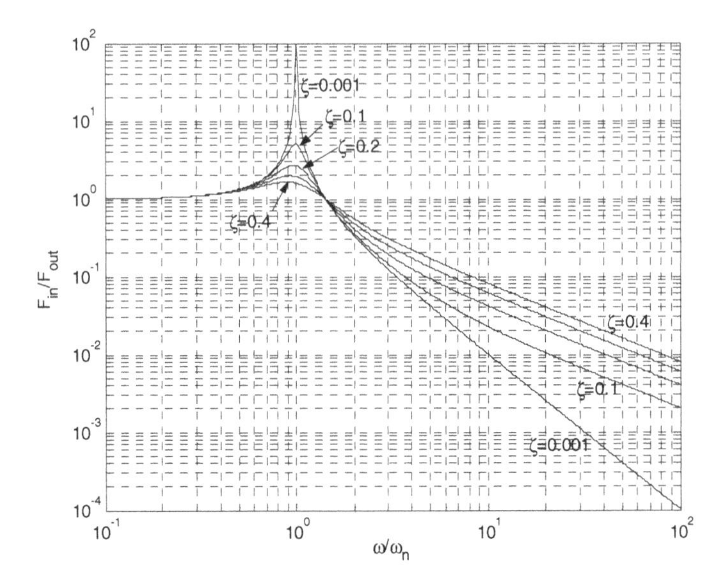

Fig. 11.50. Transfer function between the force generated by the engine (F_{in}) and the force transmitted to the vehicle structure (F_{out}). The curves refer to different damping values.

stiffness K. This last value cannot be reduced below a certain limit to avoid unwanted contacts between the power train and engine compartment during high accelerations (due to sharp obstacles, for example) or when the output torque is maximum.

The choice of the resonant frequency is therefore a compromise between the need to provide high attenuation on one hand and avoid contacts on the other. A four cylinder, four stroke engine usually has an undamped natural frequency of the engine suspension between 10 and 20 Hz.

11.4.1 Engine Suspension Mounts

To achieve a compromise between the needs of a low suspension stiffness for small amplitude oscillations and that of limiting the maximum displacements at high loads, the elastomeric mounts used for the engine suspension usually have a nonlinear force (F) to displacement (z) characteristic. As shown in Fig. 11.51, the stiffness increases considerably at the end of the displacement range. This is obtained by increasing the portion of elastomeric material is involved in the transmission of the force. This is obtained, for example, by exploiting the contact between one part of the mount and the other for displacements larger than a given limit. Additionally, the configuration of the mount is devised with different force

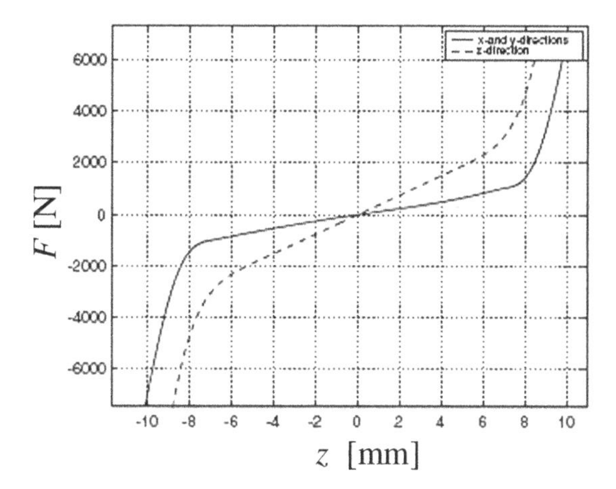

Fig. 11.51. Force to displacement characteristic along the x and z directions. The stiffness increases at the end of the displacement range.

to displacement characteristics in different directions to cope with the different acceleration values that occur in vertical, longitudinal and lateral directions.

Fig. 11.54 shows the undeformed configuration (left) with that corresponding to the maximum upwards (center) and downwards displacements (right). During deformation the rubber element is mainly deformed in shear enabling it to exploit the large dissipations that occur in the viscoelastic material under this type of loading. The nonlinear part of the characteristic starts when the displacements are sufficiently high to close one of the slots. For high loads, the upper or lower slot closes so that the rubber around the inner part of the mount is directly in contact with the outer part ring, the result being increased stiffness. Displacements perpendicular to those represented in the figure involve a completely different type of deformation in the elastomeric material, involving mainly tensile and compression deformation, hence justifying the non isotropic characteristic.

11.4.2 Role of the Damping in Engine Suspensions

Fig. 11.50 shows that for frequencies lower than $\omega^* < \sqrt{2}$, the transmissibility decreases with increasing damping; in this frequency range it is then convenient to have a high damping to reduce the amplitude of the resonance peak. Conversely for $\omega^* > \sqrt{2}$, the transmissibility increases with increasing damping; in this frequency range it is then convenient to have an engine suspension with the lowest damping possible. This is also clear from Eq. 11.189: The larger the damping ζ, the smaller the zero frequency s_z^* of the transmissibility function

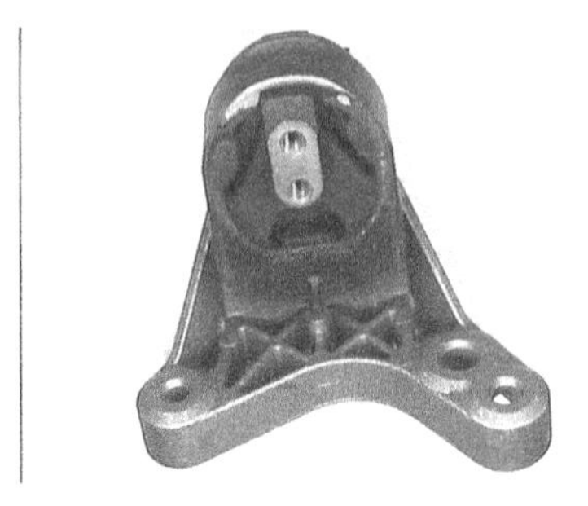

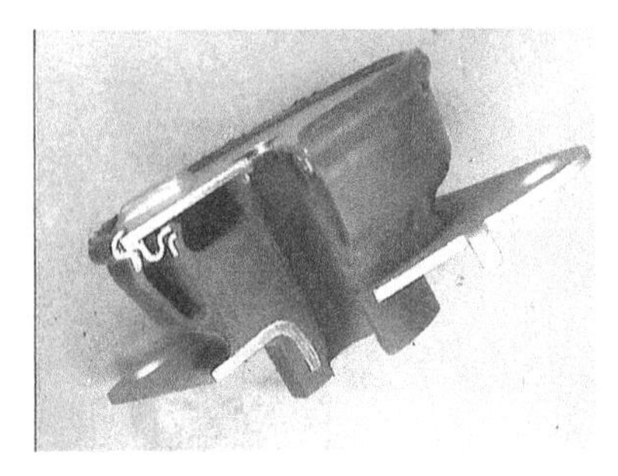

Fig. 11.52. Elastomeric mounts for the engine suspension. The shape of the rubber element determines a nonlinear and non isotropic elastic characteristic.

(a zero of a transfer function is the value of the Laplace variable s^* that reduces to zero the numerator):

$$s_z^* = -\frac{1}{2\zeta} \tag{11.191}$$

Increasing the damping factor for $\omega^* > \sqrt{2}$ increases the transmissibility.

This demonstrates that it could be convenient to have an engine suspension with a damping coefficient that varies with the frequency. The loss factor η and the elastic modulus E of viscoelastic materials change with frequency. An appropriate selection of the viscoelastic material allows a high dissipation at low frequencies to be combined with lower dissipation at higher frequencies. Nevertheless, the possibility to tune the frequency characteristics based on the materials selection alone is rather limited. Additionally, it could be convenient to devise suspension mounts with amplitude dependent characteristics that maintain a constant behavior with temperature and ageing. All this justifies the integration of an hydraulic circuit in the mounts (hydroelastic mounts).

Fig. 11.55 is the cross section of an hydroelastic mount. The upper rubber body has different functions: it transmits the static load of the engine and contributes to the stiffness and some damping of the suspension. When the upper attachment point moves towards the lower attachment it acts as a piston letting the fluid laminate from the upper to the lower chamber through one or more orifices b, i.e. acting like the valves (albeit of fixed size) of a hydraulic shock absorbers used in suspension applications.

The compliance of the rubber element b adds to the relatively low compliance of the fluid which fills the mount. The rubber membrane that separates the lower fluid from volume d also provides a small contribution to the stiffness.

The dynamic behavior of this kind of component is usually characterized in terms of a dynamic stiffness which is represented by the frequency response function between the input displacement $q(\omega)$ and output force $F(\omega)$.

$$K_{dyn}(\omega) = \frac{F(\omega)}{q(\omega)} = K(\omega) + j\omega C(\omega). \tag{11.192}$$

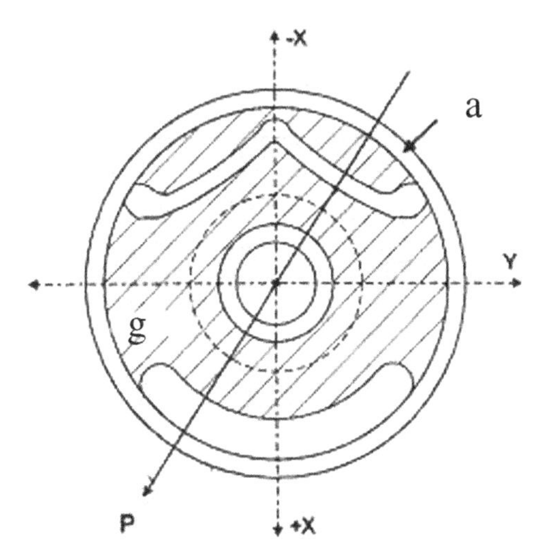

Fig. 11.53. Cross section of an elastomeric mount. The non isotropic characteristich is due to the shape of the rubber part. a) steel outer shell; g) elastomeric part.

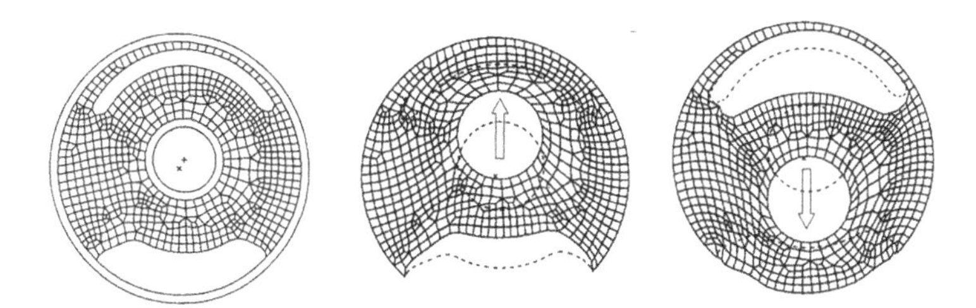

Fig. 11.54. Deformed shape of an elastomeric mount for vertical loads. The different size of the slots allows to obtain different characteristics in the up and down directions.

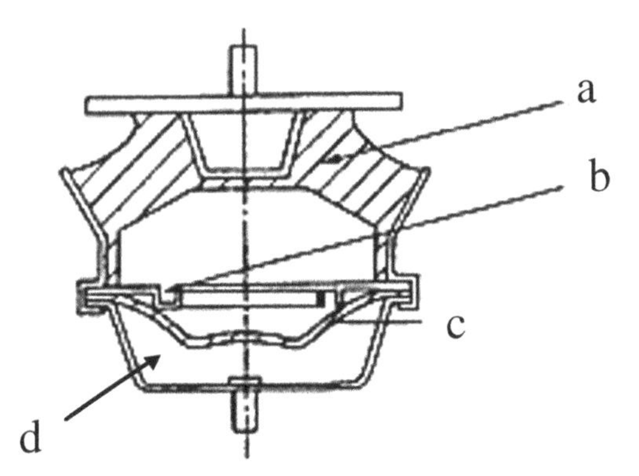

Fig. 11.55. Cross section of a simple hydroelastic engine mount. (a) primary rubber. The fluid laminates from the upper to the lower chamber throug orifice (b). The compliance of the secondary rubber (c) allows the fluid to go from one chamber to the other. The air in the lower chamber (d) can escape through relatively large holes in the lower steel element.

For a sinusoidal input displacement of constant amplitude and frequency, the real part of the dynamic stiffness $K(\omega)$ is the equivalent stiffness whereas $C(\omega)$ is the equivalent damping. Similarly to the complex elastic modulus which is used to represent the hysteretic behavior of a structural material, complex notation is meaningful only under the assumption of an harmonic motion with constant frequency and amplitude.

Fig. 11.56 represents the lumped parameters model of a simple mount as that of Fig. 11.55. K_r is the stiffness of the primary rubber, C_{rs} and K_{rs} represent a simplified model of its dissipative behavior. K_s is the stiffness of the secondary rubber that is considered to be non dissipative; C_h is the damping coefficient due to the lamination of the fluid in holes b. K_v accounts for the volumetric stiffness of the fluid (usually oil) and of the primary rubber.

For low excitation frequencies ($f < 5$ Hz), the force contribution from the dampers is negligible with respect to the contribution of the springs since for low speeds and frequencies, the pressure drop across the orifices is low. The pressure in the lower chamber is the same as that of the upper one. In this case the fluid acts as a kinematic link between the primary and secondary rubber. Taking into account that the volumetric stiffness K_v is much larger than the secondary rubber K_s

$$K_{dyn} \approx K_r + K_s, \qquad f < 5 \text{ Hz}. \tag{11.193}$$

At high excitation frequencies ($f > 20$ Hz), the deformation of the dampers is negligible compared to that of the springs. The small orifices prevent the motion of the fluid from the upper to the lower chamber. The upper chamber is therefore isolated. The stiffness in this case is due in part to the primary rubber (volumetric and shear stiffness) and to the fluid

$$K_{dyn} \approx K_r + K_{rs} + K_v, \qquad f > 20 \text{ Hz}. \tag{11.194}$$

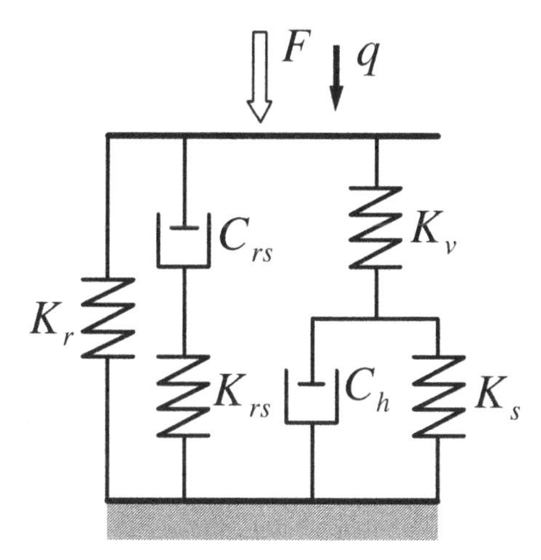

Fig. 11.56. Lumped parameters model of a simple mount as that of Fig. 11.55. K_r is the stiffness of the primary rubber, C_{rs} and K_{rs} are a simplified model of its dissipative behavior. K_s is the stiffness of the secondary rubber that is considered as non dissipative; C_h is the damping coefficient due to the lamination of the fluid in holes b. K_v accounts for the volumetric stiffness of the fluid (usually oil) and of the primary rubber.

At intermediate frequencies (5 Hz $< f <$ 20 Hz), the behavior is dominated by the dampers. In this range the dynamic stiffness is:

$$K_{dyn} \approx j\omega \left(C_{rs} + C_h\right), \qquad 5 \text{ Hz} < f < 20 \text{ Hz}. \tag{11.195}$$

While for low and high frequencies ($f <$ 5 Hz, or $f >$ 20 Hz) the contribution of the dampers is negligible, for frequencies close to the natural frequency of the engine suspension ($f \approx$ 15 Hz) their effect is dominant. This enables the attenuation of the amplitude of the transmissibility function at the resonance. Above the resonance damping is reduced in order to lower the transmissibility.

Fig. 11.57 shows the amplitude of the dynamic stiffness of the engine mount of Fig. 11.55 as function of the frequency. The behavior is dominated by the stiffness for low and high frequencies, regardless of the displacement amplitude. In the range of the engine suspension resonance (15 Hz in this case) the dynamic stiffness increases, demonstrating the role played by the damping in this range.

Above the resonance ω_n, the transmissibility of Eq. 11.189 (Fig. 11.50) decreases by increasing the ratio $\omega^* = \omega/\omega_n$. This means that for a given excitation frequency, reducing the stiffness (and, for a given engine mass, the natural frequency ω_n) reduces the transmissibility at that frequency.

The stiffness can not be reduced too severely because it is necessary to avoid excessive displacements due to the engine torque and to the accelerations. These limitations can be summarized as follows:

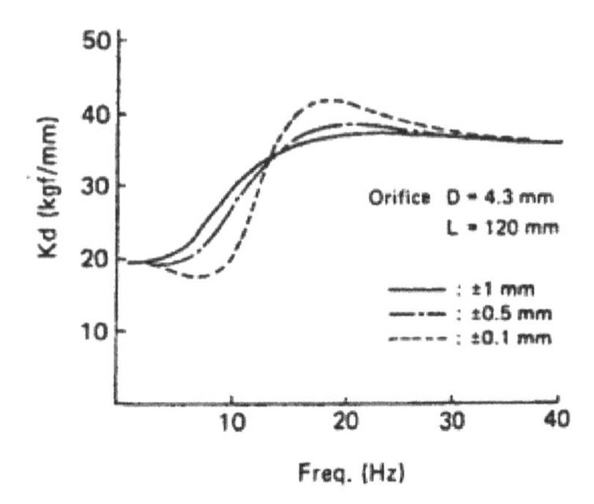

Fig. 11.57. Amplitude of the dynamic stiffness of the engine suspension mount of Fig. 11.55 as function of the frequency. The curves refer to harmonic excitations of various amplitudes and frequencies.

- low frequency, high amplitude vibrations require high dynamic stiffnesses to avoid large displacements of the powertrain;

- high frequency, low amplitude vibrations mostly related by the engine excitations above idle speed require low dynamic stiffnesses to reduce the transmissibility.

These two needs are not adequately satisfied by purely elastomeric mounts (Fig. 11.52) or by simple hydraulic mounts (Fig. 11.55), the dynamic stiffness of which is little influenced by the vibration amplitude. The increased stiffness at the end of the travel range represents only a partial solution to limit the displacement of the powertrain.

A better compromise between low stiffness at high frequency-low amplitude, and high stiffness at low frequency-high amplitude, can be obtained by means of hydraulic mounts with a decoupler (Fig. 11.58). In comparison with the simple hydraulic mount, the plate between the lower and upper hydraulic chambers is more complicated and includes a set of valves. The main parts of the valve system are the inertial valve 3 (decoupler) and the long hydraulic pipe 4 (inertial path) both of which work in parallel to connect the upper and lower hydraulic chambers. For high amplitude displacements, the decoupler is in contact with its housing and avoids the motion of the fluid around it. For small amplitudes the decoupler is free to move and allows the fluid to flow around it and pass from the lower to the upper chambers.

The inertial path is a relatively long pipe between the lower and upper chambers; its reduced cross section gives rise to a high acceleration of the fluid that passes through it. If the fluid is not deformable, the speed along inertial path is

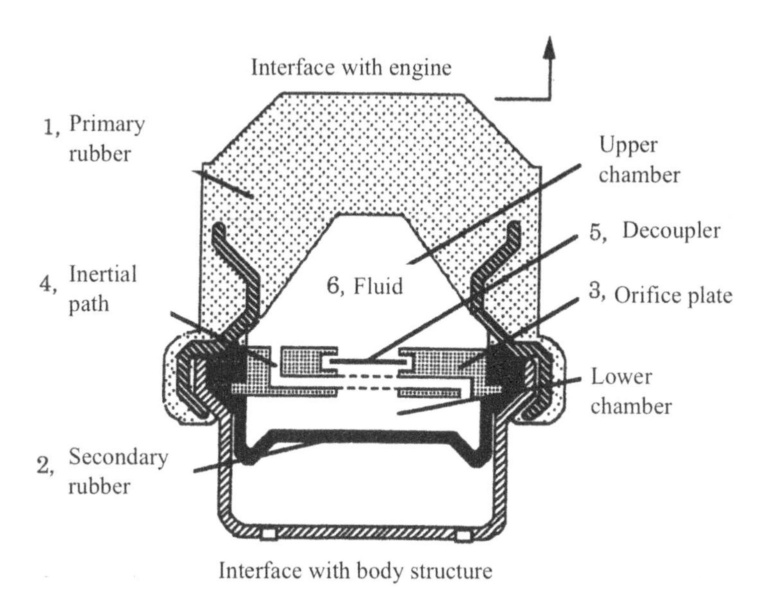

Fig. 11.58. Hydraulic mount with decoupler. For high amplitude displacements the decoupler is in contact with the housing and blocks the fluid. For small displacements the decoupler is free to move and allows the fluid to move besides it. The inertial path is a long pipe that connect the lower and upper chambers.

$$v = \frac{Q}{A_{it}}, \tag{11.196}$$

where Q is the volume flow rate and A_{it} is the cross section area of the pipe. The momentum of the fluid in the pipe is then:

$$P_{it} = \rho A_{it} l v = \rho l Q, \tag{11.197}$$

where l is the length of the pipe and ρ the mass density of the fluid.

The pressure drop p_{it} across the pipe is that needed to change its momentum is:

$$p_{it\ m} = \frac{\rho l}{A_{it}} \dot{Q}. \tag{11.198}$$

The high l/A_{it} ratio make the inertia forces dominate over the other fluid dynamic forces involved in this part of the hydraulic circuit.

The viscous losses in the same pipe induce a pressure drop which is proportional to the volume flow rate:

$$p_{it\ c} = BQ. \tag{11.199}$$

Fig. 11.59 shows the lumped parameters model of an hydraulic mount with a decoupler. The underlying assumption is that the mount is subject to high amplitude motion, so that the decoupler is blocked and does not allow the free motion of the fluid. Mass m_{it} an damping C_t represent the inertial effect; K_v and K_s are the volumetric stiffness of the oil and the rubber stiffness, respectively; K_r C_{rs}, K_{rs} are the stiffness and damping parameters of the primary rubber.

At low frequencies ($f < 5$ Hz), the contribution of mass m_{it} is negligible and the dynamic stiffness is dominated by the primary rubber K_r.

At high frequencies ($f > 20$ Hz), mass m_{it} behaves as a seismic mass, decoupling the upper and the lower hydraulic chambers. The dynamic stiffness is dominated by the primary rubber and the volumetric stiffness ($K_{dyn} \approx K_r + K_v$, $f > 20$ Hz). The experimental curves with 1 mm displacement amplitude of Fig.s 11.61 ($- \diamond -$) and 11.60 show the equivalent stiffness and damping of the mount for high amplitude motion. The damping peaks close to the resonant frequency of the engine suspension (15 Hz).

For low amplitude displacements, the decoupler is free to move and allows the fluid to move through it. Even if the decoupler and the inertial path are in parallel, most of the fluid flows through the decoupler because of the lower pressure drop. This allows to obtain a dynamic stiffness equal to that of the primary rubber up to rather high frequencies (100 Hz), as evidenced in 11.61 (0.1 mm amplitude, $- \circ -$) and 11.60.

Above 200 Hz, the decoupler inertia and the fluid around it behave as a seismic mass, preventing the motion of the fluid. The volumetric stiffness K_v and K_{rs} work in parallel with the primary rubber. At high frequencies ($\omega > K_{rs}/C_{rs}$), the displacements of damper C_{rs} are much smaller than spring K_{rs}. The result is that for low amplitude vibrations, the dynamic stiffness remains low up to the maximum excitation frequencies generated by the engine. This enables high attenuation to be achieved along the structural transmission path that connects the engine to the vehicle body.

Engine mounts with a decoupler allow a considerable reduction of the transmitted vibrations but still have an intrinsic limitation because their parameters are tuned during design and cannot be changed during operation. The availability of fluids that change properties with the application of a magnetic field (magnetorheological) or an electric field (electrorheological) enable devise engine mounts to be devised with characteristics that can be modified during operation through feedback loops based on the measurement of the engine and structural vibration.

The mount configuration of Fig. 11.62, is basically the same as an hydraulic mount. The use of an electrorheological fluid allows the characteristics of the valves to be changed by the application of an electric field. The electric field modifies the viscosity of the fluid and, therefore, the pressure drop generated by the valve. Regardless of their effectiveness, to date the high cost of these solutions has limited their application to high class vehicles.

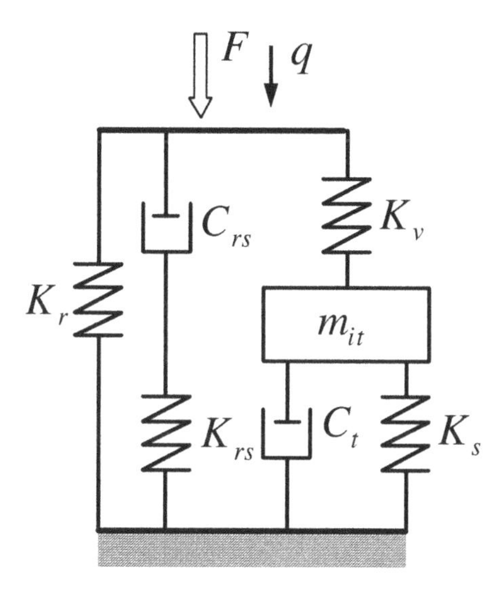

Fig. 11.59. Lumperd parameters model of an engine mount with decoupler. Mass m_{it} is the equivalent mass of the inertial path.

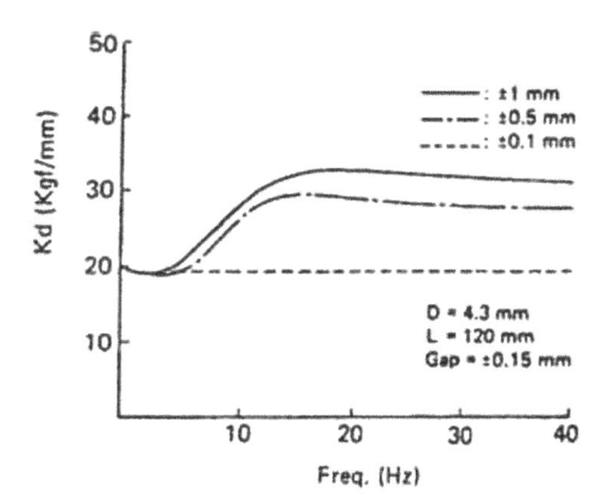

Fig. 11.60. Dynamic stiffness of an engine mount with decoupler for different displacement amplitudes. For large displacement amplitudes the dynamic stiffness increases in the interval 5÷20 Hz to limit the vibration amplitude. For low amplitudes the dynamic stiffness remains low up to frequencies of about 100 Hz.

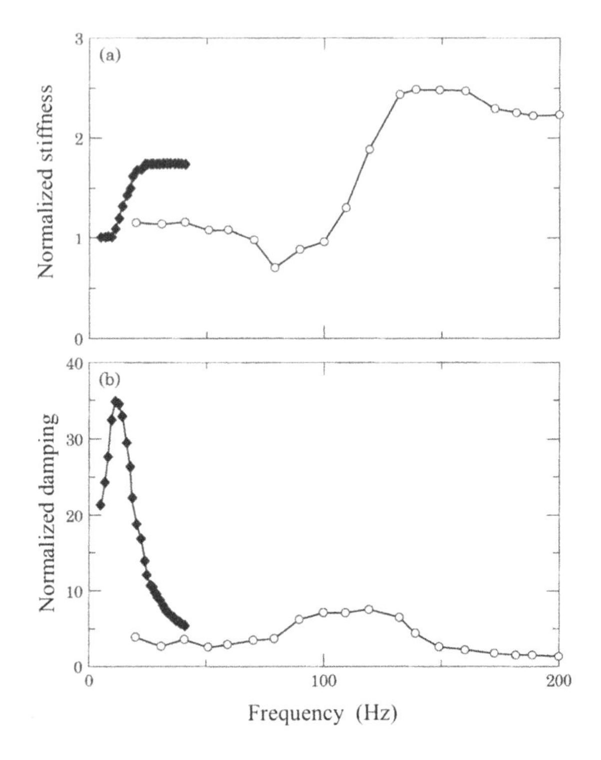

Fig. 11.61. Dynamic stiffness of an engine suspension mount with decoupler. Equivalent stiffness a); equivalent damping b). Large amplitude (1 mm, – ◇ –); and small amplitude (0,1 mm, – ○ –).

11.4.3 Engine Suspension Architectures

The main goal of the engine suspension architecture is to obtain a good compromise between the dynamic properties (transmissibility function), the deflection under quasi static and impulsive loads and the interaction with the body dynamics.

Due to its relevance, more attention will be devoted in the following to a powertrain with transversely mounted front-engine/front wheel drive.

Three mounts suspension

Fig. 11.63 shows a three mounts suspension for a transversely mounted front-engine/front wheel drive layout. Two mounts are located in front of the powertrain, one connected to the engine block, the other to the gearbox. Both of them are bolted to the main longitudinal beams of the vehicle body. The third mount is behind the powertrain, is linked to a crossbeam belonging to the frame that hosts the lower control arm of the suspensions and the steering rack.

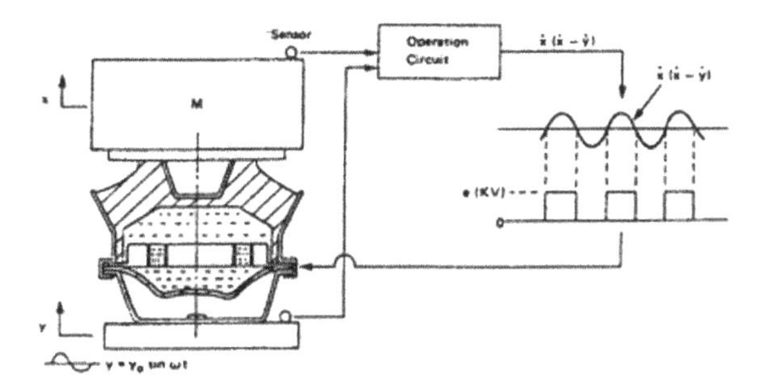

Fig. 11.62. Electrorheological (ER) engine mounts. The hydraulic configuration is similar to that of simple hydraulic mounts. The application of an electric field to the fluid that flows from the upper to the lower chambers induces a variation of the viscosity and, hence, a variation of the pressure drop.

To reduce the load on the mounts, the powertrain centre of gravity should be within the triangle defined by the three mounts. This solution is quite convenient because it allows an easy attachment to the vehicle body but the mounts are located in places where the modal displacements of the body are relatively large. This would require low stiffness to reduce the transmissibility of the engine vibrations. Nevertheless, the rather small longitudinal distance between the front and rear mounts implies large loads caused by the engine torque. This requires high stiffness to avoid too large engine displacements, although the result is relatively high transmissibility.

An additional connecting rod is sometimes added to react to the torque transmission and achieve a better compromise in the design of the three lower mounts (right of Fig. 11.63). In this case an additional longitudinal load induced by the rod must be taken into account. In some cases the mounts are connected to an auxiliary frame, connected to the body through other elastomeric elements. Advantages arise at the assembly level because the engine, suspensions and steering system are assembled in a separate group connected to the vehicle via fewer number of points. An additional benefit is the introduction of a secondary filtering stage between the auxiliary frame and the body.

Centre of gravity suspension

This solution overcomes most drawbacks of the three mounts type, and is therefore the most commonly adopted in recent vehicles. This architecture comprises two flexible mounts, aligned along the main inertia axis of the powertrain, and by one or more rods connected to the powertrain and the vehicle frame. Fig. 11.64 a) shows a solution with two rods. The weight and the inertia forces acting on the powertrain are supported by the two mounts (on connected to the engine block, the other to the gearbox) linked to the upper part of the main longitudinal beams of the vehicle frame.

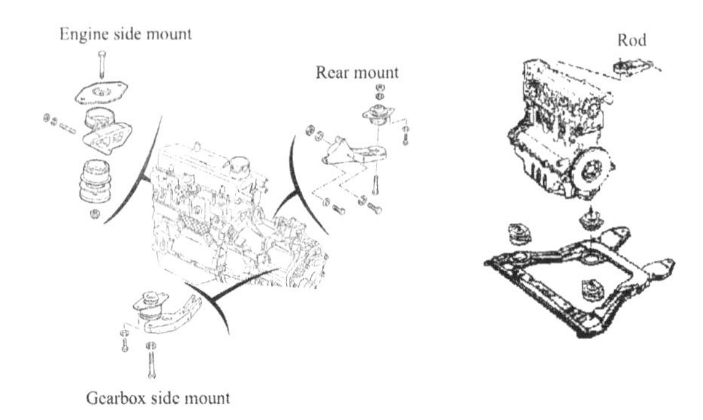

Fig. 11.63. Three mounts engine suspension for a transverse front engine and transmission. Two mounts are in front of the and one at the rear of the differential.

The alignment of the mounts along the main axis of inertia of the powertrain characterized by the lower moment of inertia enables one of the modes of vibration to involve the rotation of the engine about that axis.

The two connecting rods avoid the free rotation of the powertrain due to the torque transmitted to the wheels but their alignment allows the free motion of the engine in vertical direction. The solution with two rods (Fig. 11.64 a) allows to completely decouple the force from the torque reaction. The first is given to the two mounts, the second to the rods. The solution with just one rod (Fig. 11.64 b) leads to some loads on the two mounts when a torque is transmitted to the wheels.

The large number of different options for the same powertrain (presence or not of the air conditioning compressor, of a robotized gearbox, starter alternator,...) makes it impossible to align the mounts along the inertia axis of the powertrain in all configurations. On the other hand other issues such as the selection of attachment points to the frame involving low modal excitation must be taken into account in the design of the suspension.

Longitudinally mounted engine suspension

The most common solution for longitudinally mounted engines is constituted by four mounts arranged symmetrically about a longitudinal plane with two at the front end of the engine and two at the rear (Fig. 11.65). The engine torque is opposed by axial (tension or compression) and shear forces in the mounts, depending on their location and orientation. The elastic centers of the front and rear mounts are designed to be near to the principal inertia axis of the powertrain in order to decouple the torsional mode from the others.

The elastic center of each couple of mounts is the center of rotation of the body attached to the mounts when subject to a pure torque. If the two mounts are

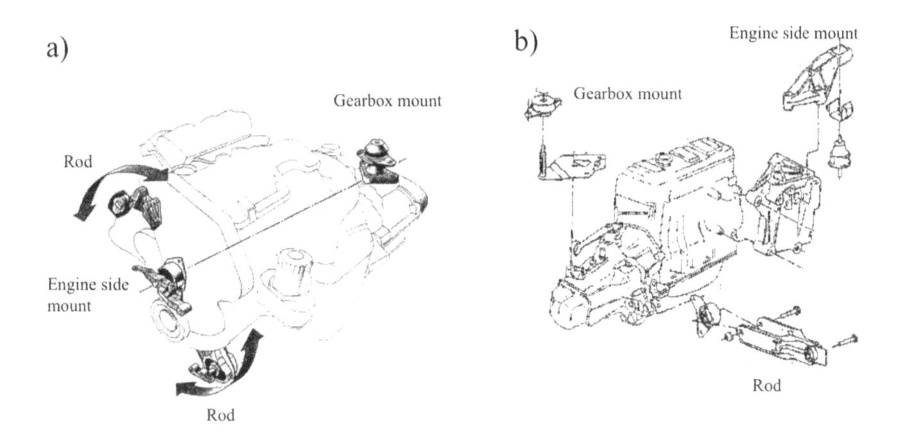

Fig. 11.64. Centre of gravity suspension for transversely mounted front-engine/front wheel drive layout. Two mounts are aligned along the main inertia axis of the powertrain of lower moment of inertia. These mounts react to the forces acting on the powertrain. Two rods react to the torque transmitted to the wheels a). If the rod is just one, the torque to the wheels induces some force on the two mounts b).

identical and installed symmetrically with respect to the longitudinal symmetry plane (Fig. 11.66), the elastic center is on the symmetry plane at a distance a:

$$a = m\frac{\tan\alpha(L-1)}{L\tan^2\alpha + 1}, \tag{11.200}$$

where $L = K_A/K_R$ is the ratio between the axial stiffness K_A of each mount and the shear (or radial) one K_R.

In this way it is possible to decouple the roll mode of the powertrain from the other vibration modes improving the dynamic behavior, in particular at idle.

11.4.4 Location of the Attachment Points to the Car Body Frame

At frequencies above the main resonances of the suspension, the powertrain can be considered to be a free rigid body, the motion of which is determined by the inertia forces and the torques acting on it. The resulting motion is a rotation about an axis depending on the inertia properties (mass, center of gravity, principal inertia axis and principal moments of inertia). The position of this axis can be estimated from maps such as that on the right of Fig. 11.67. This greyscale (or color coded) map represents, for given operating conditions of the engine, the displacements of points rigidly connected to the powertrain. Dark shading indicates the regions with displacements lower than light shaded ones. In the case of a center of gravity suspension the best location for the two mounts is that characterized by small displacements in order to minimize the deformations of

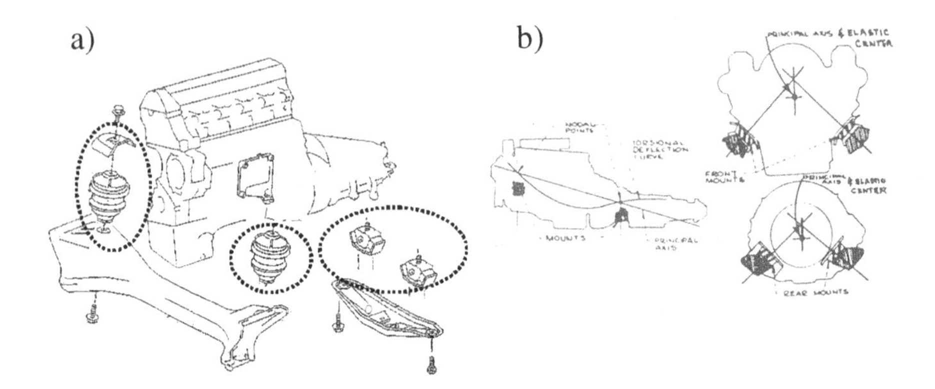

Fig. 11.65. Engine suspension for longitudinally mounted powertrain. The mounts are located so that the elastic axis of the suspension is near the principal inertia axis of the powertrain.

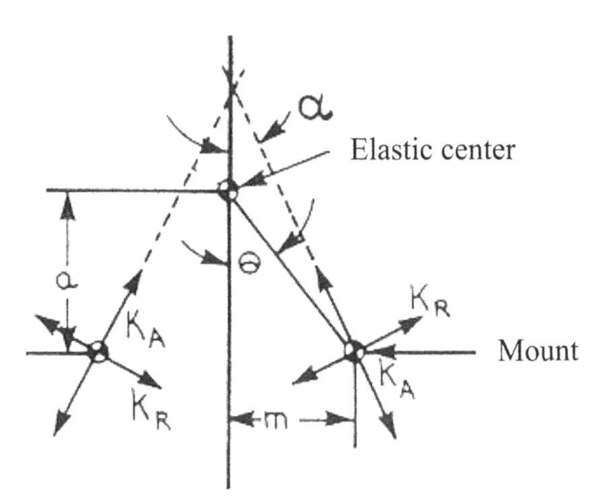

Fig. 11.66. Location of the elastic center of a couple of identical mounts of axial stiffness K_A, and radial stiffness K_R.

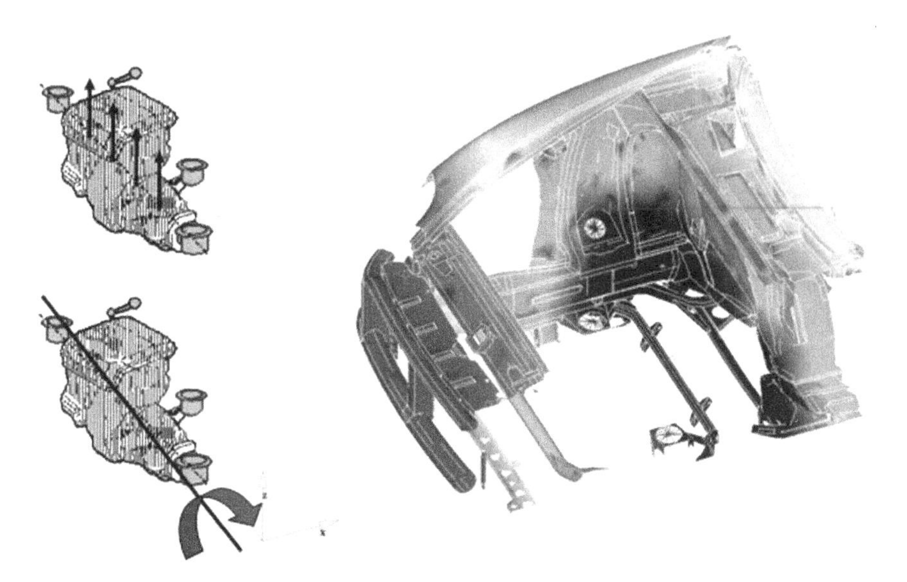

Fig. 11.67. Location of the attachment points of the engine to the frame. Right: grayscale map representing the displacement amplitude of points attached to the powertrain that vibrates under given operating conditions. Darker shade indicate displacements smaller than lighter one.

the mounts due to the engine vibrations and, hence, the forces transmitted to the vehicle body.

The information of the maps reporting the displacements of points attached to the powertrain can be compared to the maps reporting the acoustic sensitivity (Fig. 11.68), i.e. maps reporting the amplitude of the noise transmitted inside the vehicle as function of the point of application of a given force. The best points for the engine mounts are those with the optimal compromise between the lower displacement of the engine and low acoustic sensitivity. The selection of the engine suspension attachment points should take into account the need to transmit the engine torque with adequate reaction length.

Additionally, the attachment points should be easily reached by the production line assembly tools. For structural reasons the metal brackets used to connect the engine to the mounts should not be too compliant in order to avoid resonances that could increase the transmissibility of the vibrations produced by the engine.

Due to the high damping of the suspension mounts, to the complex mass distribution of the powertrain, and unavoidable compromises in the definition of the attachment points to the frame, the engine suspension mode shapes are usually quite complex: All involve combined displacement and rotation. The modes with highest displacement in the vertical direction may be strongly coupled to the

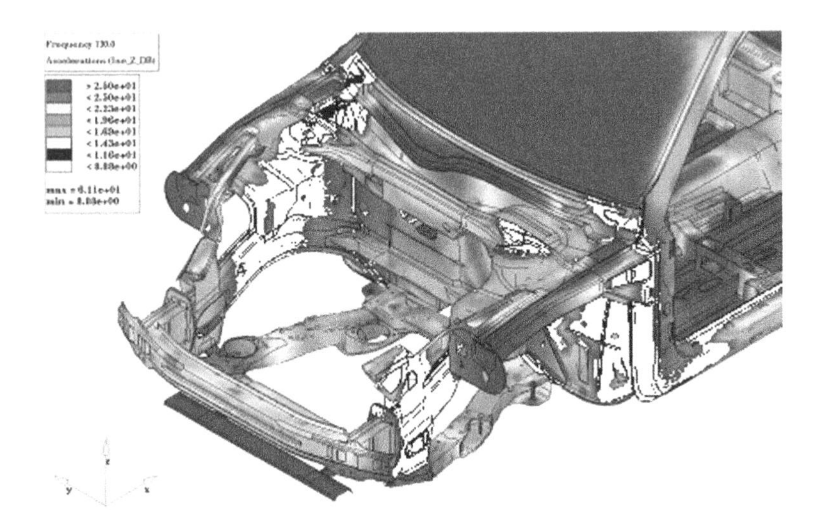

Fig. 11.68. Amplitude of the frequency response function between an input force and the sound pressure level inside the vehicle. Light grey indicates low amplitude, dark shade indicated high amplitude.

vertical modes of the vehicle body on its suspensions. These modes are excited by the road irregularities and can cause high vibrations of the body. In vehicles with front engine and transmission, such vibrations (engine shake) reach high amplitudes especially in the region of the pedals, as shown in Fig. 11.69, at a frequency of about 10 Hz.

The amplitude of the engine shake mode is controlled by the damping of the mounts. As described previously, hydraulic mounts provide high damping in the resonance region with relatively low damping elsewhere in order to reduce the transmissibility both near and far from the resonances.

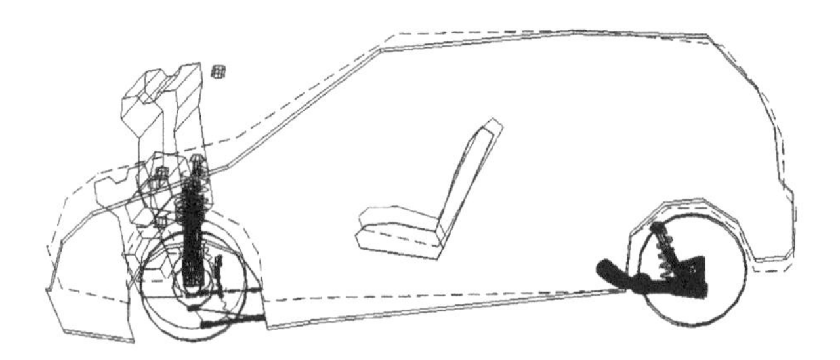

Fig. 11.69. Coupled vibration mode of the vehicle body and engine suspension. This mode usually occurs at about 15 Hz and involves the pitching motion of the body on the suspension and the vertical motion of the engine on its suspension (shake).

11.5 Acoustic Transmission and Insulation

The intensity of the noise that propagates from the sources to the occupants inside the vehicle is transmitted by the elements of the vehicle body with some attenuation. One of the key factors in determining the noise level perceived by the occupants is the acoustic transparency of the elements that delimit the inner volume from the outside. Most of the noise is transmitted through the parts with a larger noise transparency. This is similar to the case of the thermal insulation of a room where most of the heat escapes from the parts with the poorest insulation, such as the windows or thermal bridges through some parts of the walls. The noise is transmitted mostly through poorly insulated paths or gaps such as parts with less efficient acoustic treatments or direct connections between two enclosures. These gaps are difficult to avoid in a car body since they are needed for the connections such as the brake pedals, the steering column to the rack and wheels, the dashboard to the part of the air conditioning system located in the engine compartment, etc. Smaller passageways are needed for electric and hydraulic connections.

Since even small acoustic holes can lower the effective noise insulation, it is necessary to eliminate gaps already at the initial design stage.

11.5.1 Transmission Loss

The transmission coefficient τ and the transmission loss TL are two parameters that characterize the sound transmitted through a medium defined as follows:

$$\tau(\theta,\omega) = \frac{W_{tr}(\theta,\omega)}{W_{in}(\theta,\omega)}, \tag{11.201}$$

$$TL(\theta,\omega) = 10\log_{10}\frac{1}{\tau(\theta,\omega)} = 10\log_{10}\frac{W_{in}(\theta,\omega)}{W_{tr}(\theta,\omega)} \qquad [dB], \tag{11.202}$$

where $W_{in}(\theta,\omega)$ is the input sound power characterized by an incidence angle θ and angular frequency ω (rad/s), $W_{tr}(\theta,\omega)$ is the transmitted power.

The procedure for the measurement of the transmission loss TL through a partition is specified by standards such as *ASTM E90-74* and *ISO 140/3-1995* which require two contiguous reverberating chambers with relatively large volumes (64 m^3 each), a window common to the two chambers is closed by the partition that must be characterized (Fig. 11.70): The prescribed window size is 2.8 x 3.6 m^2.

One or more loudspeakers are installed in the emission chamber to generate a diffuse field that is transmitted with some attenuation to the reception chamber by the partition.

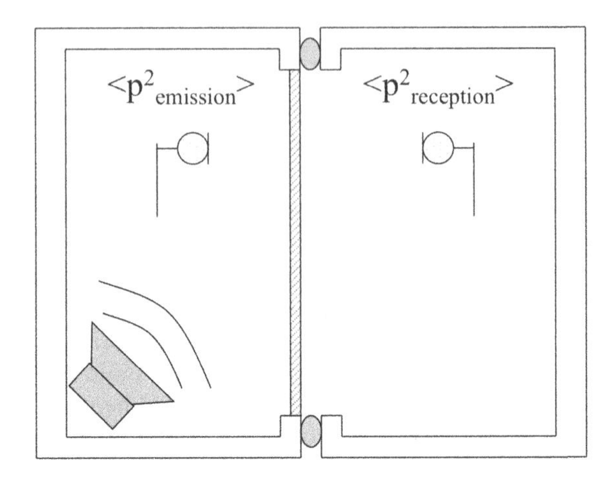

Fig. 11.70. Measurement of the *transmission loss* of a partition. The source and receiving volumes are reverberating chambers.

For a diffused field, the incident power per unit area (sound intensity) I_{random} is related to the square of the sound pressure $< p^2 >$

$$I_{random} = \frac{< p^2 >}{4\rho_0 c_0}.$$

The measurement procedure prescribes that the incident power must be measured at first in the emission chamber close to the surface of the partition of area S_W, with a temporal and spatial average $< p^2_{source} >$

$$W_{inc} = I_{inc} S_W = \frac{< p^2_{source} > S_W}{4\rho_0 c_0}. \tag{11.203}$$

The transmitted power is then measured in the reception chamber close to the surface A_r of the partition $< p^2_{rec} >$

$$W_{trasm} = \frac{< p^2_{rec} > A_r}{4\rho_0 c_0}, \tag{11.204}$$

The measurement of the incident and the transmitted power (W_{inc} and W_{trasm}) enables the determination of the transmission loss TL

$$R_{lab} = 10 \log \frac{W_{inc}}{W_{trasm}} = \; < L_p >_S - < L_p >_R + 10 \log \frac{S_W}{A_r} \quad [dB] \tag{11.205}$$

The measurement of the transmission loss is not straightforward since it requires a large and expensive setup to guarantee good acoustic insulation between

Fig. 11.71. Simplified setup to characterize the transmission loss of an automotive element. The receiving chamber is anechoic, the source is reverberating.

the two chambers. Furthermore all transmission paths must be avoided except the window used to install the specimen that must be characterized.

This setup can be simplified to characterize the small elements typical of a car. The drawback is related to some restrictions on the frequency range that can be investigated caused by the different combination of geometrical size of the chambers, sound wavelength and size of the specimen.

Fig. 11.71 shows the example of a test setup including two contiguous chambers: the emission chamber is reverberating, while the receiving one is anechoic. The window that connects them is rather small (1.6 x 2 m^2). One of the benefits of the anechoic reception chamber is that it allows the intensity maps of the transmitted noise in order to determine the parts with larger acoustic transparency.

Other alternatives are the well, including a reverberating chamber beneath the floor, and an anechoic or a reverberating one with an aperture of about 1 m^2 in between (Fig. 11.72).

Another test setup (the so-called Pisa's tower), is made by two small reverberating chambers with a 240 mm diameter test sample in between.

Acoustic transmission of plane waves through an infinite plate

The transmission loss of single or multiple layer plates can be understood considering the behavior of infinite plates of the same material for which the elastic (Young's modulus and Poisson's ratio), inertial (mass density) and geometric (thickness) properties are known.

The simpler case is that of a planar wave that propagates perpendicularly to an infinite plate made of a homogeneous and isotropic material of thickness h.

The pressure p_+ on the surface is the same at all points, resulting in the propagation of compression waves in the plate thickness. The sound pressure

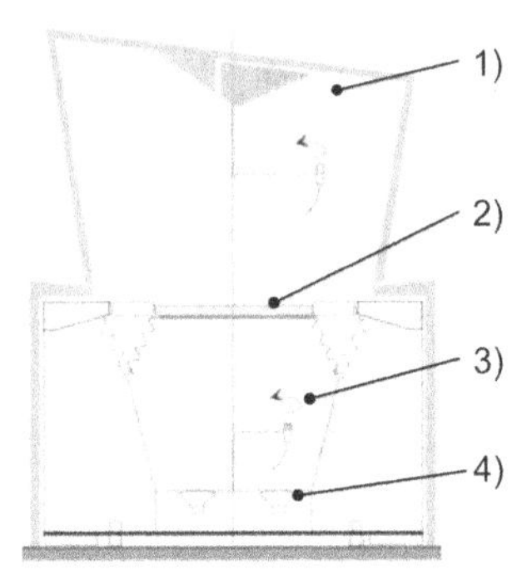

Fig. 11.72. Isokell test system to measure the transmission loss through automotive components. 1) reception chamber, 2) test specimen, 3) emission chamber, 4) loudspeakers

p_T transmitted to the reception side is the superposition of all transmitted components:

$$p_T = \frac{(1+R_0)(1+R_2)e^{-ik_m h}}{1-R_1R_2e^{-i2k_m h}} p_+, \tag{11.206}$$

where R_0, R_1 and R_2 are the reflection coefficients at the three air-plate interfaces and h is the plate thickness. The reflection coefficients are given as function of the acoustic impedance Z_m of the plate and the fluids in contact with the plate (Z_{01} and Z_{02}, Fig. 11.73)

$$R_0 = \frac{Z_m - Z_{01}}{Z_m + Z_{01}}; \quad R_1 = R_0; \quad R_0 = \frac{Z_{02} - Z_m}{Z_{02} + Z_m} \tag{11.207}$$

The losses that occur in the material can be taken into account by means of a complex wave number $k_m^{'} = k_m(1 + (1/2)i\eta)$, where $k_m = \omega/c_L$ is the wave number of the pressure waves that propagate in the plate, c_L is the speed of the sound of the longitudinal waves in the plate, and η is the material loss factor.

Recalling Eq. 11.202, the transmission loss for normal incidence is:

$$TL_N = 10\log_{10}\frac{W_{in}}{W_{tr}} = 10\log_{10}\frac{p_+^2}{p_T^2}. \tag{11.208}$$

If the gas on both sides of the plate is the same:

$$TL_N = 10\log[\cos^2 k_m^{'}h + 0.25(\frac{Z_0}{Z_m} + \frac{Z_m}{Z_0})^2 \sin^2 k_m^{'}h]. \tag{11.209}$$

Usually the thickness of the plate is very small in comparison to the wavelength of the sound that propagates in it, so that $\left|k_m^{'}h\right| << 1$. The result is the following expression of the transmission loss:

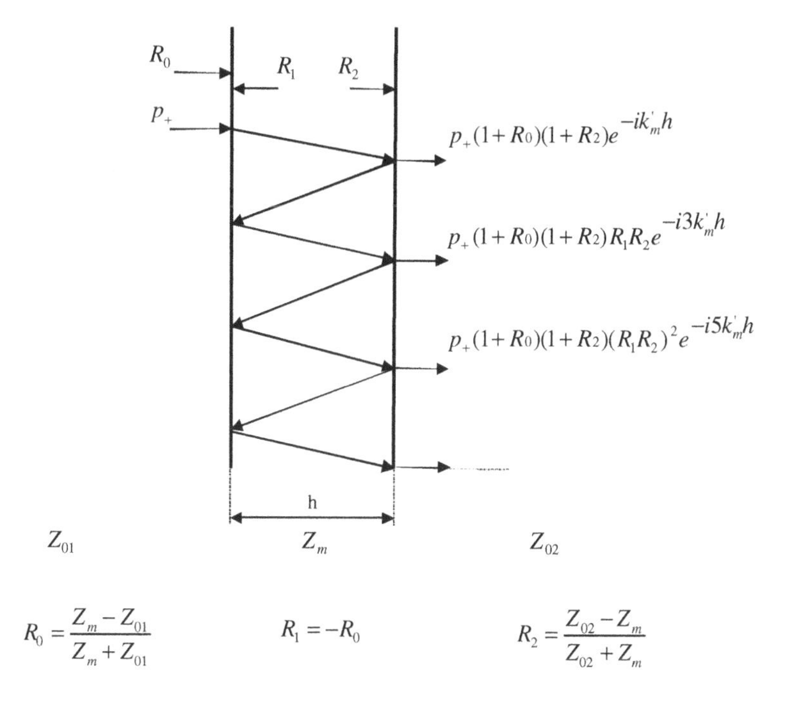

Fig. 11.73. Representation of the transmission through a plate.

$$TL_N \cong 10 \log[1 + (\frac{\rho_s \omega}{2\rho_0 c_0})^2], \tag{11.210}$$

that is usually known as the mass law for normal incidence. $\rho_s = \rho h$ is the mass per unit area of the plate and $\rho_0 c_0 = Z_0$ is the acoustic impedance of the gas that is the same on both sides of the plate. The above expression shows that the transmission loss increases of about 6 dB for each doubling of the mass per unit surface ρ_s and of the angular frequency ω.

If the acoustic wave has an incidence angle θ, the pressure on the plate surface is not the same, this produces bending waves in it. If the wavelength λ_B of the bending wave is the same as the component of the incident wavelength in the direction of the bending waves ($\lambda_B = \lambda_0/sen\theta = \lambda_{tr}$ or, considering the propagation speeds $c_{tr} = c_0/sen\theta$), the incident wave excites the bending wave. In principle under these circumstances the transmission loss should be null, in practice this does not happen because of the unavoidable dissipations: This results anyway in a dramatic reduction of the loss. The frequency corresponding to this phenomenon is indicated as coincidence frequency given by:

$$f = \frac{c_0}{1.8 c_L h \sin^2 \theta}. \tag{11.211}$$

The coincidence frequency increases with decreasing thickness h and the speed of sound in the material of the panel $c_L = (E/\rho)^{1/2}$. The smaller coincidence frequency occurs when the incident wave is parallel to the plate ($\theta = 90$ deg) is indicated as critical frequency.

The transmission loss of a plate with oblique incidence can be found considering that the bending waves can be considered as the superposition of shear and compression waves.

Considering that for thin plates the thickness is much smaller than the shear wavelength ($k_S h << 1$), the transmission loss (in dB) becomes

$$TL(\theta) \cong 10 \log \left\{ 1 + \left| \frac{\rho_s \omega}{2\rho_0 c_0 / \cos\theta} [1 - f^{*2} \sin^4 \theta] \right|^2 \right\} \tag{11.212}$$

where $f^* = f/f_c$, and the critical frequency f_c (Hz) is

$$f_c = \left(\frac{c_0}{2\pi} \right)^2 \sqrt{\frac{\rho h}{D}} \tag{11.213}$$

and $D = Eh^3/[12(1-\nu^2)]$ is the bending stiffness of the plate.

The dissipations in the material can be taken into account considering a complex Young modulus $E' = E(1 + i\eta)$. This leads to the following expression of the transmission loss (in dB):

$$TL(\theta) = 10 \log \left| 1 + \frac{\rho_s \omega}{2\rho_0 c_0 / \cos\theta} \left[\eta f^{*2} \sin^4 \theta + i \left(1 - f^{*2} \sin^4 \theta \right) \right] \right|^2 . \tag{11.214}$$

The previous expression shows that at frequencies close to the coincidence $f = f_C / \sin^2 \theta$, (or $c_0 / \sin\theta = c_B$, where c_B is the speed of the bending waves), the transmission loss has a minimum, the value of which being dominated by the damping factor η. Fig. 11.74 shows the transmission loss as function of the nondimensional frequency f^* for different incidence angles θ. For $\theta = 90°$, the minimum of the transmission loss occurs at the lowest frequency.

In a diffuse field, where plane waves propagate in all directions with the same intensity, the transmitted intensity is

$$I_{tr} = \int_\Omega \tau(\theta) I_{in} \cos\theta d\Omega. \tag{11.215}$$

The integral extends on a spherical angle Ω and $d\Omega = \sin\theta \, d\theta \, d\phi$. As I_{in} is the same for all plane waves and τ is the same for all values of ϕ, the mean transmission coefficient $\bar{\tau}$ can be defined as

$$\bar{\tau} = \frac{\displaystyle\int_0^{\theta_{\text{lim}}} \tau(\theta) \cos\theta \sin\theta d\theta}{\displaystyle\int_0^{\theta_{\text{lim}}} \cos\theta \sin\theta d\theta}, \tag{11.216}$$

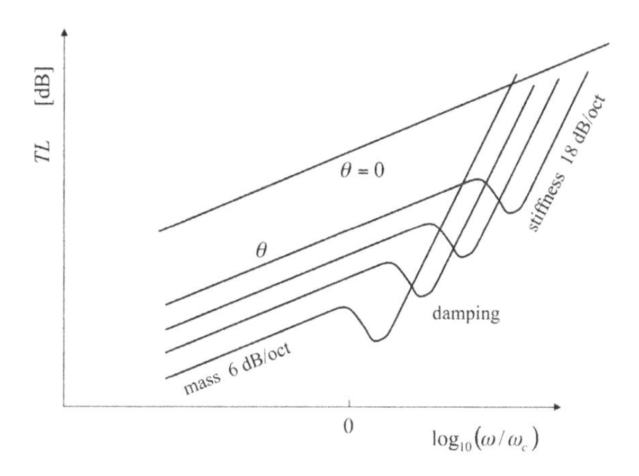

Fig. 11.74. Transmission loss as function of the frequency and of the incidence angle θ.

where θ_{lim} is the limit incidence angle of the sound field. For a random incidence $\theta_{\text{lim}} = 90$ deg. The transmission coefficient is given by Eq. 11.212, 11.214 and:

$$TL_{\text{rand}} = 10 \log(\frac{1}{\overline{\tau}}). \tag{11.217}$$

At low frequencies ($f << f_C$) the transmission loss for random incidence ($TL_N > 15dB$) is found by averaging Eq. 11.212 and 11.214 for angles θ between 0 and 90 deg:

$$TL_{\text{rand}} \cong TL_N - 10 \log(0.23\; TL_N), \tag{11.218}$$

this is usually known as mass low for random incidence.

It is common to use the mass law for field incidence, that is defined for $TL_N \geq 15dB$ as:

$$TL_{field} \cong TL_N - 5. \tag{11.219}$$

This expression can be used to evaluate a diffuse field with a limit angle of $\theta_{\text{lim}} = 78$ deg (Eq. 11.215, 11.216). The result is closer to experimental data than that of Eq. 11.218.

The different values obtained by the expressions of the transmission loss given by the mass law for normal incidence (TL_N), random (TL_{rand}) and field (TL_{field}) are compared in Fig. 11.75 to the experimental measurement obtained with a 2 mm thickness steel plate.

Transmission Loss for double layer panels

The theoretical expressions and the experimental results show that doubling the mass per unit area the transmission loss of single layer plates increases 6 dB (6 dB=20 $\log_{10}(2)$ corresponds to a factor of 2 on the amplitude). To obtain good

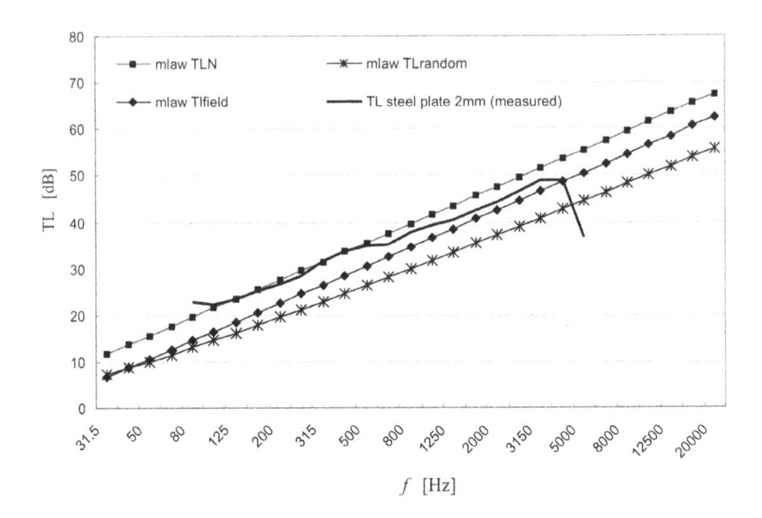

Fig. 11.75. Comparison between different expressions of the transmission loss and that obtained from experimental measurement on a 2 mm thickness steel plate.

acoustic insulation from single layer plates it is necessary to increase the mass per unit area, i.e. the thickness. The increased weight could become unacceptably large. Double layered plates allow the acoustic insulation to be improved considerably without increasing the weight. Instead of a single plate the wall is formed by two plates separated by an air layer. This solution allows the acoustic insulation to be improved in a wide frequency range, except for the so called mass-air-mass resonance.

The additional plate (plate 2) and air layer behave as a sort of dynamic damper attached to the first plate (plate 1). The resulting system has a resonant frequency given by:

$$f_0 = \frac{1}{2\pi}\sqrt{\frac{\rho_0 c_0}{d}\frac{(m_1 + m_2)}{m_1 m_2}}, \tag{11.220}$$

where d is the air layer thickness and m_1, m_2 are the mass per unit area of the two panels.

Below f_0 the stiffness of the air layer couple the motion of the two panels that move in phase as a single body. The transmission loss of the double layer panel is similar to that of a single layer panel with mass per unit area $m_t = m_1 + m_2$ (Eq. 11.210).

At frequency f_0 the mass air mass resonance makes the transmission loss of a double layer panel lower than that of a single layer panel of mass m_t.

Above f_0 the mass-air-mass system is a very efficient means to reduce the transmitted noise, the transmission loss increases 18 dB/octave (6 dB/octave=20 dB/dec are due to m_t, the additional 12 dB/octave=20 dB/dec are due to the air layer).

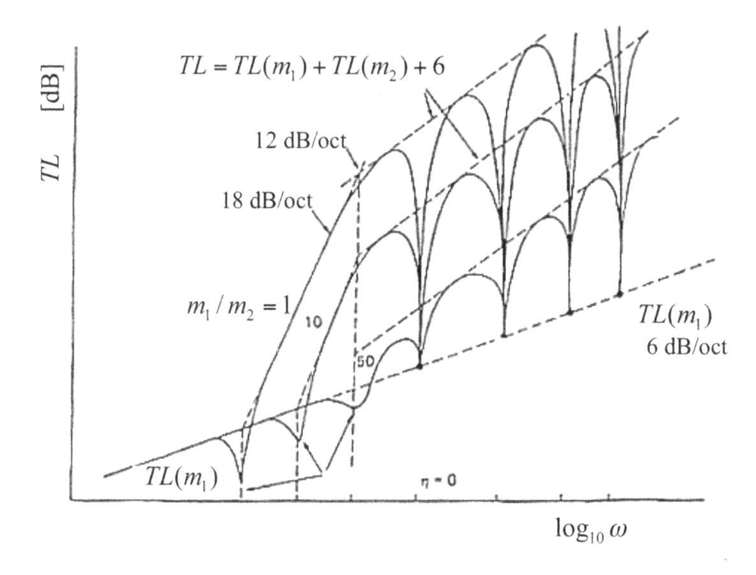

Fig. 11.76. Theoretical behavior of the transmission loss for double panel wall of mass per unit area m_1 and m_2. 6 dB correspond to a factor of 2 on the amplitude, one octave to a factor of 2 on the frequency. 6 dB/octave = 20 dB/dec.

The 18dB/octave occurs until $k\,d = \pi$ that corresponds to a resonance of the air between the two walls. At this frequency and its integer harmonics ($k\,d = n$) the air couples the plates and the transmission loss reduces to that of a single plate of mass m_t; in the intermediate frequencies $k\,d = (2n-1)/2$, the transmission loss is again that given by the contribution of the two plates and the air $TL(m_1) + TL(m_2) + 6dB$.

The transmission loss of a double layer panel is represented in Fig. 11.76 for normal incidence. The largest transmission ratio is obtained for $m_2/m_1 = 1$, the transmission loss decreases for larger values of the ratio between the mass densities.

The experimental results on double layer panels show some differences with the theoretical behavior of Fig. 11.76, this is because it is difficult to obtain a true normal field and some diffuse component is always present. Additionally, the transmission loss is usually measured in third octave bands instead of a continuous spectrum.

Practical considerations

To obtain a good transmission loss even at the medium to low frequencies typical of the lower harmonics of the internal combustion engine, the first natural frequency f_0 of Eq. 11.220 should be as low as possible. The distance d between the two panels should then be the largest possible, according to the design

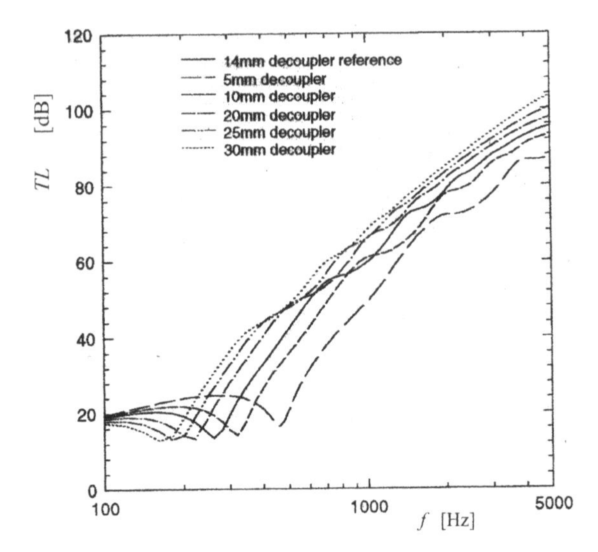

Fig. 11.77. Transmission loss of a double panel wall for different thickness of the porous means (decoupler).

specifications. Fig. 11.77 shows the transmission loss of a double panel wall for different thicknesses of the porous means (mostly air) between the plates.

The use of a layer of sound absorbing material between the two panels improves the transmission loss. The sound waves emitted by the first panel are partially dissipated in the sound absorbing layer, and reach the second panel with much less energy.

All mechanical connections between the two panels prevent the attenuation of the air or sound absorbing layer, they should therefore be avoided. If they are required for some reason, they should be very compliant and dissipative elements.

Fig. 11.78 shows the effect on the transmission loss of a sound insulating layer and of the mechanical connections in a double panel wall.

Holes or gaps establish a direct connection between the emission and the reception chambers (acoustic holes), they should be avoided as much as possible as they can deteriorate the acoustic insulation to such an extent that so as to make it useless. Unfortunately some holes are necessary to connect the volume inside a vehicle to the engine compartment or to the other surrounding spaces. For example some holes in the firewall are necessary to connect the pedals to the brake system and the powertrain, others for electrical wiring. Fig. 11.79 shows the reduction of the transmission loss as function of the surface area of the holes. Their influence is larger for sound barriers with higher transmission loss. Small holes influence mostly high frequency components (above 1kHz).

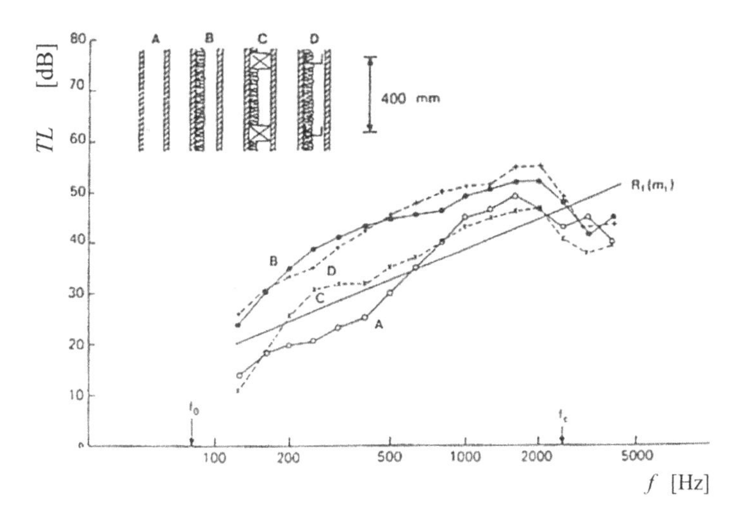

Fig. 11.78. Double panel wall: effect of the sound insulation layer and of the structural connection between the two panels. The four refer to two 13 mm thick plasterboards at a 106 mm distance. A: only air layer; B: as A + 50 mm glass wool layer; C: as B + wood spacers; D: as B + steed spacers.

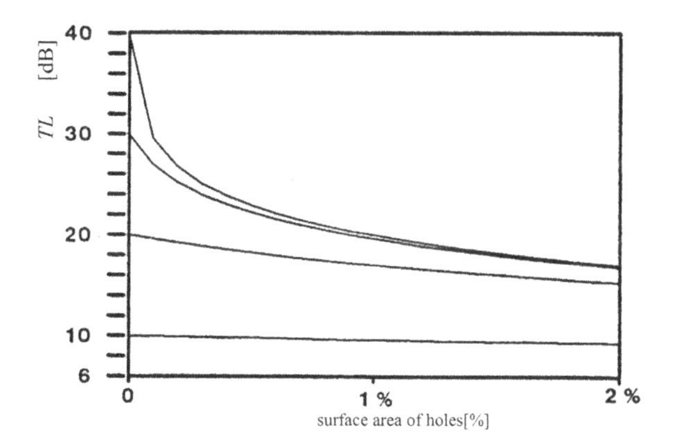

Fig. 11.79. Reduction of the transmission loss as function of the surface of holes. The curves refer to different transmission losses with no holes. The higher the curve the higher the original transmission loss of the wall.

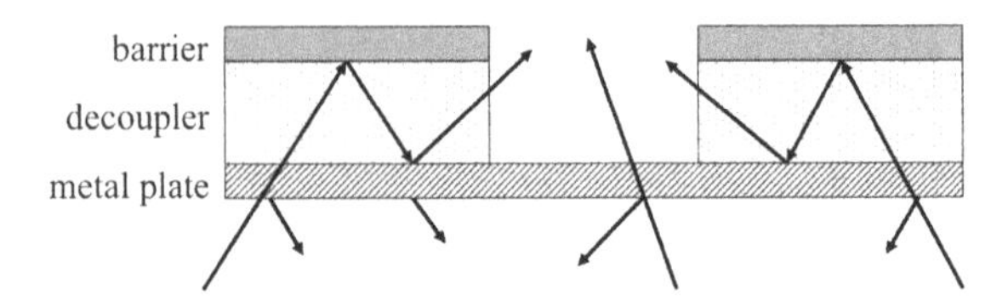

Fig. 11.80. Acoustic hole due to a gap in the acoustic insulation treatment.

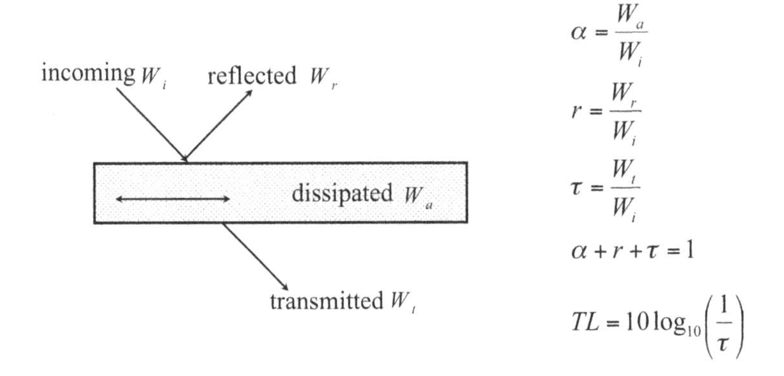

Fig. 11.81. Transmission of the sound through a barrier.

Gaps or discontinuities in the acoustic insulation that covers a panel are also acoustic holes, Fig. 11.80. The discontinuity represents a preferred transmission path for both the reflected sound and the transmitted sound.

11.5.2 Sound Barriers

The aim of the sound barriers is to reduce the amplitude of the incoming sound by means of its reflection or absorption. When the acoustic wave interacts with the barrier, the inertia of the barrier counteracts the motion of the air particles, causing a reflection of part of the energy. Another portion is dissipated as heat in the barrier itself because of the internal losses. The remaining fraction is transmitted to the opposite surface of the barrier.

(Fig. 11.81).

The double wall solution is adopted to increase the acoustic insulation of cars. The usual configuration of the sound barrier is formed by the metal sheet (for example the firewall), a layer of sound absorbing material, and a thick rubber or viscoelastic layer. As in double walls, the resulting transmission loss is larger than the sum of each individual contribution, except than at the mass-air-mass frequency.

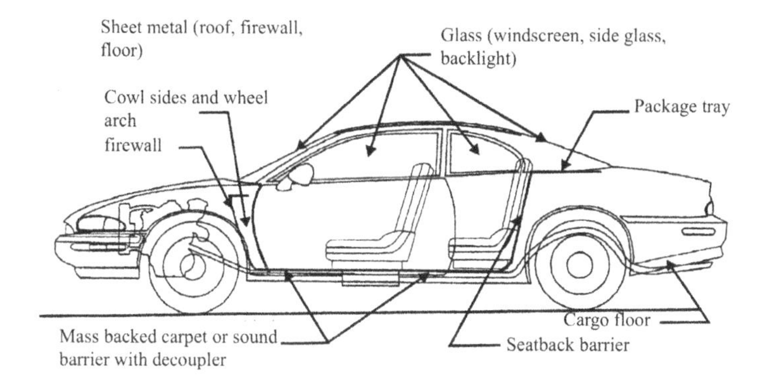

Fig. 11.82. Acoustic treatments on the car body.

La Fig. 11.82 shows the typical application of sound absorbing materials on various components of the car body. The treatments applied to body surfaces such as the body panels and the glasses reduce the noise that enters from outside; those applied on the firewall reduce the noise coming from the engine; those on the wheel arches and the floor reduce the road contact and rolling noise.

The acoustic treatment added to the firewall is usually a double walled barrier made by the steel panel, a layer of sound absorbing material and a thick viscoelastic layer (rubber or bitumen). Its thickness is about 40 mm, with a weight from 6 kg for a subcompact car (B segment), to 10 kg for a midsize (D segment). This is the most critical treatment for what the holes and uncovered portions is concerned, because of the large number of elements that are installed on the firewall.

Similarly to the firewall, a double wall treatment is also adopted for the front part of the floor. Its thickness is usually about 30 mm, covered by the finishing carpet. The weight goes from 3 kg for a B segment to 7 kg for a D segment. The rear part of the floor is usually covered by a sound insulating material and the carpet. The most critical portion of the floor is the front part of the tunnel with the holes for the gear-shift lever and the parking brake.

The glasses are rather transparent elements not only from the optical perspective but also from the acoustic standpoint. The windscreen thickness is $4 \div 5$ mm, while the side glasses $3 \div 4$ mm. Their critical frequency is about 4 kHz (Eq. 11.213), at this frequency they become almost transparent acoustically. Different solutions have been adopted to improve the behavior at this frequency, ranging from increasing the thickness to the use of laminated glasses formed by the sandwich of two glass layers and a viscoelastic one.

11.5.3 Sound Absorbing Materials

Materials with sound absorbing properties can be classified in two broad categories: porous and fibrous means. Rather typical of the automotive application are the mineral fibers (rock wool, fiber wool), polyester or polypropylene fibers (Politex, Thinsulate), cotton (shoddy), and open cells melamine or polyurethane foams (URS).

Most of the energy of the incoming sound wave is dissipated by viscous actions in the air trapped in the porous material. The acoustic wave induce high frequency vibrations the air particles that oscillate in the small gaps and channels of the porous means. The viscous forces at the boundary layer between air and solid parts dissipate the wave energy. Additionally, the compression and expansion in the irregular channels scatters the incoming wave in all directions reducing its momentum.

The contribution of the losses within the porous material are usually negligible compared to the viscous losses in the air. The only exception is that of the closed cells foams. Because the air is trapped in closed volumes, the relative motion between air and solid part is almost nulled, the dissipation is due to the relatively low losses that occur in the material. A good sound absorbing foam must then be characterized by an open cell structure, to allow the motion of the air with all dissipative mechanisms associated to it.

To understand the behavior of the sound absorbing materials it is useful to start from some basic principles such as the wave propagation equation and the definition of the mechanical impedance. The equation that describes the propagation of a sound wave is

$$\nabla^2\varphi(x,y,z,t) = \frac{\rho}{K}\frac{\partial^2\varphi(x,y,z,t)}{\partial t^2}, \tag{11.221}$$

where $\varphi(x,y,z,t)$ is the scalar potential function. It is related to the pressure field $p(x,y,z,t)$ by

$$p(x,y,z,t) = -\rho\frac{\partial^2\varphi(x,y,z,t)}{\partial t^2} \tag{11.222}$$

ρ is the mass density and K is the compression modulus. For a gas K is approximately given by

$$K = \gamma p$$

where γ is the adiabatic index. For the air $\gamma = 7/5 = 1.4$. In the following the explicit dependence from the space and time will be dropped for simplicity.

If the propagation occurs along direction x, the solution of Eq. 11.221 is an harmonic function of space x and time:

$$\varphi(x,t) = \frac{A}{\rho\omega^2}\cos[\omega(t-\frac{x}{c})+\alpha], \tag{11.223}$$

where α is the phase angle that is function of the value at the origin $x = 0$ and at time $t = 0$, ω is the angular frequency, and c is the propagation speed,

$$c = \sqrt{\frac{K}{\rho}}. \tag{11.224}$$

A is the amplitude of the acoustic pressure. Recalling Eq. 11.222

$$p(x,t) = -\rho\frac{\partial^2\varphi(x,t)}{\partial t^2} = A\cos[\omega(t - \frac{x}{c}) + \alpha]. \tag{11.225}$$

The speed of the air particles can be found considering the Euler equation (Eq. 11.2, 11.3) that relates the pressure gradient to the acceleration of the gas particles

$$u(x,t) = -\int \frac{1}{\rho}\frac{\partial p(x,t)}{\partial x}dt \tag{11.226}$$

substituting Eq. 11.223 in Eq. 11.226

$$u(x,t) = \frac{1}{\rho}\frac{\partial^2\varphi(x,t)}{\partial x\partial t}$$

so that

$$u(x,t) = \frac{A}{\rho c}\cos[\omega(t - \frac{x}{c}) + \alpha].$$

The wave number k is the number of wavelengths per unit distance, that is, $1/\lambda$ (cycle/m) where λ is the wavelength, or alternatively as $k = 2\pi/\lambda$ (rad/m), sometimes indicated as angular wave number or circular wave number or, simply wave number. Taking into account that the wavelength of a sound of angular frequency ω is

$$\lambda = \frac{2\pi c}{\omega}, \tag{11.227}$$

the angular wave number is

$$k = \frac{2\pi}{\lambda} = \frac{\omega}{c}. \tag{11.228}$$

In complex notation the potential and the pressure field of Eq. 11.223 and Eq. 11.225 become:

$$\varphi(x,t) = \frac{a}{\rho\omega^2}e^{j(-kx+\omega t)}, \tag{11.229}$$

$$p(x,t) = ae^{j(-kx+\omega t)}, \tag{11.230}$$

where j is imaginary unit ($j^2 = -1$) and $a = Ae^{j\alpha}$. The speed of the air particles is then:

$$u(x,t) = \frac{\partial^2\varphi(x,t)}{\partial t\partial x} = \frac{a}{\rho c}e^{j(-kx+\omega t)}. \tag{11.231}$$

The pressure of Eq. 11.230 and the speed of Eq. 11.231 are related by a constant coefficient Z_c

$$u(x,t) = \frac{1}{Z_c} p(x,t), \tag{11.232}$$

where Z_c is the impedance of the gas. Its value can be expressed as function of the mass density ρ and the compression modulus K

$$Z_c = \rho c = \sqrt{\rho K}. \tag{11.233}$$

The energy of the sound wave is dissipated by the viscous and thermal dissipations that occur in the air because of the motion of the air particles and the compression and expansion. In the free field propagation of audible sound waves such dissipations are negligible if the traveled distance is smaller than 10 m.

The propagation of the sound in a dissipative fluid can be described by considering a complex density ρ and a complex compression modulus K in the wave equation (Eq. 11.221). Similarly to the case of the response of mechanical systems affected by a complex stiffness, the use of the complex quantities allows to take into account of a phase lag during the propagation. The speed of sound c, the angular wave number k and the impedance Z_c given by Eq. (11.224), (11.228) and (11.233) become:

$$k = k_{\mathrm{Re}} + jk_{\mathrm{Im}}, \tag{11.234}$$

$$Z_c = Z_{c\,\mathrm{Re}} + jZ_{c\,\mathrm{Im}}. \tag{11.235}$$

if the wave is characterized by a positive term $e^{j\omega t}$ in Eq. 11.229, 11.230, the ratio $k_{\mathrm{Im}}/k_{\mathrm{Re}} < 0$. By converse, if ω is affected by a negative sign ($e^{-j\omega t}$), $k_{\mathrm{Im}}/k_{\mathrm{Re}} > 0$.

The pressure and the speed, for a wave that travels in the negative x direction are, respectively:

$$p'(x,t) = a' \exp[j(kx + \omega t)], \tag{11.236}$$

$$u'(x,t) = -(\frac{a'}{Z_c}) \exp[j(kx + \omega t)]. \tag{11.237}$$

Considering two sound waves, one traveling in the positive x direction and one in the negative one, the resulting pressure p_T and particle speed v_T is given by the superposition of the contributions of Eq.s (11.230), (11.236), and (11.237), (11.231):

$$p_T(x,t) = ae^{j(-kx+\omega t)} + a'e^{j(kx+\omega t)}, \tag{11.238}$$

$$u_T(x,t) = \frac{a}{Z_c} e^{j(-kx+\omega t)} - \frac{a'}{Z_c} e^{j(kx+\omega t)}. \tag{11.239}$$

Eq. (11.238) and (11.239) represent the general form of a unidimensional sound field of angular frequency ω. The ratio between pressure and particle speed $p_T(x,t)/u_T(x,t)$ is the general form of the impedance at coordinate x. The properties of such impedance are described in the following.

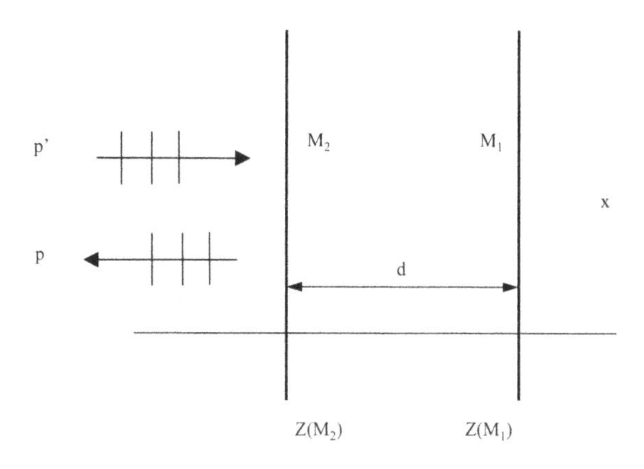

Fig. 11.83. Superposition of a forward and rearward traveling waves.

Properties of the impedance

Fig. 11.83 shows two waves that propagate along the positive and negative x directions. The impedance at coordinates x_{M_1} and x_{M_2} are

$$Z_{M_1} = \frac{p_T(x_{M_1}, t)}{u_T(x_{M_1}, t)},$$
$$Z_{M_2} = \frac{p_T(x_{M_2}, t)}{u_T(x_{M_2}, t)}.$$

Taking equations (11.238) and (11.239) into account, the time dependency of the pressure and speed disappears from the expression of the impedance Z at the two locations

$$Z(x_{M_1}) = Z_c \frac{ae^{-jkx_{M1}} + a'e^{jkx_{M1}}}{ae^{-jkx_{M1}} - a'e^{jkx_{M1}}}. \quad (11.240)$$

$$Z(x_{M_2}) = Z_c \frac{ae^{-jkx_{M2}} + a'e^{jkx_{M2}}}{ae^{-jkx_{M2}} - a'e^{jkx_{M2}}} \quad (11.241)$$

From Eq. (11.240) it is possible to obtain the ratio between the two amplitudes a/a' as function of the impedance at point x_{M_1}

$$\frac{a}{a'} = \frac{Z(x_{M_1}) - Z_c}{Z(x_{M_1}) + Z_c} e^{-2jkx_{M_1}}. \quad (11.242)$$

this expression (Eq. 11.242) can be substituted in Eq. 11.241 to get the impedance at point x_{M_2} as function of that at point x_{M_1}

$$Z(x_{M_2}) = Z_c \frac{Z_c - jZ(x_{M_1})/\tan(kd)}{-jZ_c/\tan(kd) + Z(x_{M_1})}. \quad (11.243)$$

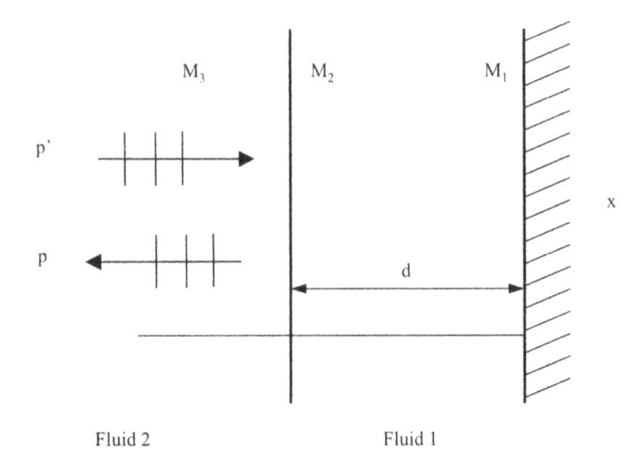

Fig. 11.84. Superposition of two pressure waves in a volume delimited by a rigid wall.

A fluid delimited at $x_{M1} = 0$ by a wall of infinite stiffness (and impedance) is represented in Fig. 11.84.

The point of coordinate x_{M_2} is located at the interface of fluid 2 and fluid 1 while the point of coordinate x_{M_3} is located in fluid 2. The impedance on the surface of fluid number 1, located at coordinate x_{M_2} is obtained from Eq. (11.243) with $Z(x_{M_1}) = \infty$:

$$Z(x_{M_2}) = -jZ_c/\tan(kd). \tag{11.244}$$

where Z_c is the impedance of a wave traveling in forward direction and k is the wave number of fluid 1. The pressure and speed are continuous at the interface between fluids 1 and 2 so that at the interface the impedance must be the same

$$Z(x_{M_3}) = Z(x_{M_2}). \tag{11.245}$$

The reflection coefficient R on the surface of a propagation means is the ratio between the pressures $p\prime$ and p of the sound waves that exit and enter the means

$$R(x_{M_3}) = \frac{p'(x_{M_3}, t)}{p(x_{M_3}, t)}. \tag{11.246}$$

as both the numerator and the denominator have the same frequency, the reflection coefficient is not function of the time. The reflection coefficient at point x_{M3} can be computed considering the impedance $Z(x_{M_3})$ and taking Eq. (11.245) into account:

$$R(x_{M_3}) = \frac{Z(x_{M_3}) - Z'_c}{Z(x_{M_3}) + Z'_c}, \tag{11.247}$$

where Z'_c is the impedance of means 2. If $|R(x_{M_3})| = 1$ the waves that enter and exit the surface at point x_{M_3} have the same amplitude. This happens when $|Z(x_{M_3})| = \infty$ or, in the opposite case, when $|Z(x_{M_3})| = 0$.

In general, if $Z(x_{M_3})$ is neither null nor infinite, the reflection coefficient has a unit value if:

$$Z^*(x_{M_3})Z_c^{'} + Z(x_{M_3})Z_c^{'*} = 0. \tag{11.248}$$

where the asterisk * indicate the complex conjugate. If Z_c' is real, $Z(x_{M_3})$ is a pure imaginary number.

If $|R(x_{M_3})| > 1$ the amplitude wave that exits the means is larger than the enters. If Z_c' is real, this happens if the real part of $Z(x_{M_3})$ is negative.

The absorption coefficient $\alpha(x)$ is defined as:

$$\alpha(x) = 1 - |R(x)|^2. \tag{11.249}$$

As the phase of the reflection coefficient $R(x)$ is not included in α, the absorption coefficient does not include some information that is present in the impedance or in the reflection coefficient. By its definition, the absorption coefficient has values between 0 and 1. It can be rewritten as:

$$\alpha(M) = 1 - \frac{E'(M)}{E(M)}, \tag{11.250}$$

where $E(M)$ and $E'(M)$ are respectively the energy flows through a plane at coordinate x of the incoming and outcoming waves.

Some considerations regarding the design of sound absorbing treatments

From this introduction on the propagation of the sound waves in different means, a number of general conclusions can be drawn regarding the design of sound absorbing treatments.

A high absorption coefficient ($\alpha = 1$, at the limit), can be obtained when the reflection coefficient R is small ($R = 0$, at the limit). This can be obtained when the surface impedance Z_s of the absorbing material is close to the impedance Z_0 of the air. The surface impedance is defined as

$$Z_s = \frac{Z_c}{\tanh(k_c d)}. \tag{11.251}$$

This can be obtained with materials comprising mostly air with a porosity index ϕ between 0.95 and 0.99.

The surface impedance is function of the thickness of the layer and from its impedance and wave number. The knowledge of these parameters is essential for the design of a sound absorbing treatment. Even if they are difficult to determine from measurements, the experimental tests on a large number of sound absorbing materials (Delany and Bazley (1970)) show that it is possible to estimate their value from a parameter called flow resistivity σ. The larger the flow resistivity the higher the characteristic impedance Z_c relative to that of the air.

If the flow resistivity is too large the sound waves are reflected from the surface, by converse, if it is too small they are not absorbed by the material. A good compromise is given by the following rule of thumb:

$$\sigma l \approx 3\rho_0 c_0. \tag{11.252}$$

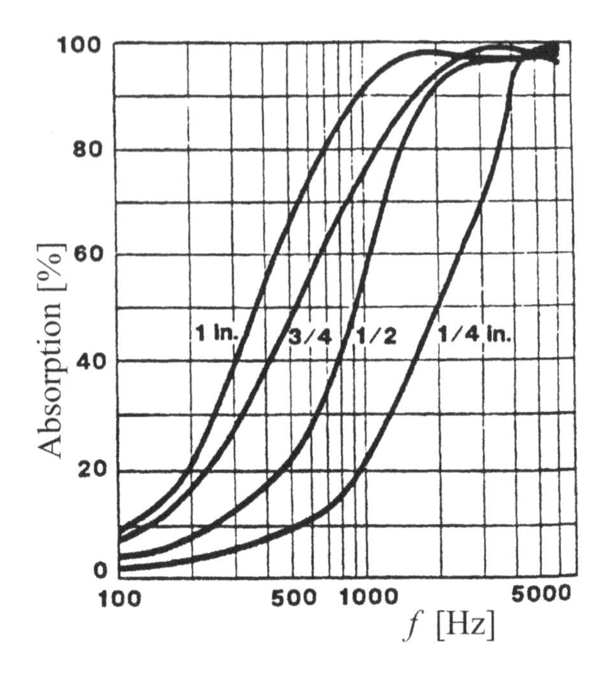

Fig. 11.85. Absorption coefficient as function of the thickness of the layer.

The largest the absorption coefficient is obtained with a thickness of the treatment equal to one quarter of the wavelength of the incident sound. A layer of a given material can give the maximum absorption only for frequencies equal and above $c/(4d)$ (Fig. 11.85).

The absorption coefficient can be improved at low frequencies adding on its surface a thin membrane with micro-holes, even if this reduces the performance at high frequencies (Fig. 11.86). The type of connection between the membrane and the underlying sound absorption material must be selected with care. This connection should not avoid to the air particles set into motion by the sound wave to enter the material. A membrane that seals completely the sound absorbing material from the incoming wave will spoil completely its performance.

Fig. 11.87 show the absorption coefficient α for a 20 mm thickness layer of different materials used in the automotive field:

Thinsulate	35% polypropilene, 65% polyester
Textile porous	recycled textile fibers
FIT	100% synthetic microfiber felt
KPF	100% synthetic fibers
Politex	100% polyester
Nitra Kombi	fibrous sandwich of different density
URS	pressure bond polyurethane chips
Melamine	fire proof open cell foam

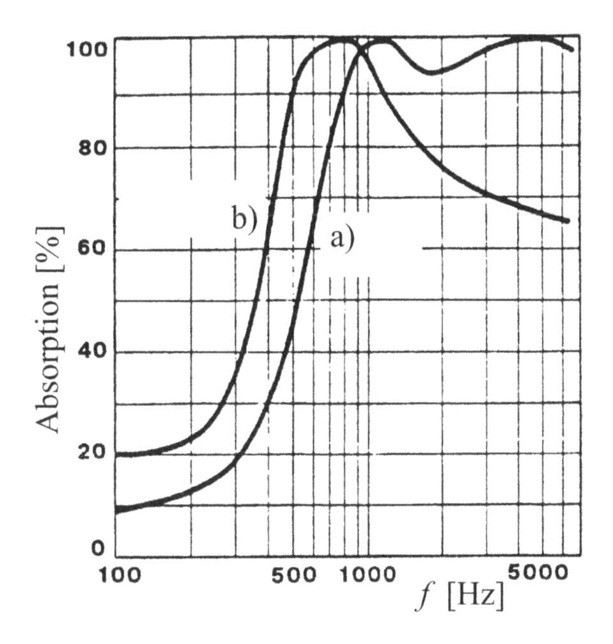

Fig. 11.86. Effect on the absorption coefficient of a mylar membrane on the surface of the sound absorbing layer. a) no membrane; b) 1.5 10^{-3} in mylar membrane.

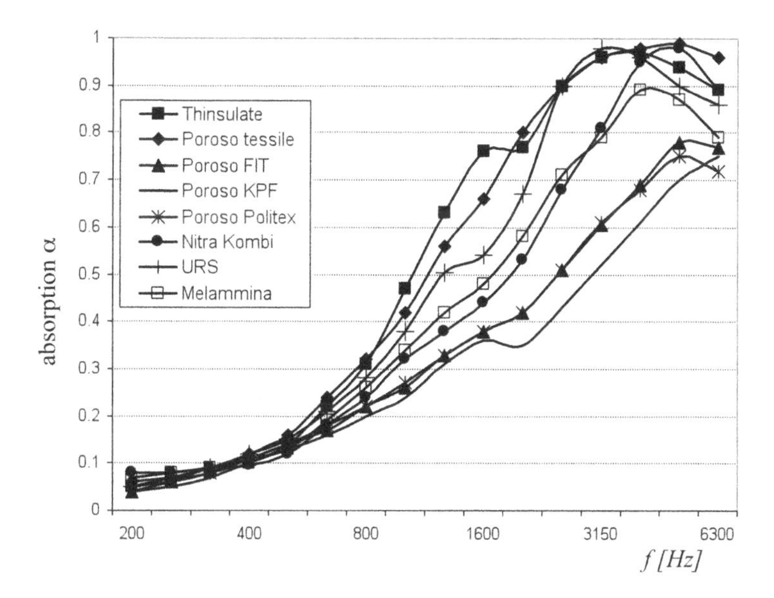

Fig. 11.87. Comparison between the absorption coefficient of different sound absorbing layers with same thickness of 20 mm.

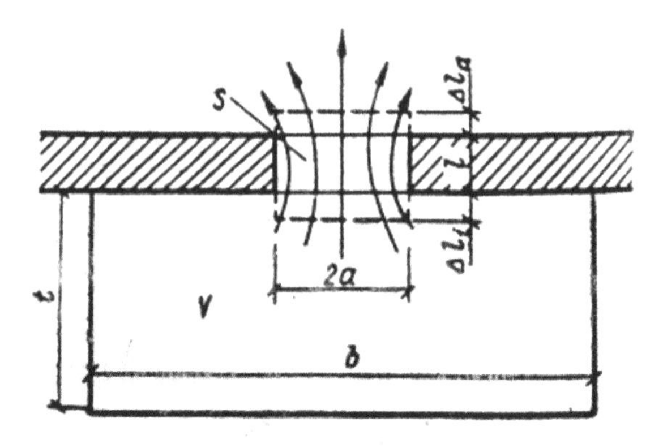

Fig. 11.88. Helmholtz resonator.

Other sound absorbing solutions

Helmoltz resonators, foil absorbers, and micro-perforated panels are other sound absorbing solutions. Compared to porous means they are effective in a narrower frequency band.

The Helmoltz resonator is an acoustic cavity with an open hole. The size of the hole is narrower than the cavity like an open and empty bottle. The mass of the air in the neck of the resonator can vibrate on the compliance of the volume of air in the acoustic cavity (Fig. 11.88). The system is then equivalent to a single degree of freedom mass-spring resonator. This resonator can be used as a spring-mass tuned damper to reduce the acoustic effects of a specific frequency (for example one of the acoustic natural frequencies of the volume). The natural frequency of the resonator can be evaluated as

$$f_0 = \frac{c}{2\pi}\sqrt{\frac{S}{Vl}}. \tag{11.253}$$

where S is the area of the cross section of the neck of length l, V is the volume of the acoustic chamber, and c is the speed of sound.

A membrane absorber is made by a thin membrane set at a distance d from a rigid wall. Similarly to the Helmoltz resonator the air cavity between the rigid wall and the membrane acts as a spring to counteract the displacements of the membrane. The resonance frequency of the oscillator in this case is

$$f_0 = \frac{c}{2\pi}\sqrt{\frac{\rho}{d\,m}}$$

where c is the speed of sound, ρ is the air density, m is the mass per unit surface of the membrane, and d is the distance between the membrane and the rigid wall. The large amplitude oscillations that occur at resonance mean great energy losses in the oscillating system. The absorption coefficient for a membrane

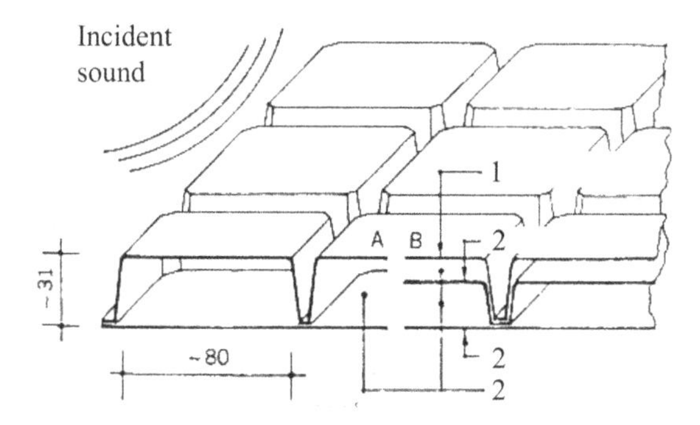

Fig. 11.89. *Foil Absorber.* 1) embossed membrane; 2) flat membrane; 3) air.

absorber thus reaches its maximum at the resonance frequency. A membrane absorber with an empty plenum will yield a high and narrow absorption peak. If the plenum is partly filled with mineral wool, the peak of the absorption will be less evident but broader.

Instead of a membrane with negligible stiffness, a foil absorber can be devised as shown in Fig. 11.89 with an embossed foil of plastic or metallic material. The foil is shaped in order to form a number of prismatic containers. The natural frequency of the system made by the each container and the enclosed air is function of the mass and stiffness properties of the container and the stiffness of the trapped air. The design parameters of the absorber in this case are

- the surface and depth of the cavities and the volume of the trapped air;
- the mass and stiffness properties of the foil and its geometry.

The combination of cavities of different dimensions allows to increase the range of frequencies where this solution is effective.

The main advantage of this solution is the resistance to the heat and to the fluids (water, lubricants, fuel), this justifies the adoption in the engine compartment.

Another solution that allows to absorb the acoustic energy is to use a micro-perforated panel of almost any kind of material (metal, polymeric, etc...) at a distance from a sound reflective surface. The design parameters of this solution are the pitch and diameter of the holes, the thickness of the plate and the distance from the sound reflecting surface. Similarly to the foil absorber this solution is suitable to high temperatures and engine fluids.

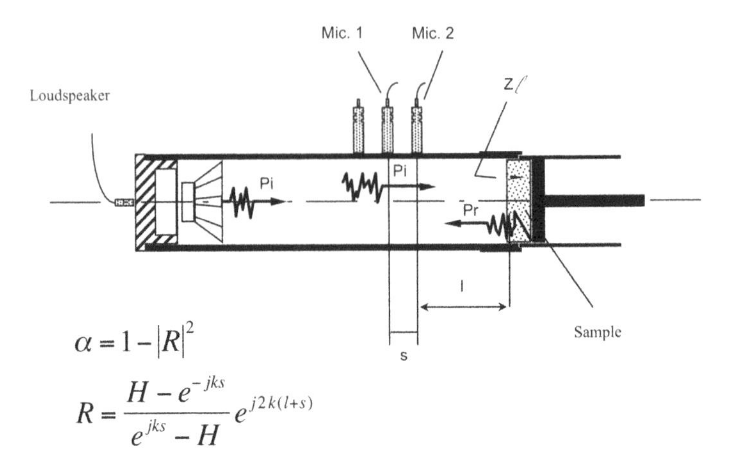

Fig. 11.90. Kundt tube test setup for the measurement of the absorption coefficient for normal incidence.

11.5.4 Measurement of the Sound Absorption Coefficient

Sound absorption for normal incidence

The reference standards are ASTM-E 1050 and DIN 52215.

Test device: impedance tube (Kundt tube), Fig. 11.90.

Samples: in the frequency range between 16÷1600 Hz the samples are cylinders with 100 mm diameter; for the 100÷6400 Hz the samples are cylinders with 30 mm diameter.

The sound absorption and the surface impedance can be evaluated from the frequency response function between two microphones installed along the tube, considering the distance s between the microphones and the distance l between one of them and the sample.

Sound absorption for random incidence

The reference standard is ISO 354-1985.

Test device: reverberating chamber with a 294 m^3 volume.

Samples: panels with 10 m^2 surface.

The sound absorption is evaluated from the measurement of the acoustic decay time.

Alpha cabin

The Alpha cabin is a reverberation chamber with linear dimensions one-third of that specified by ISO 354-1985. Its volume is 6.44 m^3 and no two walls are parallel. As a result of the size reduction of the room, the sample surface area is reduced to 1.2 m^2, an area that corresponds to that of typical hood- and roof-liners.

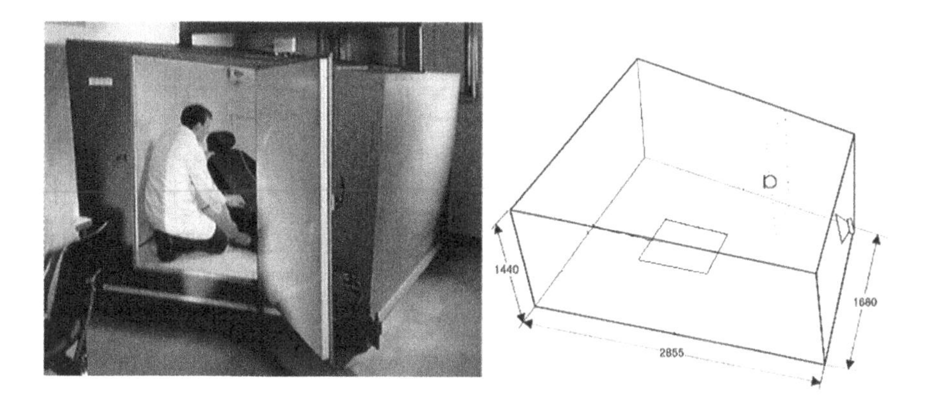

Fig. 11.91. Alpha cabin.

The measurement frequencies are also increased proportionally, so that the useful range lies between 400 and 10 kHz, which corresponds to the needs of the automotive industry.

The enclosure consists of a separate floor mounted on casters on which the upper part rests. Normal access to the cabin is through a door in one of the sides. Wall- mounted reflectors assure the sound field diffusivity necessary for the measurements.

The excitation is provided by three loudspeakers located in the corners. A microphone measures the sound pressure decay. Measurements are made in five microphone positions in a plane parallel to the floor and averaged. Movement between the positions is automatic. (Fig. 11.91).

Average absorption of a cavity

The Sabine model allows to find the mean sound absorption of an acoustic cavity is based on the assumption of the diffuse acoustic field i.e.:

- the acoustic energy flow coming at a point from all directions is the same and with random phase angle,
- the intensity of the acoustic field is the same in all points.

Sabine's model finds a large application in automotive acoustics even if the sound field inside a car is never consistent with the above assumptions.

The reverberation time T_{60} is the time required for reflections of a direct sound to decay by 60 dB below the level of the direct sound. Sabine found empirically that the reverberation time is proportional to the volume of the cavity and inversely proportional to the absorption of the room:

$$T_{60} \approx 0.161 \frac{V}{S\overline{\alpha}_d},$$

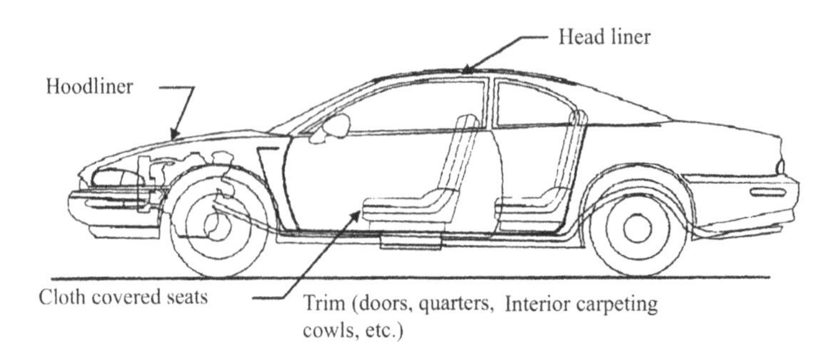

Fig. 11.92. Main sound absorbing treatments in a car.

where V is the volume of the cavity in m^3, S is the surface of the walls in m^2, $\overline{\alpha}_d$ is the average absorption coefficient of the room surfaces. A second approximation expression that takes into account of the attenuation constant m of the air (Eyring and Kuttruff)

$$T_{60} = 55.3 \frac{V}{c[4\ m\ V - S \ln(1 - \overline{\alpha}_d)]}, \tag{11.254}$$

The average absorption coefficient of room surfaces is related to the absorption coefficient $\overline{\alpha}_{di}$ of each wall of surface S_i:

$$\overline{\alpha}_d = \frac{\sum_i S_i \overline{\alpha}_{di}}{\sum_i S_i}, \tag{11.255}$$

$\sum_i S_i = S$ is the lateral surface of the cavity, not considering the objects contained in it (for example the surface of the seats).

The above expressions allow to determine the absorption coefficient of a sample of a material in diffuse field. This can be done repeating the measurement of the reverberation time with and without the sample that must be characterized. Eq. 11.254 allows then to find the average absorption coefficients and, finally, the contribution of the sample.

11.5.5 Sound Absorption Treatments in Car Body Applications

Sound absorption treatments are installed on cars to reduce the amplitude of the reflected sound waves. Fig. 11.92 shows the most common locations of these treatments. The seats account for nearly 50% of the absorption in the inner cavity, this is because they are made by materials with good absorption properties such as open cell foams and fabric. The next contribution is the roof panel, that accounts for about 25%, all other treatments make the additional 25% (door panels, floor treatment and carpet, etc.).

The sound absorption properties must be considered at the design stage even for components such as the seats or the roof panel not primarily devoted to the sound absorption. A proper choice of the textiles used for the seats and the interior panels can have a considerable influence: such fabrics should be allow the motion of the air set into motion by the sound wave. The use of heavy fabric or leather is acceptable only if micro-perforated. By converse, the velvet and rather coarse fabrics have good properties from this point of view.

Even if the main function of the roof panel is aesthetic, its considerable size makes it very important also for the absorption of the interior noise. Its base is usually a die formed panel of different kind of materials (reinforced polyurethane, for example) with a thickness from 7 to 9 mm. This panel is then covered with a finishing layer. The panel is then connected to the roof structure by bonding (low segments) or by mechanical fasteners.

As far as the engine compartment is concerned, the hood and the firewall are usually treated with fibrous tissue panels. Compatibility reasons with engine fluids, water, and mud foams suggest the use or foil absorbers for the lower engine shield.

12
Structural Integrity

There are many issues concerning the configuration of a vehicle chassis involving a range of different aspects including: ergonomics, accessibility, external shape, housing of mechanical components and suspensions. The structure is often adapted to these constraints filling the free spaces. As a consequence the body structure frequently has a complex shape.

The purpose of this chapter is to explain the structural function of the chassis components.

The starting point is the estimation of the reference loads, which are useful in the first phase of design.

Since car bodies are obtained by assembling thin sheet metal elements, methods for tension and deformation analysis of thin wall structures are recalled in the second part of the chapter. The principal failure modes, due to buckling in the elastic field, are briefly explained with respect to panels and thin wall tubular elements.

A part of the chapter is used to analyze a body structure in order to understand the function of its principal elements (longitudinal beams, cross beams, pillars, panels) and their contribution to bending and torsional stiffness. In order to analyze the contribution of each component to the global stiffness, a method based on the equilibrium conditions of the component itself is considered.

The simplified methods introduced to understand the structural function are not suited to perform quantitative analysis with the necessary accuracy. Therefore different types of finite element models, used to analyze and design car bodies, are briefly described.

L. Morello et al.: The Automotive Body, Vol. 2: System Design, MES, pp. 365–462.
springerlink.com

The last part of the chapter describes laboratory testing procedures used to determine structural characteristics of the vehicle such as torsional and bending stiffness.

12.1 Internal and External Loads

A vehicle is stressed by loads which typically are determined by the dynamic interaction between the excitation and the vehicle. During the initial stages of design the data available regarding the vehicle and its subsystems may not be enough to model such dynamic interactions. In this phase an evaluation of the loads is necessary nevertheless in order to identify an initial design. In a following stage a dynamic condition check is performed generating a more refined design.

On basis of these considerations, the objective of the following paragraph is to illustrate some load conditions useful for the initial design of the components of the structural body. Loads in the following conditions will be considered:

- parking,
- limit maneuvers,
- uneven road (uneven surface, obstacle encounter),
- internal loads due to transmission, passenger seats and seat belts,
- external loads concentrated or distributed on chassis elements.

Constant loads are applied in the following conditions: the vehicle own weight, the snow weight distributed on the roof surface and the car parked with a wheel on a curb. Instead, in all the other conditions, the loads cannot be considered to be constant.

Time constants may differ depending on the situation. Accelerating or braking to the limit of tire adhesion may only last the time needed to reach a certain speed or to stop (i.e. a time frame of few seconds).

The same can be said for cornering performed in limit conditions (similar time constants). Shorter times (fractions of a second) are involved in encountering a concentrated obstacle (curb). Moreover, during a crash, the forces applied to a seatbelt attachment points last for fractions of a second.

Subsequently the loads are assumed as if they were applied in a quasi-static way. Despite being variable in time, it is assumed again that the structure response to these loads is static, so no dynamic phenomena are involved.

Except for the seatbelt traction, all the loads described later may occur more or less frequently during the lifetime of a vehicle. Therefore components must be designed to prevent static or fatigue failures, permanent deformations or losses of functionality of the chassis.

12.1.1 Parking

Parking on flat road

A vehicle parked on a flat road, with the maximum payload possible, is not in an extreme load condition. Nevertheless flat road parking is interesting because the other conditions can be determined by calculating the variations with respect to this case. Given the mass and the position of the center of gravity, the vertical loads F_{za} and F_{zp} acting on the front and rear axles are simply:

$$\begin{aligned} F_{za} &= mg\frac{b}{p}, \\ F_{zp} &= mg\frac{a}{p}, \end{aligned} \tag{12.1}$$

where m is the mass of the vehicle, p is the wheelbase and a and b respectively are the longitudinal distances of the center of gravity from the front and rear axle. If the road is flat and the center of gravity is in the vehicle middle plane, loads F_{za} and F_{zp} are equally divided on both wheels of each axle.

Parking with a wheel on a curb

Referring to Fig. 12.1 it is assumed that the car has a wheel on a curb high enough to cause one of the other three wheels to lift off the ground. If the height of the curb is negligible with respect to the wheelbase, the vertical load on the axles does not change significantly if compared to Eq. 12.1.

With one wheel lifted, the car is supported in a statically determined way by the ground. Since the wheel contact points are not symmetric with respect to the middle plane (on which the center of gravity is assumed to lie) the structure becomes twisted. To understand which wheel leaves the ground, it is sufficient to consider the maximum torque that can be balanced by the front or rear axle.

The sketch of Fig. 12.2 shows one axle isolated from the rest of the vehicle. Applying a torque M_t determines a vertical load transfer between the two wheels with track t :

$$\Delta F_z = \frac{M_t}{t}. \tag{12.2}$$

Since the wheel-ground contact is monolateral, the maximum torque which can be balanced by the axle is that which completely unloads the wheel:

$$M_{t\,\max} = \frac{F_{zi}}{2}t. \tag{12.3}$$

where F_{zi} represents the load on the axle and the subscript $i = a, p$ denotes the front and rear axle respectively. Higher torque values cannot be balanced and would cause capsizing. When the car in Fig. 12.1 rises on to the curb, the front axle applies a torque to the rear one.

When one of the four wheels leaves the ground (in the figure the front right or the rear left) the torque cannot increase further. Therefore the maximum torque

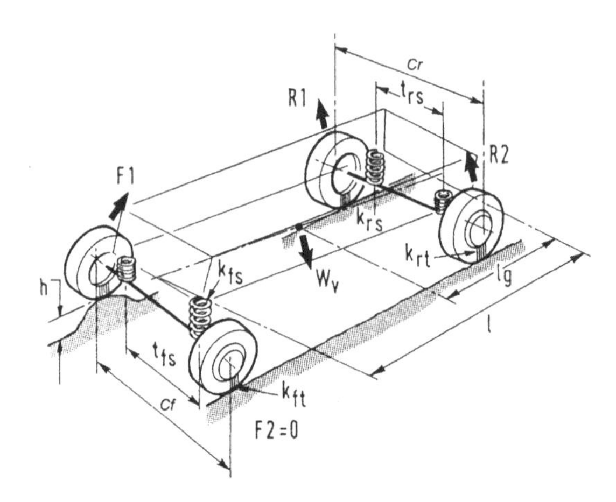

Fig. 12.1. Parking with a wheel on a curb high enough to lift one wheel off the ground, so that the car is supported by three wheels only.

during parking with a wheel on a curb is the minimum between the moments that can be balanced by the two axles:

$$M_{t\,\max} = \min(\frac{F_{za}}{2}t_a, \ \frac{F_{zp}}{2}t_p). \tag{12.4}$$

If the torque transmissible by the front axle is lower than the rear axle,

$$\frac{F_{za}}{2}t_a < \frac{F_{zp}}{2}t_p, \tag{12.5}$$

the first wheel to rise is a front one;

vice versa if:

$$\frac{F_{za}}{2}t_a > \frac{F_{zp}}{2}t_p, \tag{12.6}$$

the first wheel to rise is a rear one.

For the sole purpose of evaluating the order of magnitude of the static moment, consider a car weighing $F_F = 14$ kN, with the center of gravity located at half of the wheelbase $a = b = p/2$. In this case $F_{za} = F_{zp} = 7$ kN. If both the front and rear track are equal $t_a = t_p = 1.2$ m, the result is:

$$M_{t\,\max} = \frac{7\ \text{kN}}{2}\ 1.2\ \text{m} = 4200\ \text{Nm}. \tag{12.7}$$

The considerable magnitude of the torque requires a design intended to avoid the material yielding, the weld spots to break (static or fatigue failure) and, above all, excessive deformations.

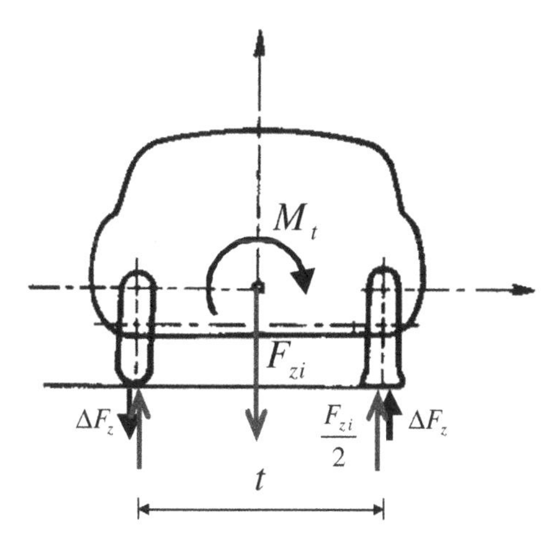

Fig. 12.2. Load transfer due to a torque M_t applied to an axle isolated from the car.

12.1.2 Limit Maneuvers

Transversal limit

When a car maneuvers a constant radius corner, the maximum lateral force ($F_{y\,\max}$) on the vehicle is limited by two conditions:

1) reaching the tires maximum lateral adhesion, or
2) the incipient capsizing.

The condition of tires maximum lateral adhesion is given by:

$$F_{y\,\max\,\mu} = \mu_{y\,\max} F_z, \tag{12.8}$$

where F_z is the total weight and $\mu_{y\,\max}$ is the maximum lateral adhesion coefficient that can be produced by the whole vehicle. Its value does not only depend on the tires features and on the road surface characteristics (dry surface, wet, snow), but is also influenced by vehicle dynamics issues, such as the features of the suspension system, its stiffness, the presence of antiroll bars, the center of gravity position, just to list the principal factors.

The incipient capsizing condition, shown in Fig. 12.3, is reached when the resultant of the inertia force and the weight force (both applied on the center of gravity) crosses the contact patch of the external wheel. In this case, the lateral load transfer cancels the vertical force on the wheels inside the corner:

$$\Delta F_z = F_{y\,\max\,R} \frac{h}{t} = \frac{F_z}{2}. \tag{12.9}$$

So a maximum lateral grip force $F_{y\,\max\,R}$ is obtained which corresponds to the incipient capsizing:

$$F_{y\,\max\,R} = \frac{1}{2} \frac{t}{h} F_z. \tag{12.10}$$

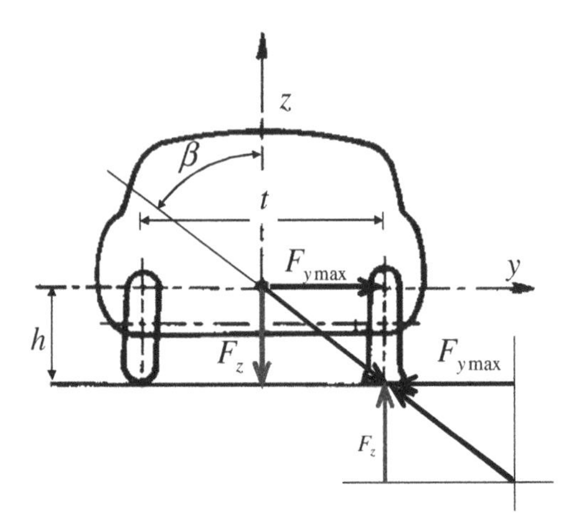

Fig. 12.3. Maximum lateral force due to the incipient capsizing

For a vehicle travelling on a road with known grip characteristics, the maximum lateral force is given by the lowest of the limit loads given by the Eq.s 12.8 (adhesion limit), 12.10 (capsizing):

$$F_{y\,\max} = \min(\mu_{y\,\max} F_z,\ \frac{1}{2}\frac{t}{h} F_z). \tag{12.11}$$

In the case of a road with good grip conditions $\mu_{y\,\max} \approx 0.8 \div 0.9$, depending on the ratio $2h/t$ between the height h of the center of gravity and the semitrack $(t/2)$, the limit of the lateral force is given by the tires or by the incipient capsizing. Considering a vehicle with track $t = 1.2$ m and a relatively low center of gravity: $h = 0.5$ m, so the ratio $t/(2h) = 1.2 > \mu_{y\,\max}$. The maximum lateral force is then imposed by $\mu_{y\,\max}$.

For a small commercial vehicle with the same track and a higher center of gravity $h = 0.8$ m $t/2h = 0.75$, the maximum lateral force is imposed by the capsizing condition.

Longitudinal limit

The maximum longitudinal force conditions can be reached during braking or accelerating to the limit of tires adhesion. For common vehicles the maximum torque produced by the brakes is much higher than the torque given by transmission, so the braking condition features higher longitudinal forces; if F_z is the total weight then:

$$F_{x\,\max\,\mu} = \mu_{x\,\max} F_z, \tag{12.12}$$

where $\mu_{x\,\max}$ is the maximum longitudinal adhesion coefficient that can be produced by the whole vehicle.

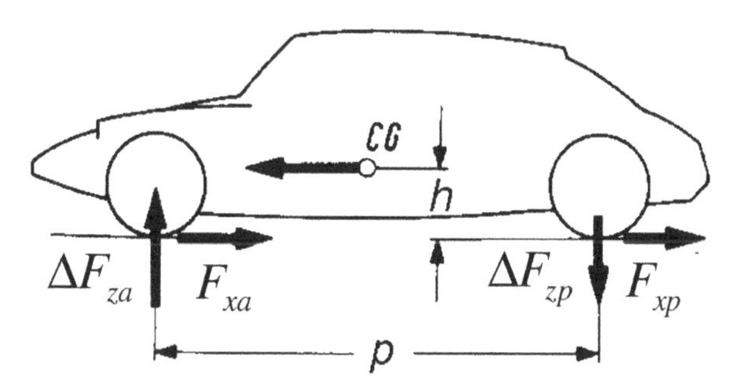

Fig. 12.4. Load transfer between the rear and front axle during braking.

Similarly to the case of limit transversal forces, the incipient capsizing condition can be taken into account. The load transfer in the braking condition is:

$$\Delta F_z = \frac{h}{p} F_{x\,\max R}; \tag{12.13}$$

the incipient capsizing is reached when the load transfer cancels the vertical load on the rear axle (Fig. 12.4):

$$\Delta F_z = F_{zp} = \frac{a}{p} F_z. \tag{12.14}$$

In this case the whole car is completely weighing on the front axle. Therefore:

$$F_{x\,\max R} = \frac{a}{h} F_z \tag{12.15}$$

The maximum longitudinal force is given by the lowest of the loads defined by equations 12.12 and 12.15:

$$F_{x\,\max} = \min(\mu_{x\,\max} F_z,\ \frac{a}{h} F_z). \tag{12.16}$$

Since usually $a > h$ the maximum longitudinal force is commonly given by reaching the limit of tires adhesion.

12.1.3 Road Unevenness

Obstacle

Road tests show that, except in the case of accidents, the maximum loads applied on the vehicle occur when encountering concentrated obstacles such as curbs and potholes. The vertical accelerations measured on the suspensions struts towers can reach values up to three times the acceleration of gravity g:

$$a_{z\,\max} = 3\ g \approx 30\ \mathrm{m/s^2}. \tag{12.17}$$

As a consequence of such a high acceleration, the suspension ends its travel so that the sprung and unsprung masses become integral. The vertical load $F_{z\ so}$

(the subscript $_{so}$ is referred to the obstacle encounter) that the ground applies on the axle, can be estimated as:

$$F_{z\ so} = a_{z\,\max} m_{a,p}, \tag{12.18}$$

where $m_{a,p}$ is respectively the mass weighing on the front or rear axle.

It has already been said that phenomena associated with encountering an obstacle are relatively complex and involve suspension and tire dynamics. Besides vertical forces, also braking forces are produced, the intensity of which can be determined using Eq. 12.18.

Considering a rather safe angle θ of 45° between the vertical and horizontal components of the force, the longitudinal components equal the vertical components:

$$F_{x\ so} = \tan\theta\ F_{z\ so} \approx F_{z\ so}. \tag{12.19}$$

Loads involved in encountering an obstacle are usually much higher than the ones involved in a braking to the limit of tire adhesion. The impulsive nature of this phenomenon makes the quasi static approach inappropriate to consider all the phenomena which occur during the transitory phase.

In order to calculate these effects, a dynamic simulation is necessary to consider the highly non linear nature of many components, such as bushings made of elastomer. These components (for example: the engine elastic mounts, suspensions bushings) may reach the end of their travel because of the high loads acting on them.

Road unevenness

Distributed road unevenness must be considered because it can lead to fatigue failures of components. Despite stressing the structure less seriously than concentrated obstacles, such as potholes and curbs, distributed unevenness can promote the initiation of fatigue cracks, which may grow and lead to mechanical failure.

Chapter 11 mentions that random road unevenness is characterized by its power spectral density. This quantity shows how the excitation involves a continuous spectrum of frequencies, illustrating that amplitudes tend to decrease at higher frequencies.

Repeated stresses, caused by a vehicle moving on an uneven road, are a source of fatigue for the structure. The evaluation of the stresses starts from considering a road profile, the features of which are expressed by the power spectral density of Eq. 11.30. By means of the two coefficients C_0 and N it is possible to choose a type of road profile (freeway, uneven profile, very uneven profile) and a speed (for example: 50 km/h). This selection enables (Eq. 11.34) to express the power spectral density in terms of temporal frequency of excitation.

Since many components have a non linear characteristics and since large displacements may arise in certain conditions, the computation of stresses requires numerical simulation in time with FEA and multibody software.

Sometimes the calculation is performed by measuring the profile of a road which is considered to be particularly significant. This measurement must take into account also the unevenness distribution along the width of the roadway so that both wheels of each axle are excited in a different way, not only inducing vertical and pitch motions, but also roll.

This road profile is then applied to the model forcing the ground-tires contact patches to have the same temporal history of vertical displacement as that measured.

If a direct measurement of the road profile is not available, it is possible to proceed by creating a road profile with the desired power spectral density, starting from a white noise.

The calculation is shown in the diagram of Fig. 12.5. The signal produced by a generator of random noise with constant power spectral density (white noise) is fed through a low-pass filter. The goal is to obtain, out of the filter, a signal with the same power spectral density as the road profile.

For this purpose the relation between the input power spectral density ($G_u(\omega)$) and the output one ($G_z(\omega)$) of a linear filter with transfer function $H(s)$ is used:

$$G_z(\omega) = |H(j\omega)|^2 G_u(\omega). \tag{12.20}$$

Assuming that the input white noise has a unitary power spectral density:

$$G_u(\omega) = 1 \tag{12.21}$$

and considering that the power spectral density $G_z(\omega)$ is that of the road profile driven at the speed V :

$$G_z(\omega) = C_0 V^{N-1} \omega^{-N}, \tag{12.22}$$

the following expression is obtained:

$$|H(j\omega)|^2 = \frac{G_z(\omega)}{G_u(\omega)} = C_0 V^{N-1} \omega^{-N} \tag{12.23}$$

Therefore:

$$H(s) = \frac{\sqrt{C_0 V^{N-1}}}{s^{N/2}} \approx \frac{\sqrt{C_0 V}}{s}. \tag{12.24}$$

After the transfer function of the filter has been determined, the vertical displacement, which output from the filter, is given as the input to a vehicle model (*multibody* + FEM) which is used to calculate the internal stresses of the chassis. It must be remembered that if $z(t)$ is the front axle vertical displacement, then the rear axle displacement is $z(t-\tau)$ where:

$$\tau = \frac{p}{V}. \tag{12.25}$$

The rear axle is subject to the same displacements as the front axle, but with a delay of τ due to the wheelbase p.

The tensions or the deformations obtained by FEM computation are then used to calculate the cumulative damage inside the structure, according to one of the hypothesis (the most simplified is the Miner hypothesis) which allows to determine the life fraction of the component that has been "consumed" because of the application of a load cycle.

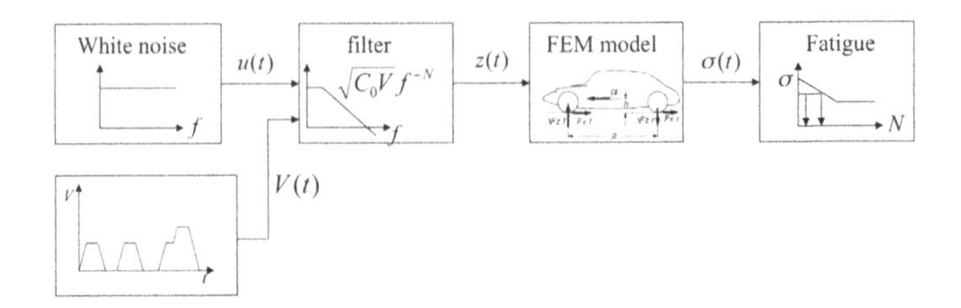

Fig. 12.5. Calculation diagram of the fatigue level reached after a travel on an uneven road.

12.1.4 Internal Loads

These loads are usually evaluated by applying static actions to *multibody* models which include the engine (or the differential), its supports, the transmission system, suspensions and all the links to the structural body.

Suspensions

The forces and couples exchanged between wheel and ground are transmitted to the chassis through the suspensions connection points. Starting from the design limit loads, previously calculated, and given the geometry and the suspensions characteristics, it is possible to obtain the loads at the suspension connection points.

As an example Fig. 12.6 shows the forces and constraints acting in a McPherson suspension on which a force with both a vertical and lateral component is exerted.

According to current engineering practice, the computation of internal and external exchanged forces is performed automatically with computational multibody codes. These programs can simulate a vehicle maneuver and calculate, besides many quantities, the forces transmitted by tires and by suspensions.

Transmission

The torque delivered by the engine causes reaction forces on the components connecting it to the structural body. Fig. 12.7 refers to a front engine and rear drive layout. $\vec{T}_{ps}$ is the torque that the gear output shaft applies to the driveshaft, $\vec{T}_w$ is the overall torque that the differential transfers to the wheels (right wheel + left wheel), $\vec{T}_{es}$ and $\vec{T}_{ds}$ respectively are the constraints reactions produced by the engine supports and by the differential supports. The constraint reaction $\vec{T}_{ds}$ must balance the resultant $\vec{T}_d$ of the torques applied to the differential by the wheels and by the drive shaft:

$$\vec{T}_{ds} = -\vec{T}_d, \tag{12.26}$$

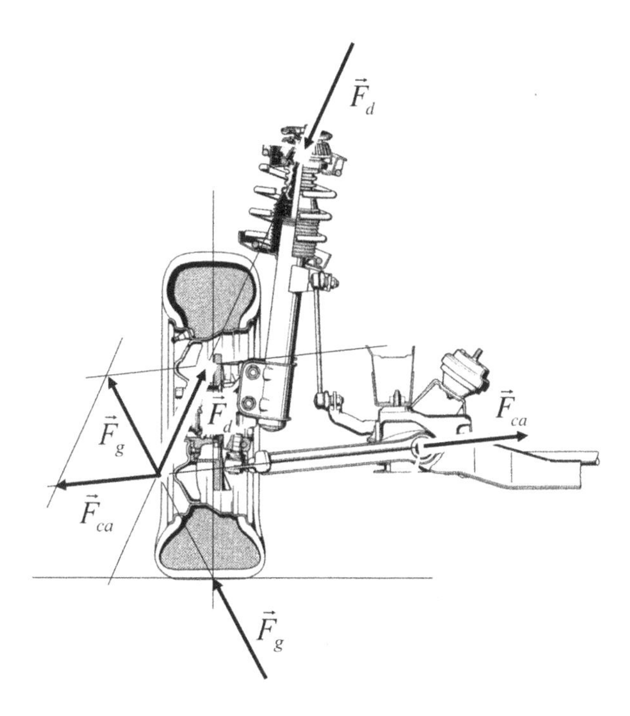

Fig. 12.6. Loads at the connection points of a McPherson suspension. The force that the ground $_g$ applies through the contact patch is translated on its line of action until it crosses the lower control arm $_{ca}$ axis. Then it is separated into components according the direction of the lower control arm axis $\vec{F}_{ca}$ and the direction through the suspension strut tower ($_d$) $\vec{F}_d$.

where:

$$\vec{T}_d = \vec{T}_{ps} - \vec{T}_{rw}. \tag{12.27}$$

The last reaction, shown by the free body diagram, is oblique to both the crankshaft and the wheels axles. The points connecting the differential to the body are positioned in order to produce the torque $\vec{T}_{ds}$, minimizing the resulting forces.

Correspondingly two of the bushings of the differential suspension are placed at the maximum distance from the line of action of the torque $\vec{T}_{ds}$. The third bushing is needed to make the suspension statically determined and to correctly support the differential respect to the forces (weight and inertia).

In the case of a four wheel drive in Fig. 12.8 the situation of the rear differential is the equal to the one in Fig. 12.7. As concerns the front powertrain composed by engine, gearbox and differential, the constraint reaction $\vec{T}_{es}$ must balance the torques of the driveshaft $\vec{T}_{ps}$ and of the front wheels $\vec{T}_{fw}$

$$\vec{T}_{es} = -\vec{T}_e; \qquad \vec{T}_e = -\vec{T}_{fw} - \vec{T}_{ps}. \tag{12.28}$$

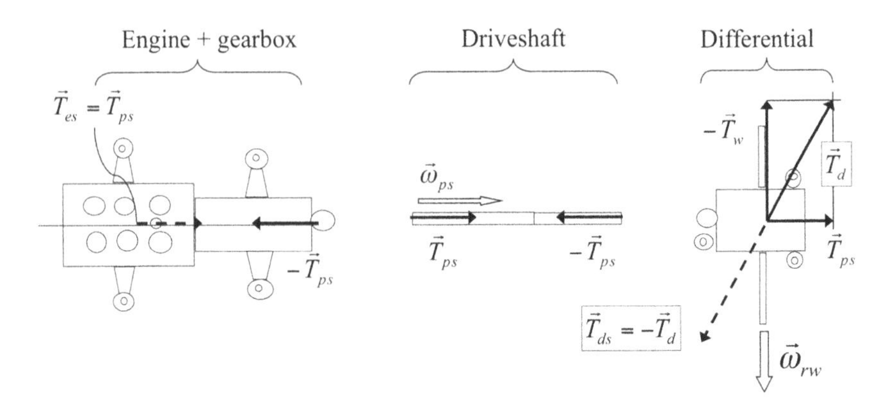

Fig. 12.7. Layout of a front engine and rear wheel drive transmission.$\vec{T}_{ps}$: is the torque that the gear output shaft applies to the driveshaft. $\vec{T}_{es}$: constraint reaction of the engine suspension. $\vec{T}_{w}$ overall torque of the rear wheels. $\vec{T}_{ds}$ constraint reaction of the engine suspension. $\vec{\omega}_{ps}$ rotational speed of the driveshaft. $\vec{\omega}_{rw}$ rotational speed of the rear wheels.

Safety belts

The local tensile strength of the seatbelts connection points is tested to verify that, under loads set by homologation rules, the permanent deformations fall within certain limits. The reader is referred to the first volume for other considerations.

Loads on the surface

Besides the loads coming from the road, other kinds of load must be taken into account such as those ones applied when pushing the vehicle or leaning against it, or due to atmospheric agents such as snow and hail.

The behavior of the bodywork with respect to loads caused by fingers or accidental contacts, is characterized using pushers which apply local loads that slowly increase up to a maximum level. After the load has been released, it is verified that no permanent deformations have occurred. The test is performed on the doors, on the hood, on the trunk door and on the side body.

Fig. 12.9 shows the result of a simulation performed with a non linear finite element model of a side door. The simulation involves a load and unload cycle performed with a sphere of 80 mm diameter. During the unloading phase, a small permanent deformation (about 0.5 mm) remains on the surface meaning that the material was subject to yield.

A similar test is made on the roof to simulate the pressure of a snow layer. In this case a distributed load is a applied, for example using bags of sand.

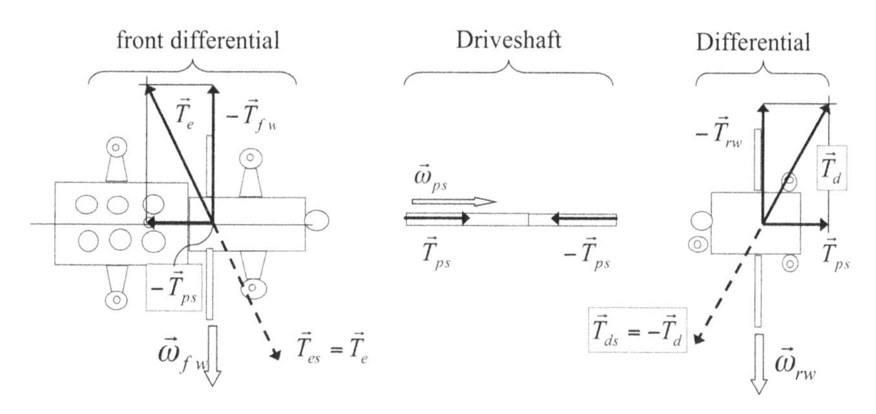

Fig. 12.8. Four wheel drive transmission. Same symbols used for Fig. 12.7. $\vec{T}_{fw}$ is the overall torque on the front wheels. $\vec{\omega}_{fw}$ is the angular velocity of the front wheels.

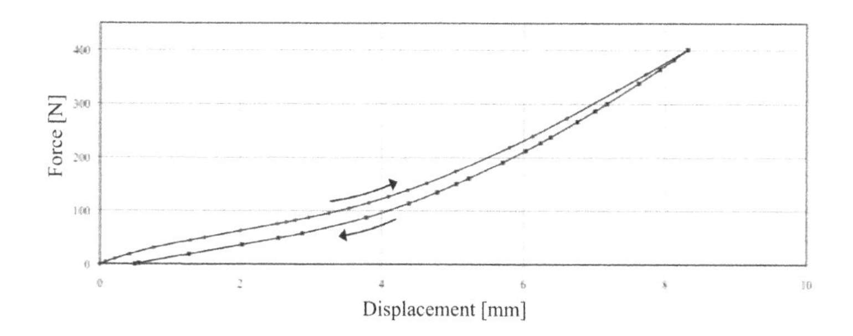

Fig. 12.9. Force – displacement diagram obtained by applying a maximum force of 400 N on the body surface with a sphere of 80 mm diameter. The arrows highlight the load and unload phases.

12.1.5 Safety Factors

As previously mentioned, as a rough approximation it is possible to consider dynamic loads as being quasi-static. In order to evaluate dynamic loads, knowing either what caused them (for example a curb) and the structure features (masses, stiffness, damping and their distribution) is a necessary requirement. So, in order to model vehicle dynamic behavior, car design must be at least defined in its basic elements.

Many data required to elaborate a dynamic model are not known in the first stages of design, so the quasi-static hypothesis is used to evaluate loads acting on the vehicle approximately, with the purpose of developing a first design approximation. The subsequent verification (and refining) steps are based on increasingly sophisticated dynamic models of the vehicle so that it is possible to consider the interaction with the external excitations.

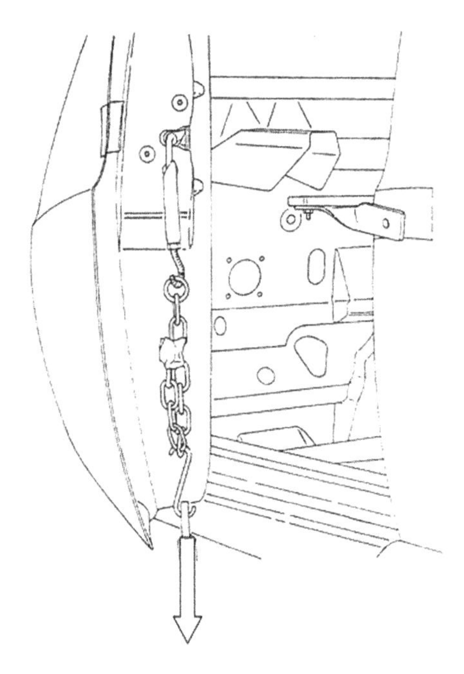

Fig. 12.10. Verification of the behavior of a door subject to a vertical load applied on the lock. In order to guarantee the functionality and the plays regularity, permanent deformations must be very low (e.g. 0.5 mm).

Table 12.1. Safety factors used to take the dynamic nature of loads into account.

Load type	**Safety factor**
obstacle	$1.4 \div 1.6$
braking to the limit of adhesion	$1.1 \div 1.8$
cornering to the limit of adhesion	$1.4 \div 1.75$
torsion	$1.3 \div 1.8$

In order to take into account the approximations introduced by the quasi-static hypothesis, an approach based on safety factors may be followed. These factors are usually based on experience and their value depends on the load type.

Table 12.1 shows the safety factors in the principal load conditions which have been obtained by comparing different authors [29], [30].

12.2 Behavior of Thin Wall Structures

The structural configuration of a chassis is often very complex due to normative and functional constraints. A vehicle chassis is characterized by a high level of static indetermination because of the complex interconnection of beams (longitudinal elements, cross elements and pillars) and panels. Therefore it is relatively difficult to perform an analytical calculation of stresses and strains unless drastic approximations are introduced. The structural analysis is normally performed numerically using the finite element method.

However, in order to provide some design criteria, it is necessary to understand the structural functionality of the principal chassis components. For this reason it is convenient to consider some basic layouts: Even though these layouts cannot provide precise quantitative information, they can prove useful to explain the structural function of the chassis components.

The comprehension of the parameters which influence chassis behavior is useful both in the outlining process, when the main configuration is selected, and during results analysis when the final design is refined.

This paragraph illustrates some simplified methods for the analysis of stress and strain in thin wall structures, such as those used in the vehicle chassis. The most common stress characteristics, that are bending, torsion and shear, will be explained.

These structures are subject to buckling because of the low thickness of their walls. Buckling phenomena will be explained for simple geometries (beams and panels).

12.2.1 Hypothesis and Definitions

A *monocoque* structure is one whose thickness is small if compared to the section dimensions, for example: a folded metal sheet so that it forms a cylinder welded on a generatrix. This structure is not well suited to supporting concentrated loads, which may cause local collapse of the monocoque.

In order to apply concentrated loads, the structure is stiffened with longitudinal elements, called longitudinal stiffeners, and transversal elements, called ribs. Such structures are called *semi-monocoques*. The aeronautic structures represent typical examples of the semi-monocoque structure. For example a wing is formed by panels stiffened by longitudinal stiffeners and transversal ribs.

Automotive structures are constituted of thin metal sheets folded and assembled in order to form open and closed sections. Longitudinal stiffeners, riveted or glued to the panels like in the aeronautic case, are not normally used. If a longitudinal stiffening element is needed, some swages or ribs may be realized on the panels, which have the same function of a longitudinal stiffener but cost less.

The connections for the application of concentrated forces are realized by adding thicker metal sheet elements. Similarly to aeronautical structures, their function is to distribute the concentrated loads on the metal panels in order to prevent local buckling.

Also the automotive frames can be studied using the hypothesis of the semi-monocoque structure.

The analysis of a semi-monocoque structure is based on the following hypothesis:

- stiffening elements to withstand traction/compression,
- panels which withstand traction and shear but not compression.

It is assumed that a compressed panel undergoes buckling so it cannot transmit compression loads. Only the strip of panel connected to a stiffening element may withstand compression loads. This is called *collaborating strip* and it has a width which is 12÷15 times the thickness of the metal sheet.

Next it is appropriate to consider beams with the following simplified characteristics:

- semi-monocoque structure,
- elastic, homogeneous, isotropic material
- long in comparison to the section dimensions,
- rectilinear axis,
- constant section.

The last two hypotheses (rectilinear axis, constant section), are not properly respected in automotive structures. Nevertheless it is possible to approximate a structure as a set of parts with rectilinear axis and constant section, joined by elements with a more complex geometry (usually nodes). Results may be used

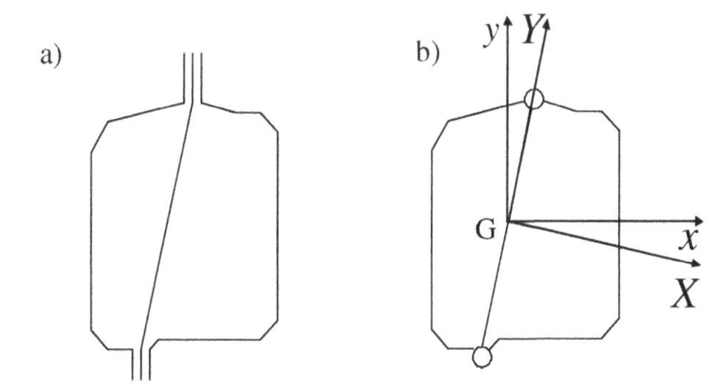

Fig. 12.11. Semi-monocoque section. a) real configuration made of three part joined by welds on the upper and lower flanges; b) schematization: the welded flanges and the adjoining sheet parts are schematized as concentrated longitudinal stiffeners that only withstand traction/compression. Panels only withstand shear.

out of the connecting elements. Similarly the curvature of beams (for example the A pillar) is often low with respect to the sections dimensions, therefore the curvature effect is also low.

Before starting the analysis of a semi-monocoque structure, its section is discretized in order to consider the different structural function of all the components.

Fig. 12.11 a) shows the section of a beam obtained joining three metal sheet elements by welding their upper and lower flanges. The same section is schematized in b) according to the hypothesis of the semi-monocoque structures. The welded flanges and the adjoining sheet parts are schematized as longitudinal stiffeners concentrated at their center of gravity. They only withstand traction/compression; their section area is indicated with S_i. Panels only withstand shear.

12.2.2 Bending

According to the semi-monocoque hypothesis, the only section elements that withstand bending are the longitudinal stiffeners, with area S_i, which are the only parts capable of producing forces perpendicular to the plane of the section.

Considering a reference system xy with the origin at the center of gravity G of the section, the area properties of the discretized section are characterized by the inertia moments

$$J_x = \int_S y^2 dS = \sum y_i^2 S_i \tag{12.29}$$

$$J_y = \int_S x^2 dS = \sum x_i^2 S_i \tag{12.30}$$

$$J_{xy} = \int_S xy dS = \sum x_i y_i S_i \tag{12.31}$$

where S_i are the areas of the sections withstanding traction and compression (longitudinal stiffeners) and x_i, y_i are the coordinates of their centers of gravity.

Based on the hypothesis, the sections of the beam remain plane when the bending moment is applied. The strain field on the section is then a linear function of the coordinates x, y

$$\epsilon = k'x + k''y \tag{12.32}$$

$$\sigma = k_1 x + k_2 y \tag{12.33}$$

The stresses distribution on the section must balance the external bending moment, with components M_x and M_y

$$\begin{aligned} M_x &= \textstyle\int_S \sigma y \; dS \\ M_y &= \textstyle\int_S \sigma x \; dS \end{aligned} \tag{12.34}$$

Substituting the Eq. 12.33 into 12.34 two constants k_1, k_2 are obtained and the distribution of stresses on the section is consequently found. (Navier)

$$\sigma = \frac{M_y J_x - M_x J_{xy}}{J_x J_y - J_{xy}^2} x + \frac{M_x J_y - M_y J_{xy}}{J_x J_y - J_{xy}^2} y. \tag{12.35}$$

Considering the discretized section, the stress in the longitudinal stiffeners is

$$\sigma_i = \frac{M_y J_x - M_x J_{xy}}{J_x J_y - J_{xy}^2} x_i + \frac{M_x J_y - M_y J_{xy}}{J_x J_y - J_{xy}^2} y_i. \tag{12.36}$$

Considering a reference system GXY , the axis of which are the principal of inertia ($J_{XY} = 0$), this expression can be simplified to:

$$\sigma = \frac{M_Y}{J_Y} X + \frac{M_X}{J_X} Y. \tag{12.37}$$

12.2.3 Torsion

The calculation of stress and strain of beams will focus on the case of thin wall beams with the following hypothesis:

- Rectilinear axis.
- Constant section.

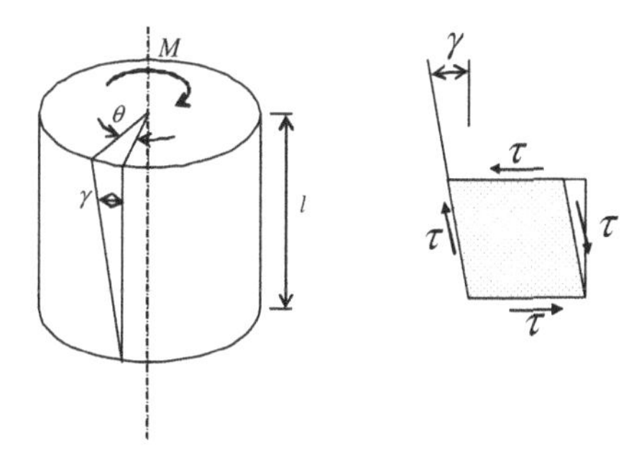

Fig. 12.12. Torsion of a cylinder with circular section and straight axis. The section is full or is closed with constant thickness.

- Elastic, homogeneous, isotropic material.
- Long axial length of the beam in comparison to the section dimensions.
- Analyzed section is far from constraints.

Circular section

A cylinder with circular section subject to torsion can be considered. Referring to Fig. 12.12 it is assumed that the beam has a full or hollow section. In the last case the wall thickness is constant and the section is closed (without openings parallel to the cylinder axis).

Due to symmetry of the structure and loads, the cylinder sections, which are perpendicular to the axis, remain plane also after deformation. After the torque is applied, two sections placed at a distance l rotate by a relative angle θ with respect to one another. As a consequence of this deformation, the material fibers, originally parallel to the axis, rotate by an angle γ. The relation between θ and γ is

$$\theta r = \gamma l \tag{12.38}$$

The material is subject to a shear deformation τ which can be determined with the shear elastic modulus G

$$\tau = G\gamma. \tag{12.39}$$

Considering the relation between γ and θ of Eq. 12.38, the result is:

$$\tau = G\frac{r}{l}\theta. \tag{12.40}$$

The tangential stresses resultant must balance the applied torque

$$M_t = \int_0^R 2\pi r^2 \tau \, dr \tag{12.41}$$

If the angle θ is equal for all the section points, then

$$M_t = \int_0^R 2\pi r^2 \, G\frac{r}{l}\theta \, dr = \frac{GJ_p}{l}\theta \tag{12.42}$$

Where J_P is the polar moment of inertia of the section. The relation between torque and torsion per unit of length of the beam θ' is then

$$\theta' = \frac{\theta}{l} = \frac{M_t}{GJ_p} \tag{12.43}$$

This result confirms the initial hypothesis that the rotation θ does not depend on the radius r.

Substituting the expression of θ of Eq.12.43 into Eq. 12.40 the distribution of tensions on the section is obtained

$$\tau = \frac{M_t}{J_p}r. \tag{12.44}$$

Eq. 12.43 and 12.44 enable the stress and strain state of a rectilinear cylinder subject to torsion to be obtained. This result is valid both for thin wall beams and not thin wall beams so long as the condition of the section symmetry is respected. It is important to underline that these results are valid only under the hypothesis that plane sections remain plane. This is true only for circular sections, so the result cannot be applied to sections of other shapes. If the section has a different shape, it does not remain plane and the sections do not simply rotate one respect to the other; instead they become warped. In these cases the relation between torque and torsion deformation is formally identical to Eq. 12.43 but, instead of the polar moment of inertia of the section, the so called modulus of torsional rigidity J_t is used:

$$\theta' = \frac{\theta}{l} = \frac{M_t}{GJ_t}. \tag{12.45}$$

Even though dimensionally it is a moment of inertia of area, $J_t \neq J_p$, its value depends on the shape of the section. Hereafter the parameter J_t will be calculated for some significant automotive cases.

Open thin wall sections

A full rectangular section of width b and thickness t subject to torque M_t becomes distorted. It is possible to demonstrate that the distribution of the tangential tensions is equal to the velocity field in a non viscous fluid which is vortically moving in a container with the same shape. Because of the mass flow rate conservation, the flow velocity is maximum in the middle point of the longest side, whereas the velocity is approximately null in the corners where the fluid is steady. Keeping this analogy in mind, the maximum value of the tangential stress is on the external surface, in the middle of the longest side of the section. (Fig. 12.13).

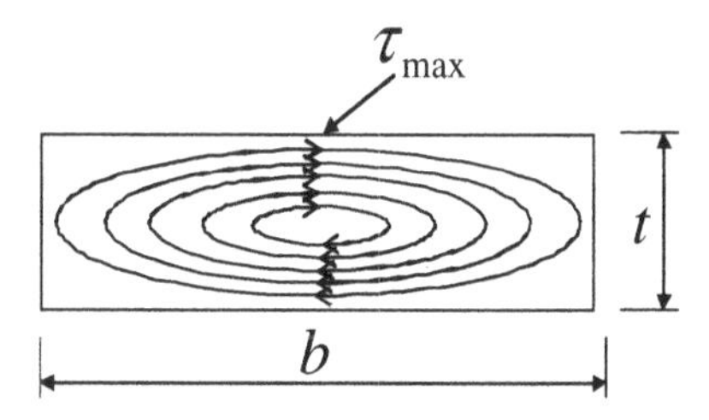

Fig. 12.13. Torsion of a beam with rectangular section: distribution of tangential tensions. The maximum value of the tension (τ_{max}) is in the middle of the longest side, on the outer surface

Table 12.2. Torsion behavior of a rectilinear beam with rectangular section with longest side b and thickness t. Coefficients α and β for the calculation of stresses and deformations.

b/t	1	1.5	2	2.5	3	4	6	8	10	∞
α	4.81	4.33	4.18	4.07	3.88	3.75	3.55	3.34	3.26	3.19
β	7.09	5.10	4.67	4.37	4.02	3.80	3.56	3.34	3.26	3.19

The maximum tangential stress (τ_{max}) and the torsion deformation per unit of length are given by:

$$\tau_{max} = \alpha \frac{M_t}{bt^2} \tag{12.46}$$

$$\frac{\theta}{l} = \beta \frac{M_t}{bt^3 G} [rad] \tag{12.47}$$

The α and β values are given by Tab. 12.2 and in Fig. 12.14 versus the aspect ratio b/t. For high values of this ratio, α and β tend to the same value, almost equal to 3.

Comparing Eq. 12.47 with Eq. 12.45 the value of J_t is obtained for a beam with rectangular section.

$$J_t = \frac{bt^3}{\beta}. \tag{12.48}$$

Equations 12.46 and 12.47 can also be used for sections obtained by folding an initial one (Fig. 12.15). This approximation improves for lower section curvature. In this way it is possible to calculate stresses and deformations in sections often used in the automotive field such as profiled C-, U- and Ω-shaped beams.

Closed thin wall sections

The hydrodynamic analogy, introduced in the previous paragraph, is useful to intuitively analyze the tangential stresses field on a closed section subject to torsion (Fig. 12.16). Since the fluid flow rate in every section must be conserved,

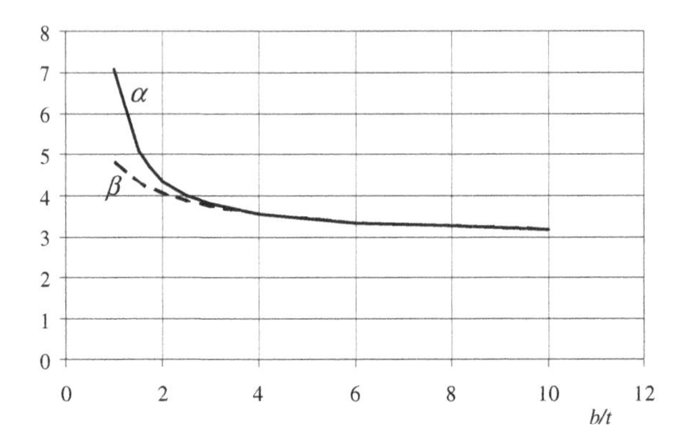

Fig. 12.14. Torsion of a rectangular section beam of width b and thickness t. Coefficients for the calculation of the maximum tension and the deformation.

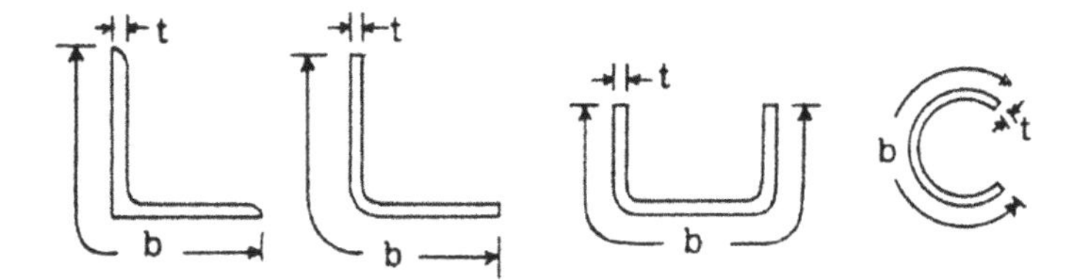

Fig. 12.15. Torsion of sections obtained by folding a rectangular one.

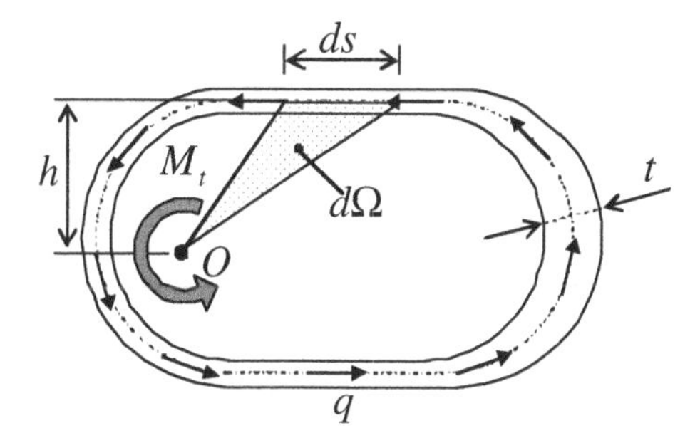

Fig. 12.16. Thin wall closed section subject to torsion. The dash dot line (-.- middle line) is traced on the middle of the thickness. Ω represents the area enclosed by the middle line.

the flow of stresses is conserved also, i.e. the stresses flow q is constant on all the points of the section.

$$q = \int_t \vec{\tau}\vec{n}\ dt = \bar{\tau}t = \text{constant} \tag{12.49}$$

where $\vec{n}$ is the unit vector normal to a basic element of thickness dt taken on a section of the flow, and $\bar{\tau}$ is the mean stress across the same thickness. If the wall thickness t is small compared to the section dimensions, it is possible to substitute the stress by the mean stress:

$$q = \tau t. \tag{12.50}$$

Bearing in mind that q =constant, the tangential stress decreases in the stretches of section where the thickness increases, and vice versa.

The shear flow may be calculated through the equilibrium of the section about any point O. The external couple M_t must be equal to the resultant torque of the tangential stresses applied on the section.

$$M_t = \oint_l h \int_t \tau\ ds\ dt \tag{12.51}$$

The integral is performed along the line traced on the middle of thickness (dash dot line placed on the middle of thickness in Fig. 12.16).

Considering that the flow q is constant and that $h\ ds = 2d\Omega$ the result is

$$M_t = q \oint 2h\ d\Omega = 2q\Omega \tag{12.52}$$

where Ω is the area enclosed by a line traced on the middle of thickness (middle line).

Torsional deformation can be calculated from the elastic energy accumulated by the material in a stretch l long and of volume v

$$U_i = \frac{1}{2}\int_v \tau\gamma\ dv \tag{12.53}$$

considering that τ are the same on all the sections and considering the stress-strain characteristics of the material ($\gamma = G\tau$):

$$U_i = \frac{1}{2} l \oint \frac{\tau^2}{G} tds \tag{12.54}$$

tangential stress can be expressed as a function of shear flow ($\tau = q/t$, Eq. 12.50),

$$U_i = \frac{1}{2} l \oint \frac{q^2}{Gt} ds \tag{12.55}$$

From Eq. 12.52

$$q = \frac{M_t}{2\Omega} \tag{12.56}$$

the following expression is obtained:

$$U_i = \frac{1}{2} \frac{l}{G} \frac{M_t^2}{4\Omega^2} \oint \frac{1}{t} ds \tag{12.57}$$

Under the assumption that the stress-strain relations are linear, the work made by moment M_t for the reached deformation θ is

$$U_e = \frac{1}{2} M_t \theta \tag{12.58}$$

For energy conservation, the work made by torque M_t must be equal to the variation of elastic potential energy accumulated inside the material.

$$U_e = U_i \tag{12.59}$$

Thus

$$\frac{\theta}{l} = \frac{M_t}{4G\Omega^2} \oint \frac{1}{t} ds \tag{12.60}$$

alternatively, considering Eq.12.52

$$\frac{\theta}{l} = \frac{q}{2G\Omega} \oint \frac{1}{t} ds. \tag{12.61}$$

Comparing Eq. 12.60 with 12.45, the torsional stiffness modulus (J_t) for the thin wall closed section is obtained:

$$J_t = \frac{4\Omega^2}{\oint \frac{1}{t} ds}. \tag{12.62}$$

If the wall thickness (t) is constant over the section, its contribution can be taken out from the integral, that becomes equal to the length $s = \oint ds$ of the middle line of section

$$J_t = 4 \frac{\Omega^2 t}{s}. \tag{12.63}$$

Eq. 12.52 and 12.60 provide the stress and deformation state of a straight axis beam with thin wall closed tubular section and are known as Bredt formulas written hereafter as:

$$q = \frac{M_t}{2\Omega} \tag{12.64}$$

$$\frac{\theta}{l} = \frac{M_t}{4G\Omega^2} \oint \frac{1}{t}\, ds \tag{12.65}$$

Fig. 12.17 shows the case of a two cells beam subject to torsion. The external torque M_t produces a shear flow in each cell. The resultant moment produced by the two shear flows must be equal to the external moment M_t. Considering the relation between flow and torque (Eq. 12.64)

$$M_t = 2q_1\Omega_1 + 2q_2\Omega_2 \tag{12.66}$$

Where Ω_1 and Ω_2 respectively are the areas enclosed by the middle lines of the left and right cells.

Eq. 12.66 is not sufficient to determine the two unknown shear flows. To calculate q_1 and q_2 it is necessary to impose that the cell rotation is the same. The condition of deformation congruency has to be imposed

$$\theta_1 = \theta_2. \tag{12.67}$$

Considering Eq. 12.61, and the fact that the axial dimension is the same for the two cells:

$$\frac{q_1}{2G\Omega_1} \oint_1 \frac{1}{t}\, ds = \frac{q_2}{2G\Omega_2} \oint_2 \frac{1}{t}\, ds. \tag{12.68}$$

Eq. 12.66 and 12.68 can be solved with respect to variables q_1 and q_2. If the two cells are identical, both in terms of geometry and material, Eq. 12.68 implies that the two shear flows are the same: $q_1 = q_2 = q$. Consequently, from Eq. 12.66: $M_t = 2q(\Omega_1 + \Omega_2)$.

The section behaves as a one cell section, as if the middle wall does not exist.

Comparison between closed and open sections

Fig. 12.18 shows two circular sections with equal dimensions. While the left section is closed, the right section is open on a line corresponding to a generatrix. The cut width is small with respect to the thickness.

The tangential stress τ_c and the deformation θ_c in the closed section can be evaluated through Eq. 12.64 and 12.65. Considering that the area enclosed by the middle line is equal to $\Omega = \pi r^2$, and that $q = \tau_c t$:

$$\tau_c = \frac{M_t}{2\pi r^2 t} \tag{12.69}$$

$$\frac{\theta_c}{l} = \frac{M_t}{2\pi G r^3 t}. \tag{12.70}$$

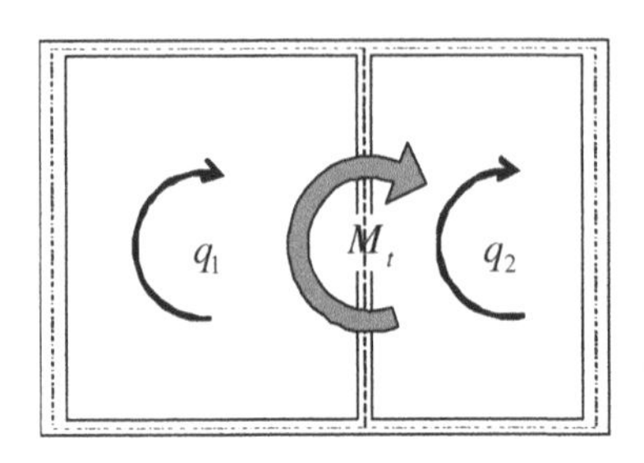

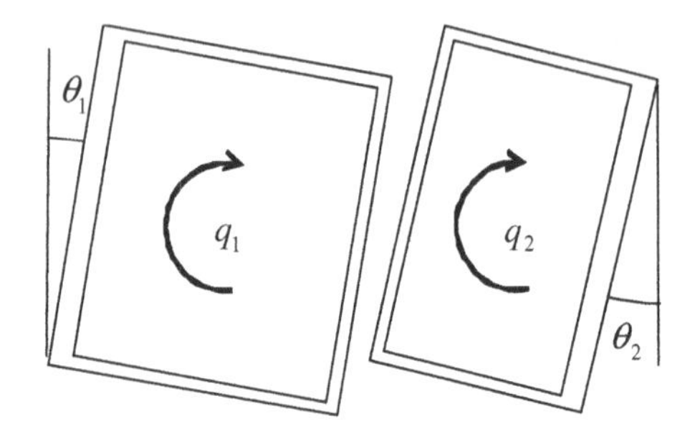

Fig. 12.17. Two cells section subject to torsion. For deformations congruency the rotation of the two cells must be equal, $\theta_1 = \theta_2$.

The maximum stress and the torsional deformation in the open section can be calculated using Eq. 12.46. Considering that $b = 2\pi r$, and that for $t << r$ the coefficient $\alpha \approx 3$, a maximum tangential stress can be determined:

$$\tau_a = 3\frac{M_t}{2\pi r t^2} \tag{12.71}$$

$$\frac{\theta_a}{l} = 3\frac{M_t}{2\pi G r t^3} \tag{12.72}$$

With equal torque and equal geometrical dimensions, the ratios between the tangential stresses and between deformations are:

$$\frac{\tau_a}{\tau_c} = 3\frac{r}{t} \tag{12.73}$$

$$\frac{\theta_a}{\theta_c} = 3\left(\frac{r}{t}\right)^2 \tag{12.74}$$

Since $r >> t$, the open section is subject to stresses and deformations much bigger than the closed section. Assuming, for example, that $r = 50$ mm and that $t = 1$ mm:

$$\frac{\tau_a}{\tau_c} = 3\frac{50 \text{ mm}}{1 \text{ mm}} = 150,$$

$$\frac{\theta_a}{\theta_c} = 3\left(\frac{50 \text{ mm}}{1 \text{ mm}}\right)^2 = 7500.$$

For the same overall dimensions and material quantity (and thus weight), a closed section beam is four orders of magnitude more rigid, and two orders of magnitude less stressed, than a open section beam.

12.2.4 Shear and Bending

Shear stress in a beam is associated with a variation of bending moment. Considering the sign conventions of Fig. 12.19, the shear stress τ is equal to the

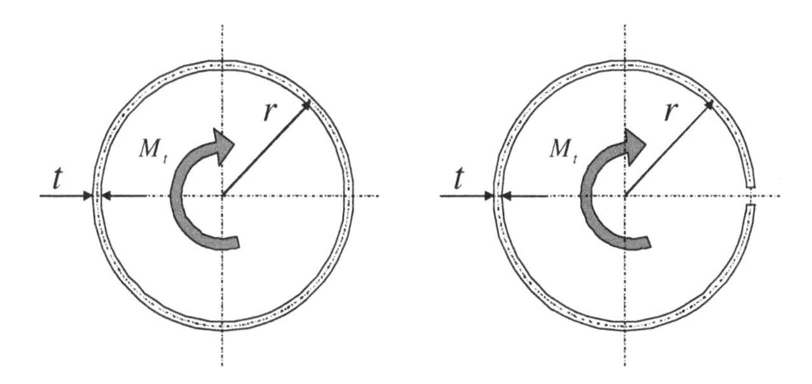

Fig. 12.18. Comparison between an open circular section and a closed circular section subject to torsion. R is the radius next to the line placed on middle of thickness. t is the wall thickness.

derivative of bending moment calculated along the x coordinate oriented along the beam axis

$$T = \frac{dM}{dx} \tag{12.75}$$

Due to the variation of bending moment, even the normal stresses σ on sections change along the axial coordinate. Referring to Fig. 12.20, an element is isolated between coordinate y and the upper edge of this part of beam, placed to $y_{\max}$ coordinate (A element, highlighted in grey). On the left side, this element is subject to normal stresses

$$\sigma = \frac{M}{I_z} y \tag{12.76}$$

where I_z is the inertia moment of the section around neutral axis z. Instead on the right side, due to the variation of the bending moment, these normal stresses are

$$\sigma + d\sigma = \frac{M + dM}{I_z} y. \tag{12.77}$$

leading to:

$$d\sigma = \frac{dM}{I_z} y \tag{12.78}$$

To guarantee the longitudinal equilibrium of the part of beam (A element), it is necessary that tangential stresses τ are applied on the lower surface such that

$$\tau(y)b(y)dx = \int_y^{y_{\max}} b(y')\ d\sigma\ dy' \tag{12.79}$$

Considering Eq. 12.78 and the relation between shear and the derivative of bending moment (Eq. 12.75):

$$\tau(y) = \frac{1}{I_z b(y)} \frac{dM}{dx} \int_y^{y_{\max}} b(y')y'\ dy' = \frac{T}{I_z b(y)} \int_y^{y_{\max}} b(y')y'\ dy'. \tag{12.80}$$

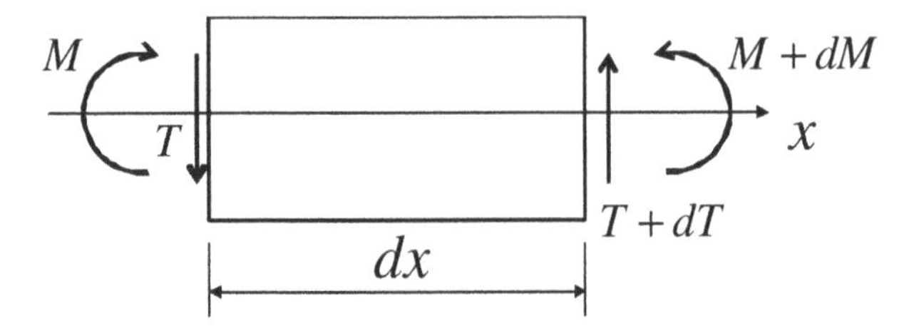

Fig. 12.19. Beam subject to shear and bending. Sign convention.

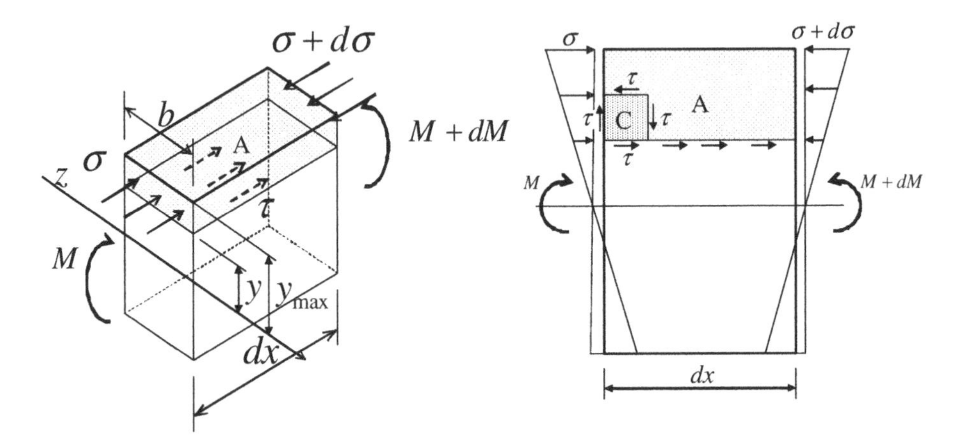

Fig. 12.20. Equilibrium of a beam element subject to shear and bending. The variation along the axis of normal stresses σ gives rise to tangential stresses τ on internal surfaces parallel to the beam axis.

For the equilibrium of the C basic element (in the right part of Fig. 12.78), the same $\tau(y)$ acts on the face of section normal to x axis for all points at a distance y from neutral axis.

In the case of a thin wall beam, the variation of normal stresses (σ) and tangential stresses (τ) over the thickness is negligible. Referring to the C section represented in Fig. 12.21, it can be assumed that:

- thickness t is constant and small with respect to the section dimensions,
- the shear stress is perpendicular to z axis.

Considering firstly the variation of normal stress over a stretch of the upper flange of length ξ:

$$\frac{d\sigma}{dx} = \frac{dM}{dx}\frac{h}{I_z 2} = \frac{Th}{2I_z} \tag{12.81}$$

where $h/2 = (h_1 + h_2)/4$. The longitudinal equilibrium of this element is guaranteed by the tangential stresses that arise at the interface with the rest of beam

$$\tau(\xi) t dx = \xi t d\sigma. \tag{12.82}$$

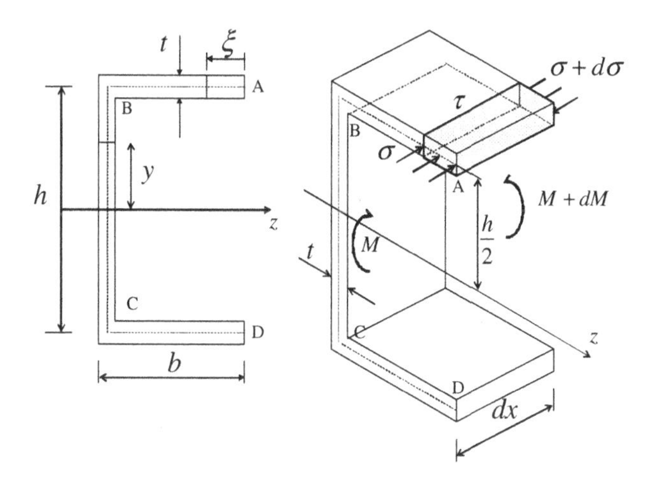

Fig. 12.21. Open thin wall section subject to shear.

Considering equation 12.81, the shear stress inside the flanges increase linearly from zero next to the edges, reaching the maximum value at the connection with central core (B and C points)

$$\tau(\xi) = \xi \frac{d\sigma}{dx} = \xi \frac{Th}{2I_z}. \tag{12.83}$$

Considering the assumption of small thickness, the shear flow can be determined: (Fig. 12.22)

$$q(\xi) = \tau(\xi)t = \xi \frac{Th}{2I_z} t. \tag{12.84}$$

The resultant force T_z of shear flow acting on the flanges is:

$$T_z = \frac{1}{2} q(b) b = \frac{Th}{4I_z} b^2 t. \tag{12.85}$$

Disregarding its distribution, the shear flow must give a resultant in the core which is equal to the shear stress $T_y = T$, acting on the section. The distribution of tangential stresses on the section is therefore equivalent to a force equal to T acting on the core and to a pair of forces T_z with a lever arm h. This system of forces is equivalent to a force equal to T, at a distance e from the core such that

$$Te_T = T_z h \tag{12.86}$$

Thus:

$$e_T = \frac{h^2 b^2}{4I_z} t. \tag{12.87}$$

The intersection between the symmetry plane and the straight line at distance e_T is called *torsion center* of section (cdt, in Fig. 12.22). A T force perpendicular to the z axis and passing through the torsion center only induces the described shear flows on the section.

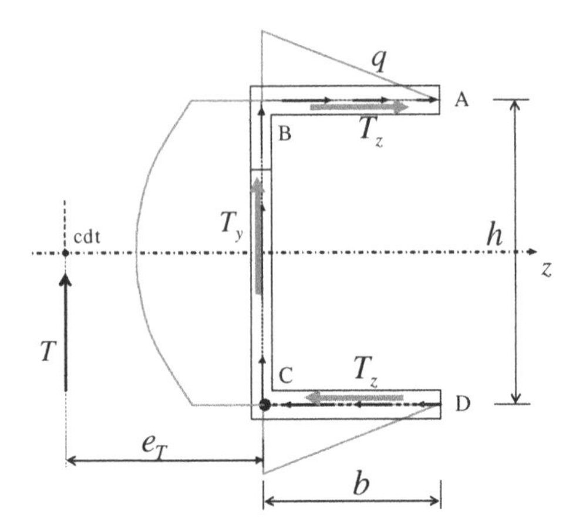

Fig. 12.22. Shear center of a C shaped section. The moment of shear force T respect to the C point is equal to the one caused by the distribution of shear flux q on section.

The effect of a force T that does not pass through the torsion center is shown in Fig. 12.23. The force T applied at a distance e (a) is equal to the same T applied to the torsion center (distance e_T) which a transposed torque $M_t = T(e - e_T)$ is summed to (Fig. 12.23 b)). The effect (Fig. 12.23 c)) is pure bending caused by T applied to the torsion center c1) which the torsion caused by M_t is summed to c2).

Stresses and deformations produced by moment M_t can be determined with Eq. 12.46 and 12.47.

The distribution of tangential stresses and shear flow on the section core (part BC) can be determined similarly to that shown in Fig. 12.20 to obtain:

$$\begin{aligned} \tau(y)t &= \int_y^{h/2} \frac{T}{2I_z} ty' dy' + bt \frac{Th}{2I_z} = \\ &= \frac{Tt}{2I_z} \left[hb + \frac{1}{2} \left(\frac{h^2}{4} - y^2 \right) \right] \end{aligned} \tag{12.88}$$

The shear stress on the same stretch is therefore:

$$q(y) = t\ \tau(y). \tag{12.89}$$

the maximum value of $q(y)$ is on the symmetry plane of section (axis z). At the edges of core ($y = \pm h/2$) the same value is reached as at the edges of flanges:

$$q(\pm h/2) = \frac{Tt}{2I_z} hb. \tag{12.90}$$

Finally the resultant of shear flows in the core is equal to T :

$$\int_{-h/2}^{h/2} q(y') dy' = T. \tag{12.91}$$

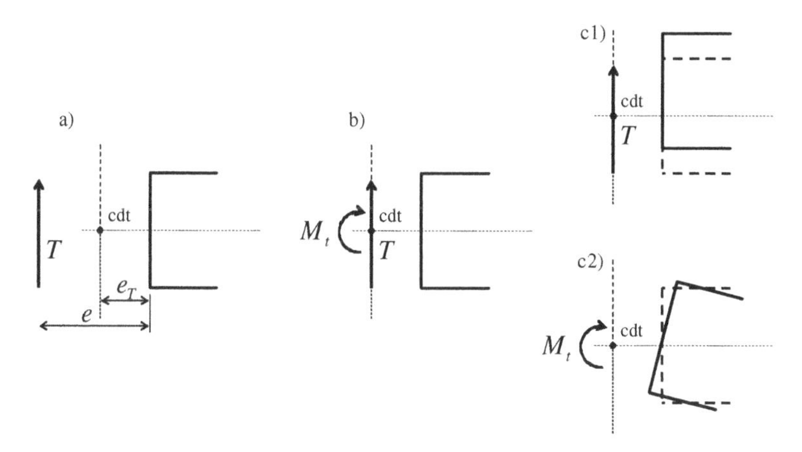

Fig. 12.23. Effect of a shear stress T not passing through the shear center (cdt). The force T applied to a distance $e \neq e_T$ a), is equal to the force T applied to the shear center, which the torque $M_t = T(e - e_T)$ is summed to b). The effect of T passing through the shear center is a pure bending c1). The torque M_t causes pure torsion.

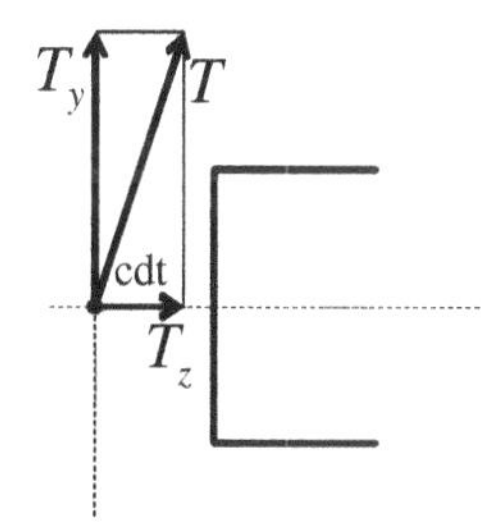

Fig. 12.24. C shaped section subject to a shear stress with components perpendicular and parallel to the symmetry plane.

If T passes through the torsion center, but has a component directed along the symmetry plane of the section (axis z), it is possible to use the superposition principle of components T_y and T_z(Fig. 12.24). T_y causes pure bending in the y direction, whereas T_z causes bending in the z direction.

The same reasoning applied to the C shaped section can be applied for more complex sections subject to shear, such as L or Ω shapes.

The assumption of a semi-monocoque structure can be applied for the analysis of thin wall beams subject to shear. For this assumption, in a semi-monocoque structure the reinforcing longitudinal elements are subject to pure traction and compression. Instead panels are subject to pure shear stress.

A semi-monocoque beam with a C section is represented in Fig. 12.25. The section is provided with four longitudinal stiffeners with equal area S. The inertia moment of section respect to the neutral axis is:

$$I_z = 4S(\frac{h}{2})^2 = Sh^2 \tag{12.92}$$

The normal stress variation caused by the bending moment variation is:

$$\frac{dF}{dx} = \frac{dM}{dx}\frac{1}{I_z}\frac{h}{2}S = \frac{T}{2h} \tag{12.93}$$

The shear flow in the three panels can be obtained through the equilibrium equations on the x direction. Considering the upper right longitudinal stiffener and a strip of panel connected to it:

$$dF = q_1 dx \tag{12.94}$$

thus

$$q_1 = \frac{dF}{dx} = \frac{T}{2h} \tag{12.95}$$

Similarly, flow q_2 is obtained considering the equilibrium in the x direction of the two upper longitudinal stiffeners and of a part of the vertical panel:

$$q_2 = 2\frac{dF}{dx} = \frac{T}{h} \tag{12.96}$$

For the flow q_3 it is possible to determine that,

$$q_3 = \frac{dF}{dx} = \frac{T}{2h}. \tag{12.97}$$

In a similar way to that identified for the section of Fig. 12.22:

$$q_1 bh = Te_T \tag{12.98}$$

The result is that the torsion center is positioned a distance equal to the half width of the horizontal flanges,

$$e_T = \frac{b}{2}. \tag{12.99}$$

Before calculating the shear stresses in a semi-monocoque beam with closed section, it is necessary to determine the resultant of the tangential forces that occurs in a curved panel (Fig. 12.26) subject to shear.

Since the shear flow q is constant on the panel, the force acting on the basic element of length $d\vec{s}$ of Fig. 12.26 a) is subject to a force with components on z and y axes that are

$$d\vec{T} = qd\vec{s} = q\vec{j}dy + q\vec{k}dz \tag{12.100}$$

where $\vec{j}$ and $\vec{k}$ respectively are the versors of y and z axes. Integrating on the whole panel one obtains that the shear forces resultant is directed as the line joining the ends of the section.

$$\vec{T} = q(Y\vec{j} + Z\vec{k}). \tag{12.101}$$

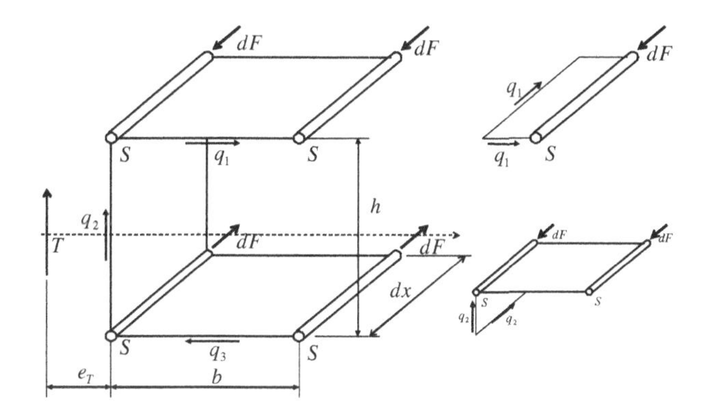

Fig. 12.25. Shear of an semi-monocoque beam with C shaped section

In order to evaluate the line of action of $\vec{T}$, it is necessary to calculate the shear force moment acting on the section with respect to any point: For convenience reference is made to point A. Denoting with r the distance between the line of action of force $qd\vec{s}$ and point A:

$$M = \int_A^B qrds = q \int_A^B 2d\Omega = 2\Omega q \tag{12.102}$$

where Ω is the area enclosed by the middle line of the panel section and the AB line.

The distance e_t between the line of action of resultant $\vec{T}$ and point A is obtained by imposing:

$$Te_t = M \tag{12.103}$$

thus:

$$e_t = \frac{2\Omega}{L}. \tag{12.104}$$

Fig. 12.27 represents a semi-monocoque beam with closed section. Since there are only two longitudinal stiffeners, the beam can only support a bending moment directed along the z axis and, therefore, a shear stress along y. To allow the beam to resist shear stresses and bending moments directed anywhere, it would be necessary to add a longitudinal stiffener not aligned with the first two.

For the rotational equilibrium around axis z, the axial force variation in the two longitudinal stiffeners is:

$$h_2 dF = Tdx \tag{12.105}$$

and thus

$$\frac{dF}{dx} = \frac{T}{h_2} \tag{12.106}$$

For equilibrium along the x direction of the upper longitudinal stiffener, it is necessary that:

$$dF = (q_1 + q_2)\, dx \tag{12.107}$$

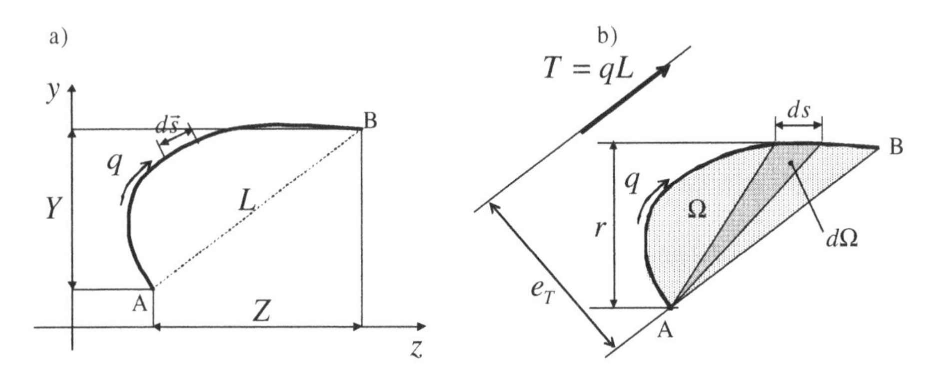

Fig. 12.26. Resultant force of a panel subject to shear. L: distance between the ends of section. Ω: area enclosed by the middle line of section and the line AB.

considering Eq. 12.106

$$\frac{T}{h_2} = q_1 + q_2. \tag{12.108}$$

To calculate the two shear flows, it is necessary to consider the equilibrium equation to rotation around the x axis. Referring to the lower longitudinal stiffener:

$$Te = 2\Omega q_1 \tag{12.109}$$

thus:

$$\begin{aligned} q_1 &= \frac{e}{2\Omega}T \\ q_2 &= (\frac{1}{h_2} - \frac{e}{2\Omega})T \end{aligned} \tag{12.110}$$

The torsion center of this section can be determined identifying the value of e for which section deformations do not occur. Considering Bredt second formula (Eq.12.65),

$$\begin{aligned} \theta' &= \frac{1}{2G\Omega} \oint \frac{q}{t} ds = \\ &= \frac{1}{2G\Omega} \left(\frac{q_1}{t_1} h_1 - \frac{q_2}{t_2} h_2 \right) = 0 \end{aligned} \tag{12.111}$$

where h_1 and h_2 are the lengths of the two panels.

Substituting Eq. 12.110 into Eq. 12.111 and setting $e = e_T$,:

$$e_T = \frac{2\Omega}{h_2} \frac{1}{1 + \frac{t_2 h_1}{t_1 h_2}}.$$

The procedure followed for the calculation of shear stresses, torsional deformations and section torsion center of Fig. 12.27 can be repeated in the case of more complex sections, including those comprising more cells. In this case, the shear

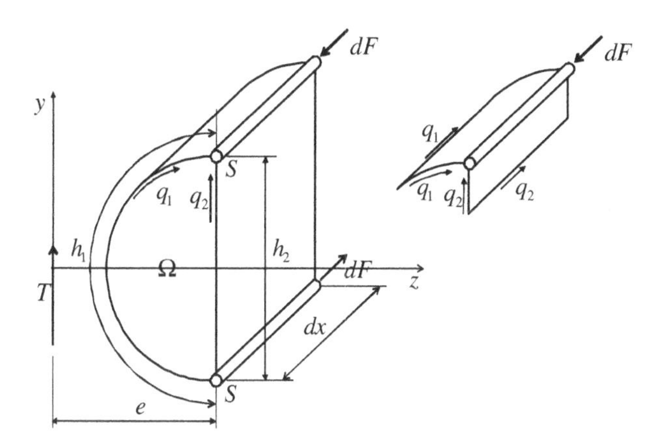

Fig. 12.27. Semi-monocoque beam with closed section subject to shear. Ω : area enclosed by curved panel. S: longitudinal stiffener section.

flows can be determined considering also the congruency conditions in addition to the equilibrium equations to translation and rotation.

Considering the case of Fig. 12.28 the bending moment variation depends on shear according to the relations

$$\begin{aligned} \frac{dM_z}{dx} &= T_y \\ \frac{dM_y}{dx} &= T_z \end{aligned} \tag{12.112}$$

The axial load variation on longitudinal stiffeners can be calculated through Eq. 12.36.

Considering Eq. 12.112:

$$\frac{dF_i}{dx} = S_i \left(\frac{T_z J_z - T_y J_{zy}}{J_z J_y - J_{zy}^2} z_i + \frac{T_y J_y - T_z J_{zy}}{J_z J_y - J_{zy}^2} y_i \right), \qquad i = 1, ..., 4. \tag{12.113}$$

As in the case with a two longitudinal stiffeners section, for which it was possible to write only one equilibrium equation along x (the other equation depends on the first one), in the case there are n longitudinal stiffeners it is possible to write only $n - 1$ independent equations. In the case of Fig. 12.112, there are three equations with five variables:

$$\begin{aligned} &\frac{dF_1}{dz} = q_5 - q_1 - q_2 \\ &\frac{dF_1}{dz} + \frac{dF_2}{dz} = q_5 - q_1 - q_3 \\ &\frac{dF_1}{dz} + \frac{dF_2}{dz} - \frac{dF_3}{dz} = q_5 - q_1 - q_4 \end{aligned} \tag{12.114}$$

where the equilibrium to rotation around the x axis is the fourth equation.

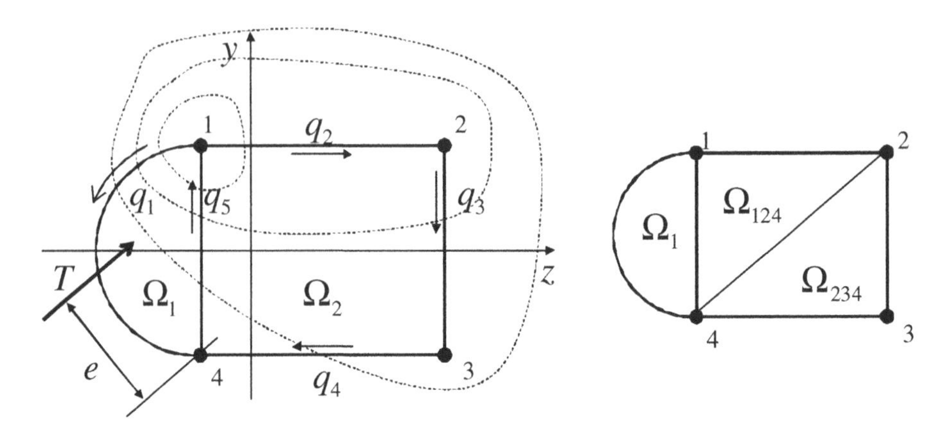

Fig. 12.28. Semi-monocoque section with two cells subject to shear and bending. Dashed lines indicate the subsystem considered to write the equilibrium equations along x.

Considering the longitudinal stiffener 4 as reference for the couples, and considering the moment generated from shear flow in a panel (Eq. 12.102):

$$Te = -2q_1\Omega_1 + 2q_2\Omega_{124} - 2q_3\Omega_{234}. \tag{12.115}$$

The last equation is the congruency equation; the torsional deformation of the two cells must be the same:

$$\theta_1' = \theta_2'. \tag{12.116}$$

Considering Eq.12.65,

$$\frac{1}{2G\Omega_1}\left(-q_1\frac{h_1}{t_1} - q_5\frac{h_5}{t_5}\right) = \frac{1}{2G\Omega_2}\left(q_5\frac{h_5}{t_5} + q_2\frac{h_2}{t_2} + q_3\frac{h_3}{t_3} + q_4\frac{h_4}{t_4}\right) \tag{12.117}$$

Eq. 12.114, 12.115, 12.117 constitute 5 linear equations depending on 5 unknown flows that can be solved to determine the state of tension of the beam. Given the flows, the first or the second term of Eq. 12.117 allows the deformation state to be obtained.

12.2.5 Buckling of Beams

Buckling is one of the more significant failure process arising due to the small wall thickness that characterizes the constitutive elements of a body structure. This kind of failure often occurs for loads which are lower than material yield or breaking; correspondingly compression instability is the type of failure that mostly affects the design process.

Buckling phenomena involve also the geometrical and constraint properties of the structure in addition to the material characteristics. A common example is a beam with straight axis, shown in Fig. 12.30.

For a compression load lower than the critical value F_{cr}, a small lateral deformation of the beam causes an elastic reaction capable of bringing the beam back to the straight axis configuration. If the load exceeds F_{cr}, the bending moment produced by the compression force with the lateral displacement overcomes the elastic reaction of beam. The system, unable to achieve equilibrium, collapses.

The Euler formula is the usual expression of the critical load F_{cr}.

$$F_{cr} = \frac{\pi^2 EI}{L_e^2} \tag{12.118}$$

where:

- E = Elastic modulus of material.
- I = Inertia moment of section around an axis perpendicular to the inflexion plane where the failure occurs.
- L_e = Effective length of beam, which is equal to the geometrical length ($L_e = l$) only if beam is hinged at its two ends. The value for different constraint conditions is shown in Fig. 12.30

Among the different inflexion planes possible, the one characterized by the lower ratio I/L_e^2 determines the critical load.

Dividing both terms of Eq. 12.118 by the area A of the section, the normal critical stress σ_{cr} is obtained:

$$\sigma_{cr} = \frac{F_{cr}}{A} = \frac{\pi^2 E}{\lambda^2} \tag{12.119}$$

where $\lambda = L_e/\rho$ is the slenderness of the beam, i.e. the ratio between the effective length and the inertia radius $I = \rho^2 A$ of the section.

Under a certain value of slenderness, the Eq. 12.119 is no longer applicable because the critical stress overcomes the yield stress (σ_y). For squat beams ($\lambda <$ 100 for steel and $\lambda < 70$ for aluminum) the Euler formula is no longer valid. In this case, expressions of critical load that also involve the yield stress are needed: The Johnson (or Rankine) expression is one of the most used,

$$\sigma_{cr} = \sigma_y - \frac{\sigma_y^2}{4\pi^2 E}\lambda^2. \tag{12.120}$$

The curves of Eq.12.119 and 12.120 are tangent for a slenderness and a value of σ_{cr} equal to

$$\lambda^* = \sqrt{\frac{2\pi^2 E}{\sigma_y}}, \qquad \sigma_{cr}^* = \frac{\sigma_y}{2} \tag{12.121}$$

Fig. 12.29 represents the ratio between critical stress and yield stress versus the slenderness, under the assumption that the beam is made of steel ($\sigma_y = 680$

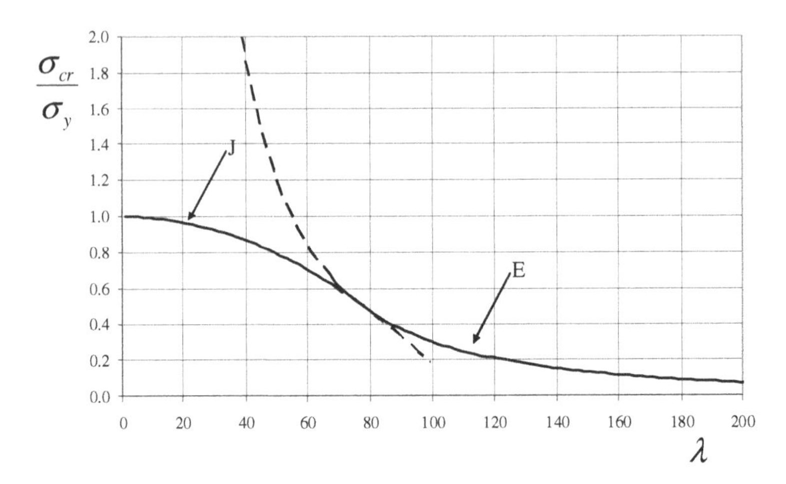

Fig. 12.29. Critical normal stress of compression (σ_{cr}) versus the axially loaded beam slenderness. $\sigma_y = 680$ MPa, $E = 2.1\ 10^{11}$ Pa. Eulerian critical stress (E curve). Critical stess for squat beams (Johnson - Rankine, J curve).

MPa, $E = 2.1\ 10^{11}$ Pa). For $\lambda > \lambda^* \approx 80$ the critical load is the Eulerian one (E curve). For $\lambda < \lambda^*$ the critical load given by the Johnson-Rankine formula (J curve) preponderates. For very low slenderness the critical stress coincides with the yield stress.

12.2.6 Buckling of Flat Panels

Similarly to beams, also panels can undergo buckling when subject to compression. Undulations are created on the panel surface during the failure. The m number of half-waves that are created along the load direction depends on the panel geometry and on the constraint conditions.

The extension of Euler theory to the flat panels leads to an expression of critical compression stress that is inversely proportional to the square of ratio b/t between loaded side and the thickness t

$$\sigma_{cr} = k \frac{\pi^2}{12(1-\nu^2)} E \left(\frac{t}{b}\right)^2, \tag{12.122}$$

where E is the elastic modulus and ν is the Poisson modulus of material.

The geometrical parameter k depends on the m number of half-waves that are created along the direction of load application (Fig. 12.31), on the aspect ratio a/b between sides of panel (a is the panel side parallel to the direction of load application, b is the one where the load is applied on) and on the constraint boundary conditions.

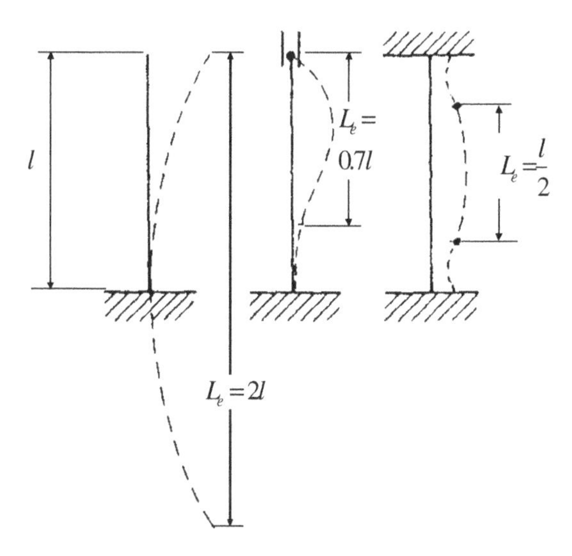

Fig. 12.30. Eulerian buckling. Effective length L versus constraint conditions. l is the geometrical length.

When all panel sides are supported,

$$k = \left(m\frac{b}{a} + \frac{a}{mb} \right)^2 , \tag{12.123}$$

i.e. the trend of the k parameter depends on a/b and m , as shown in Fig. 12.32. Each curve of the diagram represents different values of half-waves m. For a given value of a/b, each curve is characterized by a value of k and, therefore, leads to a different value of σ_{cr}. Since the plate fails at the minimum possible load, the significative value of k is obtained as the envelope between all the lines (the continuous line on diagram).

As a/b increases, the number of half-waves that characterizes the instability failure increases.

For $a/b << 1$ failure occurs for $m = 1$.

From Eq. 12.123, $k \approx (b/a)^2$, and substituting in Eq. 12.122:

$$\sigma_{cr} = \left(\frac{b}{a} \right)^2 \frac{\pi^2}{12(1-\nu^2)} E \left(\frac{t}{b} \right)^2 = \frac{\pi^2}{12(1-\nu^2)} E \left(\frac{t}{a} \right)^2 . \tag{12.124}$$

The critical stress is inversely proportional to the square of the panel length.

Instead if $a/b >> 1$, failure occurs for $m >> 1$ and $k \approx 4$ becomes independent of the aspect ratio a/b. Substituting in Eq. 12.122:

$$\sigma_{cr} = \frac{\pi^2}{3(1-\nu^2)} E \left(\frac{t}{b} \right)^2 . \tag{12.125}$$

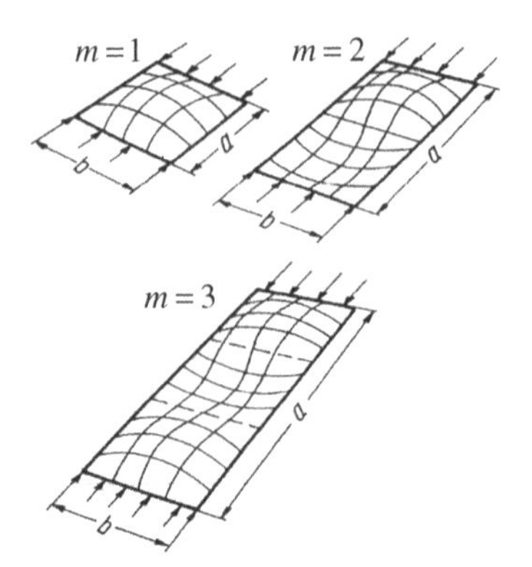

Fig. 12.31. Buckling cpmpression modes of a flat plate.

The critical stress σ_{cr} is independent of the length a of the panel and depends only on the aspect ratio of the loaded section t/b. Correspondingly the instability is termed *local.*

Fig. 12.33 shows the trend of the parameter k of Eq. 12.122 for different conditions of panel constraint. Continuous lines refer to a panel with loaded (b) sides which are supported, dashed lines refer to a panel with blocked b sides. From the diagram, it is possible to observe how parameter k increases with the increase of the constraints applied to the sides.

Also in this case it is possible to observe that for long panels ($a/b > 3 \div 4$) the value of k becomes practically constant. Therefore σ_{cr} becomes dependent only on the aspect ratio b/t of the loaded side, and the type of buckling is local.

Fig. 12.34 shows the tension distribution versus the total load on the b side of a panel with a longitudinal stiffeners frame.

Until $\sigma < \sigma_{cr}$ (Fig. 12.34 a), the load is uniformly distributed over the entire width of the panel. When σ_{cr} is exceeded (Fig. 12.34 b), the middle part of panel fails due to buckling. Compression stresses in this part of the panel are reduced to very low values. Panel portions adjacent to vertical stiffeners are instead able to support increasing compression stresses due to the influence of constraint condition made by longitudinal stiffeners on the panel. These portions of panel are usually called *collaborating strips*: Here the load can increase until reaching buckling for compression of the longitudinal stiffeners themselves. The distribution of stresses on a compression loaded panel is often simplified as shown in Fig. 12.34 c). It is assumed that middle part does not cooperate to resistance; instead the two lateral strips, of total width b_w, support the compression load and can be considered to be part of the longitudinal stiffeners.

Total width b_w of collaborating strips can be evaluated as the width that satisfy the condition,

$$\sigma_{cr} = \sigma_y. \tag{12.126}$$

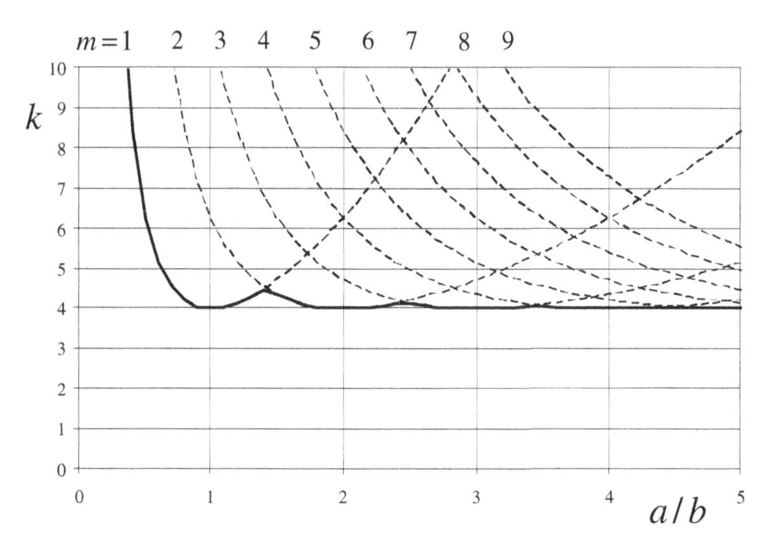

Fig. 12.32. Flat plate subject to compression. All edges supported. Value of shape coefficient k versus a/b and the half-waves number m.

meaning that the middle part that becomes unstable can be neglected, and the collaborating width of the panel is able to reach yield without failing for instability.

Assuming that the panel is long ($a >> b$) and that it is supported on all sides, from Fig. 12.33 $k = 4$. From Eq. 12.126 and from Eq. 12.122,

$$\sigma_y = \frac{\pi^2}{3(1-\nu^2)} E \left(\frac{t}{b_w}\right)^2 \tag{12.127}$$

thus:

$$\frac{b_w}{t} = \sqrt{\frac{\pi^2}{3(1-\nu^2)} \frac{E}{\sigma_y}} \tag{12.128}$$

A steel panel with E=210000 MPa, ν=0.3 and σ_y = 280 MPa will have $b_w/t \approx 50$.

An aluminum alloy panel with E=73000 MPa, ν=0.3 and σ_y = 200 MPa will have $b_w/t \approx 36$.

12.2.7 Buckling of Composite Shapes

Thin walls beams with closed or open section, often used in automotive applications, are usually considered to be a set of flat plates connected to each other. For example: a C section can be considered as the joining of three panels constrained to each other. The constraint that acts on the edges of each plate depends on the stiffness of the adjacent plates. In the case of a C section with small thickness, since the torsional stiffness of each plate is low with respect to the bending

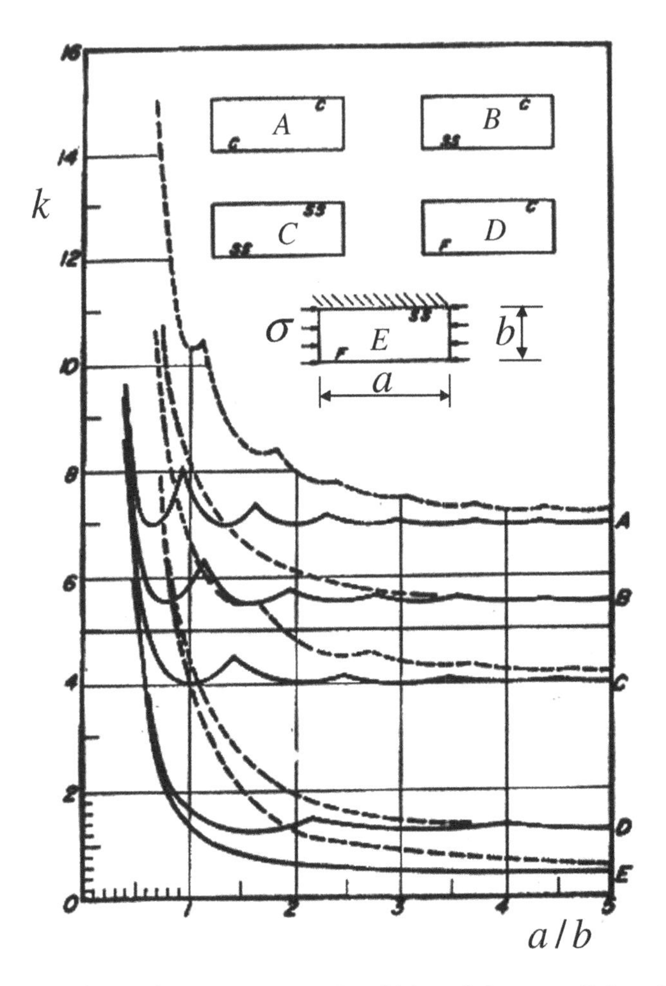

Fig. 12.33. Flat plate subject to compression. Value of shape coefficient k versus a/b and half-waves number m.Continous lines refer to b supported sides. Dashed lines to b blocked sides.

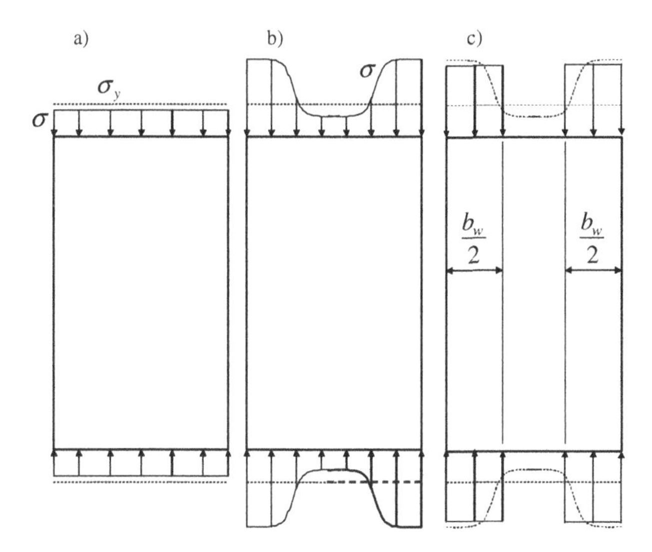

Fig. 12.34. Flat plate subject to compression. Supported edges. a) For $\sigma < \sigma_{cr}$ stresses uniformly distribute b) for $\sigma > \sigma_{cr}$ the middle part of the panel becomes stooped and its compression stress decreases respect to the lateral strips. c) Scheme of distribution of stresses with two collaborating strips.

stiffness on its plane, connection lines between one plate and another can be considered as supports (hinges).

Therefore the critical load of each part of the section can be calculated with Eq. 12.122, adopting for k the value relative to the constraint conditions imposed by contiguous panels. The critical load of the whole structure will be the minimum of the single components.

In this process, the main aspect is to correctly evaluate the constraint level that adjacent elements impose on sides of the panel under consideration. Concerning this specific issue, some approximated formulas are available in published literature that enable the critical load on thin walls sections of different shapes to be obtained that do not required the disassembly of sections into constituent components.

For Z or C sections

$$\sigma_{cr} = \frac{k_w \pi^2 E}{12\,(1-\nu^2)} \left(\frac{t_w}{b_w}\right)^2 \tag{12.129}$$

where t_w and b_w are the thickness and the width of the middle panel of section (*web*). Instead, the shape coefficient k_w, which is a function of the section dimensions, can be obtained from the diagram shown in Fig. 12.35.

Similarly, in the case of boxed sections

$$\sigma_{cr} = \frac{k_h \pi^2 E}{12\,(1-\nu^2)} \left(\frac{t_h}{h}\right)^2 \tag{12.130}$$

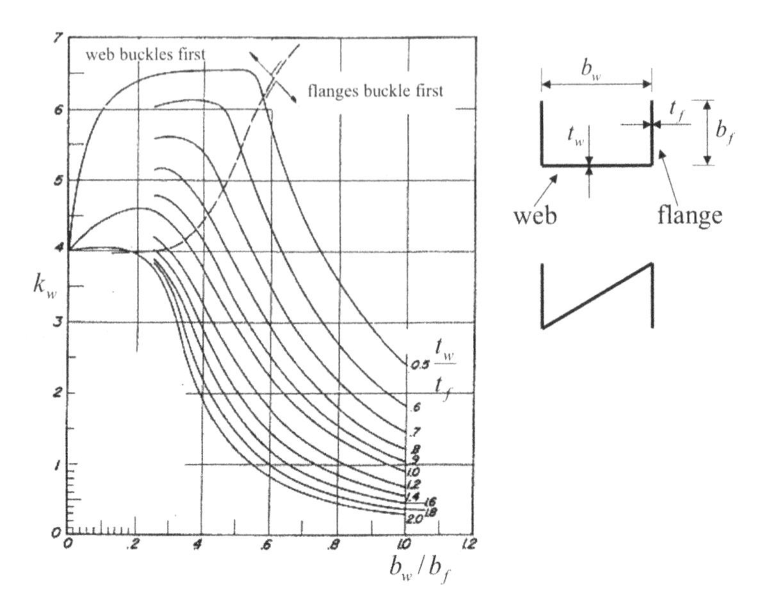

Fig. 12.35. Shape factor for the calculation of critical compression stress of C or Z sections.

where t_h and h are the thickness and the length of the longest side of the section respectively. The shape coefficient k_h is shown on diagram of Fig. 12.36.

Finally, in the case of Ω sections,

$$\sigma_{cr} = \frac{k_T \pi^2 E}{12\left(1-\nu^2\right)} \left(\frac{t}{b_T}\right)^2 \tag{12.131}$$

where t is the thickness (constant) and b_T is the length of the longest side of the section The shape factor k_T is shown in Fig. 12.37.

12.2.8 Buckling of Thin Wall Cylinders

A monocoque cylinder corresponds to a thin walled cylinder without longitudinal stiffening or transversal sections connected to the cylinder skin.

Classification:

- Short cylinders (which behave as flat panels. The edge fixings have significant importance).
- Intermediate length cylinders.
- Long cylinders, which exhibit buckling with diamond shape, for which length and edge constraints are not so important.
- Very long cylinders, which exhibit buckling like Eulerian beams.

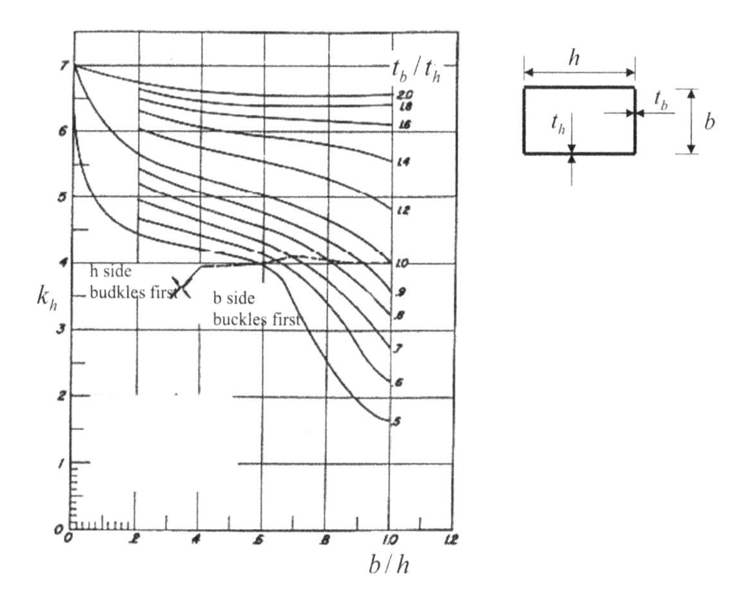

Fig. 12.36. Shape factor for the calculation of critical compression stress of rectangular thin wall sections.

In the formula for the calculation of critical stress:

$$\sigma_{cr} = \frac{k_c \cdot \pi^2 \cdot E}{12\left(1-\nu^2\right)}\left(\frac{t}{L}\right)^2 \tag{12.132}$$

t denotes the wall thickness; L is the cylinder length; ν is the Poisson coefficient;

For simply supported short to long cylinders, the shape coefficient k_c can be evaluated using:

$$k_c = \frac{\left(m^2+\beta^2\right)^2}{m^2} + 12\frac{Z^2 m^2}{\pi^4\left(m^2+\beta^2\right)} \tag{12.133}$$

where,

m is longitudinal half-waves number

$$\beta = \frac{L}{\lambda} \tag{12.134}$$

λ is the cross half-wave length

r is the cylinder radius

$$Z = \frac{L^2}{rt}\sqrt{1-\nu^2} \tag{12.135}$$

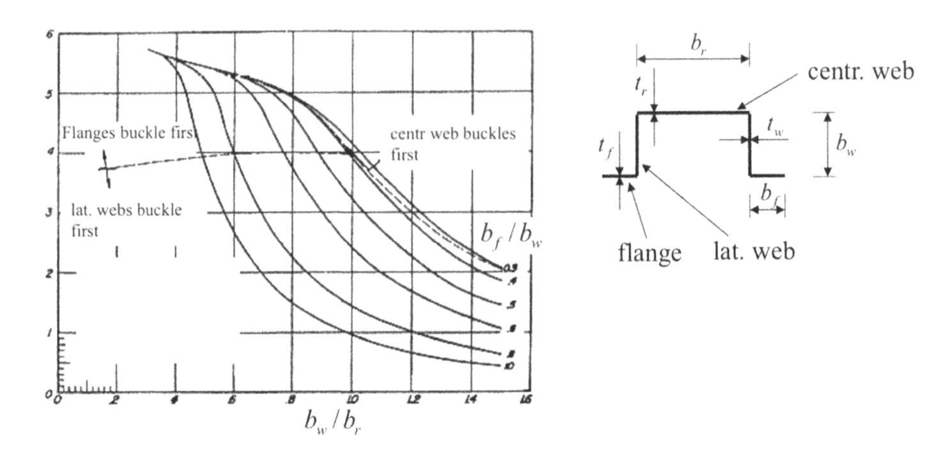

Fig. 12.37. Shape factor for the calculation of critical compression stress of Ω thin wall sections. It is supposed $t_w = t_f = t_r$.

Instead for short cylinders:

$$k_c = 4\frac{\sqrt{3}}{\pi^2}Z \tag{12.136}$$

An alternative expression that enables the critical stress of a short thin wall cylinder to be calculated is:

$$\sigma_{cr} = k_p E\frac{t}{r} \tag{12.137}$$

where the shape coefficient k_p can be obtained from Fig. 12.38 function of ratio r/t.

This formula indicates how thin wall cylinders are subject to a buckling form that involves only the dimensions of section. Therefore this is a local instability phenomenon.

12.2.9 Shear Buckling of Flat Panels

The stress state in a rectangular panel subject to pure shear (τ) is characterized by two main stress directions inclined at 45° with respect to the panel sides (Fig. 12.39).

$$\begin{aligned}\sigma_1 &= \tau\\ \sigma_2 &= -\tau\end{aligned} \tag{12.138}$$

If it is assumed that the panel is made by two sets of fibers oriented at 45°, those oriented along σ_1 are stretched, while the other fibres are compressed. When a sufficiently high load is reached, the compressed fibers collapse due to buckling. Undulations appear on the panel surface where their crest lines are perpendicular to direction σ_2.

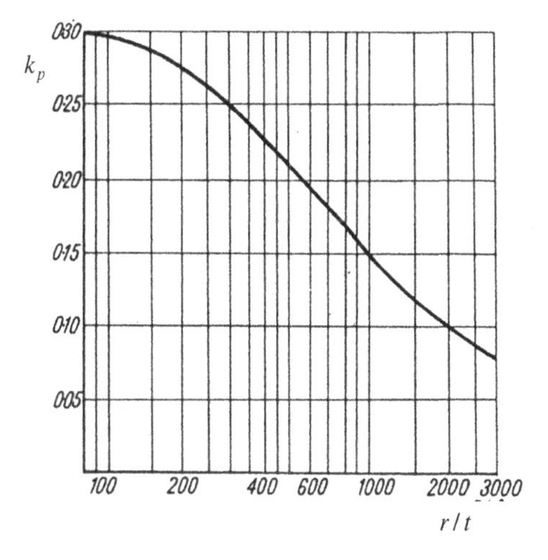

Fig. 12.38. Shape factor for the calculation of critical compression stress of a thin wall cylinder. r : cylinder radius, t wall thickness.

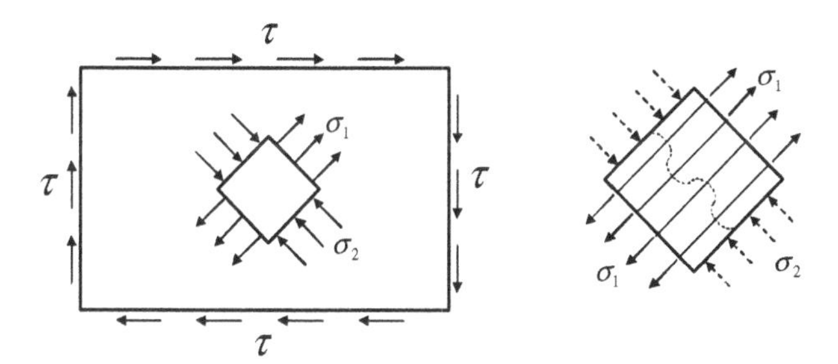

Fig. 12.39. Shear buckling. $\sigma_1 = \tau$, $\sigma_2 = -\tau$ are the main stresses. When compressed fibers collapse for buckling $\sigma_2 \approx 0$.

This type of buckling is called *shear buckling.* The tangential stress that determines the shear buckling is given by an expression similar to that used for panels subject to compression (Eq. 12.122)

$$\tau_{cr} = k_s E \left(\frac{b}{t}\right)^2 \tag{12.139}$$

where E is the modulus of elasticity of material, b is the length of the shortest side of the panel and t is its thickness. The shape coefficient k_s depends on the aspect ratio of the panel and can be obtained from Fig. 12.40.

If τ_{cr} is exceeded, the compressed fibers become curved and the load becomes negligible $\sigma_2 \approx 0$. Referring to Fig. 12.41, in order to guarantee equilibrium in the horizontal direction, the stress on the stretched fibers σ_d must be:

$$\sigma_d = 2\tau. \tag{12.140}$$

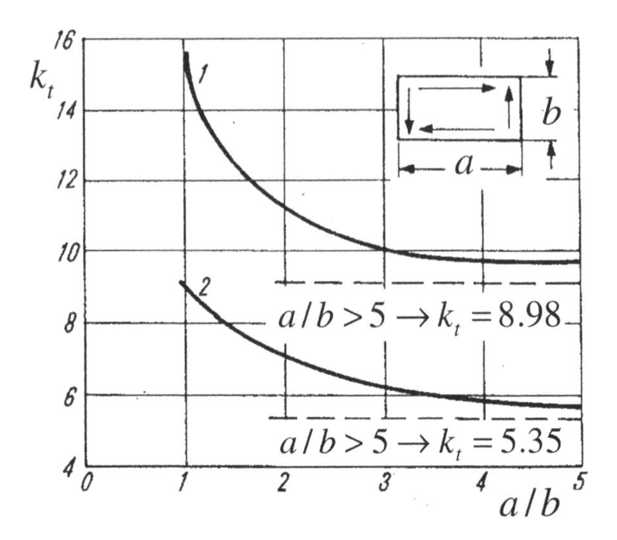

Fig. 12.40. Shape factor for the calculation of the tangential critical stress of a flat panel. 1) joint sides; 2) supported sides.

In order to ensure equilibrium in the vertical direction, normal stresses σ_n must arise on the panel sides

$$\sigma_n = \frac{\sigma_d}{2} = \tau. \tag{12.141}$$

The stress field so defined is usually denoted as a *diagonal stress field* and can be supported by the structure only if the frame of the panels is able to resist normal stress σ_n that bends each element of the frame.

12.3 Simplified Structural Models

As stated in the first volume, a structural body can be imagined as being formed through the connection of frames. In case of a closed body structure, the underbody, the sideframes and the roof, joint by structural nodes and common elements (for example sills), enclose the central portion whereas the connecting elements for the engine and suspensions protrude from the front and rear part of this central portion.

Each frame is made of longitudinal axis beams (rails) and transversal axis beams (crossmembers in the case of the underbody and roof pillars in the case of the sideframes). The space between the beams of a frame may either be left free or closed with a panel. For example: the space between the rails and the crossmembers of the floor is closed by the floor panel. In the case of the sideframe the rails and pillars surround the door frame that, clearly, cannot be closed with panels.

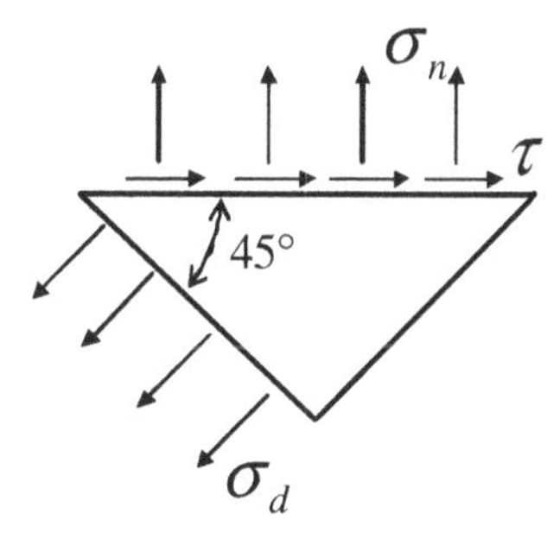

Fig. 12.41. Diagonal stress field. Because of the fact that exceeded τ_{cr}, compressed fibers do not transmit load anymore, for equilibrium the panel edges are subject to normal stress σ_n.

In order to investigate different aspects of structures built from plane elements, it is useful to introduce the concept of *structural surface* [29], [30]. A structural surface is a plane structure with the following characteristics:

- it reacts to forces applied on its plane in an arbitrary direction, but it is not able to react to forces perpendicular to its plane;
- forces can be applied to it only at corners (nodes). It is possible to apply forces along the sides only if they are directed along a line of action coincident with the side itself.

This definition does not imply a particular building shape.

Example a) of Fig. 12.42 shows a rectangle comprising four rods and a diagonal. The rods are all hinged at their ends.

In Example b) four bending resistant beams are joint on the corners (frame).

In Example c) four bars are hinged among them on the corners in order to create a frame for the panel. This panel is connected to the bars along their whole length in order to exchange shear forces that are parallel to the sides of the panel itself.

Example d) is the same as Example a) without the diagonal. The structure in this case is statically indeterminate and so it is not able to resist forces applied at the corners in any direction. Thus this is not a structural surface.

In addition to the configurations of Fig. 12.42 (a, b, c) it is possible to have mixed configurations. The frame (b) can be closed with a panel welded along the entire internal perimeter. Alternatively two corners can be joined with a diagonal. In the same way the reticular structure can be closed with a panel or its hinges can be replaced by locked ones. For technological or functional reasons, real structural surfaces in the automotive field are typically of mixed type.

As concerns the methods of applying loads, the examples of Fig. 12.42 are not all equivalent. The reticular structure a) is made of elements (rods) that can be loaded axially only at the corners.

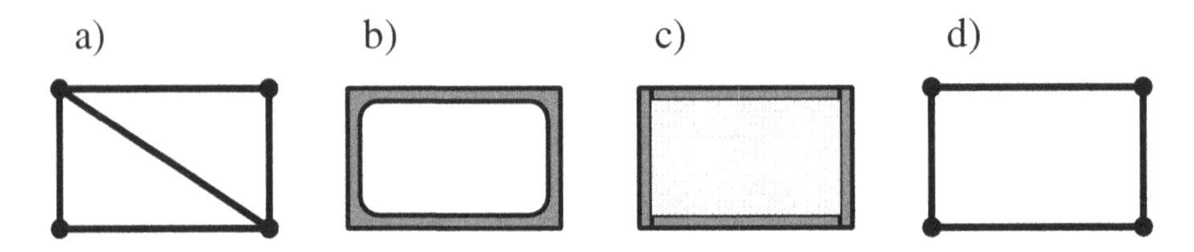

Fig. 12.42. Structural surfaces (a, b, c) and not structural (d). a) reticular structure with hinged rods and diagonal; b) frame of bending resistant beams jointed each other on corners; c) panel framed by four bars resistant to traction and compression hinged each other; d) rods hinged each other without diagonal: this structure is statically indeterminate and it is not able to resist to loads on its plane oriented in any directions.

Thus external loads can be applied only through the connection nodes. Nodes are not only connection elements but also have the function of distributing the external loads within the structure (Fig. 12.43 a). In a frame realized with a truss structure, the application points of forces (for example at the suspensions or engine connections) must coincide with the grid nodes.

The flat panel of Fig. 12.42 b) is made instead of bending resistant beams which can also support loads applied along the sides. Each beam, in this case, can redistribute the load.

The structural surface of Fig. 12.42 c) is an example of semi-monocoque structure that is a composition of a stiffening structure (the frame) and a shell (the panel) The frame transforms the external concentrated loads, distributing them as axial forces in the longitudinal stiffeners and as shear forces on the periphery of the panel.

The panel, in fact, could not support concentrated loads applied on any single spot. If, for example, a paper sheet is pulled in a direction parallel to its surface from a point on its edge, it can resist only weak loads (some N) before it breaks. On the other hand, if a distributed load is applied along an entire side, the sheet will resist much higher loads (in the order of hundreds of Newtons).

Returning to Fig. 12.43 c) the force $\vec{F}_v$ can be applied at any point along the vertical stiffener, whereas $\vec{F}_h$ can be applied at any point along the horizontal stiffener. Since the stiffeners have negligible bending stiffness, it is impossible to apply a force such as that shown in Fig. 12.43 b). The negligible bending stiffness allows the horizontal component to become a concentrated load on the panel, bringing it to breaking.

A vertical force $\vec{F}$ applied to a reticular structural surface, such as that shown in Fig. 12.44, stresses the diagonal to traction. Instead the horizontal lower rod is compressed. The calculation of these forces is made simply by considering that the rods react only to traction/compression, disassembling the external force according to the directions of the rods.

The same vertical force $\vec{F}$ can be applied to a frame structural surface as shown in Fig. 12.45. Since it is a statically indeterminate structure, the calculation of

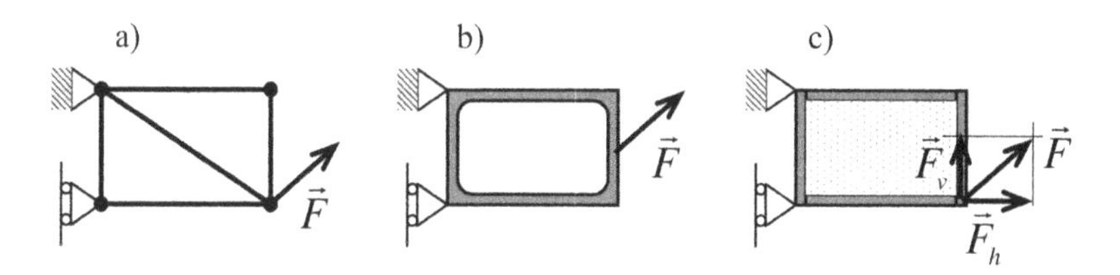

Fig. 12.43. Forces application to different types of structural surfaces. a) reticular structure: forces are applied to nodes; b) frame: forces can be applied to sides, that resists to bending on their plane; c) semi-monocoque: forces can be applied parallel to longitudinal stiffeners.

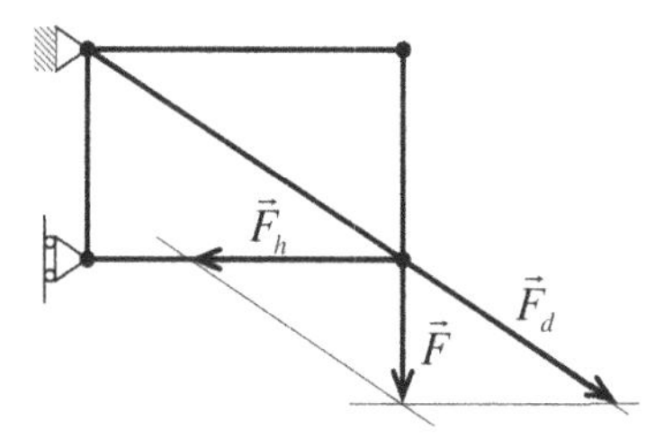

Fig. 12.44. Reticular structural surface. The diagonal is stretched, the horizontal lower rod is compressed.

beams stresses must consider also their stiffness in addition to the equilibrium of each element. In the case that both the horizontal beams have the same stiffness, and that both the vertical beams have the same stiffness, the calculation is simplified: In fact the structure results in being symmetric and loaded in an antisymmetric way.

The equilibrium of the highlighted part in Fig. 12.45 b) involves shear of the two horizontal beams. For symmetry these two beams have the same vertical stiffness; therefore the force $\vec{F}$ is divided between them equally:

$$T_u = T_d = \frac{F}{2}. \tag{12.142}$$

With the adopted sign convention, the relationship between shear and bending moment is $T = -dM/dx$; therefore the bending couple in the horizontal elements is a linear function of distance x :

$$M_u(x) = M_d(x) = -T_u x + M_u(x = 0). \tag{12.143}$$

The subscripts u, d denote either the upper beam ($_u$), or lower beam ($_d$) respectively. Because of symmetry, the curvature of the deformed shape is null in the middle section of each beam ($x = l/2$ and $y = h/2$). This leads to a null bending moment in those points (bending moment is proportional to curvature).

This condition enables the calculation of the constant of integration $M_{u,l}(x = 0) = Fl/4$; therefore:

$$M_{u,l}(x) = \frac{F}{2}l\left(\frac{1}{2} - \frac{x}{l}\right). \tag{12.144}$$

Finally, the equilibrium to rotation of the whole portion of structure highlighted in Fig. 12.45 enables to calculation of the normal stress:

$$N_u = N_d = \frac{F}{2}\frac{l}{h}. \tag{12.145}$$

Given the stress characteristics (T, N, M) in the horizontal beams, the equilibrium to translations and rotations allows the stress characteristics in the vertical beams to be determined:

$$T_r = T_l = -\frac{F}{2}\frac{l}{h}, \tag{12.146}$$

$$M_r = -M_l = \frac{F}{2}l\left(\frac{1}{2} - \frac{y}{h}\right), \tag{12.147}$$

$$N_r = N_l = \frac{F}{2}. \tag{12.148}$$

Besides the quantitative aspects, Fig. 12.45 highlights that:

- all the beams are subject to normal stress, shear stress and bending moment;
- the maximum values of bending moment are reached in the corners, requiring they be made be stiff to bending.

Fig. 12.46 shows the loads distribution inside a semi-monocoque structural surface loaded as the two previous cases.

Under the assumption that the stiffeners have the following characteristics:

- are rigid to traction/compression, and are hinged to one another,
- transmit only shear forces to the panels,

the structure become statically determined.

The equilibrium to translations of the vertical right longitudinal stiffener allows the shear force T_h on the vertical side of the panel to be obtained:

$$T_r = F. \tag{12.149}$$

The equilibrium in the vertical direction of the panel allows the shear on its vertical right side to be obtained:

$$T_l = T_r = F. \tag{12.150}$$

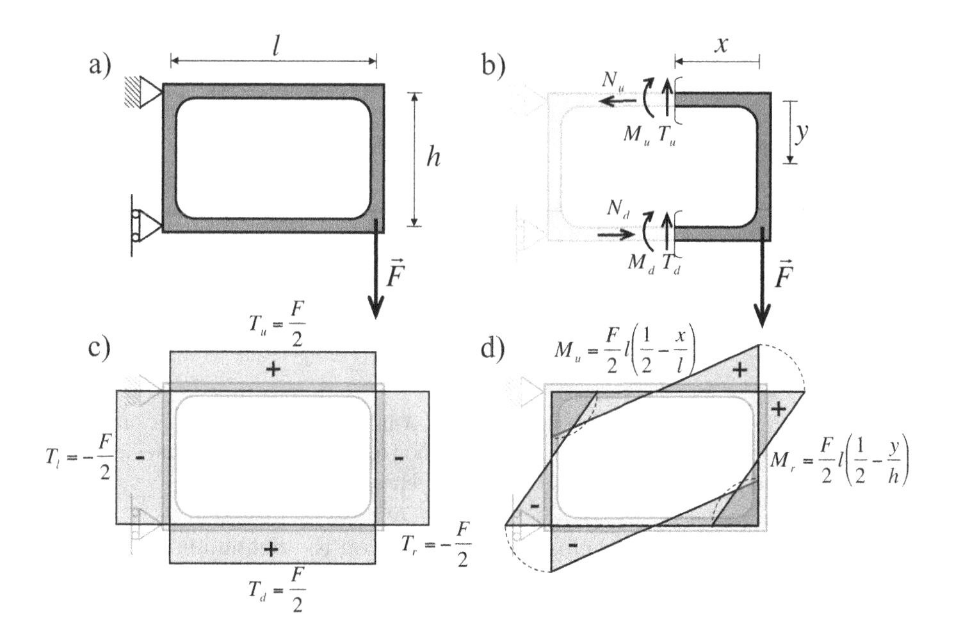

Fig. 12.45. Frame structural surface. Analysis of stress characteristcs on beams. Beams are subject to shear and bending on their plane. The most stressed points are corners.

Considering that the length of vertical sides is h, the associated shear flow is:

$$q_l = q_r = \frac{F}{h}. \tag{12.151}$$

The equilibrium in the horizontal direction and to rotation of the panel, allows the shear on the horizontal sides to be obtained:

$$T_u = T_d = F\frac{l}{h}; \tag{12.152}$$

where l is the length of horizontal sides. The corresponding shear flows are:

$$q_u = q_d = \frac{1}{l} F \frac{l}{h} = \frac{F}{h}. \tag{12.153}$$

The conclusion is that there is only one shear flow on all the panel sides. The orientation on each side guarantees the equilibrium to translation and to rotation:

$$q = q_u = q_d = q_l = q_r = \frac{F}{h}. \tag{12.154}$$

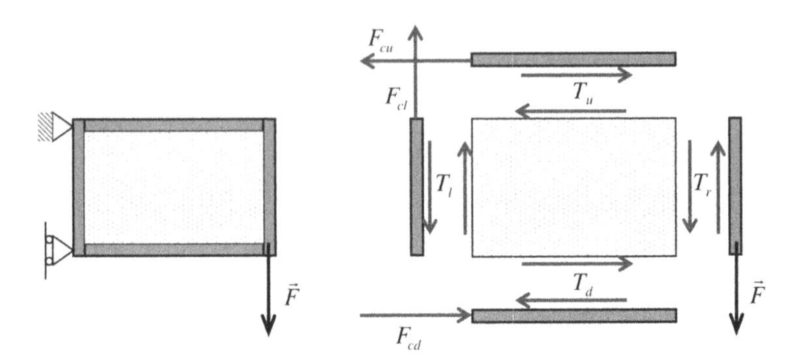

Fig. 12.46. Semi-monocoque structural surface. Analysis of forces that act on components. Longitudinal stiffeners are subject to traction/compression. The panel is subject to pure shear (assuming that it is not subject to shear buckling).

Given the shear stresses on the panel sides, those on the remaining longitudinal stiffeners can be obtained and therefore also the constraint reactions F_{cu}, F_{cd}, F_{cl} :

$$F_{cu} = F_{cd} = F\frac{l}{h}, \tag{12.155}$$

$$F_{cl} = F. \tag{12.156}$$

The upper, right and left longitudinal stiffeners are stretched whereas the lower one is compressed.

12.3.1 Box Model

The objective of this paragraph is to analyze the bending and torsion behavior of some simple box structures using the method of structural surfaces. The cases analyzed will be a box completely closed by structural surfaces, and a box open at the top. The first can be imagined as an extremely simplified example of a two volumes sedan, the second of a spider-type car.

In the previous paragraph, it is demonstrated that the distribution of stresses inside a structural surface depends greatly on its layout whereas its behavior from the point of view of external loads and constraint reactions is the same. So later it will be generally called structural surface, not considering its layout.

Closed box

Bending

Fig. 12.47 represents a closed box comprised of structural surfaces. The external load F is applied inside the box on the middle plane. This load can not be applied directly to the floor because, being a structural surface itself, it cannot support loads perpendicular to its plane. So the function of the cross-member is to distribute the load F to the two side-frames.

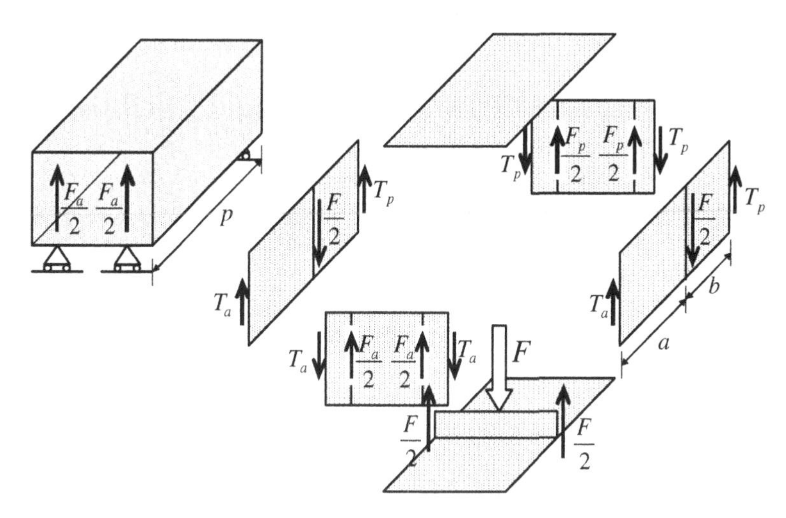

Fig. 12.47. Bending of a box built with structural surfaces. The crossmember is needed to apply the external load F. The floor and the roof are unloaded.

Fig. 12.47 on the upper left shows the general constraint scheme and the constraint reactions, and the exploded view shows the internal forces. The calculation of these forces is possible by starting from a component subject to an external given load (for example the cross-member).

The equilibrium of the central cross-member requires that a load equivalent to $F/2$ is acting on its vertical sides. These loads are produced by sideframes. To enable interfacing with the cross-member, the side-frames could have a central longitudinal stiffener. The objective is to distribute vertical forces $F/2$ on the two structural surfaces that compose the side-frame.

The equilibrium of the side-frame to vertical translations and rotation enables the reactions T_a and T_p to be obtained:

$$T_a = \frac{b}{p}\frac{F}{2}; \qquad T_p = \frac{a}{p}\frac{F}{2}. \tag{12.157}$$

These forces are balanced respectively by the front and rear faces. The equilibrium of these faces will be given by constraint reactions:

$$F_a = \frac{b}{p}F; \qquad F_p = \frac{a}{p}F. \tag{12.158}$$

In the case of bending, the roof and the floor are unloaded, so removing them from the box would not affect the equilibrium.

Torsion

Fig. 12.48 represents the same box subject to a torque. The application of load F_{at} to the right of the front surface requires a reaction equal and opposite on the front constraint. The couple of forces F_{at} is balanced by an equal and opposite couple on the rear side:

$$F_{at}c_a = F_{pt}c_p. \quad (12.159)$$

The equilibrium to translation and rotation of the front side necessitates the forces T_1 and T_2 according to the scheme shown in figure. The couple of forces T_1 and T_2 cannot be determined only through the equilibrium equation to rotation that is available: The statically indeterminate nature of the structure requires knowledge of the stiffness of other box faces, in order to evaluate T_1 and T_2. In the case that the other lateral surfaces are semi-monocoque ones, it is possible to use the first Bredt formula to calculate the shear flow on panels:

$$q = \frac{M_t}{2\Omega} = \frac{F_{at}c_a}{2hd}, \quad (12.160)$$

thus:

$$T_1 = hq; \qquad T_2 = dq. \quad (12.161)$$

At this point the actions on lateral faces can be evaluated by exploiting considerations regarding rotational and translational equilibrium. This means that opposite sides of the same structural surface are subject to equal and opposite forces. In this way a couple is generated that has to be balanced by another equal and opposite couple on the two orthogonal sides.

Considering the lateral right face must be:

$$T_3h = T_1p. \quad (12.162)$$

Given T_1 , this equation enables the determination of T_3.

The scheme of Fig. 12.48 indicates that in the case of torsion all structural surfaces are loaded. In particular, the side faces are subject to shear stress. This is explained by converging forces on opposite corners of one structural surface. On the other hand, the direction of forces applied on the front and rear faces implies that these are not subject to shear stress.

Box open on top

Bending

In the case of bending of a closed box, both the roof and the floor are unloaded. A box without the roof works, for bending loads, in the same way as the closed box of Fig. 12.47. Without the side-frames also (Fig. 12.49), the only connecting element between anterior and rear surface is the underbody. The floor, being subject to loads perpendicular to its surface, is not a structural surface acts as a frame subject to bending.

Torsion

Since the roof is missing, the upper edge of side-frames is unloaded. Consequently all forces applied to the side-frames cancel each other.

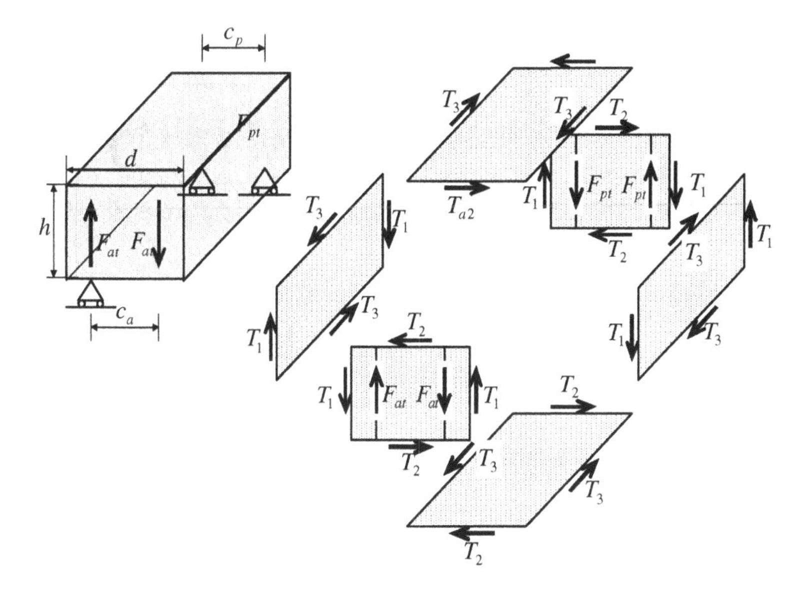

Fig. 12.48. Torsion of a closed box made of structural surfaces. All lateral surfaces are subject to shear stress.

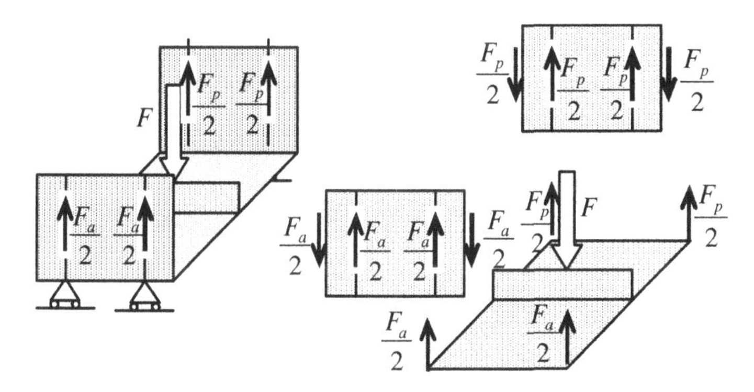

Fig. 12.49. Bending of an open box without roof and sideframes. The underbody is subject to bending.

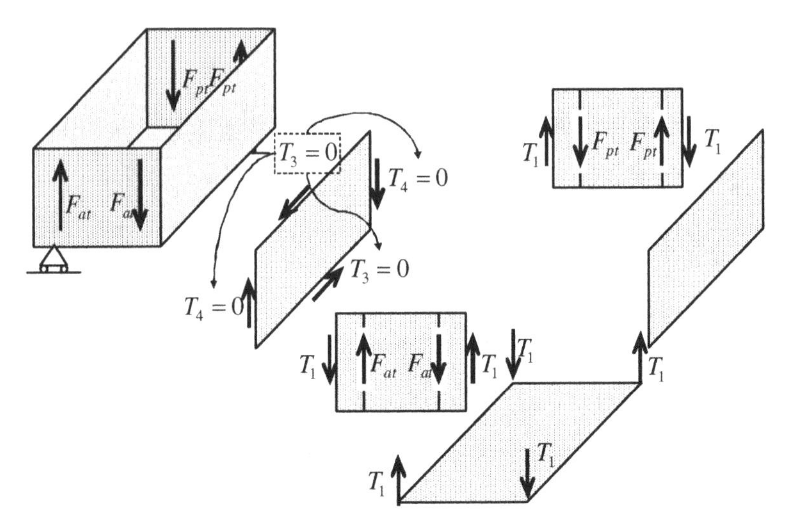

Fig. 12.50. Torsion of a box open on top. Since the roof is missing, the upper edge of sideframe is unloaded. Consequently all other forces working on sideframe cancel each other. The equilibrium of front and rear faces can only be guaranteed by the floor.

In Fig.12.50 the case of left side-frame is shown. Since no element can apply any force on the upper side, the lower one is also unloaded. Missing forces on these two sides must necessarily miss forces on the vertical sides for the rotational equilibrium to be guaranteed.

The equilibrium of the front and rear faces can only be guaranteed by the floor. Also in this case the floor is not a structural surface but a frame subject this time to torsion.

12.3.2 Underbody Configurations

The previous paragraph stated how a floor that can undergo bending and torsion is needed in the case of an open box. This paragraph aims to analyze some simple floor configurations to understand the functions of its elements and the kind of stresses it is subject to. Bending and torsion will be considered as loads.

I shaped frame

This is made of a central beam rigidly connected to two cross-members that interface with four longitudinal rails to the suspensions (Fig. 12.51). The central beam is the element that provides bending and torsional stiffness to the frame. Neglecting the cross-members and rails compliance, torsional stiffness can be evaluated from the equation of the central beam deformed shape,

$$\frac{d\theta}{dx} = \frac{M_t}{GJ_t}, \tag{12.163}$$

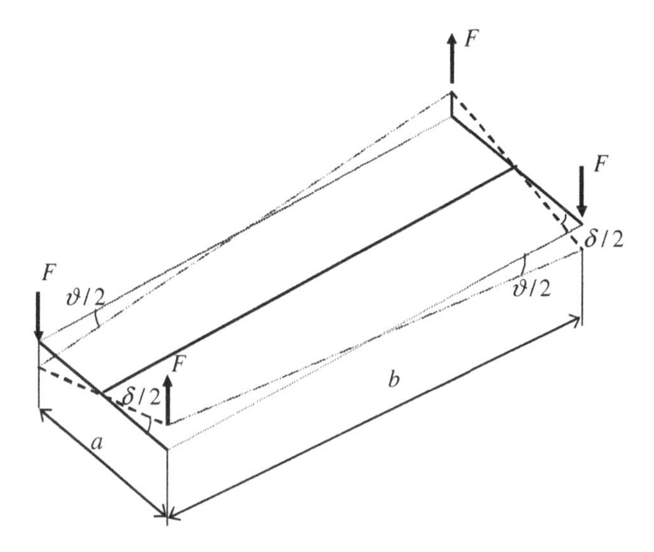

Fig. 12.51. I shaped frame. Bending and torsional stiffness are given by central tube.

where J_t is the torsional stiffness modulus of the beam section and M_t is the torque applied to the frame. If p is the beam length, the torsional stiffness K_t is

$$K_t = \frac{M_t}{\theta} = GJ_t\frac{1}{p}. \tag{12.164}$$

It is very important, in this case, to adopt a closed section beam: In fact a closed section (e.g. a closed section pipe) has a torsional stiffness modulus much higher than an open section (e.g. fissured pipe) of equivalent area.

Advantages.

- High ratio between torsional stiffness and mass.
- Ability to integrate the central beam in the *tunnel*.
- Simplicity

Disadvantages.

- The lower part of door frame is not present.
- The central *tunnel* with a closed section does not allow the installation of exhaust or transmission parts.
- The front cross-member is an obstacle in the feet area.

X shaped frame

This is made of two beams diagonally positioned and connected at the central node (Fig. 12.52). When the frame undergoes torsion, both beams are subject to bending. For symmetry, the central node does not move or change orientation. Thus it works like a rigid constraint for the semi-diagonals.

Torsional stiffness can be evaluated from the deformed shape of each semi-diagonal, under the assumption that forces are applied at the free ends while the ends at the central node are blocked.

The bending deformation of the beam portion between the central node and the free end subject to load F is

$$\frac{\delta}{2}\frac{a}{2} = \frac{(d/2)^3}{3EI}F = \frac{d^3}{24EI}F \tag{12.165}$$

where I is the inertia moment of the diagonal section around the neutral axis and $d = \sqrt{a^2 + b^2}$ is the length of each diagonal. Thus the frame torsional stiffness is

$$K_t = \frac{Fa}{\delta} = \frac{6EI}{d^3}a^2. \tag{12.166}$$

Since the stress on the diagonals is pure bending, in this case an open section shape may be adopted.

Advantages.

- Simplicity of construction.
- Light weight.

Disadvantages.

- The lower part of door frame is not present.
- This solution is hard to integrate with longitudinal elements such as the exhaust and transmission, because it requires increased ground clearance.

H shaped frame

This is made of two rails connected by only one central crossmember (Fig. 12.53). The elements that provide the bending stiffness in this case are the rails, and the cross-member defines the torsional stiffness. Neglecting the bending compliance of rails, the torsional stiffness can be evaluated from the equation of the cross-member deformed shape resulting in:

$$K_t = GJ_t\frac{a}{b^2}. \tag{12.167}$$

where J_t is the torsional stiffness modulus of the cross-member section, a is the distance between rails and b is their length.

Even in this case it is essential to adopt a closed section cross-member to obtain, with equivalent mass, a high torsional stiffness. Eq. 12.167 indicates that with equal wheelbase the torsional stiffness increases with the increase of rails distance.

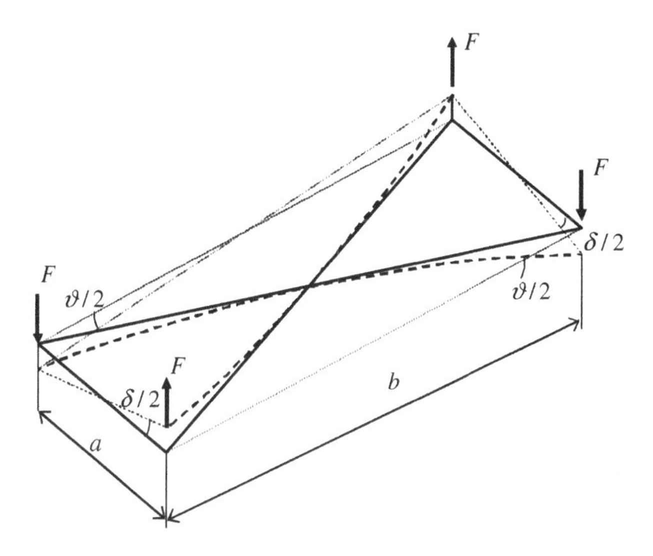

Fig. 12.52. X shaped frame. The two diagonal beams are subject to bending even if the frame is subject to torsion.

Advantages.

- Simplicity.
- The cross-member can be positioned under the seats.
- Low size in the feet area.

Disadvantages.

- Low ratio between torsional stiffness and mass.

Rectangular frame

This is made of two rails and of two cross-members to build a rectangular structure (Fig. 12.59). The elements that provide bending stiffness are rails and, partially, the cross-members. On the other hand only the two cross-members provide torsional stiffness. Assuming,

- the rear and front rails not connected at their end to the cross beams are not considered,
- the rails and cross-members are not deformed by bending,
- the torsional stiffness of cross-members is negligible compared to the rails one,
- the structure is symmetric respect to xz and yz planes,

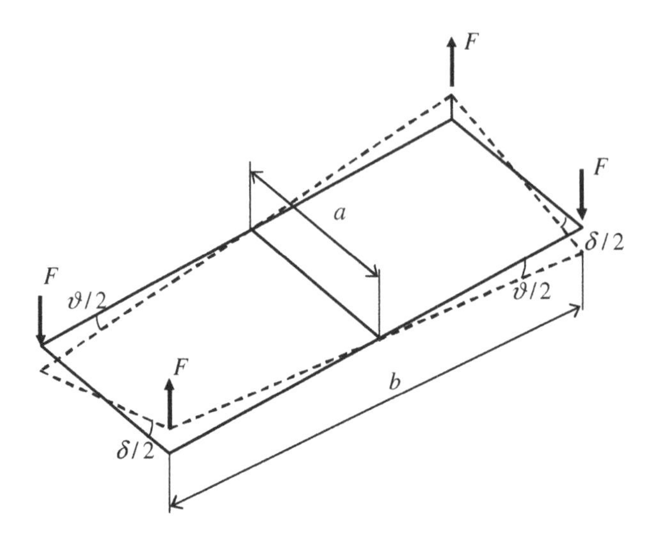

Fig. 12.53. H shaped frame. It is made by two rails and only one crossmember

after the application of a torque the two crossmembers rotate around x axis, keeping themselves straight.

The relative rotation between the two cross-members is the same that occurs between the ends of each rail. Furthermore, if the torsional compliance of the cross-members is negligible compared to the rails, there is no bending of the rails so that they result in being stressed by pure torsion (as on the I shaped frame). Denoting with J_{tl} the torsional stiffness modulus of each rail, the total torsional stiffness is the sum of the two rails contributions:

$$K_t = 2\frac{GJ_{tl}}{b}. \tag{12.168}$$

On the other hand, if rails have a negligible torsional stiffness compared to the cross-members, after the torsion they rotate around the y axis, keeping themselves straight (i.e. according to the assumptions, they are not deformed by bending).

In this way, the sections of the cross-members ends rotate by an angle ϕ as in the example of the H shaped frame. In this case the torsional stiffness is equivalent to double that of the H shaped frame because there are two cross-members with torsional stiffness modulus J_{tt} :

$$K_t = 2GJ_{tt}\frac{a}{b^2}. \tag{12.169}$$

If the cross-members are more than two, the frame is called ladder frame, which is commonly used in industrial vehicles.

The torsional stiffness of a ladder frame can be calculated in a straightforward way when both the rails and cross-members are deformable by torsion but not

Fig. 12.54. Ladder frame subject to torsion. Rails keep a straight axis. Either rails and crossmembers are subject to torsion.

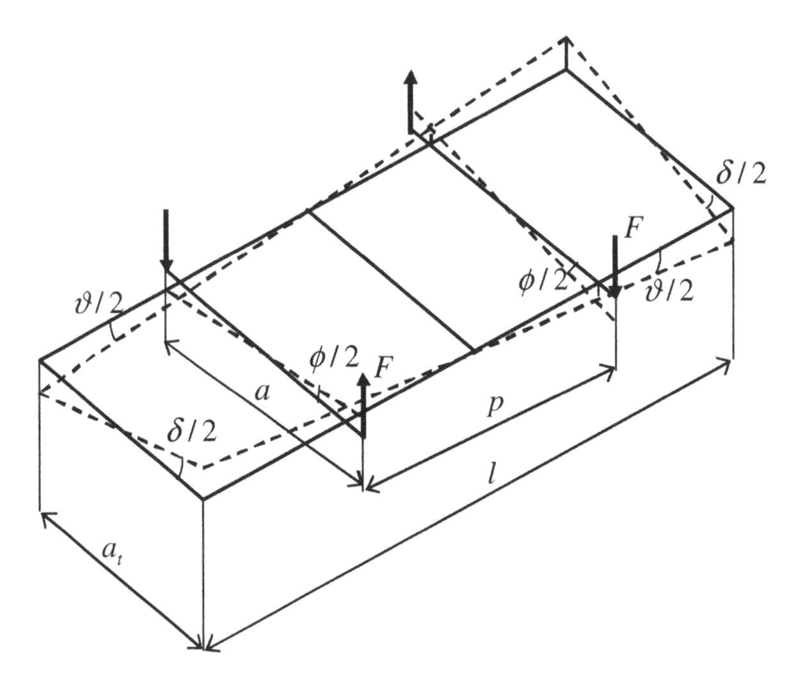

Fig. 12.55. Ladder frame subject to torsion. p distance between axles. l frame length. d distance between connection points of suspensions to the frame.

by bending. The assumption of null bending compliance is true when both rails and crossmembers are built with open section shapes.

Fig. 12.54 shows an industrial vehicle frame subject to a torsion test. Both the rails and cross-members are built with a C section. To confirm the assumption made, a high torsion of rails can be observed that instead do not demonstrate any significant bending.

In Fig. 12.55, a scheme of a ladder frame subject to torsion is represented. The model is based on the following assumptions:

- rails and cross-members have straight axis and constant section;
- rectangular modulus (i.e. parallel rails and cross-members);
- cross-members are all coincident (i.e. same section, length and material);
- rails are all coincident (i.e. same section, length and material);
- two cross-members are placed at the beginning and the end of the rails;
- the torsion center of the sections corresponds to their center of gravity;
- cross-members are placed symmetrically with respect to the middle of the rails;

- the external torque is applied to the cross-members to which the suspensions are connected. The second and the fourth cross-members are placed next to axles. The distance a_t between the rails is constant and is shorter than the distance (a) between the connection points of suspensions.

Under the application of a torque, the frame is deformed following the scheme shown in the figure, where the dashed lines represent the deformed configuration. Each rail rotates around the y axis by an angle $\theta/2$ (where θ is the relative rotation between rails). Correspondingly the cross-members rotate around the longitudinal axis (x). Cross-members rotation increases with the increase of distance from the frame middle section.

In the example shown in figure, the central cross-member does not rotate around the x axis; the first and last cross-members rotate by an angle $(\delta/2)$ more than that $(\phi/2)$ of the second and fourth cross-members.

Angles δ and ϕ can be expressed as a function of θ considering the assumption that the rails and cross-members keep their axis straight.

$$\begin{aligned} \phi &= \frac{p}{a}\theta, \\ \delta &= \frac{l}{a}\theta. \end{aligned} \tag{12.170}$$

The frame torsional stiffness can be calculated starting with the expression of the elastic potential energy E given by the sum of the contributions of the rails and cross-members:

$$E = n_t \frac{1}{2} K_t \theta^2 + 2 \frac{1}{2} K_l \delta^2, \tag{12.171}$$

where K_t and K_l respectively are the torsional stiffnesses of cross-members and rails, and n_t is the number of cross-members .

The factor 2, that multiplies the second term of energy relates to the fact that there are two rails. The elastic energy, on the other side, can be expressed as a function of the frame torsional stiffness K_ϕ and of the rotation (ϕ) between the two sections that correspond to the axles:

$$E = \frac{1}{2} K_\phi \phi^2. \tag{12.172}$$

By equating the energies given by Eq.12.171 and 12.172:

$$n_t \frac{1}{2} K_t \theta^2 + 2 \frac{1}{2} K_l \delta^2 = \frac{1}{2} K_\phi \phi^2; \tag{12.173}$$

and considering the relationship between angles ϕ, δ and θ (Eq. 12.170):

$$K_\phi = \left(\frac{a}{p}\right)^2 \left[n_t K_t + 2 K_l \left(\frac{l}{a}\right)^2 \right]. \tag{12.174}$$

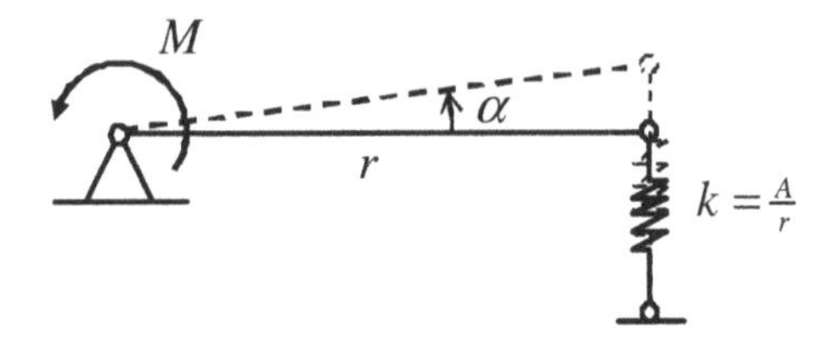

Fig. 12.56. Even if the stiffness of the spring $k = A/r$ decreases with r the torsional stiffness $K_\alpha = M/\alpha = Ar$ increases with r.

The torsional stiffness of the rails and cross-members can be expressed according to the assumption of straight axis and constant section:

$$K_t = \frac{GJ_{tt}}{a}$$
$$K_l = \frac{GJ_{tl}}{l} \tag{12.175}$$

where J_{tt} and J_{tl} are the torsional stiffness moduli of the sections of the cross-members and rails respectively. Substituting Eq. 12.175 into Eq. 12.174:

$$K_\phi = \frac{Ga}{p^2}\left(n_t J_{tt} + 2J_{tl}\frac{l}{a}\right). \tag{12.176}$$

The torsional stiffness of the frame is proportional to the stiffness modulus of the rails and cross-members. Given a certain width a and wheelbase p, the second term in the brackets shows how the stiffness increases with increasing rail length as can be understood by analyzing Fig. 12.55 and Eq.12.170.

For a given angle ϕ, the torsional deformation of the rails increases with their length l and, consequently, the energy needed for their deformation increases. The fact that the rail stiffness is inversely proportional to its length is balanced by the fact that the contribution of the rail to the torsional stiffness (Eq. 12.174) is proportional to the square of its length. Qualitatively the phenomenon is the same as the case of the hinged lever of Fig. 12.56. The torsional stiffness of the lever is proportional to the spring stiffness and to the square of the radius r :

$$K_\alpha = \frac{M}{\alpha} = kr^2. \tag{12.177}$$

Since the stiffness $k = A/r$ is inversely proportional to the radius, the stiffness K_α increases with r :

$$K_\alpha = Ar. \tag{12.178}$$

The assumptions made regarding the stiffnesses are applicable only for frames with open section cross-members and closed section rails (or viceversa), which hardly ever occurs in practice. In the case that all profiled beams have closed sections, the calculation of stress inside the frame is complicated by static

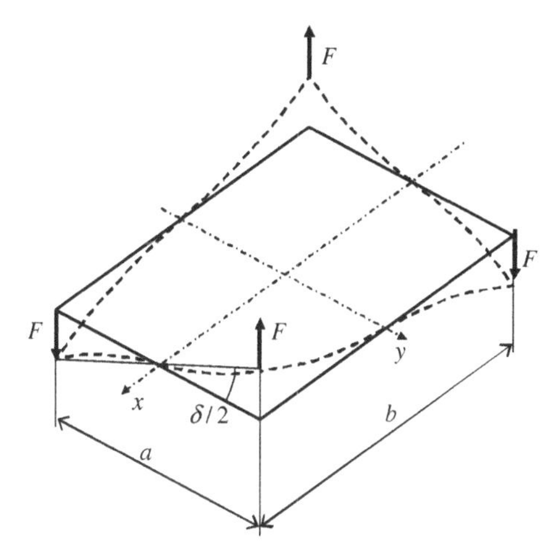

Fig. 12.57. Rectangular frame subject to torsion. The structure is symmetric respect to xz and yz planes.

indetermination, and by the fact that the bending deformation is not negligible compared to the beams torsional deformation.

The calculation of stress and deformation state is made under the following assumptions:

- rectangular frame;
- rails and cross-members with straight axis and constant section;
- torsion center coincident with the section center of gravity;
- pure torsion stress obtained by applying forces on the frame corners.

Under these assumptions, the structure is symmetric and symmetrically loaded. Its deformation agrees with its symmetry (Fig. 12.57). In particular, the rails middle sections are subject to no displacement and no torsional rotation.

To analyze the frame is possible to exploit its symmetry and analyze just one quarter, substituting the rest of the structure with hinges Fig. 12.58 A. The constraint reactions that are produced after the application of an external force F are the vertical forces F_{1A} and F_{1B} in sections 1 and 2 and the torques M_{1A} and M_{1B}. For equilibrium in the vertical direction and rotation around axes x and y,

$$\begin{aligned} F &= F_{1A} + F_{2A} \\ M_{1A} &= F_{2A}\frac{a}{2} \\ M_{1A} &= -F_{1A}\frac{b}{2} \end{aligned} \tag{12.179}$$

The structure is statically indeterminate because four constraint reactions have to be identified from three equilibrium equations. The fourth equation is obtained by applying the virtual work principle. Thus, in addition to the static indeterminate system A, an auxiliary system B is considered that becomes statically determinate following substitution of the cylindrical hinge on section 1 with a spherical hinge. Thus:

$$L_{eAB} = L_{iAB} \tag{12.180}$$

where

L_{eAB}=virtual work made by external forces of system B due to the deformations of system A.

L_{iAB}=virtual work made by internal forces of system B due to the deformations of system A.

Supposing that system B is subject to an arbitrary moment $M_{1B} = T$ on point 1 in x direction, since point 1 of system A cannot rotate around the x axis of the rail, the work made by moment M_{1B} with the deformation of system A is null:

$$L_{eAB} = 0. \tag{12.181}$$

Supposing that the structure is subject only to bending and torsional deformations (neglecting the shear compliance)

$$L_{iAB} = \int_{102} \left(\frac{M_{fA} M_{fB}}{EI} + \frac{M_{TA} M_{TB}}{GJ_t} \right) dl \tag{12.182}$$

The integral is extended to the structure included between points 1, 0 and 2.

For system A the bending and torque in the rail and cross-member are equal to:

$$\begin{array}{lll} \text{segment 1-0} & M_{fA} = F_{1A}\eta & M_{TA} = M_{1A} \\ \text{segment 0-2} & M_{fA} = F_{2A}\xi & M_{TA} = M_{2A} \end{array} \tag{12.183}$$

For system B:

$$\begin{array}{lll} \text{segment 1-0} & M_{fB} = F_{1B}\eta & M_{TB} = T \\ \text{segment 0-2} & M_{fB} = F_{2B}\xi & M_{TB} = M_{2B} \end{array} \tag{12.184}$$

Since structure B is statically determined, the constraint reactions of system B can be calculated from the equilibrium equations:

$$\begin{aligned} F_{2B} &= \frac{2T}{a} \\ F_{1B} &= F_{2B} = \frac{2T}{a} \\ M_{2B} &= F_{1B}\frac{b}{2} = T\frac{b}{a} \end{aligned} \tag{12.185}$$

Substituting the moments of Eq.s 12.183 and 12.184 into the expression of internal work L_{iAB} of Eq. 12.182 and considering the constraint reactions of Eq. 12.181, one obtains

$$L_{iAB} = F_{1A}\left(\frac{b^3}{3EI_l}+\frac{ab^2}{GJ_{tt}}\right) - F_{2A}\left(\frac{a^3}{3EI_t}+\frac{a^2b}{GJ_{tl}}\right) \tag{12.186}$$

where I_l and I_t are the inertia moments of rail and crossmember sections respect their neutral axis; J_{tl} and J_{tt} are the torsional stiffness moduli of rail and crossmember.

Equating the internal and external work (Eq. 12.180) and considering that the external work is null (Eq. 12.181):

$$C_1F_{1A} = C_2F_{2A} \tag{12.187}$$

where:

$$C_1 = \frac{b^3}{3EI_l}+\frac{ab^2}{GJ_{tt}} \tag{12.188}$$

$$C_2 = \frac{a^3}{3EI_t}+\frac{a^2b}{GJ_{tl}} \tag{12.189}$$

Eq. 12.187 can be solved together with Eq. 12.179 to obtain the constraint reactions:

$$F_{1A} = \frac{C_2}{C_1+C_2}F \qquad M_{1A} = -\frac{aC_1}{2\left(C_1+C_2\right)}F \tag{12.190}$$

$$F_{2A} = \frac{C_1}{C_1+C_2}F \qquad M_{2A} = -\frac{bC_2}{2\left(C_1+C_2\right)}F \tag{12.191}$$

The torsional deformation δ of the frame can be obtained by considering that work done by the external moment applied Fa is equal to the deformation elastic energy of the structure (Fig. 12.57). Since the frame comprises four equivalent parts, the elastic energy is equal to four times the deformation energy of portion 1-0-2 of Fig. 12.58

$$\frac{1}{2}Fa\delta = 4\frac{1}{2}\int_{102}\left(\frac{M_{fA}^2}{EI}+\frac{M_{TA}^2}{GJ_t}\right)dl \tag{12.192}$$

Substituting the constraint reactions (Eq. 12.190 and 12.191) into the stress characteristics (Eq. 12.183),

$$Fa = 2a^2\left(\frac{1}{C_1}+\frac{1}{C_2}\right)\delta \tag{12.193}$$

The torsional stiffness of the frame is the ratio between the torque applied (Fa) and the relative rotation δ between the first and last sections:

$$K_t = \frac{Fa}{\delta} = 2a^2\left(\frac{1}{C_1}+\frac{1}{C_2}\right). \tag{12.194}$$

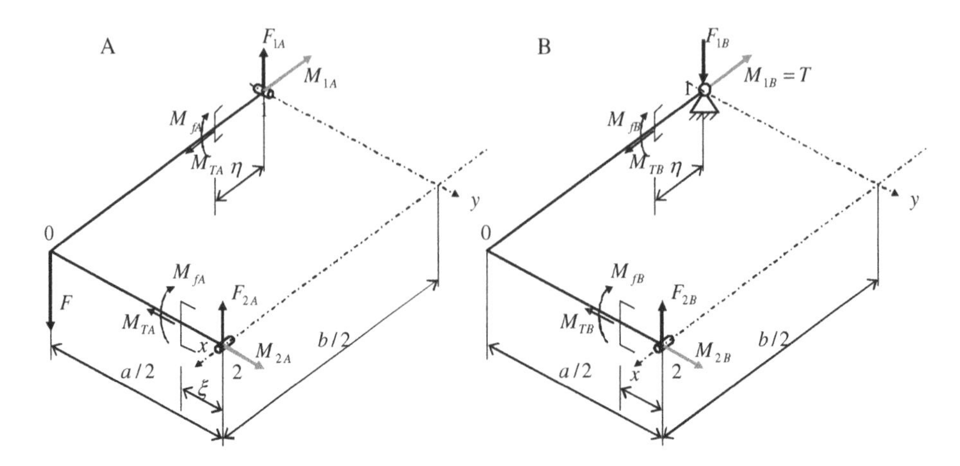

Fig. 12.58. Analysis of stress and deformation state of a rectangular frame subject to torsion.

The term in the brackets represents the ratio between the applied force F and the displacement $a\delta/2$ of its application point $\frac{F}{a\delta/2} = \frac{1}{C_1} + \frac{1}{C_2}$

$1/C_1$ and $1/C_2$ represent two stiffnesses in parallel. Each of these (Eq. 12.188 and 12.189) includes the contribution of the two rails and the cross-members.

If it is assumed that the rails and the cross-members are infinitely stiff to bending ($I_l = I_t = \infty$) and the cross-members are infinitely deformable by torsion ($J_{tt} = 0$):

$$C_1 = \infty \qquad C_2 = \frac{a^2 b}{GJ_{tl}} \tag{12.195}$$

and by substitution into Eq. 12.194,

$$K_t = 2\frac{GJ_{tl}}{b} \tag{12.196}$$

that coincides with the expression previously obtained analyzing the same limit case (Eq. 12.168).

Vice versa, if the rails are assumed to be infinitely deformable by torsion ($J_{tl} = 0$):

$$C_1 = \frac{ab^2}{GJ_{tt}} \qquad C_2 = \infty \tag{12.197}$$

and by substitution again into Eq. 12.194,

$$K_t = 2GJ_{tt}\frac{a}{b^2} \tag{12.198}$$

that coincides with Eq. 12.169.

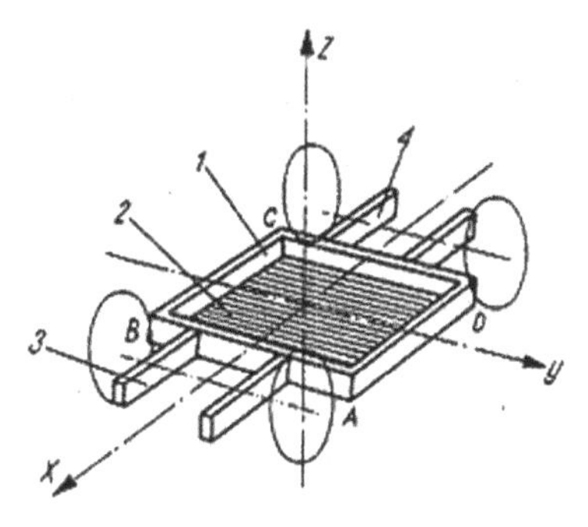

Fig. 12.59. Rectangular frame. It is built by two rails and two crossmembers placed at the ends of rails. The connection to suspensions is obtained with overhang beams connected to crossmembers.

In real cases, given a certain size (given a and b), it is necessary to design a structure with a high ratio of torsional stiffness to weight. For this purpose it is appropriate to use closed sections both for the rails and cross-members in order to enable the high torsional stiffness of both sections contribute to the overall torsional stiffness.

The advantages of rectangular frame with closed section profiled beams are as follow:

- significant free space for passengers.
- one of the cross-members can fit under the rear seats.
- the floor can be flat.
- the rails can be part of the lower portion of the sideframe.

whereas the disadvantages are:

- encumbrance in pedals area.
- difficulty to realize the front and rear overhang beams.

Integral frame

The underbody scheme of Fig. 12.60 can be considered to be the integration of a ladder frame and a rectangular frame; the first is positioned under the floor,whereas the second one surrounds it. In this case the front cross-member mainly transmits the vertical loads, so its torsional stiffness can be low; on the other hand the high bending stiffness allows the sills to undergo torsion, creating more space in the pedals area. Since the load on the front cross-member is in a vertical plane, its functions can be performed by the firewall structural surface. Instead the central and rear cross-members can be made with closed sections in order to increase the torsional stiffness.

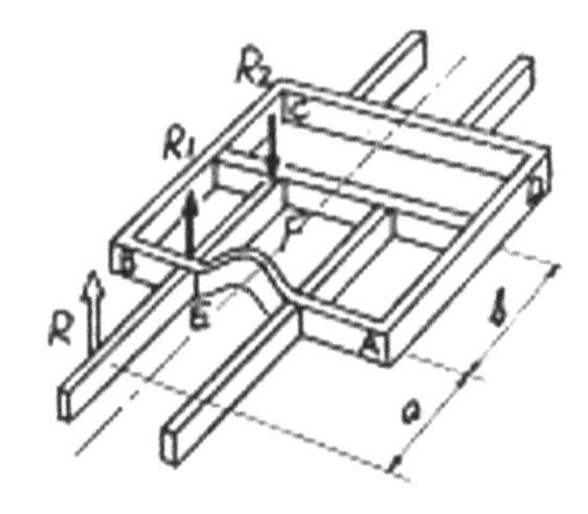

Fig. 12.60. Integral frame scheme.

12.3.3 Central Portion Model

The aim of this paragraph is to analyze the behavior to bending and torsion of the central portion of a structural body. The analysis is simplified using the method of structural surfaces in order to understand which type of stress acts on each part.

Bending

The external force F applied by the suspension (Fig. 12.61 a)) is equivalent to the system formed by force F applied on the firewall and by the couple of longitudinal forces $F_l = Fd/l$ (Fig. 12.61 a)). The firewall withstands forces F since they act in its plane (Fig. 12.61 c)). Instead longitudinal forces F_l stress the underfloor rails and sideframes. The stressed elements are the sideframes, the firewall and the floor, whereas the roof and the windshield are unloaded.

Torsion

The torque transmitted to the firewall by suspensions is transmitted to the sideframes, the windshield and the roof. Similarly to the case of a closed box, all structural surfaces are stressed. In particular the roof, floor and windshield are stressed by shear Fig. 12.62.

12.3.4 Functional Requirements for a Structural Body

Included in just a single component the structural functions with those relative to ergonomics and accessibility, a structural body has very high requirements of torsional and bending stiffness when compared to a separable frame configuration.

The need for high stiffness comes from the fact that, following the application of loads, the entire body is deformed. The deformation also involves the spaces

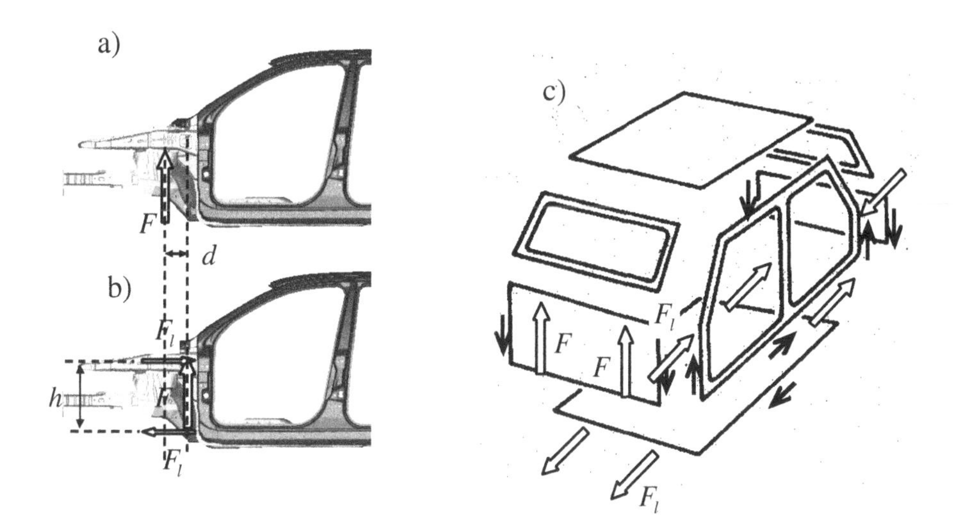

Fig. 12.61. Stress of a central portion subject to bending. All parts are modeled as structural surfaces.

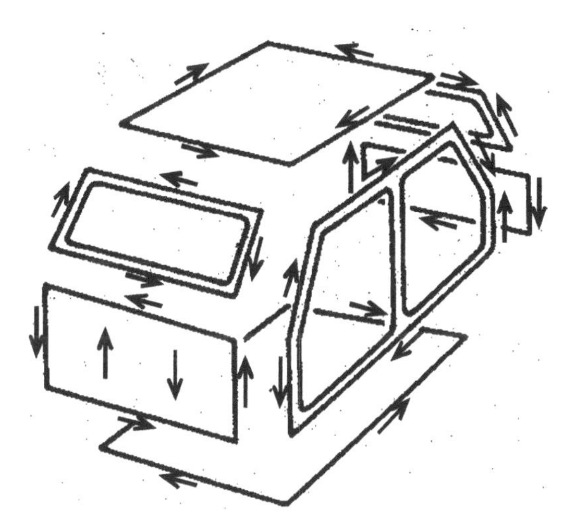

Fig. 12.62. Stress of central portion subject to torsion. All parts are modeled as structural surfaces.

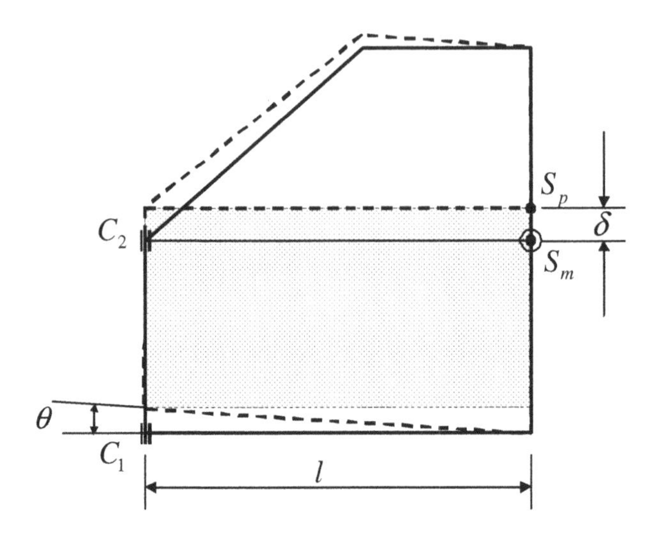

Fig. 12.63. Scheme of deformation of a door frame. Undeformed configuration (continous line) and deformed configuration (dashed line and shaded area).

where the mobile parts are positioned (doors, hood, trunk door,...) which instead tend to remain undeformed since not subject to loading. This generates a relative displacement between a mobile element and the frame that can cause some problems such as:

- creaking due to creeping of seals on moving parts,
- loss of sealing of seals and aerodynamic noise,
- loss of lock functionality.

The loss of lock functionality is due to the fact that locks used in automotive field can only tolerate displacements between two connected parts in the range $1 \div 2$ mm. At higher levels the lock may jam or even block. To avoid this, a spacer may be inserted.

Fig. 12.63 illustrates this phenomenon schematically when torsion occurs. The door frame is deformed by an angle θ represented by the dashed configuration. During distortion the door is not loaded, so its structure (C_1, C_2, S_p) is subject to negligible deformations compared to those of the door frame. Correspondingly a displacement δ occurs between the lock part connected to the door (S_p) and the one connected to the pillar (S_m). Since $\delta_{\max} \approx 2$ mm and $l \approx 1$ m then the maximum distortion is $\theta_{\max} \approx 2\ 10^{-3}$ rad.

Door frame distortion phenomena are particularly important in the case of body torsion. In order to guarantee the functionality, the torsional stiffness of the structural body must be sufficiently high (typically at least 500kNm/rad for compact cars).

12.4 Numerical Models for Structural Analysis

The number of functions of a car, the high level of integration between them in a relatively narrow space, and the needs related to the production, are just some of the factors that result in the complex configuration of the car body structure. The structure itself must satisfy an increasing number of expectations ranging from quasi static performance like the bending and torsional stiffness, to dynamic properties such as noise and vibration, to reliability under normal and misuse loading. In the past decades, passive safety requirements have introduced additional and very severe requirements that have motivated the search for new configurations and design criteria. More recently energy issues are putting pressure on reducing the weight and size without compromising the other requirements that cannot be sacrificed.

Until late 1970's the design of structural components relied on simple analytical tools such as the structural surface method and other methods for the analysis of thin shell structures. More in depth analyses were performed mostly at the experimental level on relatively simple critical components such as the suspension and powertrain attachment points. The approximations involved in such design methods required experimental tests on prototypes and pre-series vehicles. The design iteration in this case involved the expensive and time consuming building of new prototypes and, in some cases, modifications of already built production tools.

The present push to reduce the time required for putting a new model into production requires minimizing, or even avoiding entirely, the design iteration on prototypes. Over the past three decades, the availability of high computational power at a reasonable cost on one side, and of finite elements codes and their integration with CAD software on the other, has made in depth structural analysis on practically all elements of a car body structure possible. This can be undertaken starting from the earliest stages of the design process to the verification stages on the ready for production configuration.

At the preliminary stages, finite element codes with integrated optimization tools can help the selection of the configuration for given boundary conditions, design constraints and load cases. The selected configuration is then refined with an iteration that involves CAD and FEM tools considering the number of additional constraints coming from production and from other subsystems.

The types of structural analyses applied range from the linear analysis for the study of the quasi static and dynamic behavior under the assumption of small displacements in the elastic range, to the full nonlinear analysis for the study of the large deformations during crash. Finite elements models can also be integrated in the software used for the multi-body simulation of the car and of its subsystems (typical application is the study of the elasto-kinematic behavior of the suspensions).

Today the use of the finite element analysis is so pervasive in the design process that it finds application also for the production process. Nonlinear analysis codes are used for the design of the stamping process so as to minimize the time required for the optimization of the dies.

The literature regarding finite element methods is so broad that it is beyond the scope of this section to provide in depth analysis of the many important issues to be considered ranging from the elements formulation and their assembly, to the linear and nonlinear solutions. Instead the aim here is to provide just a brief summary of the main classes of finite element models for the linear and nonlinear analysis to allow the user understand the possibilities are available and the principal logic behind the solution methods.

A first part is dedicated to a brief introduction on the application of the finite elements method for structural analysis. The focus then turns to the main types of linear and nonlinear models of a car body.

12.4.1 Shape Functions and Degrees of Freedom

The application of loads and constraints to a structure induce a deformation and a stress and strain field in its material. The deformed shape at time t can be described by the displacement vector $\vec{u}(x, y, z, t)$ experienced by each point of coordinates $(x,\ y,\ z)$ of the undeformed structure

$$\vec{u}(x, y, z, t) = u(x, y, z, t)\vec{i} + v(x, y, z, t)\vec{j} + w(x, y, z, t)\vec{k}. \tag{12.199}$$

where $i,\ j,\ k$ are the unit vectors of a rectangular coordinate frames and $u(x, y, z, t)$, $v(x, y, z, t)$ and $w(x, y, z, t)$ are the displacements along each direction. Because of the continuous nature of the structure, displacement vector $\vec{u}(x, y, z, t)$ should be given in each of the infinite points of the structure. In other words, the structure has infinite degrees of freedom.

The essential step of the finite element method for the structural analysis is to split the structure into parts, called elements, delimited by a geometrically simple boundary. Each element exchanges forces and displacements with the rest of the structure by means of a small number of points called nodes. Even if the interaction between the material in an element and the rest of the structure occurs in all the infinite points of the boundary, the nodes are the only interfaces by which each element exchanges energy with the neighboring ones.

The displacement field in each element is approximated with a linear combination of the nodal displacements:

$$\mathbf{u}(x, y, z, t) \approx \mathbf{N}(x, y, z)\mathbf{q}(t); \tag{12.200}$$

column matrix $\mathbf{u}(x, y, z, t)$ (3x1) includes the components of vector $\vec{u}(x, y, z, t)$, the n nodal displacements displacements are aligned in the $n \times 1$ column vector $\mathbf{q}(t)$

$$\mathbf{u}(x, y, z, t) = \{u(x, y, z, t),\ v(x, y, z, t),\ w(x, y, z, t)\}^T, \tag{12.201}$$

$$\mathbf{q}(t) = \{q_1(t), \ldots,\ q_n(t)\}^T. \tag{12.202}$$

The $3 \times n$ matrix $\mathbf{N}(x, y, z)$ includes the shape functions. These are known functions of the coordinates of a given point that allow to approximate the displacement at that point as function of that at the nodes.

Eq. 12.200 shows that:

- the infinite degrees of freedom of the element, included in vector $\mathbf{u}(x, y, z, t)$ are approximated by the n nodal degrees of freedom of vector $\mathbf{q}(t)$;
- the shape functions are used to take the dependence on the coordinates x, y, z of each point into account. The dependence from time t is attributed completely to the nodal displacements $\mathbf{q}(t)$. The implicit assumption here is that it is possible to separate the spatial to the time dependence, as if the displacement $\mathbf{u}(x, y, z, t)$ at a point inside the element is only function of the nodal displacements $\mathbf{q}(t)$ and not of the time.

The shape functions are known functions of the position inside the element. Even if arbitrary, to some extent, they must satisfy a number of properties:

- They must have unit values at the element nodes location, so that vector $\mathbf{u}(x, y, z, t)$ (Eq. 12.200) reduces to the nodal displacement included in vector $\mathbf{q}(t)$ at the nodes.
- The shape function $\mathbf{N}(x, y, z)$ must be space differentiable up to a degree that allows to determine the strain field inside the element.
- Eq. 12.200 should be able to describe the rigid body motion of the element without involving a change in its potential energy.
- The deformed shape each element boundary should be consistent with the neighboring elements. The boundary shared by two elements should remain common to the two elements after a deformation with neither overlapping nor separation.

12.4.2 Equations of the Motion

The shape functions enable the approximation of the displacement field in the element as a function of the nodal displacements; they are then the configuration variables of the element: i.e. the variables that allow the definition of the configuration (as opposed to the state, that includes also the information about the speed) of the element at a given time. The equations of the motion of the element can be obtained using a Lagrangian approach, starting from the kinetic and potential energy.

Kinetic energy

The position $\mathbf{U}(x, y, z, t)$ of a generic point $\mathbf{P} = \{x_P, y_P, z_P\}^T$ of the element can be expressed in an inertial frame (XYZ) as

$$\mathbf{U}(x, y, z, t) = \mathbf{U}_0(t) + \mathbf{R}(t)(\mathbf{P} + \mathbf{u}(x, y, z, t)) \tag{12.203}$$

where $\mathbf{U}_0(t)$ is the position of the origin of the reference frame ($Oxyz$) fixed to the element, $\mathbf{R}(t)$ is the rotation matrix that takes into account of a different alignment between reference frames XYZ and $Oxyz$ and $\mathbf{u}(x,y,z,t)$ is the displacement vector affecting point $\mathbf{P}$ in the non inertial reference frame. The speed of point $\mathbf{P}$ in the inertial reference frame is obtained as the time derivative of Eq. 12.203

$$\dot{\mathbf{U}}(x,y,z,t) = \dot{\mathbf{U}}_0(t) + \boldsymbol{\omega}(t)(\mathbf{P} + \mathbf{u}(x,y,z,t)) + \mathbf{R}(t)\dot{\mathbf{u}}(x,y,z,t)$$

matrix $\boldsymbol{\omega}(t)=\dot{\mathbf{R}}(t)$ takes the angular speed of the element into account.

The displacement $\mathbf{u}(x,y,z,t)$ can now be expressed by means of the shape functions and of the nodal displacements of Eq. 12.200

$$\dot{\mathbf{U}}(x,y,z,t) = \dot{\mathbf{U}}_0(t) + \boldsymbol{\omega}(t)(\mathbf{P} + \mathbf{N}(x,y,z)\mathbf{q}(t)) + \mathbf{R}(t)\mathbf{N}(x,y,z)\dot{\mathbf{q}}(t) \quad (12.204)$$

The kinetic energy $\mathcal{T}$ is obtained by integrating the contribution of each infinitesimal volume dv over all the volume v of the element. In matrix form:

$$\mathcal{T} = \frac{1}{2}\int_v \rho\dot{\mathbf{U}}(x,y,z,t)^T\dot{\mathbf{U}}(x,y,z,t)\,dv; \quad (12.205)$$

where superscript ^{T} indicates the matrix transpose. Taking Eq. 12.204 into account, and dropping for simplicity the indication of the dependence from position and time of the variables:

$$\begin{aligned}\mathcal{T} = \frac{1}{2}\int_v \rho\,\big(&\dot{\mathbf{q}}^T\mathbf{N}^T\mathbf{N}\dot{\mathbf{q}} + 2\dot{\mathbf{q}}^T\mathbf{N}^T\mathbf{R}^T\boldsymbol{\omega}\mathbf{N}\mathbf{q} + \mathbf{q}^T\mathbf{N}^T\boldsymbol{\omega}^T\boldsymbol{\omega}\mathbf{N}\mathbf{q}+\\ &+2\dot{\mathbf{q}}^T\mathbf{N}^T\mathbf{R}^T\boldsymbol{\omega}\mathbf{P} + 2\dot{\mathbf{q}}^T\mathbf{N}^T\mathbf{R}^T\dot{\mathbf{U}}_0 + 2\mathbf{q}^T\mathbf{N}^T\boldsymbol{\omega}^T\boldsymbol{\omega}\mathbf{P} + 2\mathbf{q}^T\mathbf{N}^T\boldsymbol{\omega}^T\dot{\mathbf{U}}_0 +\\ &+2\dot{\mathbf{U}}_0^T\boldsymbol{\omega}\mathbf{P} + \mathbf{P}^T\boldsymbol{\omega}^T\boldsymbol{\omega}\mathbf{P} + \dot{\mathbf{U}}_0^T\dot{\mathbf{U}}_0\big)\,dv\end{aligned}$$

that is of the same form of Eq. 11.57. The three terms in the first row take into account the second order contribution of the nodal speeds and displacements. The second and third row include the first and zero order contributions, if $\boldsymbol{\omega}$ and $\dot{\mathbf{U}}_0$ are constant these terms do not contribute to the Lagrange equations. Coming back to the terms in the first row, they can be written as

$$\mathcal{T} = \frac{1}{2}\dot{\mathbf{q}}^T\mathbf{m}\dot{\mathbf{q}} + \mathbf{q}^T\mathbf{n}\dot{\mathbf{q}} + \frac{1}{2}\mathbf{q}^T\mathbf{k}_i\mathbf{q}$$

where: $\mathbf{m}$ is the element mass matrix :

$$\mathbf{m} = \int_v \rho\mathbf{N}(x,y,z)^T\mathbf{N}(x,y,z)\,dv. \quad (12.206)$$

$\mathbf{n}$ is a matrix that takes the gyroscopic and Coriolis forces into account

$$\mathbf{n} = \int_v \rho\mathbf{N}(x,y,z)^T\mathbf{R}^T\boldsymbol{\omega}\mathbf{N}(x,y,z)\,dv \quad (12.207)$$

$\mathbf{k}_i$ is a non inertial contribution

$$\mathbf{k}_i = \int_v \rho \mathbf{N}(x, y, z)^T \boldsymbol{\omega}^T \boldsymbol{\omega} \mathbf{N}(x, y, z) \, dv. \tag{12.208}$$

The above matrices are obtained for a solid element under the assumption that each elementary volume has negligible moments of inertia. Plate or beam elements do not satisfy this assumption because each elementary volume is a full thickness slice of the element, with non negligible moments of inertia. The expressions in this case are more complicated but they maintain the same overall structure.

Elastic potential energy

If the material has a conservative behavior the potential energy per unit volume is given by the area under its stress-strain curve

$$\frac{d\mathcal{U}}{dvol} = \int_0^{\epsilon} \boldsymbol{\sigma}(\epsilon')^T \mathbf{d}\epsilon'; \tag{12.209}$$

where matrices $\boldsymbol{\sigma}$ and $\boldsymbol{\epsilon}$ include all components of the stress and strain tensors arranged in a column. The non dissipative assumption for the material is evidenced by the dependence of the stress $\boldsymbol{\sigma}$ by the strain $\boldsymbol{\epsilon}$ only and not, for example, from the transformation that has led the material to that strain.

Additionally, if the material has linearly elastic behavior, the stress strain curve becomes a line and the energy per unit volume becomes the area of a triangle:

$$\frac{d\mathcal{U}}{dvol} = \frac{1}{2} \boldsymbol{\sigma}(\boldsymbol{\epsilon})^T \boldsymbol{\epsilon}, \tag{12.210}$$

in this case the stress and strain are related by the Hooke's law, through the elastic coefficients matrix $\mathbf{E}$:

$$\boldsymbol{\sigma}(\boldsymbol{\epsilon}) = \mathbf{E}\boldsymbol{\epsilon}. \tag{12.211}$$

The components of the strain tensor $\boldsymbol{\epsilon}$ can be obtained from the displacement field by means of a differential operator $\mathbf{D}$:

$$\boldsymbol{\epsilon}(x, y, z, t) = \mathbf{D} \ \mathbf{u}(x, y, z, t). \tag{12.212}$$

The displacements can now be expressed as function of the nodal degrees of freedom and of the shape functions 12.200:

$$\boldsymbol{\epsilon}(x, y, z, t) = \mathbf{N}'(x, y, z)\mathbf{q}(t), \tag{12.213}$$

where

$$\mathbf{N}'(x, y, z) = \mathbf{D}\mathbf{N}(x, y, z). \tag{12.214}$$

The strain and stress of Eq. 12.211 and 12.213 can be substituted in the potential energy density of Eq. 12.210 and, after the integration in the element volume the potential energy of the element is finally:

$$\mathcal{U} = \frac{1}{2} \mathbf{q}(t) \ \mathbf{k}_e \ \mathbf{q}(t), \tag{12.215}$$

where $\mathbf{k}_e$ is the element stiffness matrix :

$$\mathbf{k}_e = \int_v \mathbf{N}'(x,y,z)^T \mathbf{E}\ \mathbf{N}'(x,y,z)\ dv. \tag{12.216}$$

Equivalent nodal forces

The Lagrangian forces that act on the element are computed from the work $\delta\mathcal{L}$ done on the element by the external forces $\boldsymbol{f}(x,y,z,t)$, with the virtual displacement field $\delta\mathbf{u}(x,y,z)$

$$\delta\mathcal{L} = \int_v \delta\mathbf{u}(x,y,z)^T\ \boldsymbol{f}(x,y,z,t)dv. \tag{12.217}$$

The virtual displacements can be considered the result of a virtual variation of the Lagrangian coordinates, i.e. of a variation of the nodal displacements $\delta\mathbf{q}$. Taking Eq. 12.200 into account, they are given by:

$$\delta\mathbf{u}(x,y,z) = \mathbf{N}(x,y,z)\delta\mathbf{q}. \tag{12.218}$$

Substituting Eq. 12.218 in the virtual work of Eq. 12.217, the virtual work is expressed as function of a variation of the nodal displacements:

$$\delta\mathcal{L} = \delta\mathbf{q}^T \int_v \mathbf{N}(x,y,z)^T\ \boldsymbol{f}(x,y,z,t)dv. \tag{12.219}$$

The Lagrangian forces $\mathbf{f}(t)$, are the nodal forces that perform with the virtual displacement $\delta\mathbf{q}$ the same virtual done by the physical forces $\boldsymbol{f}(x,y,z,t)$ with the same virtual displacement

$$\delta\mathbf{q}^T\mathbf{f}(t) = \delta\mathbf{q}^T \int_v \mathbf{N}(x,y,z)^T\ \boldsymbol{f}(x,y,z,t)dv. \tag{12.220}$$

The Lagrangian forces are then:

$$\mathbf{f}(t) = \int_v \boldsymbol{N}(x,y,z)^T\ \boldsymbol{f}(x,y,z,t)dv. \tag{12.221}$$

Equations of the motion of the element

The element equations of the motion can be obtained by means of the Lagrange's equations. The result is formally the same as Eq. 11.69:

$$\mathbf{m}\ \ddot{\mathbf{q}}(t) + \mathbf{g}\ \dot{\mathbf{q}}(t) + (\mathbf{k}_e + \mathbf{k}_i)\ \mathbf{q}(t) = \mathbf{f}(t). \tag{12.222}$$

the mass, elastic and non inertial stiffness matrices $\mathbf{m}$, $\mathbf{k}_e$, $\mathbf{k}_i$ are symmetrical and positive semi-definite, anti-symmetric matrix $\mathbf{g}$

$$\mathbf{g} = \mathbf{n} - \mathbf{n}^T,$$

takes the gyroscopic and Coriolis forces into account.

The dissipative contributions have been neglected in the above outline of an element formulation. The are usually added at the structure level, after the assembly of the element's matrices.

Equations of the motion of the structure - elements assembly

The equations that describe the behavior of the structure can be obtained from the dynamic equations of each element. The operation that leads from the matrices of each element to the matrices of the structure is indicated as the assembly. This operation is based on the following considerations:

1. Consistency: when two or more elements share the same node, the displacements of that node must be the same for all elements. The result is that the structure is described by the collection of the displacements of all the nodes included in the discretization (usually called mesh):

$$\boldsymbol{q}_s = \{\mathbf{q}_1, \ldots, \mathbf{q}_n\}^T . \tag{12.223}$$

vector $\boldsymbol{q}_s$ includes the displacements of all the nodes of the structure. It is then the vector of the Lagrangian coordinates of the whole structure. For example, if nodes i, j belong to element a and nodes j, k belong to element b, the two elements are described by the following degrees of freedom:

$$\boldsymbol{q}_a = \begin{Bmatrix} \mathbf{q}_i \\ \mathbf{q}_j \end{Bmatrix}, \qquad \boldsymbol{q}_b = \begin{Bmatrix} \mathbf{q}_j \\ \mathbf{q}_k \end{Bmatrix} \tag{12.224}$$

as node j is in common between elements a and b, the structure formed by the two elements is described by the following degrees of freedom

$$\boldsymbol{q}_s = \begin{Bmatrix} \mathbf{q}_i \\ \mathbf{q}_j \\ \mathbf{q}_k \end{Bmatrix} . \tag{12.225}$$

2. Equilibrium and energy: due to d'Alembert's principle, the resultant of all forces (elastic, inertial and external) acting on a node must be null. If two elements share the same node, the elastic and inertia forces on that node are internal forces and null each other; the resultant is then the external force. To understand the implications of the equilibrium equations on the assembly the easiest way is to think about the element's energies. As the energies are scalar quantities, the contribution of each element add together in the energy of structure:

$$\mathcal{T}_s = \frac{1}{2}\dot{\boldsymbol{q}}_s(t)\ \boldsymbol{M}\ \dot{\boldsymbol{q}}_s(t), \tag{12.226}$$

$$\mathcal{U}_s = \frac{1}{2}\boldsymbol{q}_s(t)\ \boldsymbol{K}\ \boldsymbol{q}_s(t); \tag{12.227}$$

each element matrix is added to the the mass ($\boldsymbol{M}$) and stiffness ($\boldsymbol{K}$) of the structure in the positions corresponding to the degrees of freedom of the structure for that element

For example, if $\mathbf{M}_a$ and $\mathbf{M}_b$ are the mass matrices of elements a and b of the previous example

$$\mathbf{M}_a = \begin{bmatrix} m_{aii} & m_{aij} \\ m_{aji} & m_{ajj} \end{bmatrix}, \qquad \mathbf{M}_b = \begin{bmatrix} m_{bii} & m_{bij} \\ m_{bji} & m_{bjj} \end{bmatrix}, \tag{12.228}$$

the mass of the structure is obtained by assembling that of each element, adding each nodal contribution:

$$\boldsymbol{M} = \begin{bmatrix} m_{aii} & m_{aij} & \mathbf{0} \\ m_{aji} & m_{ajj} + m_{bii} & m_{bij} \\ \mathbf{0} & m_{bji} & m_{bjj} \end{bmatrix}. \tag{12.229}$$

The assembly does not modify the structure of the equation of the motion. The structure is then governed by the same dynamic equation valid for each element:

$$\boldsymbol{M}\ \ddot{\boldsymbol{q}}_s(t) + \boldsymbol{K}\ \boldsymbol{q}_s(t) = \boldsymbol{f}_s(t). \tag{12.230}$$

12.4.3 Finite Elements Models of Car Body Structures

The aim of the following section is to provide a brief introduction to the main types of finite element models that can be realized to study the behavior of a car body structure, in order to highlight their relative advantages and drawbacks.

Detailed models

As for any kind of structure, the finite element model of a car body may be realized considering each of its parts as a three dimensional solid, and therefore meshed with prismatic or tetrahedral solid elements. Nevertheless, the geometric complexity and the small thickness that characterizes most of the parts of a car body in this case requires the use of an extremely large number of elements and nodes. Although the increase available computational power is accelerating, this type of mesh still requires long times and could even be not acceptable for the simulation of a complete car body. The complexity of the mesh in terms of nodes and elements can be dramatically reduced by considering the small thickness that characterizes most parts a car body. This allows to adopt plate elements instead of solid ones. In this case the mesh of a thin sheet component includes just one element (and node) in the thickness, instead of the two or more node layers necessary for a solid mesh. A specific issue with car body structures is that the thin sheet parts are welded together by spot welds or other forms of welding. The welded regions can be modeled by means of specific elements available in most commercial finite element codes, and the parts close to the welded regions may be modeled by means of unilateral contact elements that react only when the relative displacements between two parts tend to overlap. This type of elements must be used with some care because of the long computational time needed to solve the nonlinear contact problem introduced by these elements.

Regardless of their size, the current trend is increasingly towards adopting detailed models in all stages of car body design for the following reasons:

- The construction of these models from the CAD geometrical descriptions is now practically automatic so that it does not require a specific background of structural modeling by means of finite elements. Even if the geometric complexity of the parts leads to problems with many degrees of freedom, this is compensated by the increment of the available computational power.

- The very close geometrical correlation between a finite element model and its CAD drawing applies also to small details. A similar level of detail also applies to the materials characteristics and interface conditions between different parts. This allows a reliable prediction to be obtained not only of the displacements but also of the local state of strain and stress in the structure. It is then possible to analyze failure modes that are very sensitive to the stress concentrations such as, for example the fatigue failure or the crack propagation. Similarly, the detailed models allow a reliable prediction of the high order dynamic behavior to be obtained including the local, high frequency modes involved in the acoustic range.

Conversely, the detailed models do have some drawbacks:

- They need a detailed definition of the geometry that may not be available at an early design stage. The incomplete definition of the geometry could require a number of iterations between CAD/CAM and FEM analyses to define the final configuration. These iterations can be simplified by dedicated optimization software, in any case this requires long times and experience to properly set the boundary conditions and the objective of the optimization. Sometimes a design approach based on the carry over from previous models reduces the unknowns at the early design stage and contributes to reduce the conception phase.

- The many parameters involved by the geometric complexity sometimes conceal macroscopic issues that are evident with experience. For example the many details that define the cross section of a pillar may conceal some global properties such as the area and the moments of inertia of the cross sections. The potential risk is to reach a solution by modifying a large number of parameters losing the sensitivity on the important parameters.

Synthesis models (beams and plates)

A completely different strategy for the construction of the finite element model of a car body structure is to try to reduce to a minimum the number of parameters involved in the model. The aim could be that of simplifying the sensitivity analysis that leads to the definition of the main structural parameters. The finite element model in this case is rather close to the analytical models used to understand the properties of simple structures such as, for example, the simple structural surface approach (see section 12.3) and the analysis of truss and frames.

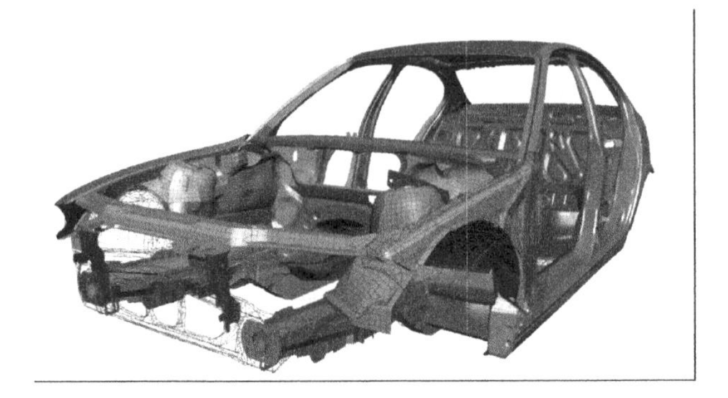

Fig. 12.64. Detailed FEM model of a car body structure. Most of the mesh is made by means of plate and shell elements.

The models of this kind are sometimes referred to a synthesis models . They allow a reasonable prediction of the deformations and of some of the stresses caused by given external loads and constraints.

The basic elements commonly used in synthesis models are (Fig. 12.65):

- Beams: longitudinal beams, cross beams and pillars can be discretized by beam elements. Each node of the mesh represents a cross section with six degrees of freedom: three displacements and three rotations in the space. The element, which connects two nodes, is characterized by macroscopic parameters of the section such as the area (A), the moments of inertia of the sections about the reference frame fixed to the element (I_1, I_2, I_{12}), the torsional moment of inertia (J_t), the location of the centre of gravity of the cross section relative to the same element reference frame, the material properties (mass density ρ, Young's modulus E, and Poisson's ratio ν). These parameters are very few compared to the large amount of information that define the geometry of the same cross section.
- Structural nodes: modeling the physical parts that connect longitudinal and cross-beams or pillars is a critical aspect of the synthesis models that deserves a careful consideration. In principle the choice is between two different options:
  - Rigid body connections: the ends of the beams that converge in a structural node share the same node of the FEM mesh. This compels the end sections of the elements that converge in that node to share the same displacements and rotations, as if the piece of structure that realizes that connection were a rigid body. The experiments show that these parts of the structure are far from being rigid bodies and their deformation accounts for a large fraction of the overall deformation. The rigid body choice then leads to a large overestimation of the overall stiffness and must be adopted with caution.

- Compliant connections: the structural node is a sort of substructure delimited by interface sections with the neighboring beams (the T shaped node between the B-pillar and the sill has three interface sections: two with the portions of the sill and one with the B-pillar). As such it is characterized by a stiffness matrix that relates the generalized forces (three forces and three torques) acting on the interface sections with the corresponding generalized displacements (three displacements and three rotations). For example, the structural node of Fig. 12.66 represents the connection between the A pillar, the longitudinal beam of the roof and the cross beam at the top of the windscreen. The stiffness of this part of structure is computed by its detailed finite element model. Each of the three interface sections has a node located at the center of gravity of the section. This node is rigidly connected (by means of a so-called "multi-point constraint" or "Rigid" element) to all nodes of the FEM mesh lying on the cross section. These interface nodes are used to connect the detailed model of the structural node to the beam elements used for meshing the rest of the structure. As the structural node has a negligible mass compared to the rest of the structure, its behavior can be accounted for by means of an 18 X 18 stiffness matrix between the three interface nodes (6 degrees of freedom each).

- Surfaces: modeled by means of plate and shell elements. A plate or shell element may have a trapezoidal shape (4 or 8 nodes) or triangular shape (3 or 6 nodes). Similarly to the beam elements, each node has 6 degrees of freedom (3 displacements and 3 rotations).

Some factors in favor of the synthesis models are:

- They can be used to evaluate global responses such as bending and torsional stiffnesses and the lower natural frequencies and mode shapes.

- The relatively small number of parameters allows to devise easily a sensitivity analysis and determine the set of parameters that match the design targets under a set of constraints.

- A change in the external surfaces can be easily dealt with by moving the nodes and connecting beams to follow the new shape. A change of the cross sectional properties or on any other parameter can be easily implemented.

- The simple structure of the model allows to study the sensitivity of each component or parameter on a given response. For example how the stiffness of each beam contributes to the overall torsional stiffness. The same can be done on other responses such as natural frequencies and mode shapes.

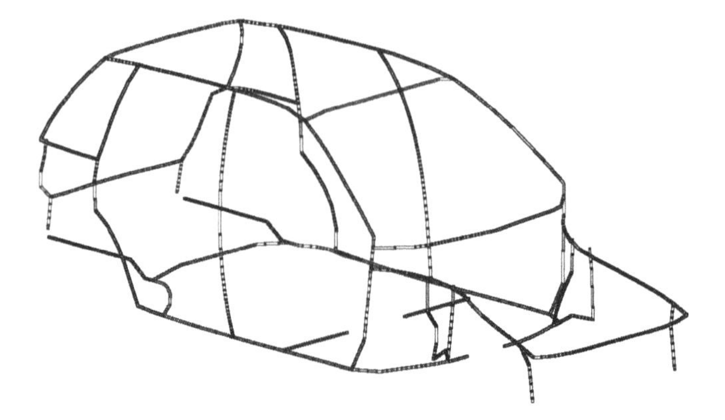

Fig. 12.65. Synthesis model of a car body structure. The pillars, longitudinal and cross beams are modeled using beam elements, the structural surfaces with plates, the structural nodes with equivalent stiffness matrices.

Instead factors against synthesis models include:

- A synthesis model cannot be used to study aspects strongly dependent on local effects such as, for example the stress and strain distribution. These effects can be investigated by means of models closer to the real geometrical configuration, so as to take local effects into account such as fillets, radii, spot welds, holes, attachment points, etc....
- The compliance of the structural nodes accounts for a large fraction of the overall deformation of the car body and therefore must be considered in the synthesis models. The proper assessment of the stiffness matrix of these parts is one of the main difficulties that affects the construction of this modeling approach. One option is to determine this information starting from the detailed FEM models of the same nodes; another is to perform experimental characterization on physical prototypes. In both cases the prerequisite is that the node is completely designed in detail. On one hand the FEM approach still allows some parametric modifications that are very difficult to perform at the experimental level. The experimental data, on the other hand, allow some of the modeling uncertainties to be solved.
- Even though synthesis models have small dimensions and can be run almost in real time, they require considerable expertise. The choices at their base are difficult to automize and the risk of obtaining unreliable results is high. Similarly to the analytical methods, their use is now limited to preliminary design phases.

Construction of a synthesis model

During the first design phases, the available information regarding the new structure may be very little. Some data concern system level parameters such as track

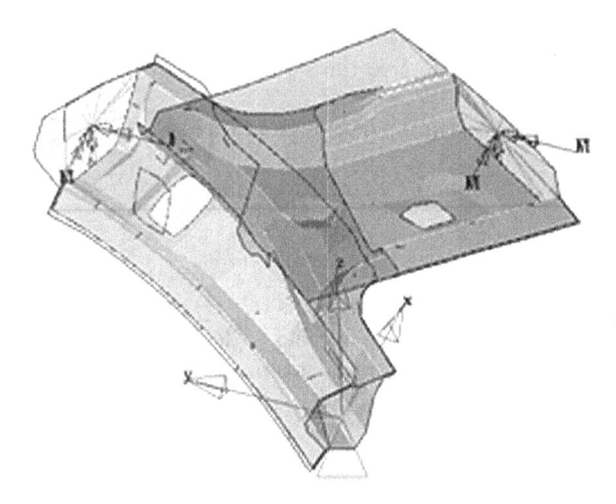

Fig. 12.66. Detail model of a structural node for the evaluation of its equivalent stiffness. All nodes on each interface cross section are connected to a single node located in the centre of gravity of the section by means of rigid body constraints (RBE elements, Rigid elements). These nodes are then used to connect to the beam elements used to mesh the rest of the structure.

and wheelbase, suspension architecture, carry over components (for example the lower frame, suspensions and powertrain). The preliminary versions of the inner and outer surfaces may have been defined by style and ergonomics along with the general shape of the apertures (doors, hood, trunk) and of the interiors. Such informations is made available in numerical form with a CAD/CAS (Computer Aided Styling) software.

The data that should be available before starting the design of the new structure by means of a synthesis model are:

- outer surfaces defined by means of a CAS or CAD software;
- detailed FEM models of the carry-over subsystems;
- sketches of the main cross sections as defined from benchmarking with previous models or competitors.

The main steps that lead to the construction of a synthesis model are:

- define the relative position of the carry over parts relative to the available surfaces;
- modify the carry-over parts considering the new track and wheelbase;
- locate the structural nodes and draft their shape;
- locate the axis of the beams between the structural nodes;

- decide the main cross sections parameters such as area and moments of inertia and attribute them to the structural members;
- locate the main structural surfaces (roof, firewall, floor, side panels);
- apply appropriate boundary conditions and loads.

Plate and shell models for design optimization

The availability of parametric drawing tools (CAD) on one hand, and of meshing software on the other, allows a large number of design iterations to be performed automatically. This enables the optimization of the structure directly on the plates and shells models, reducing the role of the synthesis models based on beams and plates.

The process is based on the definition of a set of design variables in given intervals. For each data set, the CAD software generates a drawing of the structure that is then automatically meshed and analyzed under the desired load cases to find the structural responses that must be optimized. The flexibility of the CAD software allow to explore virtually all parameters that define the drawing of the structure, for example the thickness of the cross section of a longitudinal beam, to the size of the section and its geometry (fillets, aspect ratios,...), to the location and shape of its axis.

The iterations can be manual (seldom) or automatic. The optimization process is similar to that performed on a synthesis model, but some of its limitations are avoided. The iterations are based on detailed models, critical aspects such as the unknowns related to the stiffness of the structural nodes, are solved from the beginning. It is also possible to optimize the structure considering responses that are out of the field of validity of beam and plates synthesis models, such as the stress and strain distributions, the reliability under fatigue loads, the performance during crash.

Since the iterations are performed directly on a detailed model, the required number of choices and involved assumptions is smaller, the experience of the analyst should play a less important role in obtaining reliable results.

The risk of this approach is related to the explosion of the number of design parameters that masks in some case a much more smaller number of more aggregated data (typical example: the torsional moment of inertia J_t of a cross section instead of the number of geometric dimensions that define the same cross section).

Structural optimization

The main objective of the structural optimization is to improve the design by reaching a better compromise between the performances on one side and the cost (in a broad sense, including production costs, maintenance, weight,...) on the other. This compromise means that an optimization process broadly involves two factors:

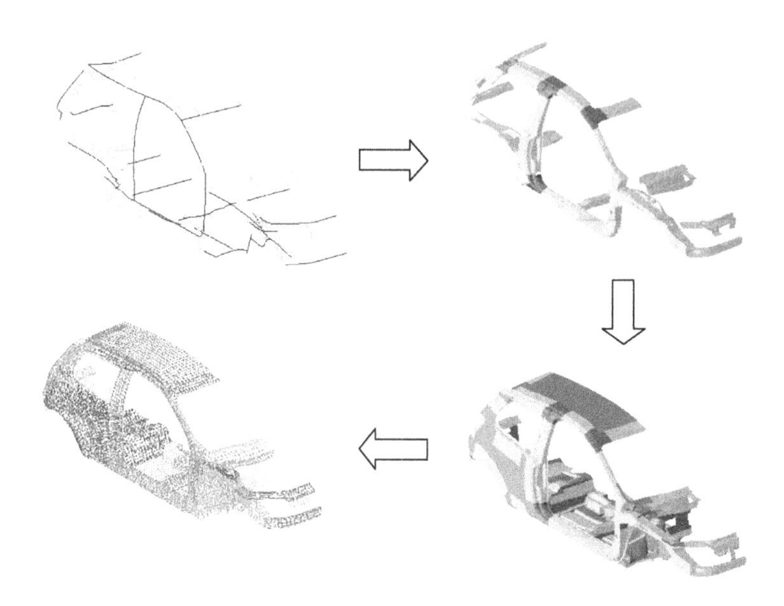

Fig. 12.67. FEM models of increasing detail.

1. the function that must be optimized, this could involve one or more parameters such as the stiffness, the behavior during crash, the natural frequencies, the weight. Since the number of parameters that must be optimized could be large, it is necessary to define a function that mixes together all of them (design objective). In the simplest form this function could be the weighted sum of all indexes. The optimization will try to optimize this function.

2. The design constraints, i.e. the boundaries that can not be exceeded. These boundaries could be set on a design variable (for example the size of a cross beam must be smaller than a given value) or on a structural response (for example a given limit stress).

It is almost standard to distinguish between two types of optimization: topologic and topographic (or shape optimization).

The aim of the topologic optimization is to identify the shape to give to a structural element in given boundaries to optimize a given objective. For example the shape to give to an engine suspension bracket in a given volume. The aim of the topographic optimization is to decide what values to give to a set of design parameters, for example: the thickness of a plate beam or the size of a rectangular cross section.

The optimization codes may be integrated in finite element softwares (this is typical of topologic optimization) or they can manage automatically multiple runs of the sequence that goes from the definition of a CAD model to its finite elements mesh and to the computation of the relevant structural responses (topographic optimization).

The structural optimization is applied in the automotive field to the design of relatively small parts to the complete car body structure.

As in all multivariable optimization algorithms, one of the key problems of structural optimization is that the functional to be optimized may have local minima. The risk is to converge on one of them not considering a more favorable point in the design space.

In principle the optimization of a structure could be performed by analytical means, provided that the response of a structure can be linked to a given set of variables by analytical means. This can be done for simple structures with reasonably simple boundary conditions such as beams and plates. In automotive applications the possibility of determining the structural response analytically is limited to very simple and idealized structures (for example frames or structures that can be modeled by the structural surfaces method). In most practical cases the geometry, load and boundary conditions are so complex that the solution can only be found numerically (using the FEM, for example). The optimization in this cases can only be performed numerically. A number of commercial softwares are currently available, sometimes integrated with the FEM (Altair Optistruct, MSC Patran/Nastran, Ansys, etc.), sometimes are software tools that perform only the optimization task by managing all steps that are individually run by specific software: from the generation of the geometric model with given data set (CAD software), to the mesh and the FEM analysis (FEM software).

Topologic optimization

The topologic optimization can be a powerful tool in the preliminary design phase when it is necessary to define the shape to give to a structure for a given set of loads, constraints and allowable design space.

The starting point is the definition of a design space, of the loads and the constraints or the interfaces with other components. The design space is virtually filled of a homogeneous material that is meshed in finite elements. The FEM analysis of this structure allows to find the regions of the design space with less importance on the structural response. For example, if the variable that must be optimized is the stiffness, the regions of small importance are those with low stresses. The material properties (Young modulus, mass density) in such regions are decreased, this operation is usually indicated as density reduction. Density equal to 1 indicates the material with nominal characteristics, null density indicates a material with negligible mechanical properties.

The finite element analysis is then repeated on the volume with the new material property until the density distribution converges.

The result is the same design space defined at the beginning but with a non homogeneous density. The density map indicates where it is convenient to eliminate the material and where it is efficient to leave it. Usually this operation is performed by plotting only those elements characterized by a material density higher than a given threshold.

The result is a shape with irregular contours because the surface of the structure is now formed by the element corners that were originally connected to deleted elements.

The obtained geometry is the base for the design of the component, that can be verified by a conventional finite elements analysis. In some case it is possible to perform a new optimization on a smaller design space.

Topographic optimization

The goal of the topographical optimization is to optimize the geometrical parameters of a given design. For example the size and depth of swages in a panel. In the simplest form these elements can be obtained by moving the nodes of a predefined mesh, alternatively they exploit the iteration of a parametric CAD with a FEM. A CAD model with a new parameters set is produced at each step, meshed, and analyzed with the FEM to find the structural response.

The main difference with the topologic optimization is that in this case the design variables are geometric parameters of a given shape. For example the goal could be to find the profile of the floor panel to optimize its stiffness and natural frequencies. The optimization software in this case modifies the profile trying to find where the introduction of a swage is more effective. Also in this case the optimization is constrained by design limits (the swages cannot be higher than a given value or they cannot be introduced in some areas).

12.5 Measurement of the Car Body Stiffness

During the operational life of a vehicle, the combinations of the loads acting on it may be very complicated. Apart from crashes and cases of misuse, most of these loads do not involve permanent deformations in the structure, thus they can be considered as the linear combination of a rather limited number of load cases amongst which the torsional and the flexural loads are perhaps the most significant.

The analysis of the torsional and flexural deflection of a the vehicle chassis can be found with different methods ranging from the numerical models (FEM) to experimental tests. The analysis of the deformed shapes enables the overall deflections to be determined together with how each structural element contributes to the deformation. It is then possible to identify the elements or the structural portions affected by higher deformations, which could be redesigned not only to increase the overall stiffness, but also to increase the reliability of the structure subject to the fatigue loads during operation.

This analysis can be repeated at different assembly stages of the vehicle: from the body in white to the fully assembled car complete with powertrain,

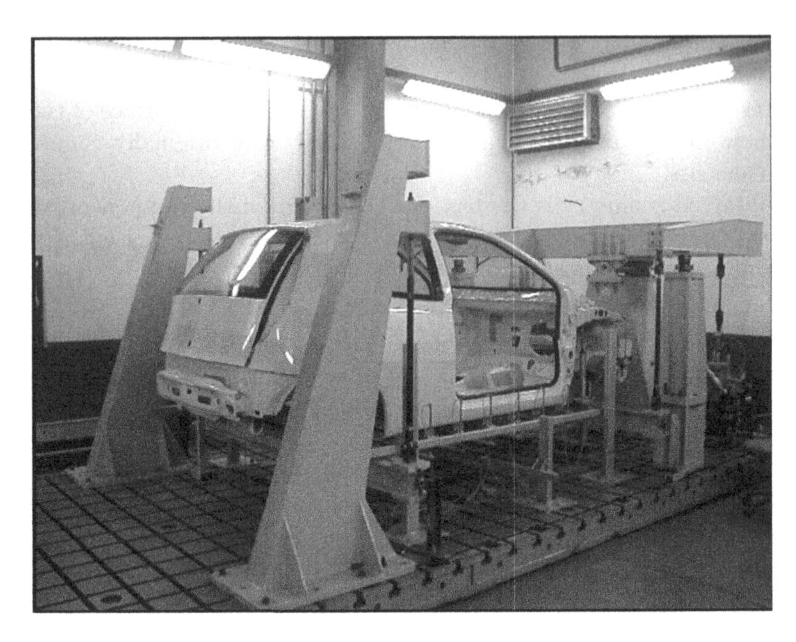

Fig. 12.68. Test bench for the measurement of the torsional stiffness.

suspensions and interiors. This allows to understand the structural contribution of elements such as the windscreen, the dashboard, the seats and the outer body panels.

The torsional and flexural load cases are usually standardized by each manufacturer in terms of test setup, load application and constraints. The torsional test replicates the load acting on the vehicle when parked with just one wheel on a sidewalk, whereas the flexural one replicates the load due to passengers, luggage and vertical accelerations.

The experimental setup used for the two load cases is almost the same so that the torsional and bending tests can be performed on the same car structure without requiring a significative changes both on the car and on the test bench.

12.5.1 Test Setup

The test setup (Fig. 12.68 and Fig. 12.69) for the measurement of the torsional and flexural stiffness includes a very stiff (relative to the car body) base structure with pillars that can be moved in different locations. These pillars are used to support the car body structure and apply the loads.

12.5.2 Vehicle Setup

Usually the test is performed at the body in white assembly stage, without the interiors (dashboard, seats), the powertrain (engine and transmission, exhaust),

Fig. 12.69. Test bench for the measurement of the bending stiffness.

the electrical and hydraulic connections, the bumpers and the wheels. Other tests are repeated at different stages of assembly to understand the contribution of "non-structural" elements such as the windscreen, the dashboard and movable parts. The suspensions are installed but all compliant elements (springs, rubber bushings) are substituted with stiff connections. Rigid struts are installed, for example, instead of the shock absorbers. The underlying idea is that the suspension must be present to transmit to the body structure the loads acting on the wheel in the same way they are transmitted during normal operation. All compliant elements are made to be stiff so as to avoid rigid body motions of the car body during the tests. When substituting the suspension components with stiff elements, it is important to avoid the introduction of additional constraints that would lead to a fictitious increase of the stiffness. Double wishbones, Mc Pherson and multi link suspensions should then be connected preserving their isostatic nature. Less evident choices are necessary for hyperstatic connections such as those between suspensions subframes and the body structure.

Just one test methodology will be described in the following because indicative of that adopted by different manufacturers.

The wheel hubs are rigidly connected to two, very stiff, cross beams connected at their ends to the test bench pillars. The connection is made using vertical rods. The vehicle body and the two cross beams constitute the test article that, constrained by at least three vertical rods, still have three rigid body degrees of freedom (motions along x and y directions and rotation about the z axis). These degrees of freedom are constrained in different ways, for example by constraining the x and y displacement of one hub (in addition to the z constraint given by the vertical rods) and the x or y displacement of another hub.

12.5.3 Bending Load

All four ends of the cross beams connected to the hubs are connected to the test bench pillars using vertical rods. Even though hyperstatic, this constraint allows the body to be hinged about two transverse axes corresponding to the axles. The vertical load is usually applied at the mid-wheelbase by means of weights applied on the sills by means of a cross beam with rubber or wood blocks in between to distribute evenly the load and avoid stress concentrations.

12.5.4 Torsional Load

The rear cross beam connected to the hubs is constrained to avoid all displacements on one end and just the vertical ones on the other. This allows this beam to rotate about the y axis. The front cross beam is constrained on one end to avoid the vertical displacement and one other displacement (either x or y). The other end of this cross beam is left free allowing its rotation about the x axis. The external load is applied by means of weights to the free end of the front cross beam.

12.5.5 Stiffness Measurement

The car body is instrumented with at least four displacement sensors that measure the vertical displacement of the four wheel hubs. Additional sensors can be installed at the attachment points of the suspensions, i.e. at the front suspension strut towers and on the longitudinal beams at points corresponding to the rear axle (Fig. 12.70), and along the sills (Fig. 12.71).

The torsional stiffness is defined as the ratio between the torque load M_t and the relative rotation $\Delta\theta$ between the front and the rear axle, evaluated at points corresponding to the wheel hubs:

$$K_t = \frac{M_t}{\Delta\theta} = \frac{M_t}{\theta_f - \theta_r} \tag{12.231}$$

where θ_f and θ_r are the rotations of the front and rear axle, respectively. To compare the torsional stiffness of vehicles with different size, it is usual to compute the ratio between the torsional stiffness and one of the characteristic lengths, the wheel base or the overall length. Another index that describes the quality of the structural design is the ratio between the torsional stiffness and the mass of the chassis.

The bending stiffness can be computed as the ratio between the applied vertical force F and the displacement Δz of the section where it is applied relative to the axles:

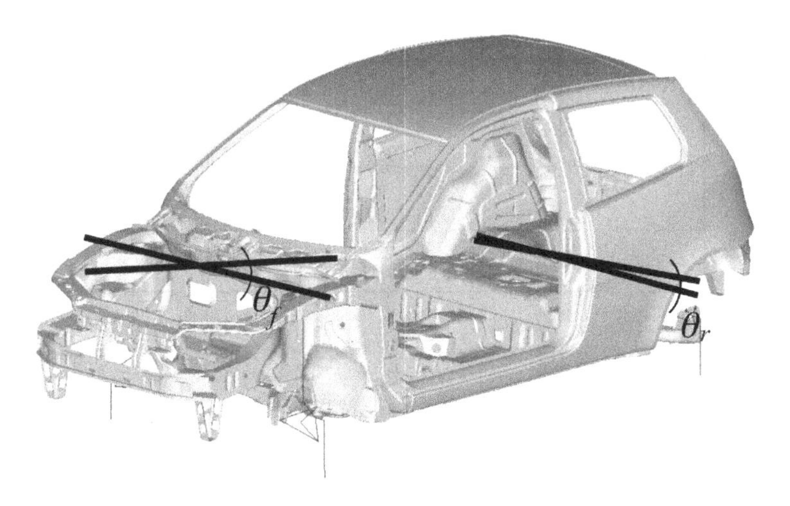

Fig. 12.70. Measurement of the torsional stiffness.

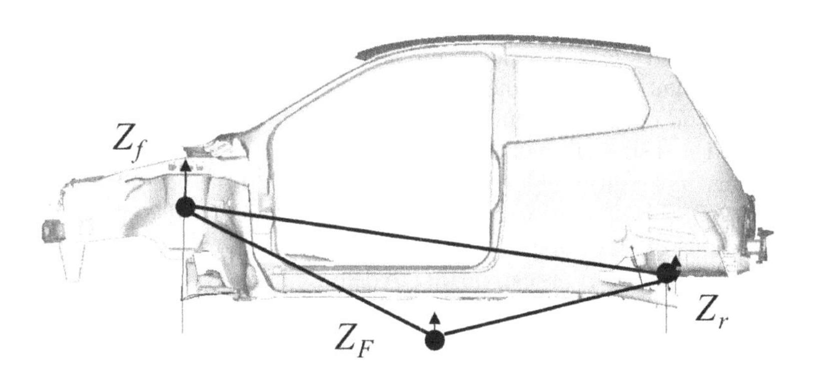

Fig. 12.71. Measurement of the bending stiffness.

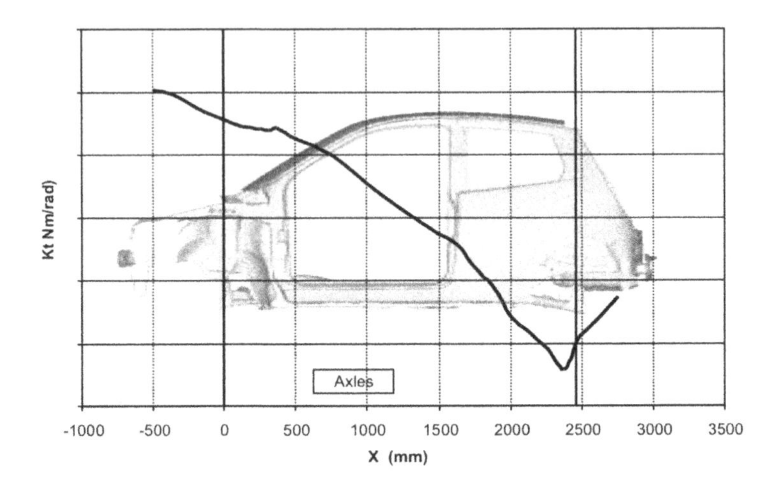

Fig. 12.72. Distribution of the torsional deformation along the longitudinal axis of the vehicle.

$$K_f = \frac{F}{\Delta z} = \frac{F}{z_F - (z_f b + z_r a) / w_b} \tag{12.232}$$

where z_F is the vertical displacement of the cross beam used to apply the vertical force, z_f and z_r are the vertical displacements of the front and rear axles, a is the longitudinal distance between the front axle and the vertical force F, b is the longitudinal distance between the force F and the rear axle. The displacements of the front and the rear axles are considered in the previous expression to eliminate the contribution of the test bench compliance.

The installation of displacement sensors along the sills or other longitudinal beams (front and rear longitudinal beams, underfloor longitudinal beams) allows to determine the distribution of the elastic deflection. These displacements can be plotted along the longitudinal axis of the vehicle to find the portions of structure that give a larger contribution to the deformation (Fig. 12.72) and Fig. 12.73). Local variations of the slope of these curves indicate portions of the structure with larger compliance. This can be due to defects during the production process or structural design problems. Sometimes discontinuities in the deflection plots may be due to local structural problems such as warping or local deformation of some of the measurement points.

The potentially large stresses that can be transmitted to the glass surfaces (bonded glasses such as windscreen and some lateral glasses) justify the installation of dedicated sensors for the measurement of the distortion of the frames where these glasses are bond. The simpler solution in this case is to install high resolution displacement sensors on the diagonals of such frames. Similarly, also the frames that host movable elements such as doors, hood, trunk door can be

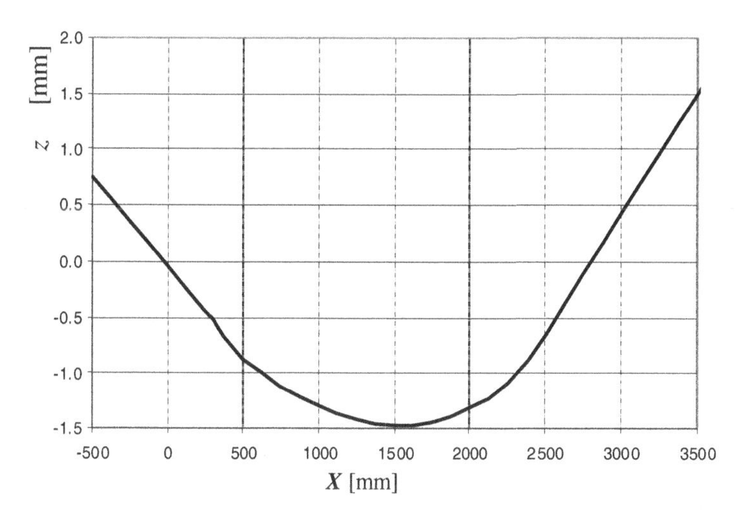

Fig. 12.73. Distribution of the bending deformation along the longitudinal axis of the vehicle.

instrumented to measure the distortions caused by torsional and bending loads. The aim in this case is to avoid excessive deformation that could cause noise (squeaks), leakage of the seals, and malfunctioning of the lock and hinge system. Usually such deformations are subject to internal design requirements specific of each manufacturer.

The large structural contribution that can be provided by glasses and other elements added to the body in white structure justifies the measurement of the deformations at various assembly stages. Starting from the body in white stage the stiffness increases by adding glasses interiors and doors, the larger contribution being that of the glasses, as shown in Tab. 12.3.

Table 12.3. Torsional and bending stiffness at different assembly stages. BIW: body in white, FG: fixed glasses; I: interiors, doors, dashboard

	Torsional [kNm/rad]			Bending [Nm/rad]		
Vehicle model	BIW	BIW + FG	BIW+ FG+I	BIW	BIW + FG	BIW +FG+I
Fiat Punto MY94 5D	573	701	796	630	640	670
Renault Clio MY90 5P	540	740	770			
Fiat Uno MY90 5D	342	404	478	430	445	475
Ford Fiesta MY89 5D	420	515	655	350	355	385
Autobianchi Y10 MY85	445	556	678	635	675	760
Citroen Ax MY87	455	635	690	455	500	570
Peugeot 205 3D	390	500	588	320	355	385
Fiat Punto MY94 3D	578	728	834	595	620	630
Fiat Uno MY90 3D	336	423	486	445	460	480
Peugeot 106 3D	567	730	820	570	590	640
Opel Corsa 3D	410	540	690	490	500	510
Nissan Micra 5D	300	370	510	480	490	
VW Polo MY94	550	660	790	410	420	
VW Polo MY82 3D	360	380	435	490	500	525
Renault 5 3D	340	480	530	425	450	475
Fiat Palio HB 5D	428	579	683	580	600	620
Lancia Y	583	735	835	680	690	720

13
Passive Safety

13.1 Biomechanics

13.1.1 Biomechanical Approach

The safety concept that will be developed in this chapter takes into consideration the mitigation of injuries to the occupants of a vehicle that my result from a collision against an external obstacle or another vehicle. In this way this safety concept assumes both the possibility that a collision may occur and the probability of injuries arising due to this collision.

Accepting this assumption implies that the safety concept is associated with probability. Thus, rather than pursuing the design of an absolutely safe vehicle, the approach taken is to consider vehicle safety issues regarding certain types of accidental events. Moreover, with regards to these different accidental events, the approach is not to seek damage avoidance but instead a reasonable mitigation of damage to the vehicle and, above all, injury to its occupants.

Originally the passive safety of a car was associated with the maintenance of cockpit integrity. In this way the car body was considered to be a shell which can protect the occupants but which can deform, in defined situations of static and dynamic loads, within certain geometric limits. In this way, the car body can offer a good degree of protection for the occupants in case of impact.

This approach to safety is here identified as the *geometric approach*. Due to this approach, a series of prescriptions were created, some of which are still in force today. For example, during a front impact test on a car, the backwards movement of the steering wheel must be less than 12.7 cm, while, in the case of side impacts, intrusion must not exceed specific geometric values.

L. Morello et al.: The Automotive Body, Vol. 2: System Design, MES, pp. 463–558.
springerlink.com

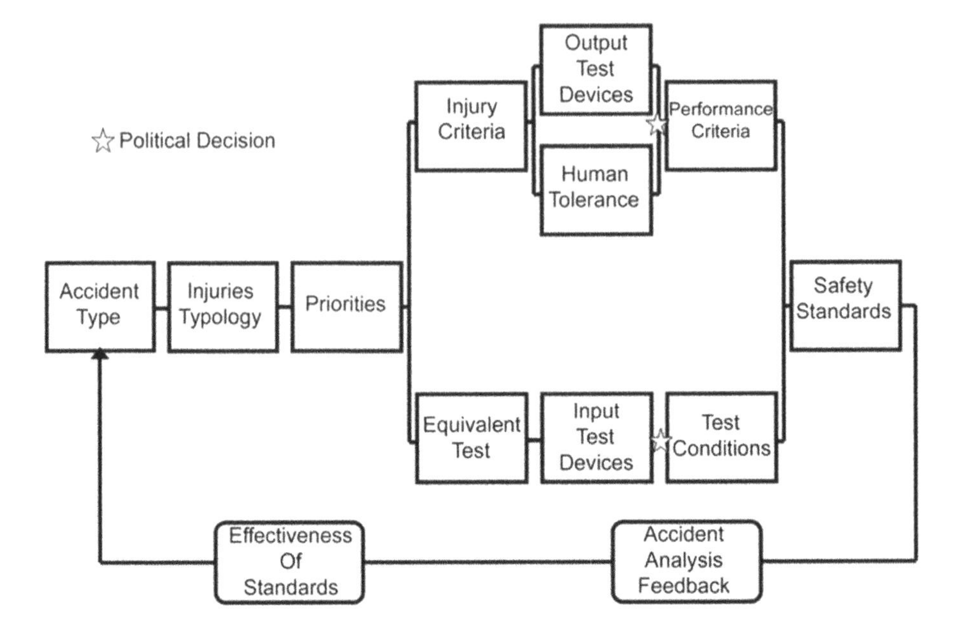

Fig. 13.1. Schematic representation of the biomechanical approach.

Several years later it became apparent that this approach is incomplete since it is not linked with the human tolerance limits as a consequence of trauma due to accidents. Thus the need became evident to evaluate the safety of a car with *biomechanical criteria* requiring the need to verify, in case of accident, that the stress suffered by the occupants are lower than human tolerance limit. This became known as the so-called *biomechanical approach*. The logical scheme to define the safety *standards* according to this approach is multi-disciplinary, as shown in Fig. 13.1.

Before examining this representation in more detail, it is necessary to remember that, according to European law, during an accident, all the occupants of a car are exposed to the level of risk related to that type and severity of impact. However only those who correctly use the restraint systems provided will benefit from rules imposed by this regulation. Moreover, it is necessary to clarify the meaning of the following terms that are shown in Fig. 13.1.

Lesion: in the case of an accident, lesion is a physiological change arising from a mechanical stress. The detection of lesions and the attribution of different severity limits concerns the first aid doctors.

Injury criterion: this is a physical parameter (acceleration, force, deformation, etc.) which establishes the correlation between the extent of damage to the human body, in particular with respect to the severity of the lesion for the body segment of interest. If, for example, for a belted passenger involved in a frontal impact, the severity of thoracic lesion (rib fractures) increases with crushing, the crushing of the thorax is an injury criterion for the thorax. To identify injury criteria and attribute human tolerance levels, it is necessary to make tests with corpses. Thus this type of activity is mainly done within Forensic Medicine Institutes.

Performance criterion: this is the value of the injury criterion which is obtained from the biomechanical process and must not be exceeded. It is measured with a dummy positioned inside a car in an impact test. The selection of performance criteria concerns the legislator.

The biomechanical approach combines three typologies of activities and knowledge which contribute to the formulation of a regulation relating to biomechanics specifications:

- accident analysis which must define the priority of intervention and determine the efficiency of the specification;
- the definition of test conditions, in order that the car body is submitted to the same stress levels as in a real accident
- the definition of performance/protection criteria which are expressed in biomechanics terms.

At this stage it is appropriate to examined each of these three different aspects in greater detail.

Accident analysis

The detailed analysis of an accident, which is conducted by multi-disciplinary team, has to establish the fundamental information to orientate the studies. In particular, it is necessary to identify the types of priority accident, their severity, and the subdivision for body segment. The results of this analysis are summarized in Fig. 13.2

It is necessary to determine:

- in which types of accident injuries or deaths occur (classified, for example, in terms of frontal, side, rear impacts or rollover, etc.);
- which parameters characterize the severity of the impact (eg. vehicle speed, damage to the vehicle, etc.);
- how the occupants are injured (for example the body segment, and the cause of the injuries, etc.).

The accident analysis must provide the priority of action and verify, over time, the validity of the regulation issued.

Test conditions

The definition of an equivalent test with respect to the real accident under consideration is fundamental. Usually an equivalent test involves the definition of an

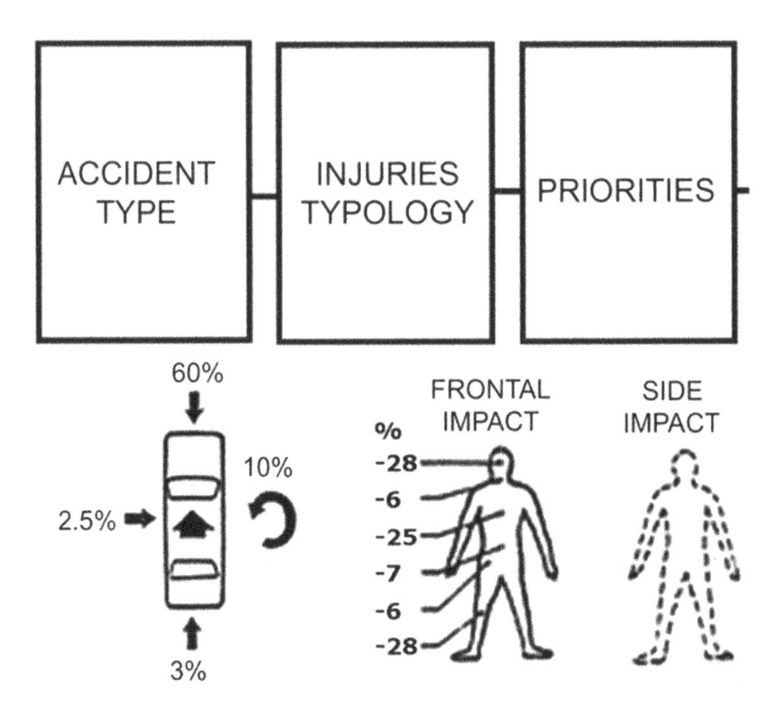

Fig. 13.2. Schematic representation of accident analysis.

impactor or an obstacle, and the definition of impact zones and angles. In this way it is posible to define the test instrument for the vehicle or the component, the test conditions, etc. as shown in Fig. 13.3.

The determination of test condition severity (speed of impact, barrier mass, etc.) is the responsibility of the legislator who must decide on the limits required to adequately protect the occupants.

Performance criteria

When the body segment which must be protected has been defined for each type of accident, it is necessary to determine for each body segment measurable physical parameters (IC = injury criterion) which are linked to the injury severity (IS = injury severity). For example, if increasing the bending of the chest increases the number of ribs broken, then this can be an injury criterion for the chest, as shown schematically in Fig. 13.4.

Once the injury criterion has been defined, it is necessary to proceed to determine the maximum values that can be applied on the human body (H = Human tolerance level). These values have to be correlated with an injury scale (AIS = Abbreviated Injury Scale).

In parallel it is necessary to develop a conformity anthropomorphic instrument which can measure the injury criterion during the impact.

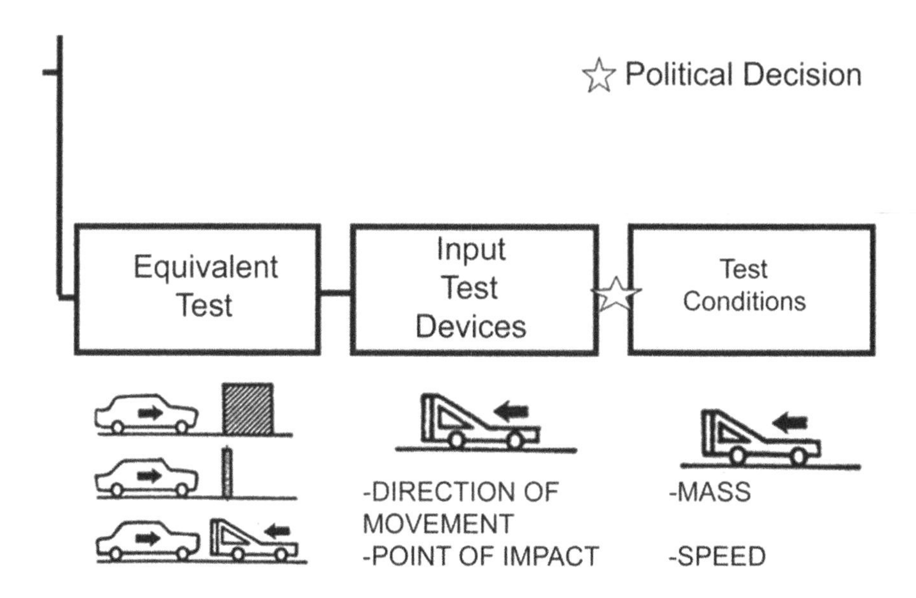

Fig. 13.3. Test conditions.

These instruments may be complex (eg. dummies, which are made by different body segments with the appropriate instrumentation in order to measure the different injury criteria such as, for example, chest crushing) or relatively simple (eg. an impactor with the anthropomorphic characteristics of head used to verify the energy dissipation of the dashboard).

These are typical measurement instruments used to evaluate the passive safety degree of a car under examination. It is important to highlight that the definition of acceptable values, the performance criteria, explained in Fig. 13.4, are the responsibility of the legislator having established which of those exposed to risk the legislation intends to protect (for example which percentile of people is included into the regulations) and which injury levels can be deemed acceptable. These two aspects are summarized with the ID = Injury Distribution for the different test conditions considered.

Safety standards

The output of the biomechanical approach comprises the test conditions and the performance criteria, used to define the safety standards which are the regulations in matters of passive safety.

Before concluding this introduction on biomechanics and examination of the injury criteria for the different body sectors, it is necessary to address the injury mechanism and injury scale.

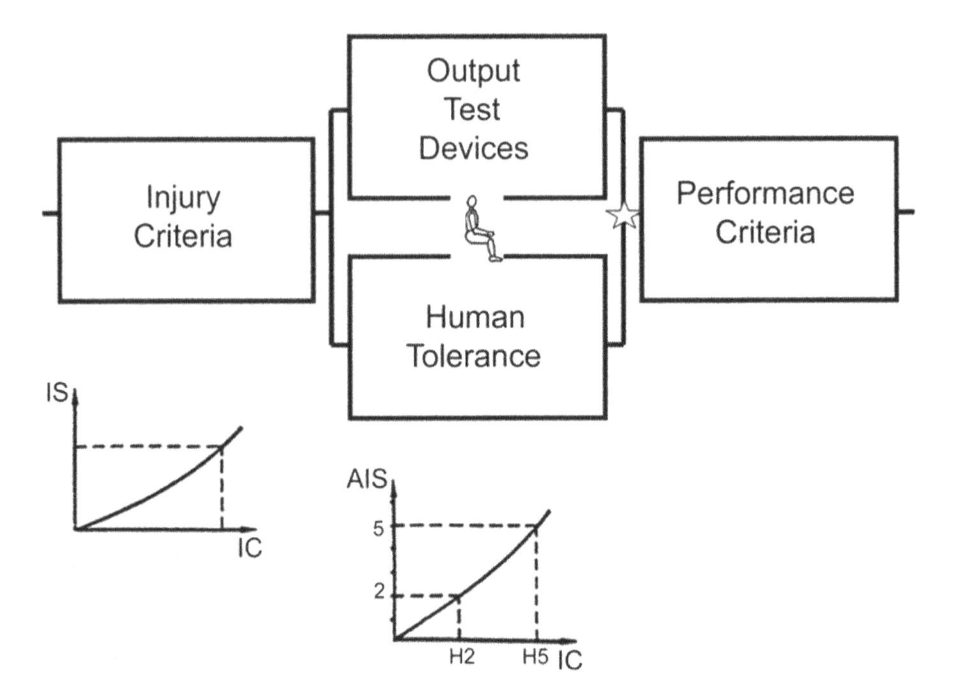

Fig. 13.4. Performance/protection criteria.

Injury mechanism

The logic of the biomechanical approach assumes that, for each body segment, an absolute injury criterion can be determined independently of how the injury occurs, as shown in Fig. 13.5. This is not possible when the injuries can be created by different mechanisms; a significant example will be examined subsequently which concerns injury to the neck.

Today it has been understood that for a body segment it is necessary to determine as many criteria as the causes of injury. The interaction between these mechanisms will determine different values for each single injury criterion. How to identify the consequences of interactions between different injury mechanism is one of the main issues currently being addressed in biomechanics.

Injury scale

The injury scale can be divided into three main types.

- *Anatomic scales*: these are applicable to injuries which result in a noticed anatomic alteration and describe the injuries in term of anatomic collocation, injury type, and severity. Anatomic scales classify the injuries themselves and not the results of injuries, the most common being the Abbreviated Injury Scale (AIS).

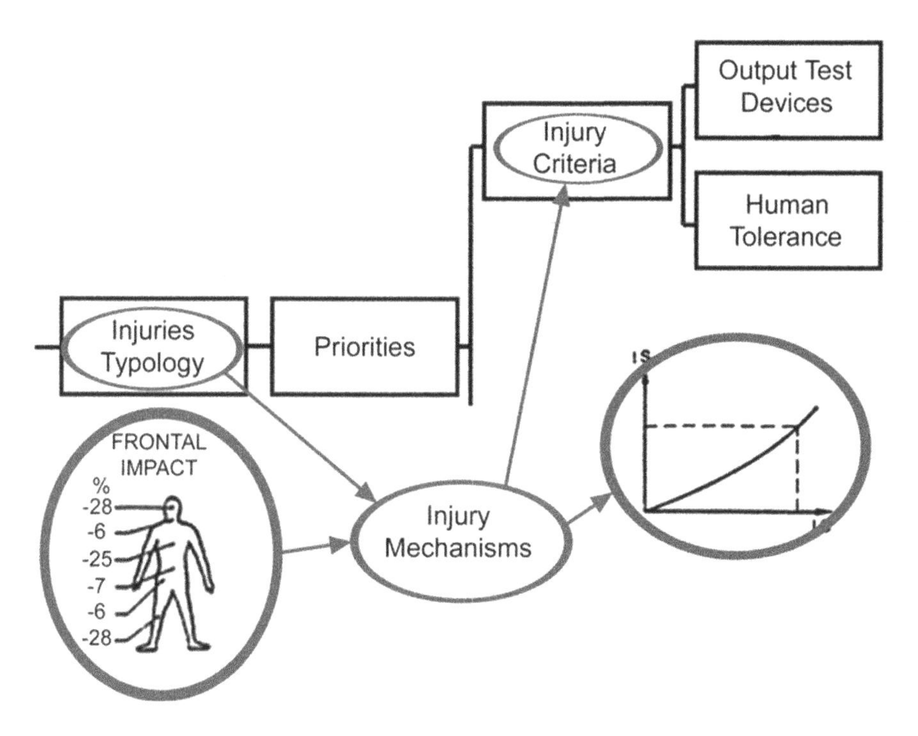

Fig. 13.5. Injury mechanisms.

- *Physiological scales*: these describe the physiological state of the injured person, basing on the functional changes due to the injury. This state, and thus the values of injury severity, changes during the treatment and thus differs from the anatomic scale where a single injury level is given to a particular injury. An example of physiological scale is the Glasgow Coma Scale, which classifies the injury severity to the brain based on three indicators: eye opening, motor response and verbal response.

- *Scales for particular severity criteria*: these scales do not classify the severity of the injury itself but are oriented towards the social costs or long term consequences. As an example of such scales, the Injury Impartment Scale (IIS) assigns to each injury, already classified with the AIS scale, a probable level of functionality loss of the affected body.

The physiological and severity injury scales can be applied only to living persons. The anatomic scales, as the AIS, are the only scales that can be used to determine the injury severity of corpses and thus, the definition of injury criteria for each body segment.

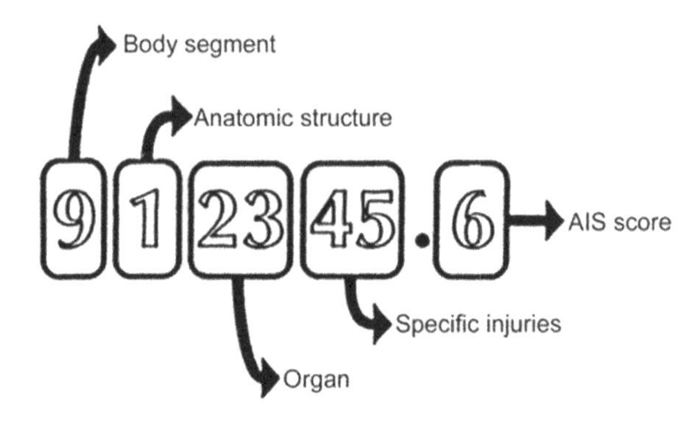

Fig. 13.6. AIS code.

Abbreviated Injury Scale

The AIS scale identifies an injury with a numeric code comprising seven numbers. The last number is the entity of lesion for each body segment which is affected by the impact. This last number has six degrees of classification (from 1 to 6). Level 1 denotes an injury just beyond the threshold of pain, such for example, a contusion or a hematoma, while level 6 indicates a mortal injury; level 5 indicates a injury with a 50% probability of being fatal; 2, 3 and 4 correspond to progressive intermediate levels of injury. The Fig. 13.6 shows the composition of AIS code.

The use of AIS code represents a universal method used in the automotive field. It is able to provide objective and repeatable evaluations for the traumatic consequences of vehicle occupants in case of real impact, and can be used for a single injury.

If there is not a documented anatomic lesion, but there is information regarding the loss of consciousness, the AIS value is defined as the time that the injured person is unconscious:

less than 1 h	AIS 2,
between 1 and 6 h	AIS 3,
between 6 and 24 h	AIS 4,
over 24 h	AIS 5.

Injury Severity Score (ISS)

The ISS is a global severity index for persons with many injuries. It is based on the evaluation of single injuries with the AIS scale and is defined as the sum of squares for the highest AIS scoring in the most affected three body region. Conventionally the value of the sum equal to 75 corresponds a fatal injury:

$$ISS = AIS_1^2 + AIS_2^2 + AIS_3^2.$$

Evidently, if a lesion with AIS equal to 6 has occurred, the ISS is set automatically to 75.

13.1.2 Injury Criteria

In this section the injury criteria adopted for the different body segment by the European and United States regulations will be discussed.

Head

The HIC (Head Injury Criterion) is the injury criterion universally known for the head. This indicator is known also as HPC (Head Performance Criterion, a definition used by the regulations for pedestrian impact cases). It expresses an injury criterion based on the resultant acceleration on the centre of gravity of the head.

The expression of HIC is:

$$HIC_{\max} = \left\{ \left[\frac{1}{t_2 - t_1} \int_{t_1}^{t_2} a(t) \right]^{2,5} (t_2 - t_1) \right\} < 1.000,$$

where $a(t)$ is the temporal progress of resultant acceleration, expressed in [g], while $t1$ e $t2$ (with $t2 > t1$) are two instants which define the width of a temporal window along the time interval encompassing the acceleration impulse measured on the head.

On the $a(t)$ curve, a mobile temporal window with variable width is applied (along the time axis). For each window configuration, a correspondent HIC value is obtained: the maximum values obtained in this way are the HIC. It is compared with the limit value of 1000, that is the human tolerance value (AIS<3) and the performance criterion.

Moreover it is important to specify that the integral contained in the definition of HIC yields an average value over the $t2 - t1$ window: In this way the influence of peaks is reduced.

At this point, in order to better understand the limit value of HIC, it is appropriate to introduce a short historical note on the HIC.

The original studies on the HIC were conducted in the 1930 at Wayne State University in Detroit. In these first investigations, the response of the skull during impact against a rigid wall was evaluated (ie. resulting in a high level of acceleration for very short time). In particular, to represent the injury level for the head, the presence or absence of linear fracture in the skull were evaluated. This fracture was used as damage indicator and as index of the risk to life. With these first data, the first part of the curve shown in Fig. 13.7, demonstrating a typical hyperbolic trend, were obtained.

The first results were added to over time with data from experimental tests performed in the 1950s on animals and volunteers (with lower levels of acceleration over a longer time interval). With these tests it was possible to obtain a series of points on the second part of the curve shown in Fig. 13.7.

The points that constitute this curve, when plotted on a semi logarithmic graphs, correspond quite well to a straight line. Making the best fit of these points, it is possible to obtain a very simple formulation:

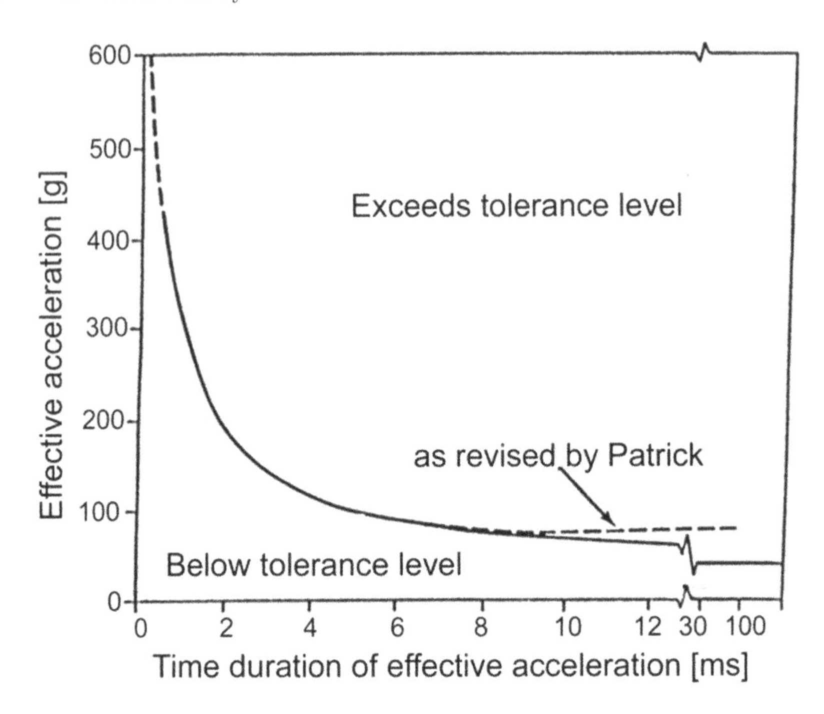

Fig. 13.7. Injury criterion for the head.

$$a^{2.5}t = 1.000.$$

In summary, the limit value of HIC = 1000 is tied to a straight line equation that represents the trend line of the border zone between the absence and presence of linear fracture of the skull. This straight line is positioned on a logarithmic diagram with time on the abscissa and acceleration on the ordinate. This was obtained with experimental test mentioned previously.

So it is important to bear in mind that the HIC has meaning only when an impact with the head and interior components or an air-bag occurs.

Currently, for the calculation of the HIC, mobile windows with width of $t2 - t1 \leq 36ms$ are used. The maximum value of this indicator is obtained in these ranges of amplitude (with time interval typical of impact head phenomenon).

A practical interpretation of the 36 ms limit can be obtained noting that if the acceleration is assumed constant for the whole time interval $t2 - t1$ (which is not very realistic but is relatively simple), the interval of 36 ms and a HIC of 1000 correspond to an acceleration of 60 g.

Recently, in the U.S., the NHTSA (the Governmental Department responsible for traffic security) has proposed to reduce to HIC $15 \leq 700$ the performance criterion for the head of H3 dummy 50%ile male. This value corresponds to a 31% probability of an injury with $AIS \leq 2$ occuring. The reduction of Δt from 36 to

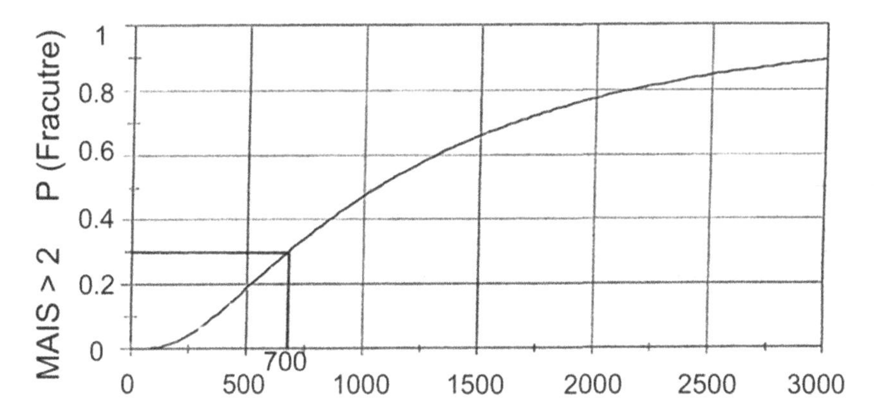

Fig. 13.8. Curve obtained with 54 head impact samples with HIC value known. The data are referred to the 50%ile male.

15 ms can be justified by the fact that in none of the 54 head impacts has the Δt corresponding to the maximum HIC exceeded 2 ms. So HIC 15 corresponds to HIC 36 when a head impact occurs.

Correspondingly as concerns the reduction from 1000 to 700 for the HIC value, the value of 700 has only statistical meaning. The curve plotted in Fig. 13.7 establishes the border value (1000); thus it is not necessarily true that the maximum or minimum distances from this curve correspond to major or minor injury levels.

Furthermore since 2003 the NHTSA has extended the biomechanical specification to the entire family of H3 dummies. Since experimental biomechanical data are not available, the NHTSA has used scaling technique to derive values from those of the 50%ile male (see Fig. 13.8).

Today there is a clear need to not limit the biomechanical approach to safety to just the 50%ile male but to extend it to all those exposed to risk (children 1, 3, 6, years; female 5%ile, male 95%ile).

The problem being encountered is the lack of specific biomechanical experimental data. Thus recently there has been a move towards using scaling techniques integrated with technical evaluations.

This technique, starting from the known performance criteria for a dummies of the family (e.g. H3 50%ile male), enables the unknown performance criteria for the other dummies to be calculated. However it is necessary to know the scale factor that associates several fundamental entities.

Neck

As concerns the injuries to the neck, it is necessary to bear in mind that the results of accident analysis represent a relatively low proportion of serious injuries (with AIS level >3) and a particular high number of minor injuries. The Japanese

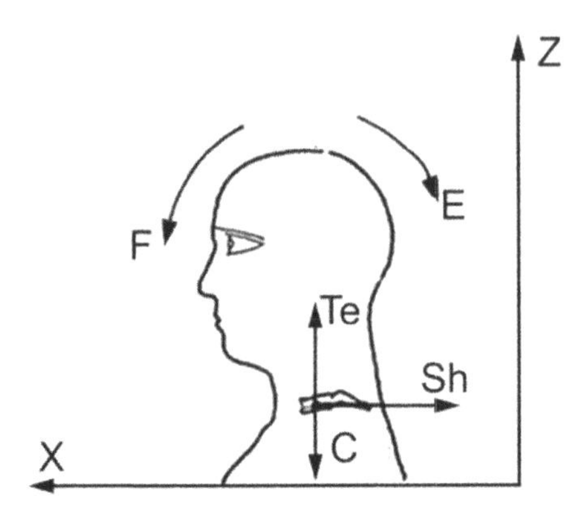

Fig. 13.9. Neck injury criteria.

study of Koshiro Ono (IRCOBI 1993), indicates that, during a rear impact, 93% of neck injuries were classified as AIS level 1. The serious injuries to the neck occur only during high severity impact and are usually associated to serious injuries to other body segments (eg. head, chest, etc.) while the minor injuries to the neck occur mainly with low severity impacts and are often not associated to other injuries.

The neck injuries, despite not being a particular priority from the perspective of injury seriousness, have remarkable importance from the economic standpoint due to their high numbers. For this reason, also the reduction of medium and minor injuries have significant importance.

As concerns the injury criteria, the situation has not yet been consolidated and today there are no regulations or dummies outside of the research arena.

Nevertheless it is possible to classify three different types of neck injuries:

- Criteria linked to single injury mechanisms.
- Criteria based on the interaction of injury mechanisms.
- Fluidodynamic criterion.

Criteria tied to single injury mechanisms

The neck is stressed on the xz plane even with not in axis impacts. Early in the 1970s, five injury mechanisms were examined, which are shown schematically in Fig. 13.9.

- Bending and extension, due to forward and backward head rotation.
- Tension and compression, with loading along the z axis, due to forward and backward rotation of the head.
- Shear, with loading along the x axis, due to relative movement between head and chest.

In these cases, the injury criterion used is the NIC (Neck Injury Criterion); as defined by European directive D 1996/97, the shear force Fx and the axial force Fz has to respect some limit curve, while the bending moment My must not exceed 57 Nm in extension.

As concerns the bending moment, currently no limit is applied (although 190 Nm is being postulated as a future regulation limit).

The definition of values for human tolerance for these injury mechanisms go back to 1970s, based on some experimental tests; these values are accepted as performance criteria in the aforementioned directive and in the 208 standard for U.S. regulations.

As concerns the accepted limit value of force, these curves are cumulative, ie. for each value of force, the curves indicate the maximum time for which they may be applied; the curves are ordered in terms of decreasing magnitude as shown in Fig. 13.10.

For example, if the shear force is $>$ 1.5 kN for a Δt $>$25 ms and if the maximum force value is $<$3.1 kN, the performance criterion is not passed.

Plotting the performance criteria on the force/moment plane, the conformity area for a 50%ile male dummy is obtained, as shown in Fig. 13.11.

In practice the current specifications are respected if the dummy measures a tension $<$ 3300 N and an extension moment $<$ 57 Nm, a compression force lower than 4000 N and bending moment lower than 190 Nm.

The point on the graphs defined through the force and moment values have to be located inside the area bounded by these limit values. It is important to remember that this criterion does not consider the combined effect of forces and moments.

The following considerations may be drawn in conclusion:

- The values of injury criteria defined in the 1970s are no longer considered to be reliable.
- The simultaneously application of axial force and moment reduces the human tolerance level.

Criteria based on the interaction of injury mechanisms

It has been demonstrated experimentally that the simultaneously application of axial force (tension or compression) and moment (flexion or extension) reduces human tolerance. For this reason the conformity area force/moment on $x - z$ plane has been modified as shown in Fig. 13.12.

This concept has been applied with U.S. Std. 208, where, in addition to the peaks values that must not be exceeded, the Nij injury criterion is introduced.

This indicator corresponds to a ratio of the measured entities with respect to the corresponding critical values for the human body:

$$N_{ij} = \left(\frac{F_Z}{F_{ZC}}\right) + \left(\frac{M_{OCY}}{M_{YC}}\right),$$

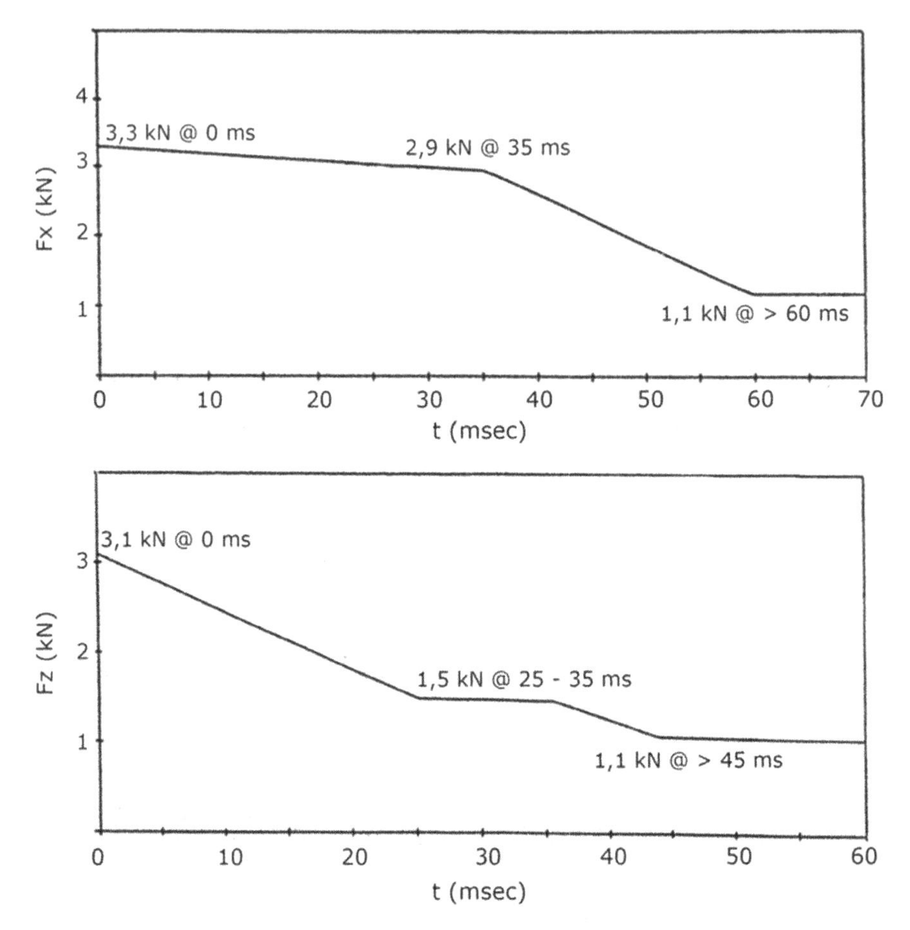

Fig. 13.10. Neck, trend of performance criteria as a function of the time. Shear criterion above, tension criterion lower.

where F_{ZC} and M_{YC} are the critical values, which can assume different values depending on the type of force or moment exerted during the event.

During the impact, the axial force F_Z can be in tension or in compression, while the moment at the occipital zone (M_{OCY}) can be either of bending or extension.

In all four different loading conditions can be described by Nij:

- tension-extension(N_{te});
- tension-bending (N_{tf});
- compression-extension (N_{ce});
- compression-bending (N_{cf}).

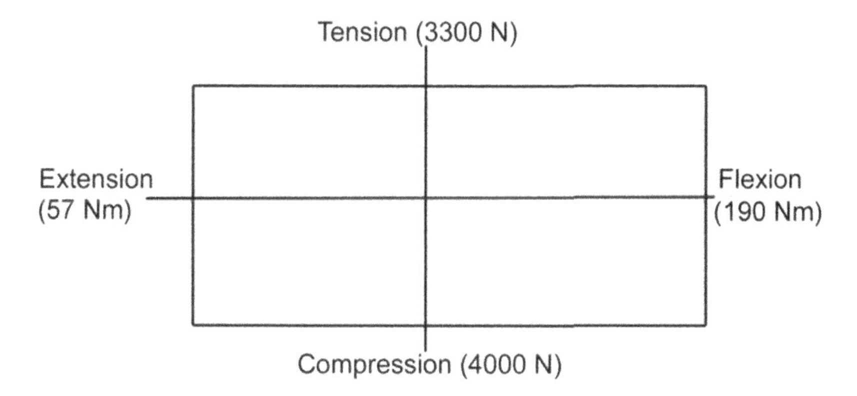

Fig. 13.11. Conformity area on plane xz for performance criteria linked to a single criterion.

In correspondence with each temporal points, only one of the four loading conditions can occur; consequently the correspondent Nij value is calculated while the other three loading conditions are considered to have a value equal to zero. In every moment during the impact, none of the four Nij conditions must exceed 1.0.

The kinematics of the neck during rear impact (whiplash) can be divided in four phases, as shown in Fig. 13.13.

- Phase 1, corresponding to the normal drive position at the beginning of the impact.
- Phase 2, as the chest starts to move forward, pushed by the backrest: due to inertia, the head nominally remains in its original position; the neck assumes an S shape.
- Phase 3, the head rotates towards the rear, the neck curves backward.
- Phase 4, the head reaches the maximum rotation and the neck the maximum curve; in this phase the tension and extension stress are maximum.

Whereas the previous injury criteria associates the injury mechanisms with Phase 4, the fluidodynamic criterion, which has been developed by researchers at Chalmers University of Technology in Sweden, associates the injury mechanisms with Phase 2 .

The cervical spine (the upper part of the back), when passing rapidly from phase 1 to phase 2 and from phase 2 to phase 3, exhibits a section variation of spine canal with an inversion of blood flux direction. As a consequence, an over pressure occurs in some points of main nervous system and the peripherical one, possible resulting in micro injuries on the peripherical nervous tissue.

The NIC injury criterion considers also the fluidodynamic criterion which has been determined via experimental tests on pigs. In practice, the analytical

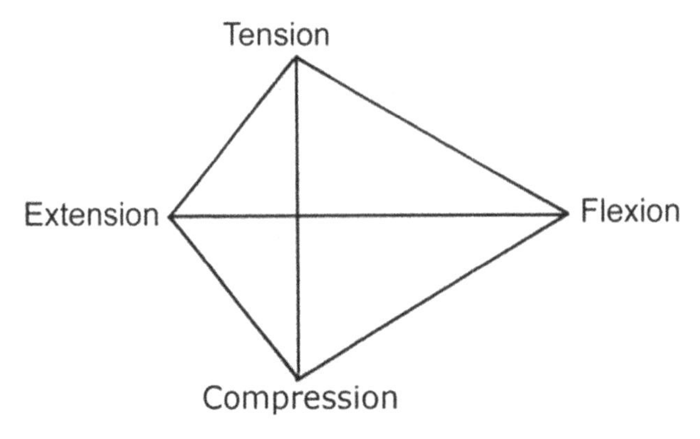

Fig. 13.12. Conformity area reduced for the simultaneously application of axial forces (tension or compression) and moments (flexion or extension).

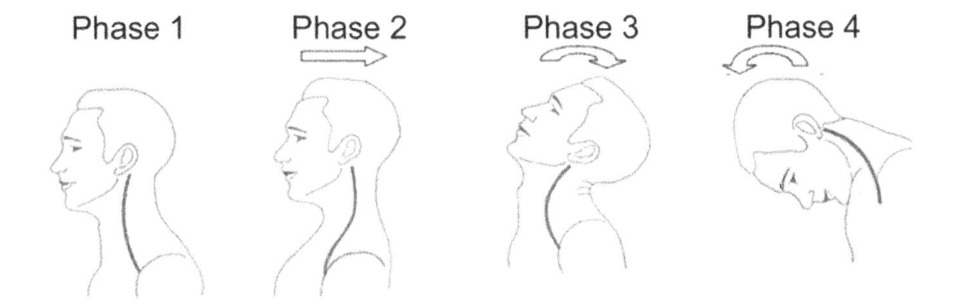

Fig. 13.13. The whiplash phenomenon.

expression of NIC determines that, for a defined subject, the maximum pressure of main nervous system is a function of the acceleration and of the square of relative speed between the head and the neck.

A first evaluation of the performance criteria, with a correspondence human tolerance level equal to AIS<2 is NIC $< 15\ m^2/s^2$. The performance criterion has to be evaluated with a RID (Rear Impact Dummy) dummy: Today it is still undergoing testing.

Thorax

Conventionally studies into thoracic injuries are divided into frontal and side impact case studies.

Frontal impact

The thoracic injuries can be divided in two types:

- injuries to internal organs, where the most frequent and most severe are those affecting the arteries linked to the heart;
- rib fracture, which in some cases can cause perforated internal organs.

Since the first type of injuries are due to mass inertia, it is logic to consider an injury criterion based on acceleration. The second type of injuries are due to thoracic deformation, it is logic to consider an injury criterion based on thoracic deformation which is difficult to measure on corpses causing a lack of representativeness of real thoracic deformation with a single measurement. For these reasons, many researchers have considered a single criterion based on acceleration considering the two injury mechanisms simultaneously.

The injury criterion determined was the application length (in ms) for acceleration. In this context, the Std. 208 specifies a maximum acceleration value of 60 g for a maximum time interval of 3 ms.

This limit derives not only from tests made on corpses but also from tests on volunteers (stunt-drivers, soldiers) and include data relative to accidents derived from free falls. The specified value of 60 g for the acceleration of the centre of gravity of thorax is mainly the result of Mertz's research and was proposed in 1971. This value has been implemented as performance criterion AIS ≤ 3 for the 50%ile male.

Successively, the crushing criterion for the sternum was defined with different human tolerance values depending on whether the load is concentrated (applied by means of the thoracic part of three point seatbelt, for example, as is the case in Europe) or distributed (applied by means of contact with the bag of the airbag, for example, as is the case in the U.S). These criteria were implemented as follows:

- for AIS $\leq$3 for 50%ile male: $\leq$76.2 mm distributed load (U.S.);
- for AIS $\leq$3 for 50%ile male: $\leq$ 50.0 mm concentrated load (Europe).

Subsequently the viscous criterion was introduced, having been developed at the end of the 1980s by two researchers working at General Motors. This criterion considers that the human body, exhibiting viscoelastic characteristics, reacts on equal penetration, with higher loads increasing the speed of penetration. This implies increased effect on the human body for the same intrusion, therefore more severe injuries.

The injury criterion refers to injuries of soft tissue of the thorax (heart, lungs and blood vessels). These injuries, as has been demonstrated during studies on corpses, apart from the high deformation values of the chest, are also caused by high impact speed with very low deformation of the thorax (as occurs, for example, during the impact of a bullet on a bulletproof vest).

An effective risk index for the thoracic index is thus the maximum value of the temporal function of the product of instantaneous speed of deformation of the thorax with its instantaneous compression. The compression of thorax can

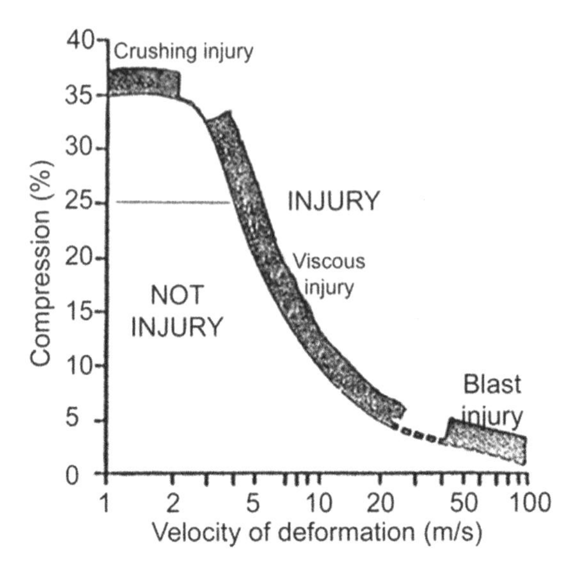

Fig. 13.14. Injury criterion for the thorax.

be considered with respect to two main modes due to frontal and side impact, giving rise to two VC indices.

The VC defined for the case of frontal impact is:

$$VC_{\max} = \left[1.3 \frac{D(t)}{0,229} \frac{dD(t)}{dt} \right],$$

Where D(t) is the sternum deformation in [m].

This expression differs from that of side impact only with respect to the reference constant. The maximum value of VC(T) is the injury criterion expressed in [m/s].

In case of frontal impact, for example, experimental studies were performed on animals and corpses. These studies showed that a resultant impact in a value of VC equal to 1.3 m/s has a 50% probability of causing bad thoracic injury.

For a VC value equal to 1 m/s, this injury probability drops to 25% and this is the VC value imposed as the imit by regulation D 96/97.

Plotting all the available experimental results (percentage thoracic compression vs. deformation speed), as shown in Fig. 13.14, it is possible to define the separation curve between injuries (AIS ≥ 3) and no injuries (AIS < 3). From these values it is possible to define the human tolerance value and the performance criterion for a H3 dummy, 50%ile male, with K = 0.299 m.

Side impact

The critical nature of side impacts is not only determined by their frequency and gravity, but also by the limited distance between interior part of the side of the car and the occupants.

Studies for the determination of injury criteria for thorax for side impact have been developed in independent way in Europe and in U.S.; this has led to the definion of two injury criteria and two different dummies.

In Europe the thorax tolerance for side impact was studied by using fresh corpses falling horizontally onto materials with various stiffness, positioned on dynamometric plane surfaces. Different physical parameters were taken into consideration as possible injury criteria; of these, the deflection was found to be the best parameter in terms of correlation with injury severity.

Thus in Europe a dummy (Eurosid) has been developed which is used to measure the thoracic deflexion in the y direction.

The thoracic deflection injury criterion was derived from experimental data for AIS $< 3 \leq 30\%$ of thoracic flexion, 42 mm of thorax flexion for 50%ile male Eurosid dummy.

In the U.S, to address the determination of crushing, standard instrumentation was initially used, with twelve accelerometers positioned on defined zones of chest both on corpses and dummies: Using this instrumentation, a large number of experimental tests were performed on corpses and dummies, both using sleds and impactors.

The analysis of the experimental obtained data led to the definition of the Thoracic Trauma Index TTI:

$$TTI = 1,4AGE + 0.5\,(RIBY + T12Y)\,\frac{MASS}{Mstd},$$

where AGE denotes the age of the test subjects, RIBY is the maximum absolute value of acceleration of ribs in the lateral direction, T12Y is the maximum absolute value of twelve thoracic vertebra in the lateral direction, MASS is the weight of the subject, and *Mstd* is a standard mass value equal to 75 kg.

As can be seen from this expression, this index represents an injury criterion based on the average acceleration values in the y direction measured on the impacted ribs and the spine. The formulation of TTI was obtained applying a regressive analysis technique on experimental data. The specific benefit of the TTI is that it enables the risk of injury for particular population exposed to the risk to be established. In fact, in the algorithm, the mass and the age of the subject are included.

The U.S. standard for car occupant protection in case of side impact requires the use of a US-SID dummy with the following performance criteria for the thorax:

- $\leq$85g (for car with four doors);
- $\leq$90g (for car with two doors).

An important critical aspect is that the TTI reflects a pure statistical approach rather than a physical situation, ie. no physical basis exists to relate the TTI with the thorax injury mechanisms.

Finally, it is also necessary to consider the viscous criterion as the injury criterion for the side impact. In fact, the analytical expression for side impact is the same of that used for the frontal impact but with different constants.

Also for this injury criterion a specific dummy has been made (BIOSID) capable of measuring the deflection (X(t)) of the ribs during side impact; as for frontal impact:

$$VC_{\max} \leq 1m/s.$$

The VCmax criterion, developed with the BIOSID dummy, has been introduced in Europe as an additional specification to the eurosid dummy.

In this case the expression of VC is:

$$VC_{\max} = \left[\frac{D(t)}{0,140}\frac{dD(t)}{dt}\right],$$

where D(t) is the deformation of ribs in [m].

Abdomen and pelvis

As for the thorax, the injuries relative to abdomen ad pelvis are studied by dividing into frontal and side impact cases.

Frontal impact

The types of injuries occurring in this type of impact are usually caused by an incorrect restraint of the pelvis with the abdominal part of the safety belt. This phenomenon occurs when the abdominal part of the safety belt climbs the iliac crest of the pelvis (a phenomenon known as submarining) in such a way that the restraint is applied on the soft part of the abdomen.

For this injury mechanism, an injury criterion in terms of the maximum force applied on the safety belt has been defined. To this value, a penetration value has been assigned; human tolerance has been fixed with the following values:

$F_{max} < 3{,}5$ kN;
penetration < 39 mm.

This criterion has not been implemented in any regulations, whereas the efforts to limit this type of injury have focused on the origin of the problem, the target being to avoid the climbing of the safety belt over the iliac crest.

Side impact

The injuries to the abdomen, during side impact, are due mainly to protruding side elements inside the vehicles (for example the armrest). Due to the intrusion make by the impact, these components can penetrate the side part of the human body. On the basis of a limited number of fall corpse tests, on elements with different stiffness which penetrate on the abdomen, a maximum force injury criterion (external measurement) has been defined which has been associated with the percentage of abdominal penetration. Human tolerance has been fixed as:

$F_{max} < 4{,}5$ kN;
penetration $< 28\%$ (40 mm for 50%ile M.).

This criterion has been implemented by European legislation, where the performance criterion has been adapted to the instrumentation of Euroside dummy:

F_{max}(internal evaluation) ≤ 2.5 kN;
penetration < 40 mm.

As concerns injury to pelvis in the side impact, mainly of concern is the rupture of the symphysis pubis (joint between the two iliac bones). Also in this case the injury criterion is represented by the maximum value of the force applied. Human tolerance for the 50%ile male is:

$F_{max} \leq 10$ kN.

Also this criterion has been implemented by European legislation and the performance criterion has been adapted to Euroside dummy, with:

Fmax ≤ 6 kN.

Rotula – Femur

The injuries to upper part of leg during frontal impact are usually generated by the impact between the knee and the dashboard. This is a consequence of the forward sliding of lower part of occupant (submarining).

The injury criterion for the femur is essentially a value of the maximum compression load that must not be exceeded during the impact and reflects the experimental results obtained with tests on corpses which focuses on determining the level of force which causes a femur fracture.

Over time different limits have been applied: the initially proposed limit of 5.43 kN (1200 lb) has been incremented progressively to reach the current level of 10.19 kN (2250 lb) in Std. 208.

The maximum force would establish a limit value for the femur fracture even though, for this value of force, knee, femur and articulation femur-hip fractures can generally occur. Furthermore, since it is not possible to establish for certain that the load on the femur will cause the fractures illustrated above, this value is assumed to be a global index.

In fact, if the loads were evaluated on each part, the fracture loads of the different parts would yield different results (in the evolution of dummies, the way to evaluate the loads also in other point over that the femur is being researched).

In the European regulation, the injury criterion on femur is given in terms of the limit curve force-cumulative time.

Also in this case, as concerns the performance criteria, a difference between U.S. and European regulations exists. This difference is due to the fact that in the U.S., wearing the safety belt is not required in all States.

U.S.: performance criterion $F_{max} \leq 10$ kN for dummy H3 50%ile male.
U.E.: performance criterion $F_{max} \leq <$ 9,07 kN for dummy H3 50%ile male.

In the U.S. since 2003 also a performance criterion for the 5%ile female has been introduced:

$$F_{\max} \leq 6,8kN.$$

This value has been obtained by applying scaling techniques starting from a H3 dummy 50%ile male.

Tibia

The European regulations for occupant protection, in the case of frontal impact, impose three performance criteria of biomechanical nature referring to the tibia of H3 dummy 50%ile male. However, since these criteria have never been validated biomechanically, it is not possible to say anything about injury criteria and human tolerance values.

The three criteria are:

- compression tibia force, $F_z \leq 8$ kN;
- Tibia Index, TI ≤ 1.3;
- relative movement tibia/femur ≤ 15 mm.

The injury criterion of tibia, the TI (Tibia Index), is expressed by the following formula:

$$TI = \left[\frac{M_r}{(M_c)_r}\right] + \left[\frac{F_z}{(F_c)\,r}\right].$$

This index is calculated at the vertex and the base of each tibia; however the axial compression force can be measured (Hybrid III dummy) in only one of the two points (lower end) and the value obtained is used to calculate the tibia index on both the vertex and the base.

Instead the two bending moments M_X and M_Y, are measured in the two points: According to D 96/79, TI must not exceed the value of 1.3 (mentioned above) at both extremities.

The TI, which is the main criterion, represents the first case of a criterion based on the interaction of two injury mechanisms (bending moment and axial compression force) being implemented by a European regulation. It has not yet been implemented in any U.S. regulations.

13.2 Simplified Models for Crash

13.2.1 Impulsive Model: With Full Overlap

The effect of the main parameters affecting the crash of a vehicle can be studied in a first approximation by following a simplified approach that starts from the consideration that the crash occurs over a very short time, in the order of 100 ms. This leads to the so-called impulsive model. The forces between the impacting (bullet) vehicle (BV) and the impacted (target) vehicle (TV) are so large that all other forces that usually act on the two vehicles are negligible compared with those that are developed during the crash. Thus loads that are usually dominant during normal or limit driving, such as those produced by the tires and aerodynamics, can be neglected. During the crash, the system constituted by the two vehicles can then be approximated as being isolated, ie. with negligible external forces acting on them. The total momentum of the system of the two vehicles can therefore be considered to be a constant.

The sketch of Fig. 13.15 refers to the case of a full-overlap crash, with the speed vectors aligned in the same direction passing through the centre of gravity of the two vehicles. The upper part of the figure refers to the start of the impact (time t_1), and the lower one to the end (time t_2). Before and after the impact the two vehicles are assumed to move at constant speed whereas during impact both of them decelerate.

If d is the distance between the centre of gravity of the two vehicles, whose speeds are V_{BV} and V_{TV}, the relative speed V_{REL} is

$$V_{REL} = V_{TV} - V_{BV}. \tag{13.1}$$

subscript $_1$ denotes the start of contact, when the speed of the two vehicles starts to change, whereas subscript $_2$ denotes the end.

A rear crash usually implies that, before the crash, V_{BV} e V_{TV} have the same sign; instead for frontal impact they have the opposite sign. By convention V_{REL} is negative when the distance d decreases with time; instead it is positive when the distance increases. The dashed line reported in Fig. 13.16 represents the qualitative evolution of the relative speed during a frontal crash, whereas the solid line represents the relative distance d. At the beginning of the crash, d is a maximum; then, in a first phase, it reduces to a minimum (time t_i); in the final phase, it increases again until the end of the impact. This last effect is due to a degree of elastic restitution of the two structures. The amount of deformation is

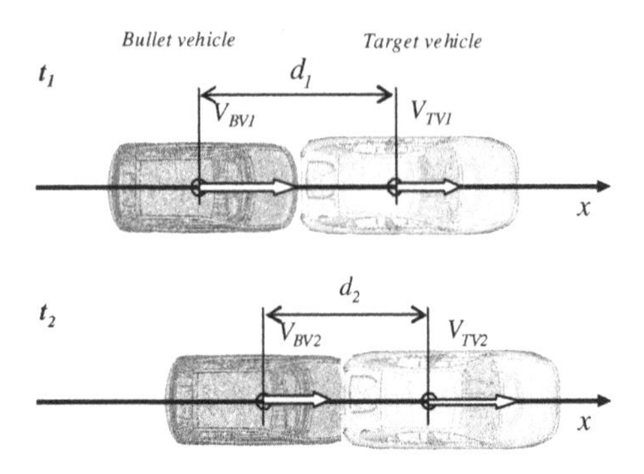

Fig. 13.15. Full overlap crash between two vehicles. The crash starts ad time t_1 and ends at t_2.

given by the difference between the distance at a generic time during the crash $d(t)$ and at the beginning (d_1):

$$S(t) = d(t) - d_1.$$

The coefficient of restitution r is usually defined as

$$r = -\frac{V_{REL2}}{V_{REL1}} = \frac{V_{TV2} - V_{BV2}}{V_{BV1} - V_{TV1}}; \tag{13.2}$$

Thus, in a perfectly inelastic collision, the relative speed after the impact is null and $r = 0$; instead, in a perfectly elastic collision, the relative speed after the impact is the $r = 1$; in a real case its value falls in the range between elastic and inelastic $0.05 < r < 0.20$.

As the system is isolated, the total momentum of the system is constant during the impact

$$m_{BV} \cdot V_{BV1} + m_{TV} \cdot V_{TV1} = m_{BV} \cdot V_{BV2} + m_{TV} \cdot V_{TV2}. \tag{13.3}$$

The plastic deformations dissipate part of the kinetic energy of the two vehicles, so that:

$$E_{C1} > E_{C2} \tag{13.4}$$

where

$$\begin{aligned} E_{C1} &= \tfrac{1}{2}\left(m_{BV}\ V_{BV1}^2 + m_{TV} V_{TV1}^2\right) \\ E_{C2} &= \tfrac{1}{2}\left(m_{BV} V_{BV2}^2 + m_{TV} V_{TV2}^2\right). \end{aligned} \tag{13.5}$$

The energy dissipated during the crash can be expressed as function of the speed of the centre of gravity of the system of the two vehicles and the coefficient of restitution r.

The coordinate x_G of the centre of gravity of the two vehicles is

$$m x_G = m_{BV} x_{BV} + m_{TV} x_{TV}. \tag{13.6}$$

where $m = m_{BV} + m_{TV}$ is the total mass.

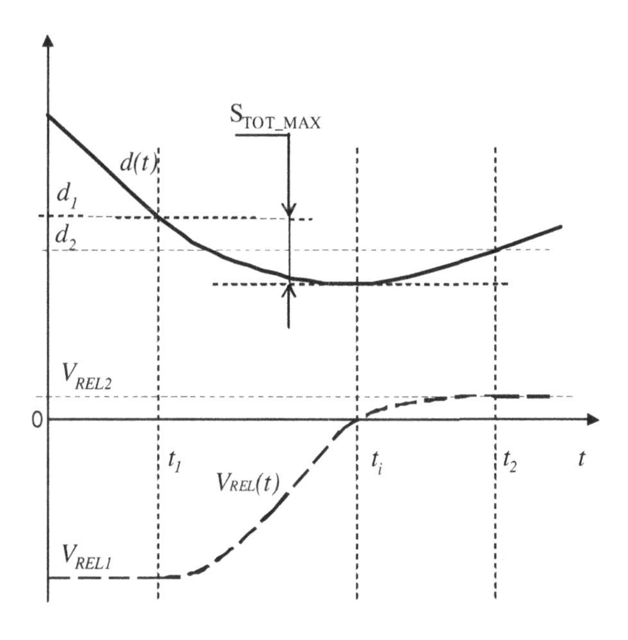

Fig. 13.16. Speed during the full overlap crash of Fig.13.15 .

The speed of the centre of gravity (V_G) is obtained as time derivative of Eq. 13.6

$$mV_G = m_{BV}V_{BV} + m_{TV}V_{TV}. \tag{13.7}$$

Since the system is isolated, the momentum is constant (Eq. 13.3), and so the speed of the centre of gravity (V_G) is also constant:

$$mV_G = m_{BV}V_{BV} + m_{TV}V_{TV} \tag{13.8}$$

Taking Eq. 13.1 into account, V_G can be expressed as function of the relative speed V_{REL} between the two vehicles

$$mV_G = m_{BV}V_{BV} + m_{TV}\left(V_{REL} + V_{BV}\right) \tag{13.9}$$

$$mV_G = m_{BV}\left(V_{TV} - V_{REL}\right) + m_{TV}V_{TV}. \tag{13.10}$$

This enables the speed of the bullet and the target vehicle to be expressed as a function of the speed of the centre of gravity and the relative speed

$$V_{BV} = V_G - \frac{m_{TV}}{m}V_{REL}; \tag{13.11}$$

$$V_{TV} = V_G + \frac{m_{BV}}{m}V_{REL}. \tag{13.12}$$

The kinetic energy before and after the crash is obtained by substituting speeds V_{BV} e V_{TV} of Eq. 13.11 and 13.12 in Eq. 13.5. Taking into account that the speed of the centre of gravity is constant,

$$E_{C1} = \frac{1}{2}m\left(V_G^2 + \frac{m_{BV}m_{TV}}{m^2}V_{REL1}^2\right)$$
$$E_{C2} = \frac{1}{2}m\left(V_G^2 + \frac{m_{BV}m_{TV}}{m^2}V_{REL2}^2\right) \tag{13.13}$$

Finally, the relative speed after the impact can be expressed considering the definition of the coefficient of restitution of Eq. 13.2. The dissipated energy is then

$$E_D = E_{C1} - E_{C2} = \frac{1}{2}\frac{m_{BV}m_{TV}}{m}(1-r^2)V_{REL1}^2. \tag{13.14}$$

This energy is that of a body of mass M_{eq} and speed V_{eq} that dissipates all kinetic energy during the impact

$$M_{eq} = \frac{m_{BV}m_{TV}}{m}; \tag{13.15}$$

$$V_{eq} = V_{REL1}\sqrt{1-r^2}. \tag{13.16}$$

As expected, the dissipation is maximum for a perfectly inelastic impact i.e. for $r = 0$. By converse, no dissipation occurs for a perfectly elastic collision ($r = 1$).

The frontal crash against a rigid barrier of infinite mass (Fig. 13.17) can be considered as a limit case by considering a target vehicle of infinite mass $m_{TV} = \infty$. In this case the relative speed is the same as that of the impacting vehicle $V_{REL} = V_{BV}$ and the speed of the centre of gravity is null $V_G = 0$. The distance d_1 at the beginning of the crash is that between the foremost part of the bumper and the centre of gravity of the vehicle. Considering that $m_{TV} = m = \infty$, from Eq. 13.14 the energy dissipated during the crash is that of the vehicle times factor $1 - r^2$.

The focus until now has been more on establishing a link between the motion before and after the impact, without considering what happens during the time interval between t_1 and t_2. To address this issue two of the most important parameters are the transmitted force and the acceleration. The force transmitted during an impact against a barrier can be measured by instrumenting the barrier with load cells, the result being a tormented sequence of peaks during the folding of the structures followed by abrupt valleys caused by bending and collapse of some parts with other contributions due to the impact of the parts in the engine compartment.

Similar behavior can be found measuring the acceleration on a part of the vehicle that is not directly involved in the crash, for example the seat rail. Fig. 13.19 shows the qualitative shape of this acceleration (curve a). The curve b represented on the same diagram is obtained by the best fit interpolation of the experimental data (curve a) by the McMillan curve which is simply a semi-empirical expression that represents the low frequency content of the acceleration:

$$a(t) = C\frac{t}{t_2}\left(1 - \frac{t}{t_2}\right)^{\beta} \tag{13.17}$$

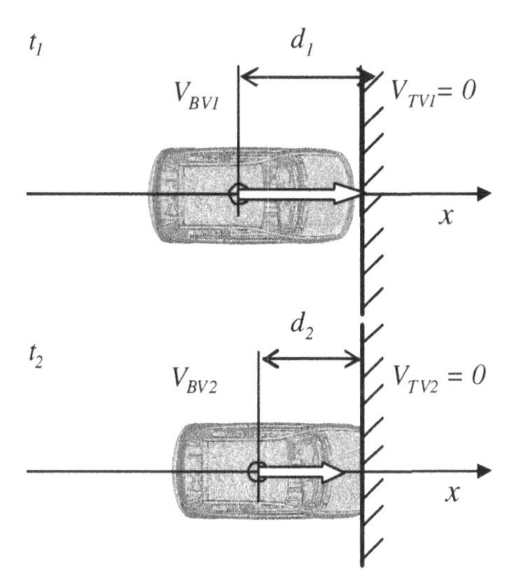

Fig. 13.17. Crash against a rigid barrier of infinite mass.

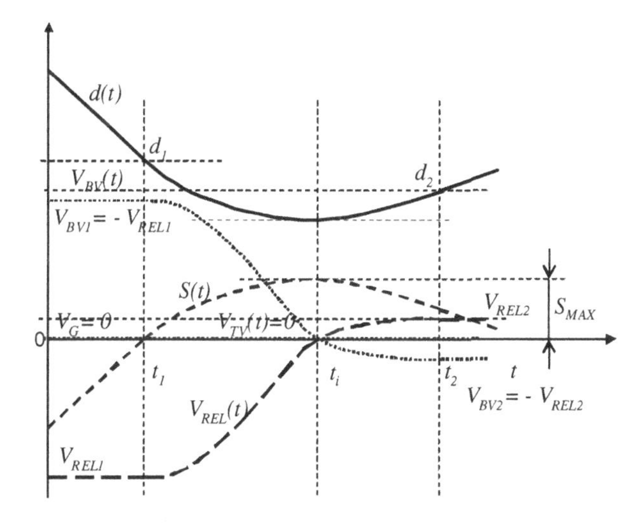

Fig. 13.18. Speed and distance during a crash against a rigid barrier.

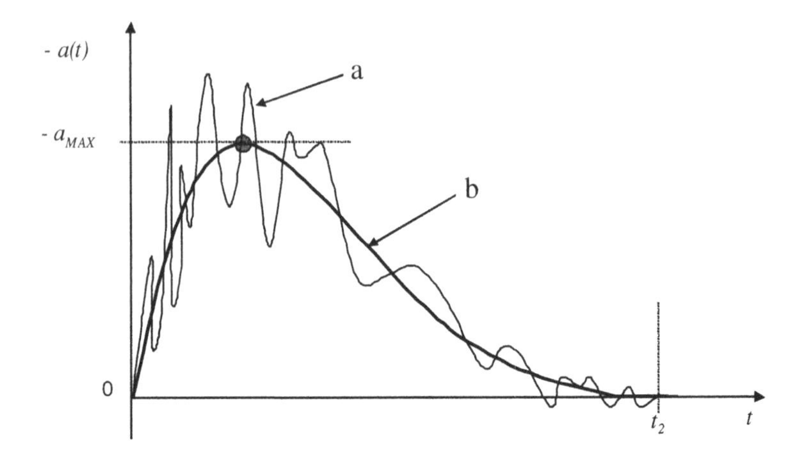

Fig. 13.19. Qualitative behavior of the acceleration measured on the seat rail during a frontal crash against a barrier. Experimental measurement a); Mc Millan semi-empyrical curve b).

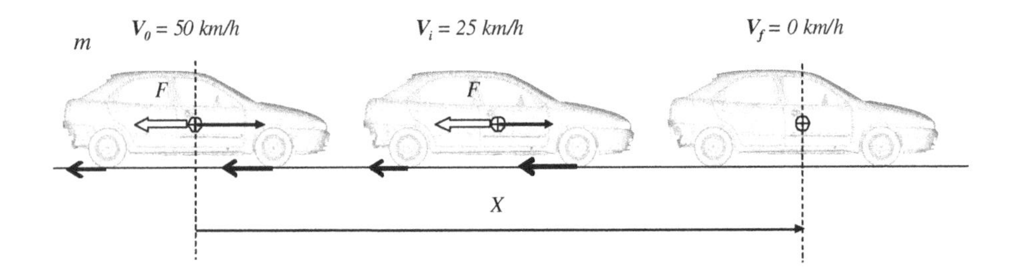

Fig. 13.20. Deceleration from 50 km/h to full stop by braking.

coefficient C is the amplitude factor and exponent β it the shape factor which can be identified by curve-fitting the experimental data with Eq. 13.17.

13.2.2 Role of the Restraint System

To understand the role played by the restraint system during a crash, it is useful to refer to two different decelerations of the same vehicle from the same initial speed, say 50 km/h. In the first case the driver acts on brake pedal until the vehicle stops in a distance X (Fig. 13.20). In the second case the vehicle crashes against a rigid wall (Fig. 13.21).

The kinetic energy of the vehicle at the beginning and at the end of the deceleration is the same in the two cases; nevertheless the process of reaching a full stop differs significantly. In the first case, the use of the brakes generates forces with resultant F between the tire and the road surface that act to reduce the vehicle speed. Due to Newton's second law, the variation of the momentum

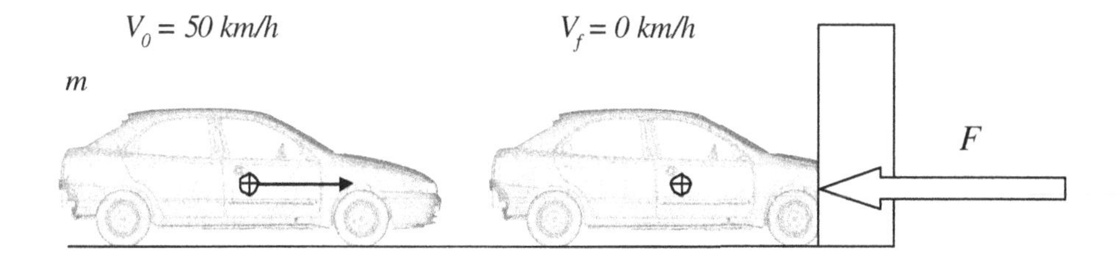

Fig. 13.21. Deceleration from 50 km/h by crash against a rigid barrier.

is equal to the impulse, ie. to the time integral of the resultant force F between the start t_0 and the final time t_f of the deceleration

$$m\,(V_0 - V_f) = \int_{t_1}^{t_2} F(t)\;dt \tag{13.18}$$

As these forces move with the vehicle, the corresponding work (force times distance) reduces the kinetic energy until the vehicle comes to a full stop.

In the second case, the vehicle crashes against a fixed barrier reaching the full stop almost instantly. Similarly to the previous case, the time integral of the force transmitted from the barrier to the vehicle is the same. The difference is that since the time interval is much shorter, the force will be correspondingly higher.

The first difference between the crash and the normal braking lies in the time interval needed to stop the vehicle from a certain speed. Looking at Eq. 13.18, since the impulse is the same, normal braking involves a relatively long time and, therefore, forces lower than in the case of a crash which occurs in very short time interval.

The second difference corresponds to how the kinetic energy of the vehicle is dissipated. During normal braking, the forces between the tires and the road surface dissipate all kinetic energy. In contrast, during the crash against a rigid barrier, the force acting on the vehicle does not dissipate energy as the surface of the barrier is not moving. The kinetic energy will then be transformed by the vehicle with the plastic deformation of its front structure.

The same considerations valid for the vehicle can be applied to the occupants. During normal braking each occupant is subject to a number of forces transmitted by the seat, belts, feet and all other elements of the body in contact with some part of the vehicle. These forces move their points of application over the braking space. Similarly to what happens at the vehicle level with the tire slip forces, the forces are exerted on the occupant so as to reduce the kinetic energy. If the kinetic energy is not transformed into internal energy of the occupant body, but elsewhere, no damage to the occupant will occur in this process.

During a crash, if the occupant is restrained, energy is dissipated by the forces developed by the restraint system. In this case there may not be a substantial difference in the energy dissipation with respect to the case of the normal braking, apart from the entity of the forces applied to its body.

A completely different picture emerges if the occupant is not restrained. In this case when the vehicle reduces its speed by plastic deformation of its structure, the occupant continues its motion until impacting the interior surfaces (dashboard or the steering wheel for the front occupants). The very short time involved in a crash (0.1 s) makes any conscious or unconscious reaction by the occupant impossible. Impact against the interior surfaces will occur at an almost unchanged speed (e.g. 50 km/h). The vehicle at the same time could be already completely stopped. The result is that the high stiffness of the interior surfaces stop the occupant in a very short time and with very little deformation of the surface itself. The energy of the occupant is transformed in this case into internal energy, the result being significant injury.

Correspondingly it can be easily understood that the main role of the restraint system is to reduce the kinetic energy of the occupant with forces that act during the deformation of the vehicle. The motion of their points of application avoids the transformation into internal energy of the body and the consequences in terms of physical injury.

13.2.3 Speed-Time Diagrams

Diagrams that represent the speed during the crash as function of the time have proven to be a useful tool to understand the role of different elements involved such as the restraint system and the shock absorption structure.

Apart from the speed, other quantities can be represented on these diagrams, such as the acceleration and the distance. If the speed is measured with respect to a reference system fixed to the ground, the acceleration is the slope of the diagram, while the distance relative to the ground is represented by the area between the time axis and the speed curves. Additionally, in the case of objects in relative motion, the relative displacement is the area between their speed curves.

Fig. 13.22 shows the speed time diagram for the case of a normal braking with no crash. At time $t = 0$ the speed is 14 m/s (~50 km/h) and the braking action starts. The intensity of the braking action is sufficient to produce a constant 5 m/s^2 (0.5 g) deceleration that leads to a full stop in

$$\frac{14 \text{ m/s}}{5 \text{ m/s}^2} = 2.8 \text{ s}$$

The braking distance is represented by the area under the speed diagram. As the deceleration is constant, this area is that of a triangle

$$\frac{1}{2} 14 \text{ m/s } 2.8 \text{ s} = 19.6 \text{ m}$$

If the restraint system were rigid, the speed of the occupant during the deceleration would be the same as that of the vehicle and the two speed-time curves

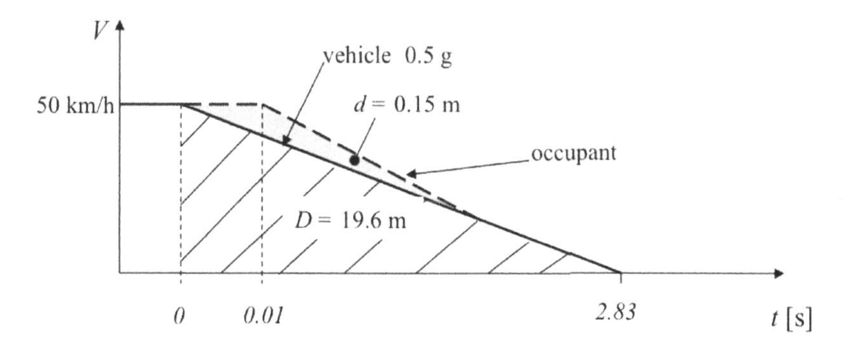

Fig. 13.22. Speed-time diagram during braking.

will overlap. Conversely, if the restraint system acts with a delay of 0.01 s, the speed curve of the occupant and that of the vehicle would not overlap. The area between the two curves is the displacement d that the restraint system allows the occupants to have with respect to the vehicle frame.

Fig. 13.23 shows the speed-time diagram for a crash against a rigid barrier with an unrestrained occupant. If the deformation involves only the front part of the car body structure, the displacement of the central portion (safety cell) will be equal to the amount of deformation of the front part, for example 0.625 m. Additionally, if: 1) the same speed before the impact of 50 km/h as for the previous case, and 2) the deceleration during the crash is constant, the deceleration during the crash can be determined by letting the area under the speed-time diagram equal to the amount of deformation of the front part

$$\frac{1}{2} 14 \text{ m/s } \Delta t = 0.625 \text{ m}$$

This leads to a duration of the impact of

$$\Delta t = 0.09 \text{ s}$$

and an acceleration of

$$a = \frac{14 \text{ m/s}}{0.09 \text{ s}} = 157 \text{ m/s}^2 = 16 \text{ g}$$

The solid line in Fig. 13.23 represents the speed of the central portion. The unrestrained occupant (dashed line) maintains its speed of 14 m/s until it strikes against the steering wheel or the dashboard. If the initial distance between the occupant and the steering wheel (or dashboard) is the same as the deformation of the front part of the vehicle (0.625 m), the occupant will impact these elements when the vehicle speed is null. The total distance covered by the occupant relative to the ground is 1.25 m (0.625 m+0.625 m).

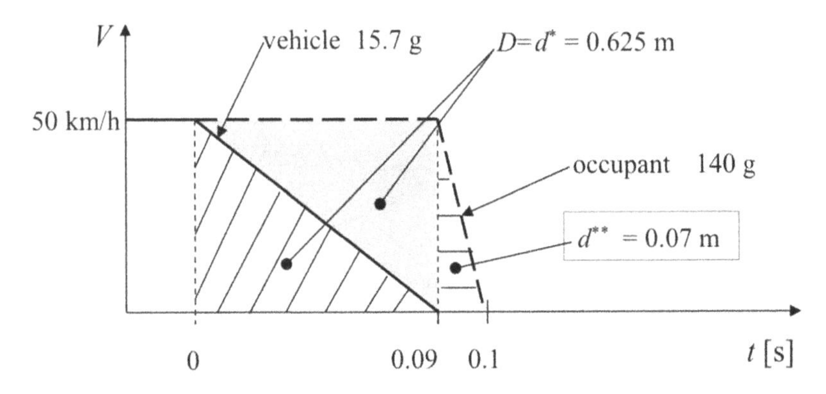

Fig. 13.23. Speed time diagram during a crash against a rigid barrier. Unrestrained occupant.

The same considerations used to estimate the deceleration of the vehicle can be repeated now for the occupant. The assumption is that the surface hit by the occupant is rigid and that the deformation of the body is 7 cm. The result is that the time required to stop the occupant is 0.01 s, corresponding to an average acceleration of 140 g. This amount of deformation, together with the short time and the very high acceleration, implies very severe or fatal injuries.

Fig. 13.24 shows the speed-time diagram for the same crash with a restrained occupant. If the restraint system allows a backlash distance of $d^* = 0.156$ m the occupant continues to travel at constant speed for 0.045 s from the beginning of the crash. If the restraint system is able to stop the occupant in the following 0.045 s, the acceleration is

$$\frac{14 \text{ m/s}}{0.045 \text{ s}} = 280 \text{ m/s}^2 = 28.5 \text{ g}.$$

In this case the occupant and the vehicle stop at the same time, after 0.09 s from the start of the crash. During the phase when the restraint system is active, the occupant moves relative to the vehicle by 0.156 m (distance d^{**} of Fig. 13.24); total displacement of the occupant relative to the ground is then 0.937 m (0.625 m is the deformation of the vehicle, 0.156 m backlash distance, 0.156 m restraint phase).

If at the beginning the distance between the occupant and the steering wheel (or the dashboard) is 0.625 m, at the end of the crash a distance of 0.312 m remains. This can be exploited so as to reduce the occupant acceleration, as shown by the dotted line in Fig. 13.24. The result is a significant reduction of the forces applied by the restraint system and, as a consequnce, of the potential injuries to the occupant.

Fig. 13.25 shows a restrained occupant during a crash. Assuming that the occupant does not have an increment of internal energy (relevant injuries), its

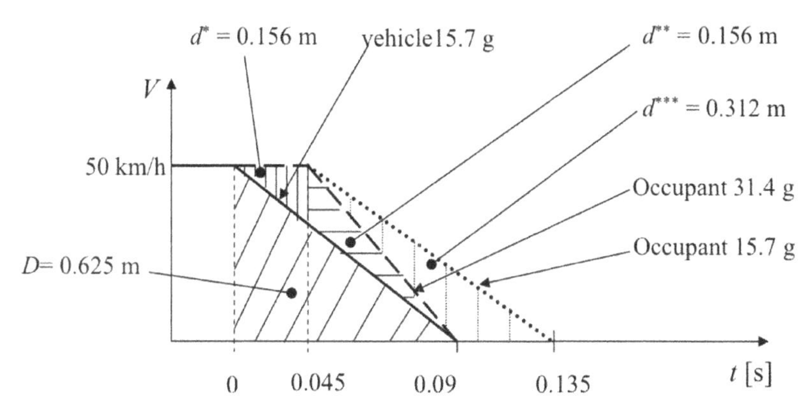

Fig. 13.24. Speed time diagram during a crash against a rigid barrier. Restrained occupant.

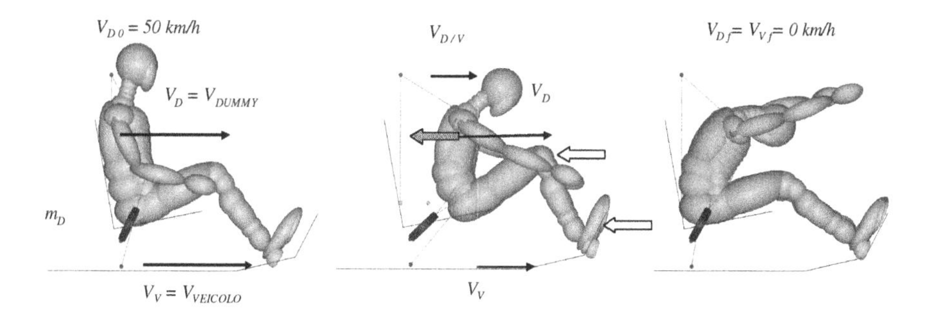

Fig. 13.25. Restrained occupant during a crash.

initial kinetic energy must be completely dissipated during the crash by the vehicle. This energy dissipation can be split into two contributions:

- Ride down energy (E_{RD}), the work done by the forces applied to the restraint system and to the seat by the vehicle with the displacement of the attachment points relative to the ground. Another contribution is the force that the vehicle applies to the occupant through the forces applied directly to the occupant, for example the feet. All this energy is transferred by the occupant to the vehicle (speed V_V) that eventually dissipates it in plastic deformation.

- Restraint system energy (E_{RS}), the work done by the forces applied to the occupant by the restraint system and the seat with the relative displacement between the occupant and the vehicle structure (speed V_{DV}).

The ride down energy can be calculated as the time integral of the power that the resultant of the inertia force acting on the occupant makes with the vehicle speed V_V

$$E_{RD} = \int_{t_1}^{t_2} m_D a_D V_V dt, \tag{13.19}$$

where m_D is the mass of the occupant (subscript D indicates dummy) and a_D is its acceleration. As the occupant is not a rigid body, this integral should be split in several contributions (head, upper body, legs, arms,...), one for each part of the body.

Similarly, the restraint system energy is the time integral of the power that the resultant of the inertia force acting on the occupant makes with the occupant ot vehicle relative speed V_{DV}

$$E_{RS}= \int_{t_1}^{t_2} m_D a_D V_{DV} dt. \tag{13.20}$$

In conclusion, with this knowledge regarding the role of the restraint system, together with some considerations about the injury criteria, it is possible to outline some general guidelines regarding the restraint system itself which should be designed to.

- maximize the distance traveled by the occupant under the action of the restraint system.

- minimize the backlash, enabling the occupant acceleration to be reduced.

- minimize the movement of the occupant body joints, the deformation and the deformation speed since the severity of injuries increases as these parameters increase.

- maximize the interface area with the restraint system so as to reduce the interface pressure and, hence the extent of injury.

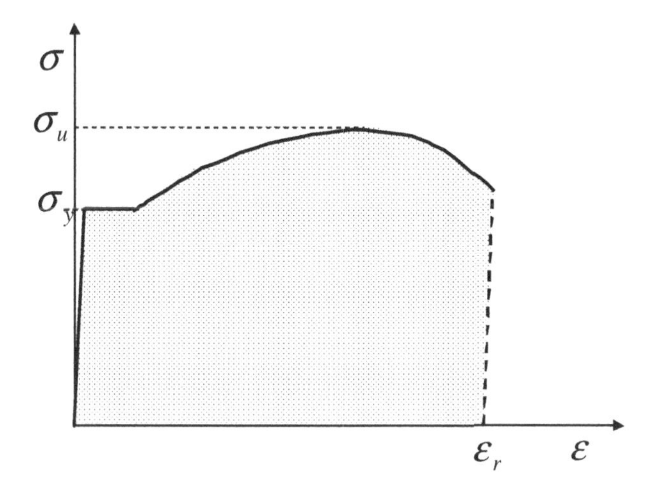

Fig. 13.26. Qualitative stress strain curve for a deep-drawing steel. The area under the diagram represents the energy dissipated to bring the material to a tensile break.

- apply the restraint forces on bones (femur, hip, chest, shoulders, head) rather than soft tissue areas since higher loads can be resisted and such areas of the body are less affected by the deformation speed that is potentially very harmful to softer tissue.

13.3 Introduction on Impact Energy Absorbers

An impact energy absorber is a system that converts the kinetic energy involved in a crash into another form of energy. In principle the conversion could be conservative, so that the energy could be stored in some form, for example as a pressure in a gas or in elastic energy in a spring. Alternatively it could be dissipated by some mechanism.

The considerable amount of energy involved even in a low speed crash on one hand, and the need to limit the weight and the size of the vehicle on the other, justifies the current adoption of energy absorbers based on the dissipation of the kinetic energy, exploiting the plastic deformation of a structural material.

The energy per unit involved in the tensile failure of a cylindrical specimen of a ductile material ε_d (J/m^3), is equal to the area under its stress-strain curve. Fig. 13.26 shows qualitatively this curve for a deep-drawing steel (Fe P04, for example). The large elongation at break ($\epsilon_r = 0.4$) and the small difference between the yield strength σ_y =228 MPa and the ultimate strength σ_r = 330 MPa justifies an elasto-plastic approximation of the strain-stress behaviour for such materials.

Neglecting the elastic energy accumulated in the sample, the density of the energy dissipated before breakage is given by:

$$\varepsilon_d \approx \sigma_y \epsilon_r = 132 \text{ MJ/m}^3. \tag{13.21}$$

The kinetic energy of a vehicle with a mass $m = 1200$ kg moving with speed $v =$ 56 km/h is

$$E_c = \frac{1}{2}mv^2 = 145 \text{ kJ}. \tag{13.22}$$

An impact energy absorber able to bring all plastic materials that forms its structure (Fe p04, for example) close to breakage would require a volume of material V_{abs}

$$V_{abs} = \frac{E_c}{\varepsilon_d} = 0,0011 \text{ m}^3, \tag{13.23}$$

that corresponds to a mass of 8.6 kg.

The high energy density dissipated in plastic materials suggest to exploit the plastic deformation to design the structures devoted to dissipate the kinetic energy during the impact. To increase the structural efficiency the geometry must be such that most of the material of the absorber is brought to the plastic range during the deformation.

The solutions adopted usually exploit the complex deformations that occur during the compression instability of thin walled structures. These deformations are often indicated as crushing to distinguish them from the rather simple deformation that occurs during global compression instability of slender structures (buckling).

Some of the most commonly used crush mechanisms exploited in shock absorbers are:

- tubular structures with constant cross section loaded along their main axis
- tubular structures with tapered cross section loaded along their main axis
- tubular structures loaded perpendicularly to their axis,
- honeycomb materials
- metallic or plastic foams
- sandwich materials

Fig. 13.27 shows two possible deformation modes of a quarter of a beam with rectangular cross section during crushing. In the case of Fig. 13.27 a), increasing the axial deformation, segment AC decreases, while CD increases of the same amount, so that the length of corner ACD remains constant. As the deformations of a surface at mid thickness are negligible, this mode of deformation is therefore said to be inextensional.

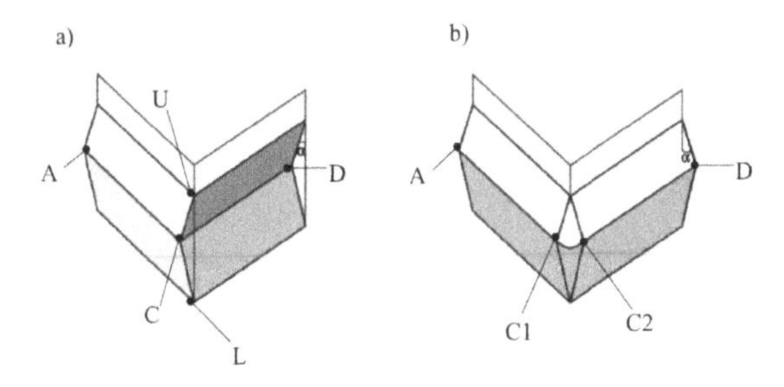

Fig. 13.27. Deformation mode during the compression of a beam with rectangular cross section. a) inextensional mode; b) extensional mode.

Conversely in Fig. 13.27 b) the surface at mid thickness is deformed so that corner A-C1-C2-D increases its length by increasing the compression deformation. Specifically, the extensional deformation is concentrated in segment C1-C2. This mode of deformation is therefore said to be extensional.

In the inextensional mode (a), the plastic deformation is due to the propagation to the left of the plastic hinge (U-C-L). Due to the effect of the thickness, the movement of the hinge line across the material causes a large bending of the plate, with associated plastic deformations. An observer on the plastic hinge will see the plate flow from left to right perpendicularly to the corner. The plate that flows through the corner is first bent in one direction and then in the reverse direction, so that at the end it becomes flat again. This means that each portion of the material is subject to a cycle that proceeds from plastic extension to plastic compression or vice-versa.

Fig. 13.28 shows the inextensional deformation mode of a plate with a non negligible thickness. As the plastic hinge U-C1-C2-L moves to the left in a fixed plate, strip A-C-D is deformed as if it were drawn on a toroidal surface (Fig. 13.28 a destra). As mentioned previously, the effect is bending that is first produced and then removed. Together with the movement of the plastic hinge through the plate, the compression of the beam causes a larger bending about corner A-C-D, so that angle U-C-L decreases. Fig. 13.28 shows that at corner U-C1-C2-L there is an additional extensional deformation along with the inextensional one.

13.3.1 Beams with Rectangular Cross Section

The first type of instability that occurs during the compression of a thin walled beam with rectangular cross section is usually that of the panels that constitute its sides. The instability of the panels triggers the process that leads to the global instability of the beam (buckling Fig. 13.30 b), or a local instability. The last case is characterized by a sequence of crushing and folding actions (folding, Fig. 13.30 a). The axis of the beam is still straight but its length is reduced.

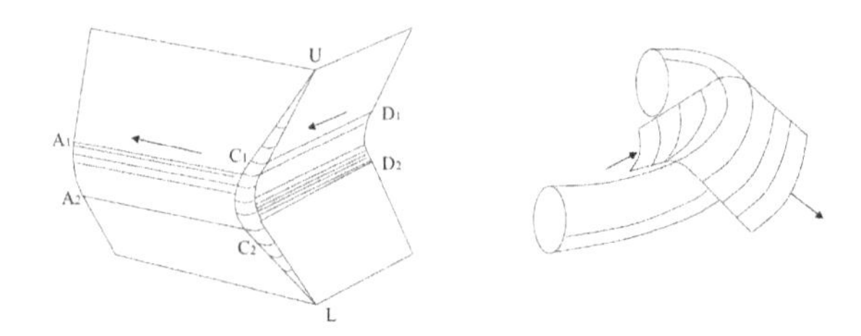

Fig. 13.28. Deformation with moving plastic hinge. During the compression plastic hinge U C_1 C_2 L moves to the left. The plate moves on the hinge as if it were drawn on a toroidal surface.

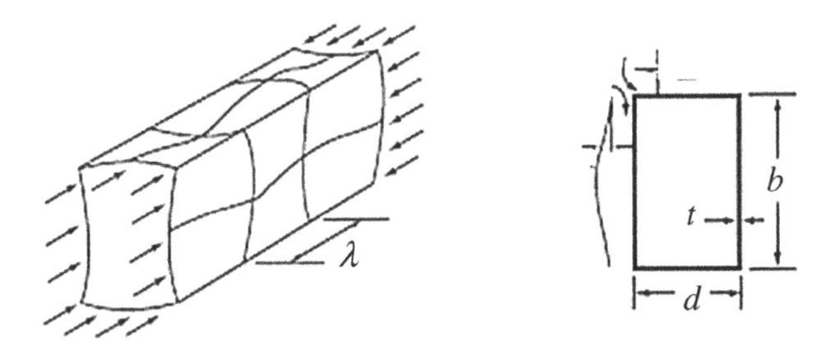

Fig. 13.29. Elastic instability of a thin walled beam with rectangular cross section. Each face is a plate that induces boundary conditions on the neighboring ones so that each corner can be considered as a hinge.

From the energy dissipation perspective (Fig. 13.31) the folding mode is the most convenient since it leads to plastic deformation a large portion of the material. Conversely, during the global instability, the straight axis beam is bent to form an angle (as a sort of knee, Fig. 13.30 b). As the plastic deformation is limited to the knee, dissipation is much lower.

The compressive stress that leads to the local instability of a beam with rectangular cross section (critical stress, σ_{cr}) can be evaluated considering each face as a panel. The two wider panels first reach the instability, as they have a smaller thickness to width ratio (t/b). The other two can be considered to be constraints that prevent the displacements perpendicular to the plate, while allowing the rotations about the corners, albeit with some effort. If l is the length of the beam in the load direction, b and d the sides of the cross section, and t is the small thickness ($t << b$, d), the critical compression can be computed as

$$\sigma_{cr} = k\frac{\pi^2 E}{12(1-\nu^2)}\left(\frac{t}{b}\right)^2, \tag{13.24}$$

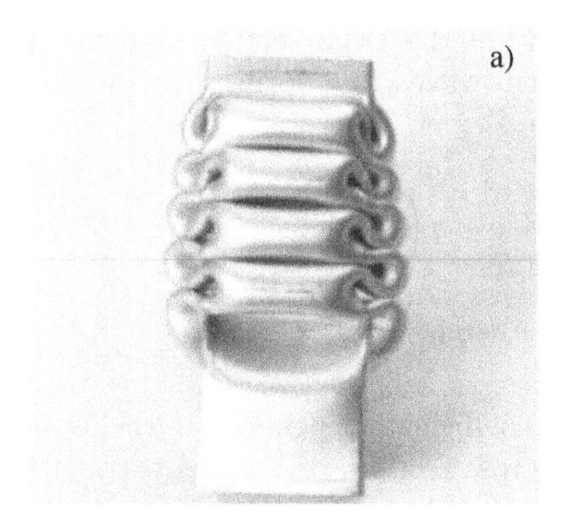

Fig. 13.30. Compression instability with large deformations. a) Axial folding, note that the axis of the beam remains straight. b) Global instability, the plastic deformations are limited to the kee and to the ends. The beam axis forms an angle about the knee.

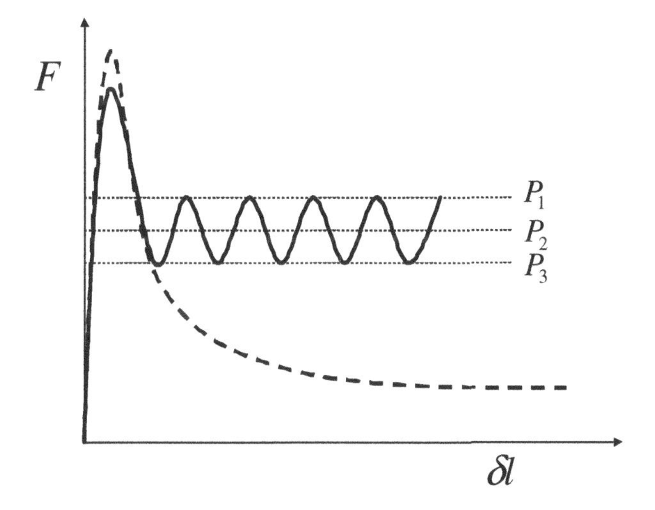

Fig. 13.31. Force-displacement characteristic: qualitative comparison between a stable crushing with folding (solid line) and a global instability (dashed line).

where k is a function of the large side of the cross section, and of the ratio $\lambda = l/m$ between the number m of half-waves that form after the instability along about the loading direction (side l):

$$k = \left(\frac{mb}{l}\right)^2 + p + q\left(\frac{l}{mb}\right)^2$$
$$= \left(\frac{b}{\lambda}\right)^2 + p + q\left(\frac{\lambda}{b}\right)^2, \tag{13.25}$$

where p and q are two non dimensional coefficients depending on the aspect ratio b/d of the cross section. For a square cross section, or a panel with with hinge boundary conditions along all sides: $p = q = 1$.

The critical stress has a minimum for a value of λ that can be found by letting $\partial\sigma_{cr}/\partial\lambda = 0$.

The result is

$$\lambda = \frac{l}{m} = \frac{b}{q^{1/4}}. \tag{13.26}$$

For a square cross section ($q = 1$), the half-wave length is the same as the side $\lambda = b$ of the cross section. For rectangular cross sections, the shorter sides constrain the longer ones so that, when one buckles inside the beam, the facing one buckles out of it. The section distorts moving its centre of gravity. As the axis of the beam is no longer a line, global instability may occur. The global instability can be triggered by the local instability, especially for thickness ratios t/b small enough to induce local instability in the elastic range ($\sigma_{cr} < \sigma_y$).

The critical stress can be raised above the elastic range by increasing the thickness t ($\sigma_{cr} > \sigma_y$). In this case Eq. 13.24 must be modified to take plastic deformation into account.

The structure of the equation is the same as for a rectangular cross section with the secant and the tangent modulus (E_s and E_t) instead of the elastic modulus E.

$$\sigma_{cr} = \frac{\pi^2 E_s}{9}\left(\frac{t}{b}\right)^2\left\{\left(\frac{1}{4} + \frac{3}{4}\frac{E_t}{E_s}\right)\left(\frac{mb}{l}\right)^2 + 2 + \left(\frac{l}{mb}\right)\right\}. \tag{13.27}$$

The half wavelength can be found similarly the elastic case:

$$\lambda = b\left(\frac{1}{4} + \frac{3}{4}\frac{E_t}{E_s}\right)^{1/4}. \tag{13.28}$$

The values of λ are smaller than in the elastic case, as in the case of a perfectly plastic material, with $E_t = 0$;

$$\lambda = b\left(\frac{1}{4}\right)^{1/4} = 0.707b. \tag{13.29}$$

The half wavelength of mild steel is between the elastic and the perfectly plastic value:

$$\lambda \approx 0.8b. \tag{13.30}$$

For rectangular cross section the half wave length is also function of the aspect ratio of the cross section through parameter q

$$\lambda = b \left[\frac{1}{q} \left(\frac{1}{4} + \frac{3}{4} \frac{E_t}{E_s} \right) \right]^{1/4}; \tag{13.31}$$

For mild steel, this leads to

$$\lambda \approx 0.8 \frac{b}{q^{1/4}}. \tag{13.32}$$

Above the critical stress, the lateral displacement of the panels are no longer small compared to the cross section: So the assumptions at the basis of the elastic or the plastic instability are no more valid and the nonlinear effects cannot be neglected in the computation of the stress and the deformations.

Another effect is that, because of the large deformations, the stress is no longer constant on the cross section. Close to the corners, since the stress is higher and can be near, or above, the yield stress, the central part of the panels is less severely loaded due to the buckling. This effect is similar to what occurs in the effective width of flat panels loaded in compression above the critical elastic stress.

13.3.2 Stable Crush: Mechanics of the Deformation

The aim of this section is to describe the kinematics involved by the folding mechanism during a stable crushing of a beam as represented in Fig. 13.30 a).

Fig. 13.32 shows the basic mechanism for a part of a beam with rectangular cross section. The two symmetry planes of the undeformed beam remain so during the crush enabling a simplification of the graphical representation by considering only one quarter of the cross section with an initial length equal to half the wavelength of the elastic instability mode.

After the instability, the reference line $A_0C_0D_0$ rotates about corner UC_0L. Increasing the axial deformation this corner becomes a plastic hinge that moves in the direction of the bisector of angle α. During deformation, line ACD maintains a constant length with good approximation ($A_0C_0D_0 = A_1C_1D_1$). The displacement of A and D on the two vertical symmetry planes is therefore the same ($A_0A_1 = D_0D_1$).

The part of the plate swept by the plastic hinge UCD is subject to an inextensional deformation. As shown in Fig. 13.28, this implies a bending of the plate followed by unbending which restores its flat surface.

Fig. 13.33 shows the the movement of the plastic hinges in a cylindrical beam. Increasing the axial deformation plastic hinges UCL move apart at a larger distance, as indicated by the increasing distance between points C.

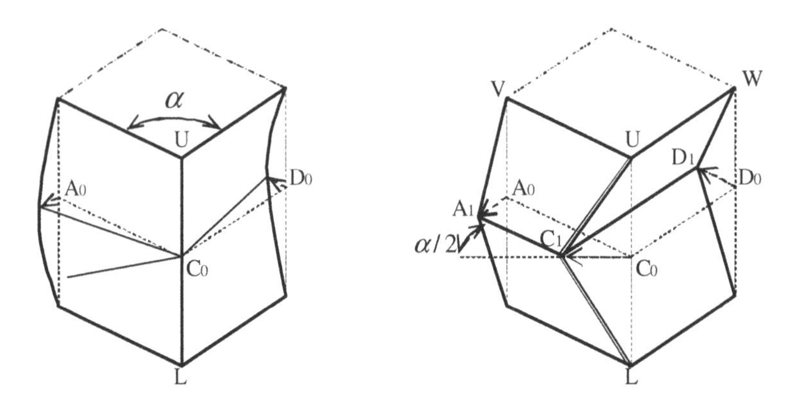

Fig. 13.32. Sketch of the kinematics involved in folding. As the length $A_1C_1D_1$ remains constant this is an inextensional mode.

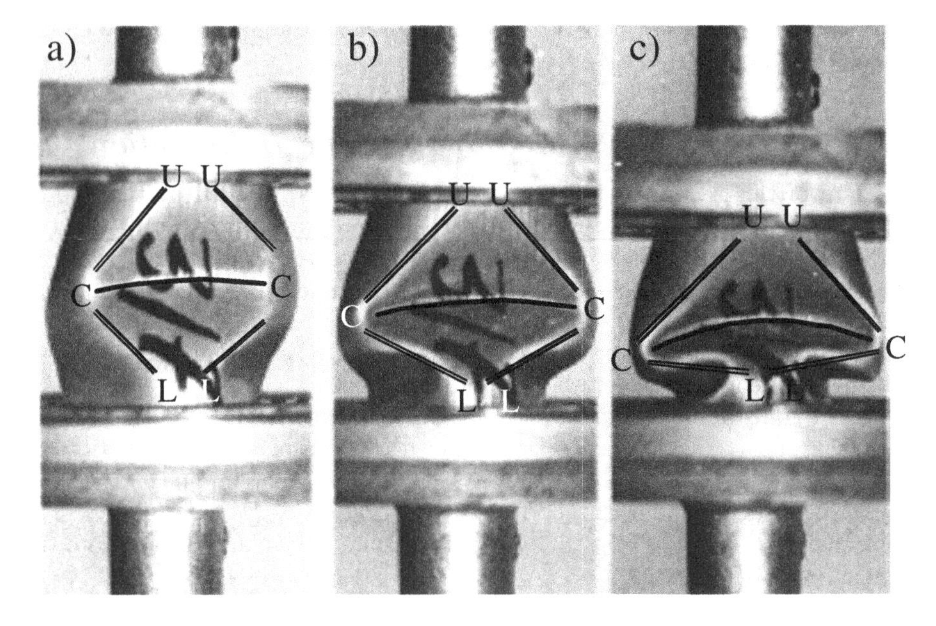

Fig. 13.33. Example of the growth of a fold on a cylindrical beam. The lines evidence the diverging movement of the plastic hinges UCL.

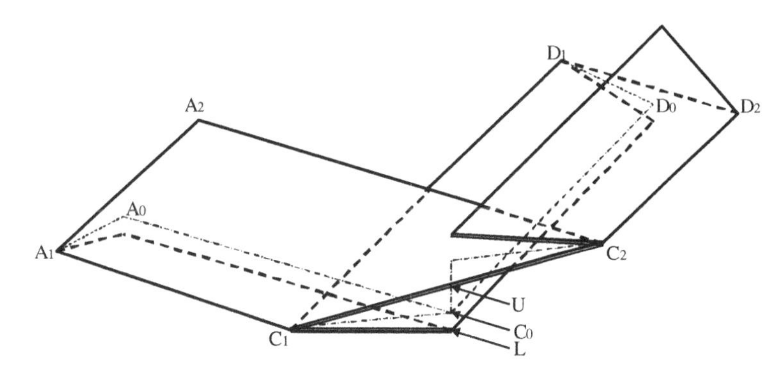

Fig. 13.34. Kinematics of neighboring folds. Before crushing line UC_0L was a straight corner. The outwards folds form alternatively on the left and the right face.

Because of continuity, the folding kinematics represented in Fig. 13.32 has implications on the neighboring parts. Considering two neighboring faces, if one moves outwards, the other will move inwards. For this reason the outwards motion of surface A_1C_1UV induces an inwards deformation of the surface above line VU. Similarly, surface C_1D_1WU induces an outwards deformation of the surface located above line UW.

If the amount of deformation is large enough to bring points U and L into contact, the first fold closes. The deformations in the neighboring parts trigger a second fold and the process repeats periodically. This is evidenced in Fig. 13.34, that represents a longer portion of the structure of Fig. 13.32 with an axial deformation large enough to generate two folds.

Fig. 13.36 shows (top view) the growth of two folds in a squared section beam figure (a) refers to the first fold wile (b) to the second. The inwards displacement l_f of the plastic hinge C_1D_1 (Fig.s 13.32 and 13.33) transforms the section from square to rectangular with a shorter side equal to:

$$b' = d - 2l_f. \tag{13.33}$$

The folding amplitude l_f is related to the half wave length λ, that, in turn, is function of the side d of the square section. For square sections, the first fold forms in longer side (Fig. 13.37), the distance b' can be so small so as to have contact between facing folds. In this case, the bending moment of inertia of the cross section can be so small that the beam could fail for global instability. This occurs for aspect ratios $d/b > 0.584$.

The force to displacement characteristic during the crushing of a square or rectangular section beam is shown in Fig. 13.38. If the thickness is such that the instability of the beam sides is in the plastic range, the stress corresponding to

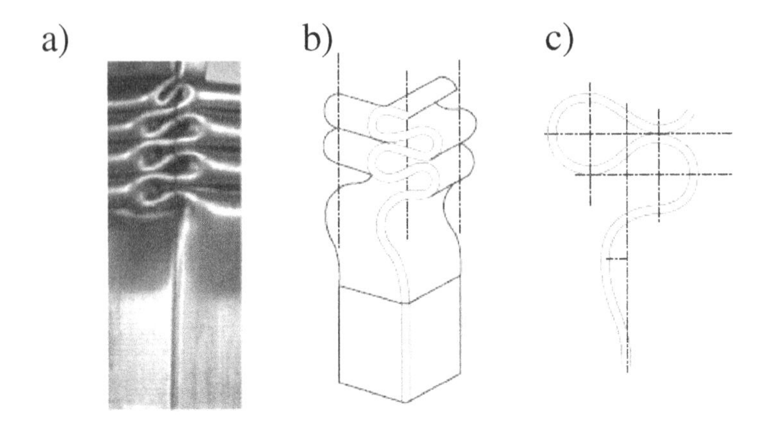

Fig. 13.35. Folds during stable axial crush. a) Experiment, b) sketch, c) section perpendicular to a face with contact between two folds.

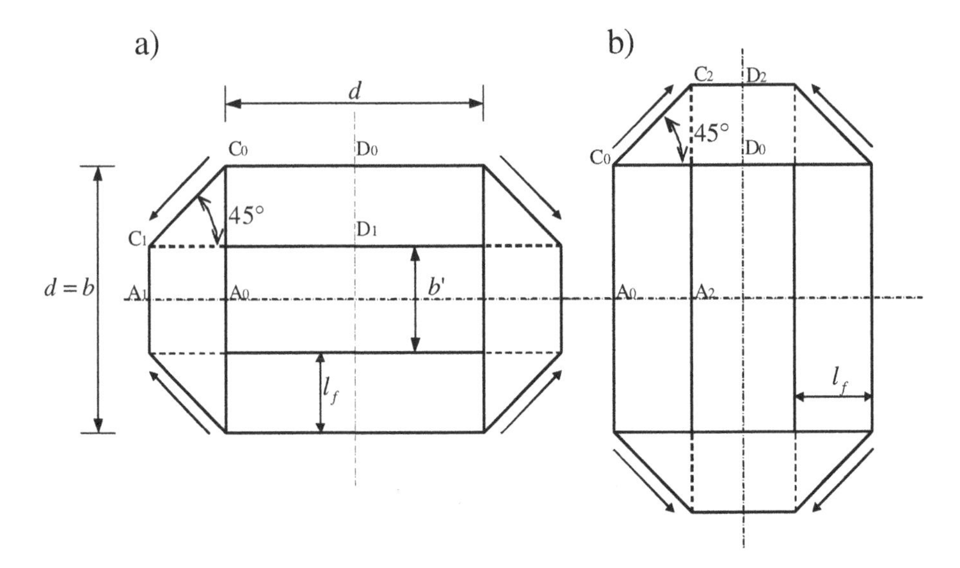

Fig. 13.36. Folding kinematics of a beam with square cross section ($d = b$). a) growth of the first lobe: side CD moves inwards; b) growth of the second lobe: side CD moves outwards.

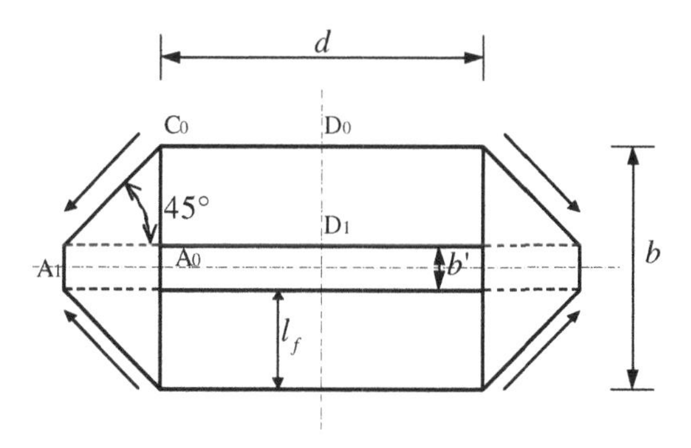

Fig. 13.37. Fold kinematics of a beam with rectangular cross section. If $d/b > 0.584$, there can be contact between facing inwards folds, leading to global instability.

the peak load C can be evaluated from the yield stress σ_y, and the elastic critical stress (σ_{cr}, Eq. 13.24):

$$\sigma_{\max} = \frac{P_{\max}}{A} = \sigma_y \left(\frac{\sigma_{cr}}{\beta\sigma_y}\right)^n . \tag{13.34}$$

n is function of the cross section shape, for a rectangular cross section $n = 0.43$. Parameter β is related to the thickness ratio t/b and from the ultimate stress σ_u of the material.

$$\begin{array}{ll} t/b \le 0.016 & \beta = 1; \\ 0.016 \le t/b \le 0.035 & 1 \le \beta \le \dfrac{\sigma_u}{\sigma_y}; \\ t/b \ge 0.035 & \beta = \dfrac{\sigma_u}{\sigma_y}. \end{array} \tag{13.35}$$

For small thickness ratios $t/b < 0.016$, the instability occurs in the elastic range and $\beta\sigma_y = \sigma_y$.

Conversely, for large thickness ratios $t/b > 0.035$ and $\beta\sigma_y = \sigma_u$.

The first part of Eq. 13.34 shows that $\sigma_{\max}$ is actually a mean stress on the cross section. In fact it is evaluated as the ratio between the maximum compression force ($P_{\max}$) and the cross section area A. Substituting Eq. 13.24 in Eq. 13.34:

$$\sigma_{\max} = \sigma_y \left\{\frac{k_p (t/b)^2 E}{(1-\nu^2)\beta\sigma_y}\right\}^n , \tag{13.36}$$

where k_p is the *crippling coefficient*, its value is reported in Fig. 13.40 as function of the aspect ratio $\alpha = d/b$ of the cross section. Fig. 13.41 shows the comparison between the maximum stress in the plastic range as given by Eq. 13.34, and the elastic critical stress of Eq. 13.24. For small thickness ratios t/b the elastic instability occurs at loads lower than the elasto-plastic instability. Increasing the thickness ratio t/b, the elasto-plastic instability prevails. An important point is that, when the crush occurs in the elastic range, its stability is rather poor and the beam will likely eventually fail for global instability (crumbling), with a very

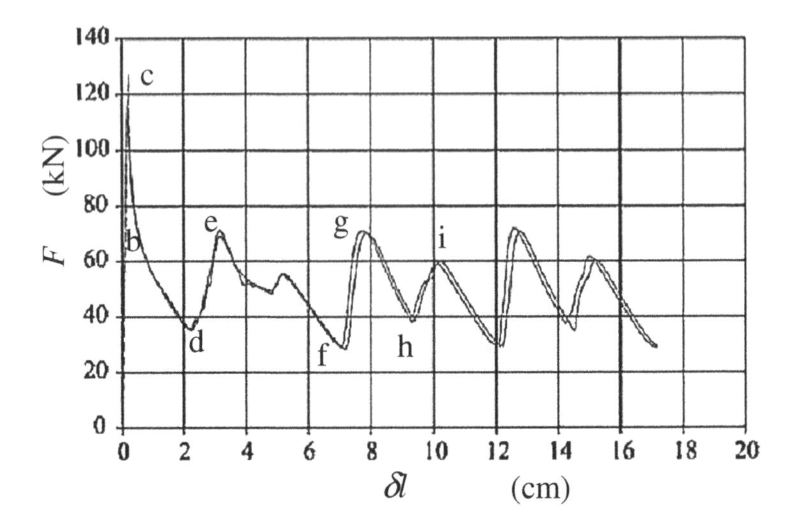

Fig. 13.38. Force to displacement characteristic during the crushing of a square or rectangular section beam. The letters indicate following phases of the crush. *b* beginning of the elastic instability; *c* max load; *d* first fold is closed; *e* compression instability and start of second lobe; *f* second fold is closed,...

low energy absorption. Conversely, the crushing in the elasto-plastic range is much more stable and is associated to a larger energy absorption. It is therefore important to design the beam thickness in order to move the critical load in the elasto-plastic range. From this viewpoint also the yield strength could play a role. Fig. 13.41 shows that for ratios $t/b \sim 0.015$ the selection of a material with larger yield stress can lead from a stable elasto-plastic instability to an elastic instability with the possibility of crumbling (global instability).

Non compact sections have a critical stress in the elastic range. With reference to Fig. 13.41, if $\sigma_y = 160$ MPa non compact sections have $t/b < 0.0085$.

Compact sections have a critical stress in the elasto plastic range. If $\sigma_y = 160$ MPa their thickness ratio must be $t/b > 0.0085$.

Considering Fig. 13.38, after peak (c) the first fold appears and increases its depth, the corresponding load has a sharp reduction as the effect of the fold is to increase the compliance of the structure. When the first fold closes at point (d), the stiffness of the box beam returns to almost the original value. The load increases again until a new instability occurs in the part of the panel not involved in the first folding (e). The load reduces again until the new fold closes (f).

In compact box structures, this process can repeat several times in a stable sequence of load increases and reductions. As the process is irreversible, the energy represented by the area under the force-displacement characteristic is dissipated. Conversely, in non compact structures, the folding sequence is not stable and the structure fails for global instability.

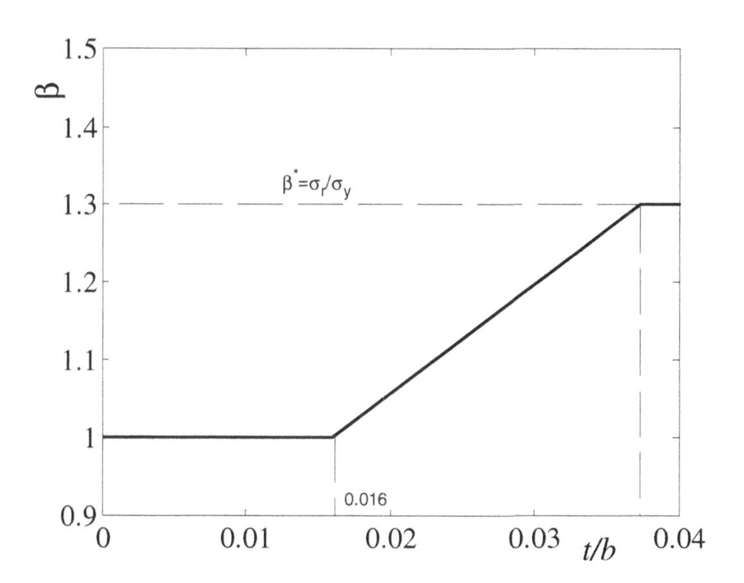

Fig. 13.39. β coefficient as function of the thickness ratio. If $t/b < 0.016$ the crush is in the elastic range, for $t/b > 0.035$ the crush is in the plastic ragne.

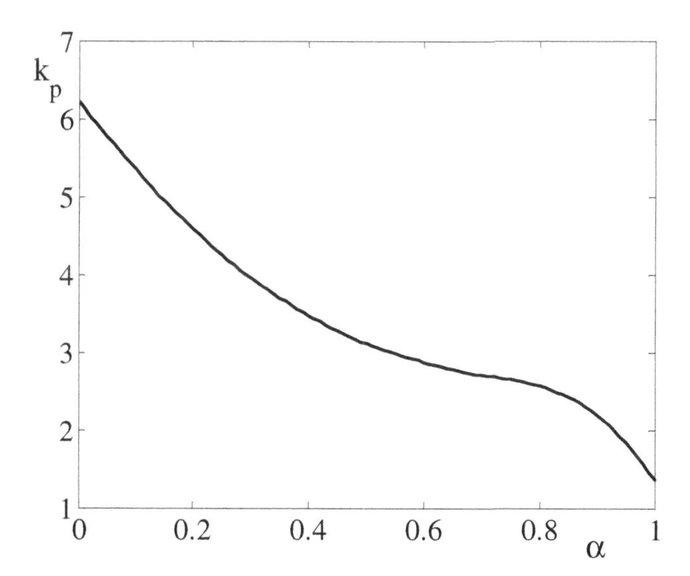

Fig. 13.40. Crippling coefficient k_p for box beam structures (rectangular cross section) as function of the aspect ratio $\alpha = d/b$ between the smaller (d) and larger side (b) of the cross section.

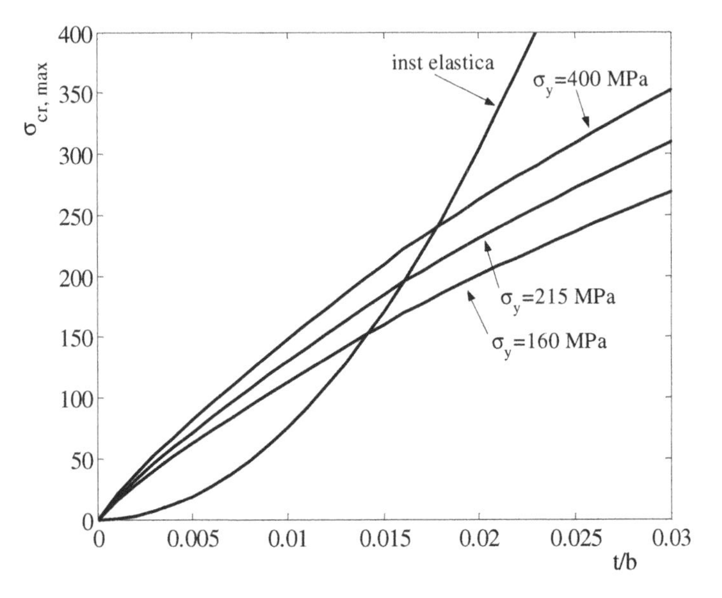

Fig. 13.41. Comparison between critical stress (crippling strength) obtained considering a crush in the elastic range and in the elasto-plastic range. For a given yield stress (σ_y) the elasto-plastic crush (stable) occurs when its stress is lower than the elastic one. This happens for large enough thickness ratios ($t/b > 0.016$).

Experimental measurements show that the force corresponding to the first instability (point c) is larger than the following ones. This implies the unwanted effect of larger accelerations at the beginning of the impact which can be avoided by adding crush triggers on the box structure (*triggers* [31]) the role of which is to reduce the load corresponding to the first instability, and permit the crushing sequence to feature an almost constant amplitude of the load oscillation.

To design a shock absorber, it is important to estimate the amplitude of the load oscillation and its mean value [32] which can be detemined by an expression very similar to Eq. 13.36 (that gives the maximum load of point c), except that the value of the geometrical coefficient k_i is different affecting the terms in the bracket

$$\sigma_i = \sigma_y \left\{ \frac{k_i (t/b)^2 E}{(1-\nu^2)\beta\sigma_y} \right\}^n , \qquad i = 1, 2, 3. \tag{13.37}$$

Even in this case, for a box beam with rectangular cross section, $n = 0.43$. Coefficients k_i can be found as function of the aspect ratio $\alpha = d/b$ of the cross section in diagrams such as Fig. 13.42. The comparison of Fig. 13.40 and 13.42 shows that $k_p > k_i$. As expected, the stress corresponding to the first peak is larger than the following peaks (points e, g, i of Fig. 13.38).

As shown by Wierzbicki and Abramowicz [33], the mean load of beams with polygonal cross section (force F_2 in Fig. 13.38) can be estimated by an approach based on the kinematics of the plastic deformations. The same approach can be used to evaluate the mean load of box structure with square cross section of side b, thickness t and made by an elasto-plastic material:

$$F_2 = 9.56\ \sigma_y t^{5/3} b^{1/3}. \tag{13.38}$$

This expression can be used as an alternative to Eq. 13.37.

13.3.3 Triggers

The maximum load reached at point c of the force-displacement characteristic of Fig. 13.38 may be considerably larger than that of the following peaks (points e, g and following, corresponding to strength σ_1 Eq. 13.37) and of the mean load (σ_2). The peak load reached at point c leads to a series of drawbacks:

- a peak of the acceleration in the main part of the vehicle and the occupants;
- the risk of global compression instability of the supporting structures; to avoid this it may be necessary to oversize them, resulting in an excessive load when they come into play as absorbers. The outcome is again a large acceleration of the occupants.

Fig. 13.38 shows that the deformation and the energy corresponding to point (c) are relatively low. To reduce the accelerations, for a given absorbed energy, the load should be constant. The first load peak of the characteristic should then be reduced at the same level of the following peaks. This is obtained by

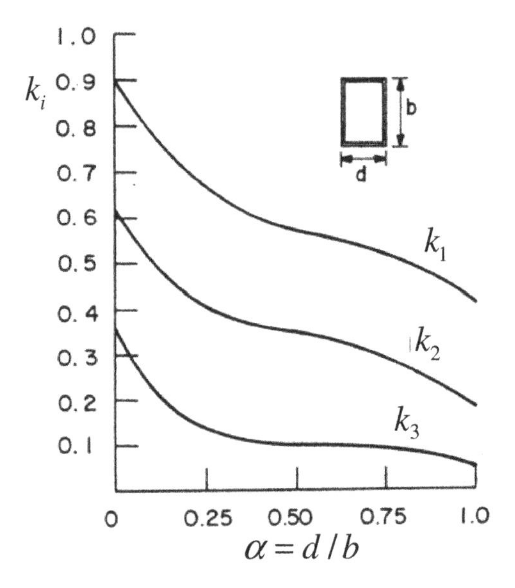

Fig. 13.42. Geometrical coefficient that allow to estimate the maximum (k_1), mean (k_2), and minimum load (k_3) during the elasto plastic crush of compact box beams given by Eq. 13.37.

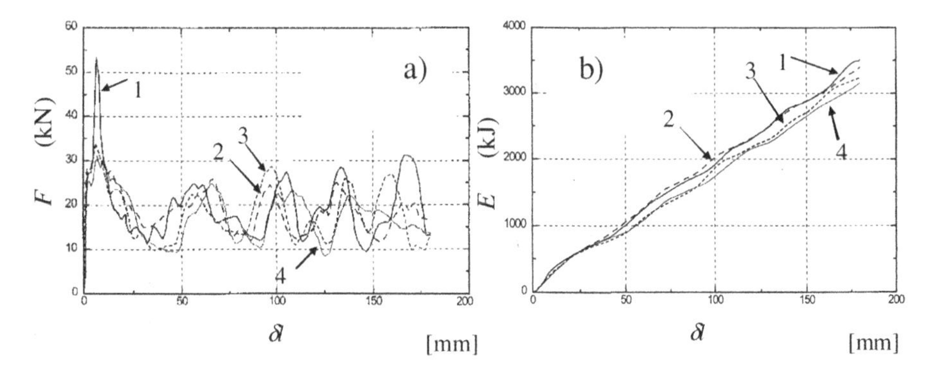

Fig. 13.43. Effect of the *triggers*. a) on the force-displacement characteristics: 1) no *triggers*. 2) e 3) different types of triggers. b) Energy-displacement characteristic.

including some imperfections (usually referred to as *triggers* or *crush initiators*) on the surface of the beam which act to reduce the first critical load below that of the flat panel in which they are included. The result is that the first folding occurs at a lower load, without altering substantially the subsequent process and the associated dissipation.

Fig. 13.43 ([31]) shows the effect of different types of triggers on the force-displacement curves of aluminum box structures with 50 mm side, 2 mm thickness square section. The comparison of curve 1) with no triggers and curves 2, 3, 4 (triggers) shows that the triggers reduce considerably the first peak, maintaining almost constant the mean load (diagram a) and the absorbed energy as function of the displacement (diagram b).

Fig. 13.44 shows different types of crush initiators obtained by stamping or punching ([34]). Solution a) (bead) is a type of notch that may involve all or part of the side panel. Notches b) (diamond notch) involve the corners, small spherical domes (c) may be stamped on the walls. Bead d) replicates the shape of the panel after its instability.

The position, type and size of the initiators are subject to optimization. As a first approximation, initiators may be located at the point of maximum deflection of the half wave λ that will grow following the first instability of the side panel. In Fig. 13.32 this location corresponds to line UW. In box structures with square cross section, this position can be approximated by Eq. 13.28, or ([33]):

$$\lambda = 2H_{fold} = 0.983\sqrt[3]{\left(\frac{a+b}{2}\right)^2 t}. \qquad (13.39)$$

This location is correct if the end section is hinged along its contour so as to leave the panels free to rotate about their sides. If the end section is welded to a

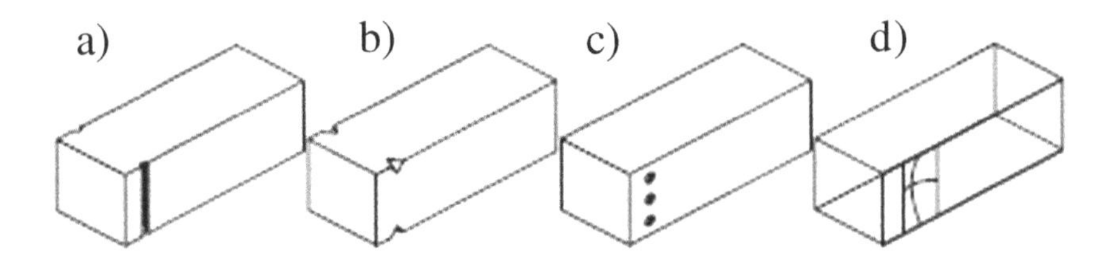

Fig. 13.44. Examples of triggers or crash initiators.

rigid element, this cannot occur: In this case the half wave-length increases and the location of the triggers should take this into account.

13.3.4 Global Compression Instability of Beams

Until now the focus has been on stub beams with a local instability that occurs at loads lower than Euler (or global) instability. If the stub beam can be considered to be compact, the crushing is stable with a sequence of foldings and high energy dissipation.

Long beams (for which the length is high compared to their cross section) may fail for global instability with a considerable reduction of the absorbed energy. The crush strength of Eq. 13.36 must be corrected in this case to take the effects of the length into account:

$$\sigma'_{\max} = \sigma_{\max} - \frac{\sigma^2_{\max}}{4\pi^2 E}\left(\frac{l}{\rho}\right)^2 , \tag{13.40}$$

where ρ is the radius of inertia of the cross section about the normal to the bending plane and $\sigma_{\max}$ is given by Eq. 13.36.

Fig. 13.45 shows that for box beams (rectangular cross section), the effect of the length is appreciable for length to section side ratios $l/b > 10$ and that for $l/b > 50 \div 70$ the Euler instability becomes the dominant failure mode. This must be taken into account during the design phase in order to avoid the global instability of beams involved in the transmission of the crash loads.

13.3.5 Bending Instability

The presence of inaccuracies in the shape and alignment of the sections, combined with the axial loading, induce unavoidable bending components that usually cause the global bending instability to be the dominant crushing mode, instead of the stable axial mode. The result is lower energy dissipation, so that the design of the shock absorber is aimed at avoiding global instability.

The bending instability produces a plastic hinge in a single portion of the beam while the rest remains almost undeformed. The type of plastic hinge and dissipation mechanism is related to the geometry of the beam section. Sections

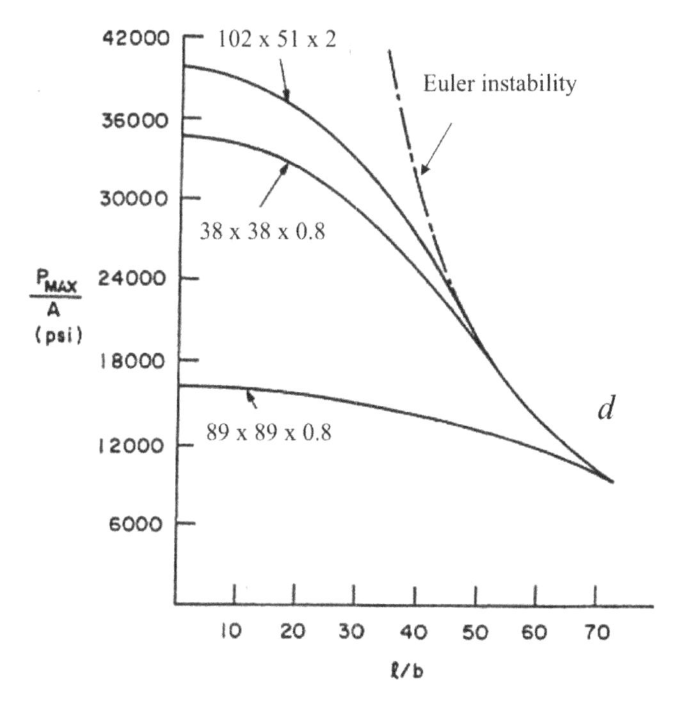

Fig. 13.45. Critical load as function of the length of the beam. The collapse is in the plastic range only for short lengths.

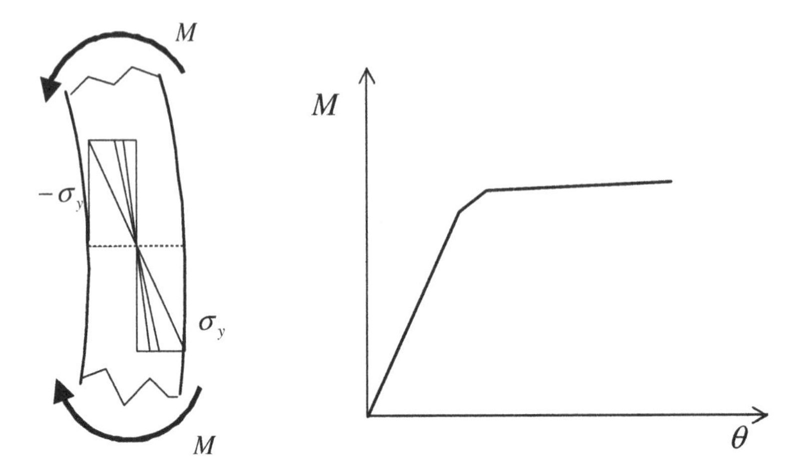

Fig. 13.46. Stress distribution (left) and bending torque as function of the relative bending angle θ (right) for a beam with full depht cross section made of an elastic-perfectly plastic material.

with large thickness or with full depth material start to yield on the outer surface. The yielded portion increases by increasing the load, until all material is plasticized.

Fig. 13.46 shows the stress distribution in the plastic hinge section. The limit elastic load corresponds to a triangular stress distribution with the yield stress σ_y on the outer surface. If the load is increased above this value, the material starts to yield. Increasing the load, a larger portion of the cross section is plasticized, from the outer surface going inwards towards the elastic axis (the line on the cross section characterized by null stress). The bending torque increases during this process, even if not so considerably, as the stress increases at a smaller and smaller distance from the elastic axis. The maximum bending torque occurs when all material of the cross section is at the yield stress either in tension or compression. Fig. 13.46 represents a sketch of the stress distribution on a cross section under the assumption that the material is elastic - perfectly plastic.

In thin walled beams, the mechanism that leads to the plastic hinge is dominated by the geometry of the cross section, rather than only by yielding, as it occurs in full depth or thick sections. The collapse starts with the instability of the panels loaded in compression.

Similarly to what has already ben mentioned regarding the crushing of thin walled beams and panels, this first instability can be found from Eq. 13.36. The instability of the panels reduces the moment of inertia of the cross section, the result being a further reduction of the bending moment M that can be applied. As shown qualitatively in Fig. 13.47 [35], [36], the bending moment decreases by increasing the bending deflection θ. The reduction is more marked for smaller thickness to width ratios of the compressed panel (t/b). Reducing this ratio (t/b),

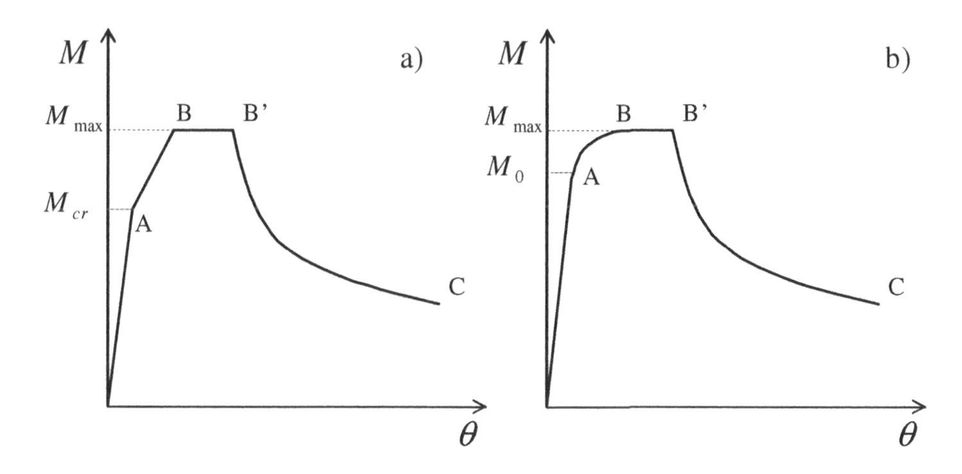

Fig. 13.47. Bending moment M to bending deformation θ characteristic for a thin walled beam. AB: instability of the compressed side. Diagram (a) refers to the case of an elastic instability. Diagram (b) to that of a plastic instability. Beyond point B' the bending load M decreases by increasing the deformation (θ).

the critical stress σ_{cr} decreases and, correspondingly, so does the stress in the compressed side during buckling (Eq. 13.34).

Fig. 13.47 shows qualitatively this behavior for a thin walled beam subject to bending. The left diagram (a) refers to an elastic instability, the right (b) to a plastic one. The instability of the compressed panel occurs in segment AB. Increasing the load, the deflection of the compressed panel increases, until the load reaches the maximum strength of the beam, represented by segment BB'. Increasing the deformation beyond point B', the required load decreases.

Similarly to what occurs with axial crushing, for non compact sections ($t/b < 0.0085$), the instability of the compressed panel occurs in the elastic range and the dissipated energy is lower than for compact sections ($t/b > 0.0085$), characterized by a plastic instability of the compressed side. In the latter case, the compressed panel maintains a certain load capacity even after the instability, allowing the dissipation of a not negligible amount of energy.

Similarly to what has been mentioned regarding the effective width of compressed panels, after the instability, the load capacity of the central portion decreases because of the buckling. The part closer to the edges maintains a higher load capacity that arrive at the yield stress. The effective width b_e can be found by assuming that the yield stress acting on the effective width produces a load equal to the maximum average stress $\sigma_{\max}$ acting on the whole panel width:

$$\sigma_y b_e t = \sigma_{\max} bt; \tag{13.41}$$

The effective width is then:

$$b_e = \frac{\sigma_{\max}}{\sigma_y} b. \tag{13.42}$$

The maximum average stress $\sigma_{\max}$ can be evaluated from Eq. 13.36,

$$b_e = b\left\{\frac{k_p(t/b)^2E}{(1-\nu^2)\beta\sigma_y}\right\}^n . \tag{13.43}$$

Alternative expressions have been introduced for the effective width.

For this quantity, Kekman states that:

$$b_e = b(0.7\sigma_{cr}/\sigma_y + 0.3), \tag{13.44}$$

where σ_{cr} is the elastic critical stress given by Eq. 13.24 in which the geometrical parameter k is expressed as:

$$k \approx 5.23 + 0.16\frac{d}{b}. \tag{13.45}$$

Instead the American Steel Institute indicates [37] that:

$$b_e = 1,9t\sqrt{\frac{E}{\sigma_y}}\left(1 - 0.415\frac{t}{b}\sqrt{\frac{E}{\sigma_y}}\right). \tag{13.46}$$

The maximum bending moment $M_{\max}$, before the elasto-plastic collapse can be evaluated by three different expressions, depending on the value of the critical stress σ_{cr} compared to the yield stress σ_y.

$$M_{\max} = \begin{cases} M'_{\max} \\ \left[M'_p + (M_p - M'_p)(\sigma_{cr} - \sigma_y)\right]/\sigma_y \\ M_p \end{cases}, \tag{13.47}$$

where

$$\begin{gathered} M'_{\max} = \sigma_y td\left[2b + d + b_e\left(3\frac{a}{b} + 2\right)\right]/3(d+b), \\ M'_p = \sigma_y td(d + b/3), \\ M_p = \sigma_y t\left[b(d-t) + (d-2t)^2/2\right]. \end{gathered} \tag{13.48}$$

Fig. 13.30 b) shows the configuration of the beam during the bending instability. Only a relatively small portion of the material is subject to plastic deformation while all the rest remains in the elastic range. The absorbed energy is therefore significantly lower than that needed for the axial crush with folding.

As the energy involved in the elastic deformation is smaller than that dissipated in plastic deformation (Fig. 13.47 b), the first part of the characteristic can be neglected.

Despite the complexity of the buckling phenomenon, the geometry of the plastic hinge is relatively simple. Fig. 13.48 shows the yield lines that appear during the bending failure of a beam with rectangular cross section. The bending moment acts about the width of the cross section.

The bending moment can be evaluated from the energy conservation principle by assuming that the power involved by the bending moment M with the deformation speed $\dot{\theta}$ is the same as the power $\dot{W}_{int}$ dissipated in the beam:

$$M\dot{\theta} = \dot{W}_{int}. \tag{13.49}$$

The power $\dot{W}_{int}$ is dissipated in the yield lines:

$$\dot{W}_{int} = \dot{W}_A + \dot{W}_D + \dot{W}_{AD} + 2\left(\dot{W}_{FE} + \dot{W}_{EL} + \dot{W}_{AE} + \dot{W}_{AL} + \dot{W}_{LM}\right). \quad (13.50)$$

At yield lines A and D there is a material flow through a toroidal surface that implies the elongation of lines AD, EA, AG and the shrinking of lines AL, BK. This is similar to what occurs in Fig. 13.28. Lines AE, AL, AG,... are then moving hinges while EF, EL, LM, GH,... are fixed.

The power dissipated by a fixed plastic hinge of length l and with a rotating speed $\dot{\psi}$ (rad/s) about its axis is

$$\dot{W}_f = M_p l \dot{\psi}, \quad (13.51)$$

where M_p is the bending moment of a perfectly plastic plate of thickness t and unit width:

$$M_p = \sigma_y \left(\frac{t}{2}\right)^2. \quad (13.52)$$

Conversely, the power dissipated by a moving hinge line is

$$\dot{W}_m = 2M_p \frac{\dot{A}_r}{r}, \quad (13.53)$$

where r is the radius of curvature of the plastic hinge and A_r is the area that is swept by the hinge during its motion.

The rotating speed $\dot{\psi}$ and the length of the moving hinges is then expressed as function of the deformation speed $\dot{\theta}$ under the assumption of a given kinematics of the part subject to plastic deformations. This approach is then usually referred to as kinematic method for the plastic collapse. The power dissipated by the yield lines is expressed in the form:

$$\dot{W}_{int} = \Gamma(\theta)\dot{\theta}, \quad (13.54)$$

From Eq. 13.51 and Eq. 13.54 it follows

$$M = \Gamma(\theta). \quad (13.55)$$

Since the function $\Gamma(\theta)$ is a rather intricate function it is not explicitly reported here. Usually the computation is performed using dedicated software tools such as DEEPCOLLAPSE, SECOLLAPSE e CRASH-CAD.

13.3.6 Collapse of Tubes with Circular Cross Section

Tubes with circular cross section can be used as shock energy absorbers by exploiting different modes of collapse [38] :

- inversion;
- splitting;

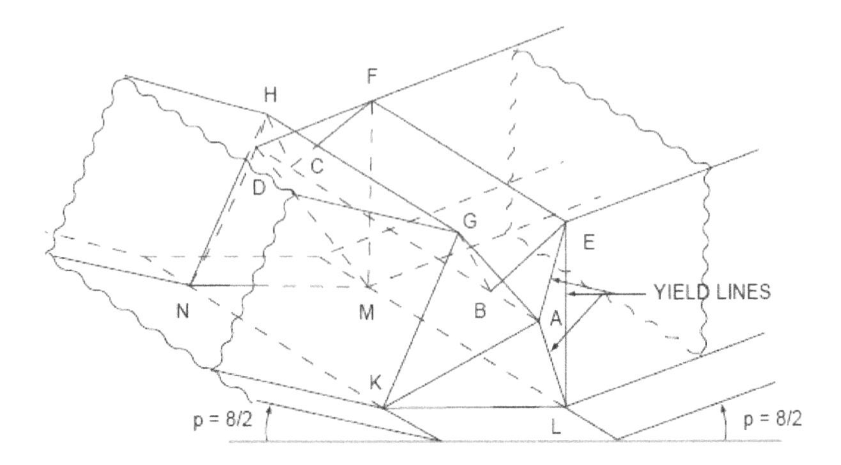

Fig. 13.48. Cinematica del cedimento per instabilità flessionale. Le linee indicano le cerniere plastiche fisse e mobili.

- lateral indentation;
- lateral flattening;
- axial crushing.

The inversion involves turning inside-out or outside-in of a thin circular tube made of ductile material, as shown in Fig. 13.49. During the inversion, the material elongates (or shrinks) in the circumferential direction while being bent on a diametral plane. This bending first produces a curvature with radius r and before restoring the straight configuration.

This mode is sometimes referred to as the invertube; its specific characteristics are the constant load and the high energy density. Conversely there is high sensitivity to the die radius r. Low values of r trigger a folding mode of collapse (as for box beams), whereas if r is too large the material splits on diametral planes (Fig. 13.50).

If the tube splits into strips, the energy disspation is due to the large deformations involved in bringing the material to the ultimate stress. Also in this case the load is reasonably constant.

Lateral indentation occurs when a tube hinged at the ends is loaded in a transverse direction (Fig. 13.51). The lateral load first produces a dent with no relevant implications on the straight axis. The axis then bends into a V shape. As the yielding is rather limited and is concentrated near the loaded section, the amount of dissipated energy is low. For this reason, this mode is seldom used as an energy absorber. Nevertheless, this is the dominant mode when a

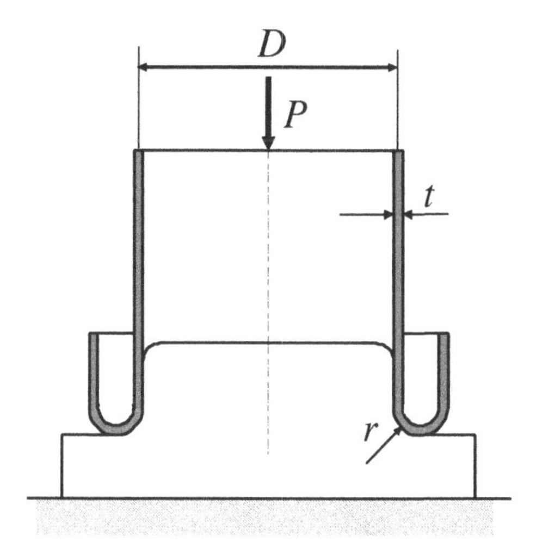

Fig. 13.49. Inversion of a tube with circular cross section (invertube).

Fig. 13.50. Tube splitting. If the die radius r of Fig. 13.49 is smaller than a given value the tube tears apart in strips. The strips can be flat or they can wrap.

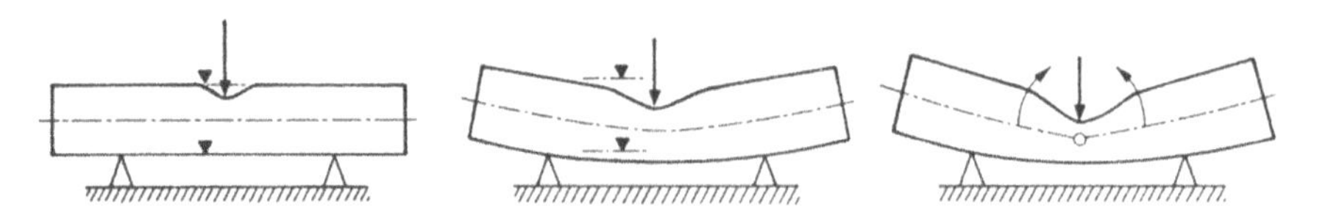

Fig. 13.51. Lateral indentation. As yielding occurs in a small amount of material, the dissipated energy is small.

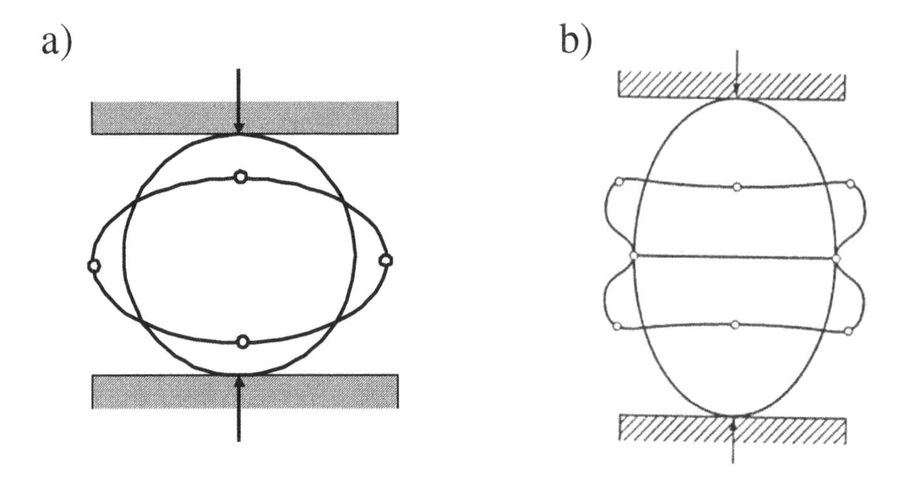

Fig. 13.52. Lateral flattening. The small circles evidence the plastic hinges. a) no tension bracing; b) with tension bracing.

beam impacts laterally, for example when the bumper beam impacts against an obstacle (a pole).

The lateral flattening of a cylinder with circular or elliptic cross section, (Fig. 13.52), enables an amount of energy larger than for the lateral indentation to be dissipated. The dissipated energy density increases in arrangements where each module induces boundary conditions to the others such that the number of plastic hinges increases. The cylinder of Fig. 13.52 a) is characterized by four plastic hinges on two diameters (the loaded one and that perpendicular to it). The arrangement of Fig. 13.52 b) involves a larger number of hinge points per each cylinder which increases the amount of dissipated energy.

13.3.7 Axial Collapse of Circular Tubes

Cylinders with circular cross section can have two axial collapse modes with high energy dissipation. They are characterized by different values of the diameter (D) to thickness ratio (t).

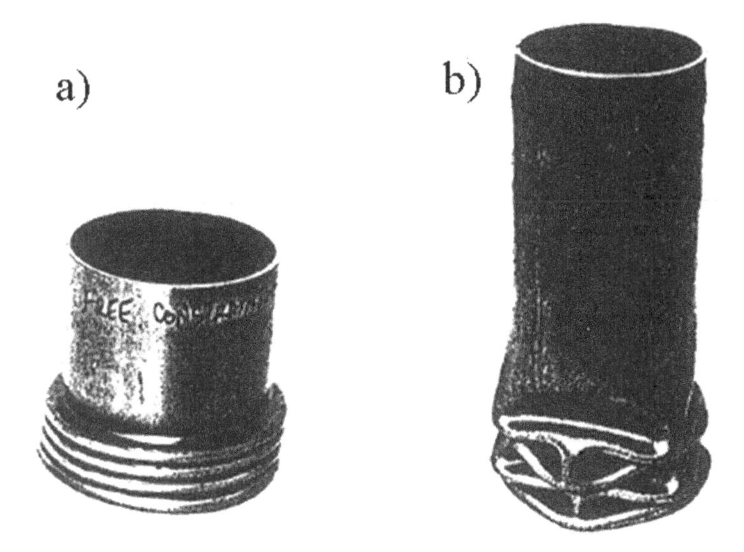

Fig. 13.53. Axisymmetrical collapse (concertina) a); and non axisymmetrical (diamond) collapse b) of a cylinder with circular cross section.

For relatively small ratios $D/t < 50 \div 100$, the cylinder crushes in a sequence of axisymmetric bellows formed by fixed plastic hinges. The result is a sort of "concertina" (concertina mode). Each hinge appears as a local axisymmetric instability mode; when a hinge is completely packed, a new hinge can develop and the process repeats again. As the diameter of each bellow is not the same as that of the cylinder, the concertina mode involves extensional hinges .

For larger values of the diameter to thickness ratio $D/t > 100$, the deformation is not axisymmetric. A number of plastic hinges appear on the surface in the diagonal direction to form diamond-like shapes (diamond mode). The plastic hinges in this case are inextensional.

Concertina mode

Once the first peak is overcome, the axial load initiates a series of oscillations as function of the axial displacement, as represented in Fig. 13.31. The mean load $\bar{P}$ during this phase can be approximated by a kinematic approach, introduced by Alexander in the 1960s [39]. The basic assumption is that the work that must be exerted to form one convolution is given by two contributions (Fig. 13.54): that required for bending the plastic hinges (W_1) and that for stretching in circumferential direction the material between joints (W_2). Additionally, the plastic hinges are assumed to be in pure bending, while the material between two hinges is in pure extension or compression. Experimental results have demonstrated the validity of these assumptions [40].

Considering these assumptions, increasing the axial deformation hinges c_1 and c_3 rotate of an angle $d\theta$, while hinge c_2 of twice this value. The work for this deformation is:

$$dW_1 = 2dW_{c1} + dW_{c2}. \tag{13.56}$$

The work for the deformation of hinges c_1 and c_2 is obtained from the bending moment M per unit circumferential length of the plastic hinge:

$$dW_{c1} = M\pi D d\theta, \tag{13.57}$$

$$dW_{c2} = M\pi(D + 2h\sin\theta)2d\theta; \tag{13.58}$$

and substitution in Eq. 13.56 gives:

$$dW_1 = 2M\pi D d\theta + 2M\pi(D + 2h\sin\theta)d\theta = 4M\pi(D + h\sin\theta)d\theta. \tag{13.59}$$

For a narrow beam with rectangular cross section, the material will be deformed in essentially plane stress conditions. Considering the material as rigid-plastic, half the thickness is subject to the tensile yield stress σ_y and the other half to compression $-\sigma_y$. The bending moment per unit length in plane stress conditions is then:

$$M = \sigma_y \frac{t^2}{4}. \tag{13.60}$$

Due to the cylindrical symmetry of the deformation, each strip of the buckling cylinder can be regarded as a part of a very wide beam. This means that the material is subject a plane strain rather than a plane stress with zero strain in the circumferential direction $\epsilon_c = 0$. Additionally, if the thickness is small, the stress perpendicular to the thickness ($\sigma_t = 0$) is negligible everywhere, the stress and strain matrix relation is then given by:

$$\begin{pmatrix} \epsilon_a \\ \epsilon_c \end{pmatrix} = \frac{1}{E} \begin{bmatrix} 1 & -\nu \\ -\nu & 1 \end{bmatrix} \begin{pmatrix} \sigma_a \\ \sigma_c \end{pmatrix}, \tag{13.61}$$

where σ_a and σ_c are the axial and circumferential stresses, while ϵ_a and ϵ_c are the corresponding strains.

During an increment $d\theta$, the variation of the circumferential strain is negligible, and from Eq. 13.61 it is possible to find:

$$\sigma_c = \nu\sigma_a. \tag{13.62}$$

As σ_a and σ_c are principal stresses, the yield can be found considering the Von Mises equivalent stress

$$\sigma_y = \frac{1}{\sqrt{2}} \sqrt{(\sigma_a - \sigma_c)^2 + (\sigma_a - \sigma_t)^2 + (\sigma_c - \sigma_t)^2}. \tag{13.63}$$

Substituting Eq. 13.62 in 13.63 results in:

$$\sigma_y = \frac{\sigma_a}{\sqrt{2}} \sqrt{2(1 + \nu + \nu^2)} \approx \frac{\sqrt{3}}{2} \sigma_a; \tag{13.64}$$

the axial stress in the plastic hinges is then:

$$\sigma_a = \pm \frac{2}{\sqrt{3}} \sigma_y \tag{13.65}$$

Finally the bending moment per unit circumferential length in plain strain conditions is:

$$M = \frac{2}{\sqrt{3}} \sigma_y \frac{t^2}{4}. \tag{13.66}$$

This enables the contribution to the increment of work due to the plastic hinges to be calculated:

$$dW_1 = \frac{\pi}{\sqrt{3}} \sigma_y t^2 (2D + h \sin \theta) d\theta. \tag{13.67}$$

The increment of work dW_2 to extend the material between two hinges can be calculated from the circumferential strain. Considering a fiber at equal distance between hinges c_1 and c_2, the circumferential length is equal to $D + h \sin \theta$. An increment of $d\theta$ corresponds to a change in the circumferential length and hence a circumferential strain

$$d\epsilon_c = \frac{\pi \left[D + h \sin\left(\theta + d\theta\right)\right] - \pi \left[D + h \sin \theta\right]}{D + h \sin \theta} = \frac{\pi h \cos \theta d\theta}{D + h \sin \theta}. \tag{13.68}$$

If the material is yielding also in circumferential direction, the increment of work per unit volume corresponding to $d\epsilon_c$ is

$$dw_2 = \sigma_y d\epsilon_c; \tag{13.69}$$

taking into account the volume of material between the two hinges $(2\left(D + h \sin \theta\right) th)$

$$\begin{aligned} dW_2 &= 2\left(D + h \sin \theta\right) th \sigma_y d\epsilon_c = 2\left(D + h \sin \theta\right) th \sigma_y \frac{\pi h \cos \theta d\theta}{D + h \sin \theta} \\ &= 2\pi \sigma_y t h^2 \cos \theta d\theta. \end{aligned} \tag{13.70}$$

The total work for collapsing one convolution (fold) is obtained by integration of angle θ from 0 and $\pi/2$:

$$W = \int_0^{\pi/2} (dW_1 + dW_2) \tag{13.71}$$

$$= \sigma_y \int_0^{\pi/2} \left[\frac{\pi}{\sqrt{3}} t^2 (2D + h \sin \theta) + 2\pi t h^2 \cos \theta\right] d\theta; \tag{13.72}$$

This (internal) work is equal to that done by the mean load $\bar{P}$ with the axial displacement $(2h)$, corresponding to the collapse of one convolution, leading to:

$$\bar{P} = \pi \sigma_y t \left[\frac{t}{\sqrt{3}} \left(\frac{\pi D}{2h} + 1\right) + h\right]. \tag{13.73}$$

The half wavelength, which is still unknown, can be found by minimizing the work necessary for the deformation:

$$\frac{\partial \bar{P}}{\partial h} = 0; \tag{13.74}$$

and hence

$$h = k\sqrt{Dt}, \tag{13.75}$$

where

$$k = \sqrt{\frac{\pi}{2\sqrt{3}}} = 0.952. \tag{13.76}$$

The semi wavelength of Eq. 13.75 and 13.76 is substituted in Eq. 13.73 to determine the mean collapse load when the convolutions form out of the diameter of the cylinder (subscript $_o$)

$$\bar{P}_o = \pi\sigma_y t\left(2k\sqrt{Dt} + \frac{t}{\sqrt{3}}\right). \tag{13.77}$$

If the convolutions form internally (subscript $_i$), instead of externally, to the cylinder (Fig. 13.54)

$$\bar{P}_i = \pi\sigma_y t\left(2k\sqrt{Dt} - \frac{t}{\sqrt{3}}\right). \tag{13.78}$$

During experimental tests, the diameter of the undeformed cylinder is between the inner and outer diameters of the collapsed convolutions ([41], [42]); the corresponding mean load can be obtained as the average between the values given by Eq.s 13.77 and 13.78

$$\bar{P} = 2\pi k\sigma_y t^{3/2}\sqrt{D} = K\sigma_y t^{3/2}\sqrt{D}. \tag{13.79}$$

This corresponds to the assumption that the mean diameter of the is the same as the undeformed cylinder. Parameter K can be computed from the value of k given by Eq. 13.25:

$$K = 2\pi k = 5.984, \tag{13.80}$$

which is slightly lower than that determined experimentally:

$$K_{\mathrm{exp}} = 6.2. \tag{13.81}$$

The difference between the experimental and the numerical value is lower than 4%, justifying the approximations at the basis of the analysis, especially those concerning the mean radius of the convolutions and that giving the semi-wave lengty h.

Other authors have introduced alternative expressions for the mean crushing load of cylinders, obtaining good correlation with experimental results, eg. Abramowicz ([43]):

$$\bar{P} = \sigma_y t\frac{6\sqrt{Dt} + 3.44t}{0.86 - 0.57\sqrt{t/D}}. \tag{13.82}$$

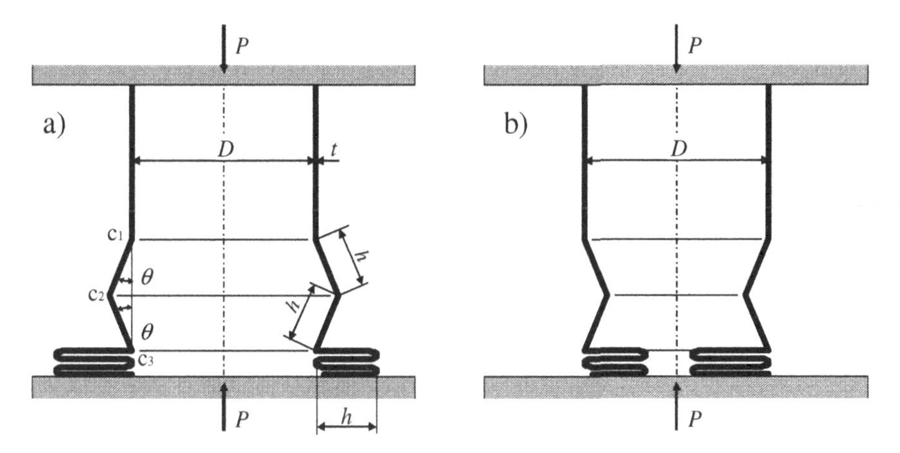

Fig. 13.54. Kinematics of the axial collapse of a circular tube: concertina mode. a) convolutions with inner a) or outer diameter b) equal to the undeformed diameter.

Diamond mode

The mean load during the axial collapse in the diamond mode has been evaluated by Pugsley [44] using a kinematic approach similar to that adopted by Alexander for the concertina mode. The mean load is expressed as function of the number n of diamonds along the circumference. This number increases with increasing the diameter to thickness ratio D/t:

$$\bar{P} = 2.286\ n^2 \sigma_y t^2. \tag{13.83}$$

Other Authors propose expressions with an explicit dependance from the D/t ratio, for example,

$$\bar{P} = 18.15\ (D/t)^{1/3} \sigma_y t^2. \tag{13.84}$$

13.3.8 Effects of the Strain Rate

The stress-strain curve of a plastic material is usually obtained during experiments at very low strain rates. The tests show that for a high deformation speed, the effect of the strain rate on the characteristic cannot be neglected, which should be taken into account in computations related to energy absorbers because of the high deformation speeds.

A relevant effect is that the yield stress increases with increasing strain rate: A semi-empirical expression that gives this dependency is:

$$\sigma_{yd} = \sigma_y \left(1 + \left(\frac{\dot{\epsilon}}{c} \right)^{1/\rho} \right), \tag{13.85}$$

where σ_y is the quasi static value of the yield stress, and coefficients c and ρ are functions of the type of material. For mild steels for deep drawing, for example $c = 40.4\ \mathrm{s}^{-1}$ and $\rho = 5$.

During the axial collapse of a cylindrical tube, the mean strain rate $\dot{\epsilon}$ increases with the deformation speed of the cylinder and decreases with the diameter D (as a first approximation).

In the concertina mode, the strain rate can be approximated as:

$$\dot{\epsilon} = \frac{V}{2D\left(0.86 - 0.568\sqrt{t/D}\right)}$$

During the diamond mode:

$$\dot{\epsilon} = 0.74\frac{V}{D}. \tag{13.86}$$

The strain rate effect is large for strain hardening materials such as high strength steels, while is rather small for ductile materials (mild steel for deep drawing and low resistance aluminum alloys); in this case, the yield strength could increase by about 10%.

Another effect of the strain rate is that increasing the deformation speed causes the inertia forces to increase. This effect is rather small until a kind of threshold is reached. Below the threshold, the convolution forms when the previous one is completely collapsed (progressive crush); above the threshold, all convolutions appear almost simultaneously (simultaneous crush).

13.3.9 Structural Foams

Structural foams have found a vast application in the automotive field for crash protection being used for bumpers, dashboards and side impact protections. Their strain stress characteristic during compression makes them attractive for crash energy dissipation (Fig. 13.55) ([45], [46] [47]). This characteristic exhibits a relatively small linear part (5% max deformation) followed by a flat section with nearly constant stress. During this phase the cells that constitute the material begin to collapse by elastic buckling, plastic yielding or brittle failure, and this collapse proceeds in the material at a nearly constant stress. When most of the cells have collapsed with opposite walls in contact (densification) the stress exhibits a sharp increase.

Apart from the type of base material (polyurethane, polypropylene or polystyrene, for example), the mechanical properties of the foam are influenced mainly by the density which depends on the amount of air trapped in the material.

The density has a large influence on the stress-strain curve. Fig. 13.55 shows that increasing the density the stress corresponding to a given deformation increases but the densification occurs at lower strain.

The energy per unit volume (W) absorbed during the deformation is equal to the area under the stress strain curve. Most of the energy is dissipated in the plateau, where the stress is nearly constant. The energy can be computed as

$$W(\epsilon) = \int_0^{\epsilon} \sigma(\epsilon')d\epsilon'. \tag{13.87}$$

For a given amount of energy that must be dissipated, the maximum force should be minimized to reduce the acceleration of the occupants. Fig. 13.56 shows three different conditions with the same energy per unit volume. The material with lower density (ρ_1) reaches the densification with high stress values, that with higher density remains in the flat region but it is characterized by a maximum stress (σ_3) higher than the material with intermediate density, that is compressed at almost the end of the flat region. The material with intermediate density is then the most convenient one because it minimizes the force a given amount of dissipated energy.

Another parameter that quantifies the effectiveness of a foam is the ratio between the energy absorbed up to a maximum stress σ and the stress itself. The efficiency E is then defined as

$$E = \frac{w(\epsilon)}{\sigma(\epsilon)}. \tag{13.88}$$

The curves of Fig. 13.57 show the efficiency to be a function of the maximum stress for the same material of Fig. 13.55 and 13.56. The diagram shows for a given maximum stress the density that allows the efficiency to be maximized. For example, the intermediate density ρ_2 has a maximum efficiency in the range $1 \div 1.7$ MPa.

As a simple design procedure, consider a cylindrical absorber with a given cross section A that must stop a body of mass m with a kinetic energy T.

1. If the maximum allowed acceleration is a_{max}, the corresponding maximum force is:

$$F_{\text{max}} = ma_{\text{max}}. \tag{13.89}$$

2. The maximum force F_{max} and cross section area A, enable the maximum design stress σ_d to be determined. The foam density should be selected so as to maximize the efficiency E^* at stress σ_d. The corresponding energy per unit value is

$$w = E^* \sigma_d. \tag{13.90}$$

3. The energy density w together with the kinetic energy T that must be dissipated enablesthe V of the shock absorber to be determined:

$$V = \frac{T}{w}. \tag{13.91}$$

4. and, hence the axial length l:

$$l = \frac{V}{A}. \tag{13.92}$$

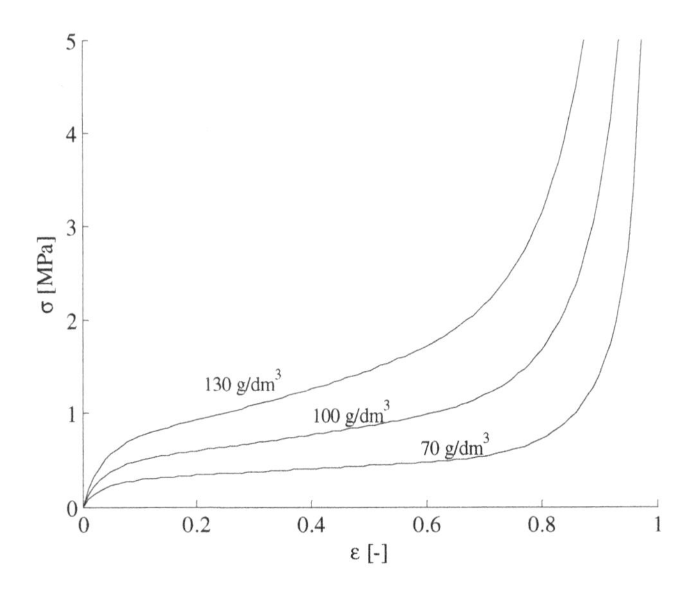

Fig. 13.55. Stress strain curve for a polypropylene foam (EPP [47]). The curves indicate different values of the density ρ.

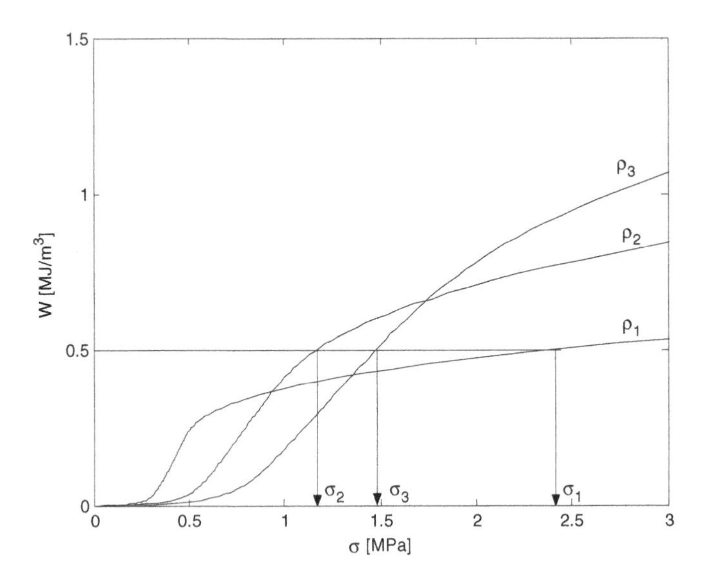

Fig. 13.56. Deformation energy per unit value from the curves of Fig. 13.55. ρ_1 =70 g/dm^3, ρ_2 =100 g/dm^3, ρ_3 =130 g/dm^3.

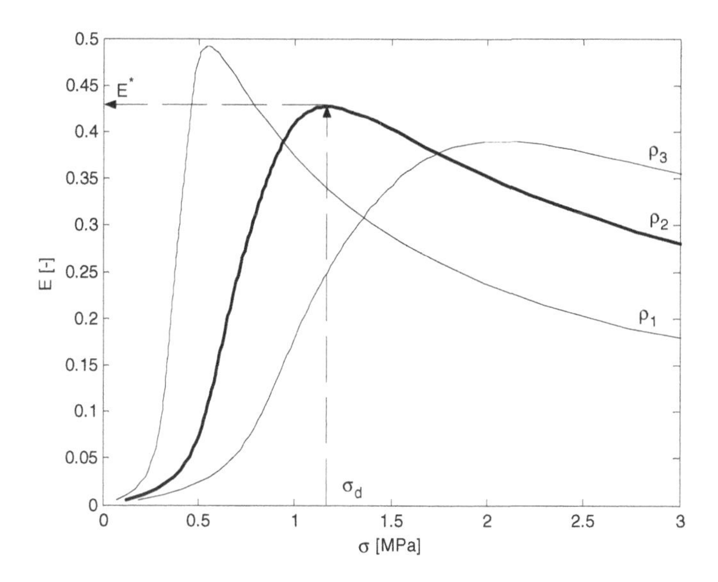

Fig. 13.57. Efficiency of the structural foam ($E = w/\sigma$) as function of the maximum stress for different densities. $\rho_1 = 70$ g/dm^3, $\rho_2 = 100$ g/dm^3, $\rho_3 = 130$ g/dm^3.

13.4 Front Structure Configuration

The design of a vehicle for crash is an iterative process of refining starting from results of sub-division of general objectives into specific targets (*target deployment*); this starting point is the output of the structuring design stage.

This way of proceeding applies the various aspects of impact behaviour and can be considered as the initial analysis of different potential solutions which are appropriate to the achieve the objective (*target*) and the subsequent refinement of the best compromise which results (considering even all the restrictions from other disciplines). This way involves and links the two principal aspects of passive safety: the biomechanical response of passengers and the structural response of vehicle.

Since today the quality of the vehicle design from the *crash* perspective is evaluated through the biomechanical response of dummies used in experimental *crash testing*, it is clear how the design has to develop from this aspect; therefore it must occupy a central position in this structuring stage which means being already able to activate the numerical instruments of biomechanical research on acceleration profiles estimated with simplified models of various types of impact. At this stage, impact models of first approximation are adopted. Subsequently the level of simulation details increases progressively up to the simulation of the impact behaviour of the complete vehicle. It is important that the different types of impact fixed by regulation are considered simultaneously for the various typologies of possible crashes (frontal, rear, ...).

Working in this way allows to have a procedure of project structuring able to already identify possible incompatibility with respect to the initial requirements, allowing a correct identification of objectives of subsystems being aware of their possible limits / criticalities.

The definition of a new vehicle requires specifying its configuration and its general performances in terms of expected final results (reflects the result of marketing research): of these specifications, those that most directly concern passive safety are:

- the biomechanical response of passengers, expressed both in terms of regulation limits to respect for homologation and in terms of results to be achieved during the rating tests.
- type of impact tests to be satisfied (through rigid or deformable barrier, at different speeds, with total or partial vehicle-barrier superposition, etc.).
- type of retraining systems that must be adopted (seatbelts with tensioner, with or without load limiter, airbags, collapsable steering column, etc.).
- geometry and configuration of interior layout (seats, dashboard, steering wheel, passengers postures, ...)
- the mass of the vehicle and macro-architecture (vehicle segment, type of structure and chassis, layout of the engine compartment, etc).

A certain vehicle architecture produces a corresponding acceleration profile depending on the structural response. Such acceleration profiles can be described by simplified curves, related, within limits, to classic design parameters such as deformation distances and collapse loads of elements adopted for impact energy absorption.

Body acceleration profiles are the entry parameter for evaluating the biomechanical response of passenger (for example, through codes like RADIOSS, MADYMO and OPTIMUS), through the interface retaining system - cockpit.

Through the iteration that starts with the selection of acceleration profile and reaches the biomechanical response, it is possible to identify the target load-displacement curve for the vehicle; these are the specifications for designing the impact absorption system (that again depends on the type of archtecture considered).

These identified specifications enable the size of sections and structures to be defined with first approximation models that initially consider only the absorption system of the impact energy (front rails). Subsequently, the use of FEM numerical simulations applied to complete vehicle models permit increasingly precise evaluations. These stages of evaluation have to be made at each step of the design process to indicate any potential criticalities from the outset.

Successively, the design of the energy absorption structure will be analysis with respect to a frontal and a rear impact.

13.4.1 Target Acceleration Profiles

The definition of target acceleration profiles and, consequently, of the load-displacement characteristics of a vehicle can be made using simplified acceleration profiles that can be evaluated for various typologies of impact test.

In fact at this stage it is appropriate to considering simultaneously different types of impact (USA FMVSS, US/EURO NCAP,...). The same type of structure produces different acceleration profiles and thus different biomechanical responses as a consequence of impacts on different types of obstacles . The final solution will be the result of a compromise between configurations that optimize the different types of impact.

Two examples of high speed impact types are, respectively:

- high speed frontal impact into a rigid plane obstacle *full overlap* (or rather with a barrier that covers all the front part of the vehicle, as in the case of USA FMVSS 208 standard, or the US NCAP rating) or,
- frontal impact into deformable barrier *offset*, ie. with a displacement respect to the longitudinal symmetry plane of vehicle (XZ) that produces a partial covering (*overlap*) of its front part by the obstacle (for example the old AMUS rating test).

In the first case, the vehicle structure is deformed symmetrically during the impact with the obstacle, involving all the front part. Since all the elements of the absorption system of frontal impact work in parallel, the deceleration profiles that are recorded in this type of test are generally very strict. In this case the retaining system must limit injuries in the upper part of the occupant (head, neck and chest). Instead intrusions in the cockpit are reduced because the whole front structure of the vehicle works from the beginning of the impact.

Since the initial impact speed (in the case of US NCAP and AMUS tests) and therefore the initial kinetic energy to dissipate is the same, the collision into an obstacle with partial coverage of the front part of the vehicle (for example 50% in the case of AMUS impact) means make the structure is made to works only partially in terms of energy absorption. The fact that structural elements located on one side do not absorb energy means that the part of structure impacted is forced to dissipate the kinetic energy through bigger deformations, with a consequent increase in the risk of cockpit intrusions. Despite less strict acceleration profiles, the impact with *offset* is the more critical in terms of deeper intrusions with high risk of occupant injury especially for lower limbs.

To better understand these aspects, it is possible to approximate the load-deformation characteristic of frontal part of vehicle with a line (13.58). To characterize this structure, deformations can be imposed using a press and the corresponding loads measured. Increasing the deformation, the load increases

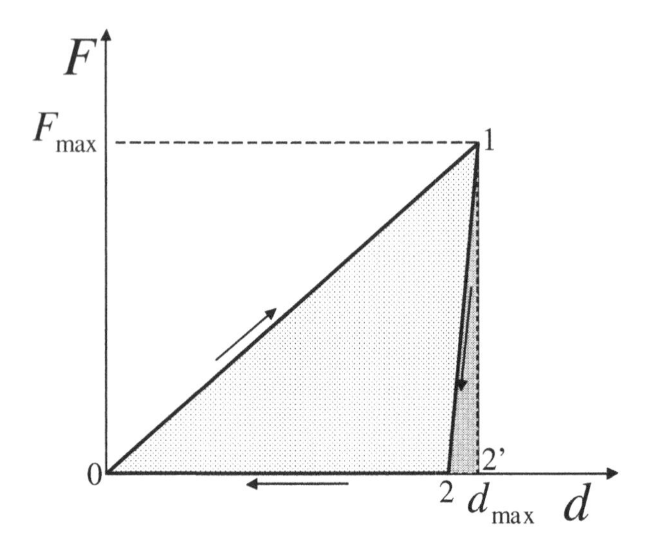

Fig. 13.58. First approximation of the load-deformation characteristics of structure with respect to impact energy absorption.

before reaching a maximum value $F_{\max}$, corresponding to the maximum deformation $d_{\max}$ (0-1 part of characteristic). Decreasing the deformation from point 1, there is a phase of elastic return before (1-2), before the load retruns to zerol.

The 0-1-2 area represents the dissipated energy during the load-unload cycle whereas the 1-2-2' area represents the restored elastic energy. Assuming that the elastic energy stored during deformation is negligible, ie. that the 1-2 line of characteristic is vertical, the dissipated energy can be expressed as function of the maximum load and maximum deformation:

$$E_d = \frac{1}{2} F_{\max} d_{\max}. \tag{13.93}$$

It must be emphasized that, even considering potential energy, energy E_d is dissipated and not restored during the unloading phase.

Similarly, the ratio between the maximum load and maximum displacement can be considered to represent a type of dissipative "spring", since the energy needed to deform it being completely dissipated rather being than stored as in a normal spring.

Clearly this characteristic is a rough approximation of what happens on real structures applied to imapct absorption, and is simply used to simplify the calculations and emphasize the basic physical aspects of phenomenon.

Fig. 13.59 represents the impact into a rigid barrier with total superposition with the obstacle (*full overlap*) and with partial superposition (*offset*) for the same vehicle; in the latter case, the overlap percentage is:

$$\frac{w_{offset}}{w} \tag{13.94}$$

where w is the width of the car body while w_{offset} is the width of the obstacle that overlaps the car body.

Since a smaller part of vehicle is involved in the impact with partial superposition, for equal deformation a smaller force is exerted. The load-deformation characteristics in the two cases are shown in Fig. 13.60. K_{offset} stiffness on *offset* impact and the one on *full* impact(K_{full}) are in first approximation proportional to the overlap percentage, as follows:

$$K_{offset} = a\frac{w_{offset}}{w}K_{full} \tag{13.95}$$

The a coefficient takes into account the fact that during an impact with *offset* even the part of structure not directly involved by the obstacle is subjected to large deformations, hence dissipating energy. The structure has an overall stiffness which is higher than that resulting with the only w_{offset}/w ratio. The stiffness increase due to the parts not directly involved with the obstacle depends on how the frontal part of the vehicle is made (in the propagation of deformation, the engine and the rest of the underhood layout of the engine compartment play a significant part). For a vehicle with frontal trasversal engine, one can assume a value $a = 1.3$.

Referring to Fig. 13.59 , and supposing that the structure is designed to absorb a *full* impact into a rigid barrier at a V_{full} speed with a d_{offset} deformation, the K_{full} stiffness of the complete structure is given by:

$$\frac{1}{2}mV_{full}^2 = \frac{1}{2}K_{full}d_{full}^2 \tag{13.96}$$

thus

$$K_{full} = m\left(\frac{V_{full}}{d_{full}}\right)^2. \tag{13.97}$$

The maximum deformation during the *offset* impact at a V_{offset} speed is given by

$$\frac{1}{2}mV_{offset}^2 = \frac{1}{2}K_{offset}d_{offset}^2 \tag{13.98}$$

Considering Eq. 13.95, the deformation of *offset* impact is obtained:

$$d_{offset}^2 = \frac{m}{a\frac{w_{offset}}{w}K_{full}}V_{offset}^2. \tag{13.99}$$

As an example of a *full* impact, the following starting data can be considered:

$$\begin{array}{lll} V_{full} & 56 & \text{km/h} \\ d_{full} & 500 & \text{mm} \\ m & 1000 & \text{kg} \end{array} \tag{13.100}$$

from which it is possible to obtain:

$$\begin{array}{ll} E_{full} & 121\ \text{kJ} \\ F_{\max full} & 484\ \text{kN} \\ K_{full} & 968\ \text{N/mm} \end{array} \tag{13.101}$$

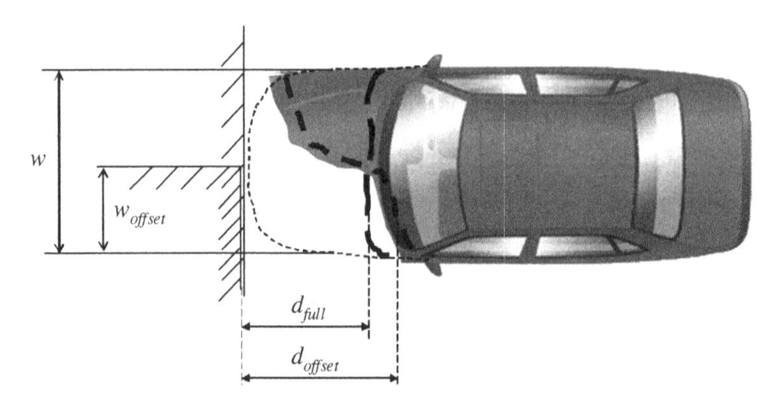

Fig. 13.59. Impact scheme with total superposition (*full overlap*) and with partial superposition (*offset*).

The same structure, in an *offset* impact in the following conditions:

$$\begin{array}{l} \frac{w_{offset}}{w} \; 0,5 \\ V_{offset} \; 55 \quad \mathrm{km/h} \end{array} \tag{13.102}$$

has a maximum deformation and a maximum force that depend on the design choices. Assuming $a = 1.3$ it is possible to obtain:

$$\begin{array}{ll} E_{offset} & 117 \text{ kJ} \\ K_{offset} & 629 \text{ N/mm} \\ F_{\max offset} & 383 \text{ kN} \\ d_{offset} & 609 \text{ mm} \end{array} \tag{13.103}$$

It can be observef that the *offset* impact is characterized by a lower maximum force and a higher deformation respect to the *full* impact; As a consequence it is less critical in terms of acceleration, whilst the cockpit intrusions are more critical.

The linear load-deformation characteristic is just a first approximation model of the behaviour of structures used to absorb the energy of impact. As stated previously, the structure is takes advantage of the collapse in the plastic field of thin wall tubular structures. Such structures exhibit an almost rectangular load-deformation characteristic. By positioning in series tubular elements that collapse with progressively higher loads, it is possible to obtain a step load characteristic with each load level corresponding to the collapse of a different parts of the structure. Such a step characteristic offers the important advantage of a lower maximum load for the same absorbed energy, thus decreasing the acceleration levels of the vehicle occupants.

Fig. 13.61 shows an example of a load-deformation characteristic of a rear part of a vehicle both in the case of *full overlap* impact and in the case of 50% *offset* impact; both cases refers to an impact into a rigid plane barrier.

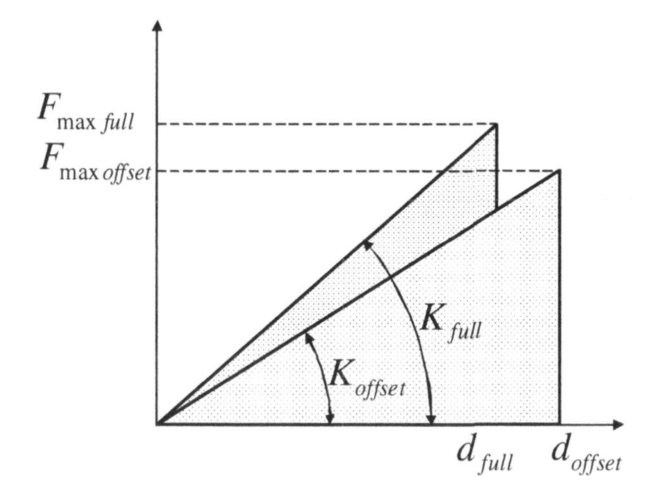

Fig. 13.60. F: Full and partial frontend impact forces

The first load level corresponds to a deformation of 100 mm and represents the insurance impact. The next level of 400 mm is designed to be sufficient to stop a vehicle in a *full overlap* impact. The last level of 100 mm is made to absorb the remaining energy in case of 50% *offset* impact.

The useful lengths for plastic deformation of the structure depend on the type of architecture, ie. the layout of the vehicle considered and the type of impact. Once the lengths have been decided, the parts of the structure that take part in the impact are sized taking account also the other functions, apart from crash resistance, that the vehicle must meet.

Instead the structure of the cockpit must remain as undeformed as possible in order to protect the occupants. In particular, mechanical parts deforming the firewall must be avoided to minimise the risk of injury to the occupants.

It can be noted how, for the same architectures, similar values of absorbed impact energy are exhibited despite the vehicles having different style and/or segment. Therefore these absorption levels can be considered to represent a standard for the rough design of the structure.

At this stage it is appropriate to consider a frontal impact into a rigid plane barrier full-overlap (US NCAP) and offset (AMUS).

From the point of view of understanding the phenomena, tests against a fixed rigid obstacle offer the advantage of causing damage due only to the vehicle itself. The energy that the structure must dissipate is equal to its initial kinetic energy.

To estimate the behaviour in collisions involving a deformable obstacle it is necessary to consider its load-deformation characteristic; in fact a part of the impact energy is dissipated by the obstacle itself. In the case of two vehicles impacting, the situation becomes significantly more complex because each can

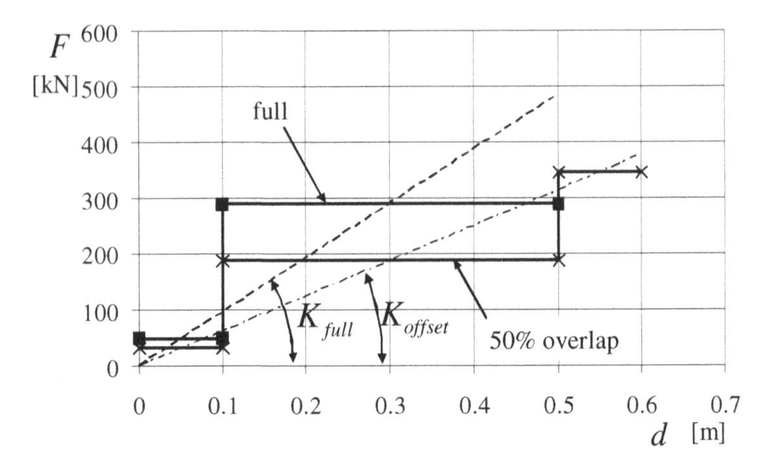

Fig. 13.61. F:frontend forze puntoni

exhibit different load-deformation characteristics and, during the impact, the speeds of both vehicles change, according to the laws of dynamics.

Fig. 13.63 shows the typical three *load lines* configuration, that is usually adopted on most vehicles today. This configuration permits a rational arrangement of the engine compartment, high energy absorption, and a near cancellation of cockpit intrusions.

For the two types of impact considered, since the obstacles are rigid and initial speeds are almost identical (56 km/h for US-NCAP and 55 km/h for AMUS), the levels of energy to be dissipated are nominally the same in the two situations.

The elements absorbing the energy are shown on Tab. 13.1. In the case of US-NCAP impact, the amounts reported refer to the entire front part (so, for example, 60% relative to the main front rails (5) is given by the sum of two contributions of 30% that each provide in accordance with the symmetry of full-overlap impact). Instead in the case of AMUS impact, dissipation is for the most part due to the elements that are in the half of the frontal part involved in the impact with the obstacle.

It can be noted how most of the energy is dissipated in both cases by the low load lines, ie. by the front rails (crashbox 3, bumper crossmember 4, main front rails 5) and the mechanics frame (6) which, together with the bumper crossmember (4) and the engine group, contributes to the propagation of damage to the side not impacted.

Although the load line made comprising the upper front rails (2) does not absorb a high amount of energy, it is fundamental since, being located above the vehicle centre of gravity, serves to avoid pitch motion of the vehicle during impact. The effect is to stabilize the vehicle and help prevent bending collapse in the cockpit with the breaking of the roof and of the central floor section.

Table 13.1. Amounts of absorbed energy during impacts into a rigid barrier.

element	absorbed energy	
	full overlap (≈US-NCAP)	50% offset (≈AMUS)
main front rails (5)	60%	40%
upper front rails (2)	6%	10%
mechanics frame (6)	20%	30%
crash box (3)	10%	5%
bumper crossmember (4)	4%	15%

Fig. 13.62 shows the deformation subjected to by the front structure in the case of insurance impact made on the left side of vehicle.

The *crash box* that is subjected to the impact is almost completely deformed axially, while the two front rails do not exhibit relevant deformation. In this way, the repair of damage consists only on the substitution of the front crossmember group, through the dismounting of the two flanges that connect it to rails. In addition, if the mechanical groups are appropriately mounted behind the flanges that support the *crash box*, they should not be damaged by the impact.

13.5 Testing on Vehicles

13.5.1 Dummies for Impact Test

The first anthropomorphic dummy was realized in 1949 for the US Air Force to evaluate the responses on humans due to accelerations imposed on the spine by ejection seats. From the middle of the 1960s dummies started to develop for applications in the automotive industry to evaluate the biomechanical response during the various types of impact a vehicle can be subjected to. Today these dummies play a fundamental role in vehicle saftey evaluation being the basis of the homologation and rating testing of vehicles.

A dummy is an anthropomorphic mechanical system comprising metallic masses, spring, dampers, articulations and polymeric coverings, that simulate the response of human body in the considered impact conditions. Inside the dummy there are sensors that enable the measurement of physical quantities related to the biological damage that occurs on the real occupant in the same impact conditions.

To be a reliable instrument for measurement during impact tests, it is necessary that a dummy exhibit the following characteristics:

- *biofidelity*: referring to injury criterion; (for example it has a good biofidelity if the chest crushes like the human chest when crushing is the injury criterion);

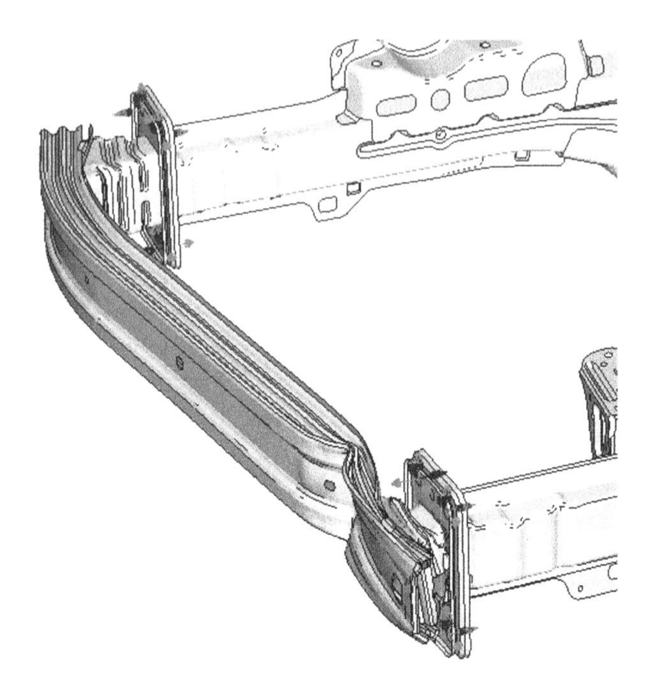

Fig. 13.62. Bumper crossmember deformation after an insurance impact on the left side of vehicles. The deformation interests mainly the *crash box* and not the front rail.

- *repeatability*: the same dummy, subject to the same stresses many times, must have equal responses;
- *reproducibility*: different dummies of the same type, subject to the same stresses even in different laboratories, must have equal responses;
- *sensitivity*: necessary in two respects: a) the dummy must be highly sensitive to the variation of harshness of impact (eg. test speed) and/or to the countermeasures adopted (eg. the variation of stiffness of the impacted part); b) on the other hand it must not be sensitive to the environmental variations (eg. temperature);
- *long-life*: the dummy parts must exhibit physical characteristics which do not vary over time with use;

To ensure these characteristics, some calibration and installation specifications of dummies are necessary. It can be noted that the performance criterion can differ from the limit of human tolerance as long as the relation between the two values is known.

The definition of correlations between physical quantities detectable on dummies and the corresponding biological damage levels on the human body, as well

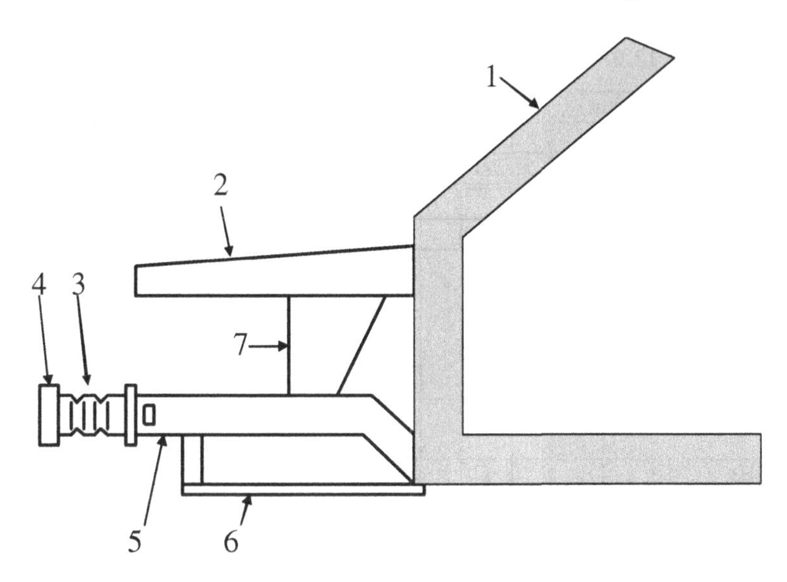

Fig. 13.63. Front structure configuration. 1) Undeformable cockpit; 2) upper front rails; 3) *crash box*; 4) bumper crossmember; 5) rails; 6) mechanics frame.

as the adjustment of the same dummies, derive from studies on cadavers (or PMHS, Post Mortem Human Subjects) conducted in a limited number of specialized centres in this kind of tests (eg: Wayne State University – USA, INRETS – F, Heidelberg – D).

As shown on table in Fig. 13.2, the characteristics of dummies depend on the type of impact they are used for and the occupants they represent:

- age, if children;
- gender (male/female) and percentile, if adults;
- impact type designed for (frontal, side, pedestrian);
- complexity: depending on the use, for certification or analysis and/or research activities;
- standard regulation.

Frontal impact dummy

The 50%ile Hybrid III (Fig. 13.64) represents the most common dummy for frontal impact tests, reproducing the size, inertial characteristics and dynamic stiffnesses of an average US male adult. The development of this dummy was undertaken made by General Motors in the 1970s: over the period 1971 to 1976, four generations of different dummies were created, leading to Hybrid III (following Hybrid I, Hybrid II and ATD 502).

Table 13.2. Dummies typologies for impact tests and their use.

type of imp.	front			side		rear	ped.	
gender (m/f)	m		f	m	f	m	m	
percentile	95	50	5	50	5	50	50	
Hybrid III	x	x	x					standard dummy for front impact
EU-SID 1				x				standard dummy for side impact
EU-SID 2				x				improvement of EU-SID 1, employed in Euro-NCAP tests
SID IIs					x			most updated dummy for side impact
US-SID				x				prescribed by 214 standard
BIOSID				x				rather sophisticated dummy
SID FMVSS 201				x				US SID body, Hybrid III head
THOR		x						improved evolution of Hybrid III
WORLDSID				x				improved evolution of EU-SID 2, under development
TNO 10		x						no instrumentation used for ECE16 tests
OCATD5			x					used to verify bag deactivation
PED.							x	
POLAR II							x	developped in cooperation with Honda
TRID						x		Hybrid III 50th percentile with improved neck for backlash simulation
RID2						x		Hybrid III based with new neck and torso for whiplash tests
BIORID						x		Developped by Volvo, Saab, Autoliv and Chalmers Univ., very sophysticated model of the backbone and neck
RID3D						x		Based on RID2 for whiplash evaluation
Virtual pregnant dummy			x					Virtual dummy developped by Volvo to evaluate injuries on woman and fetus

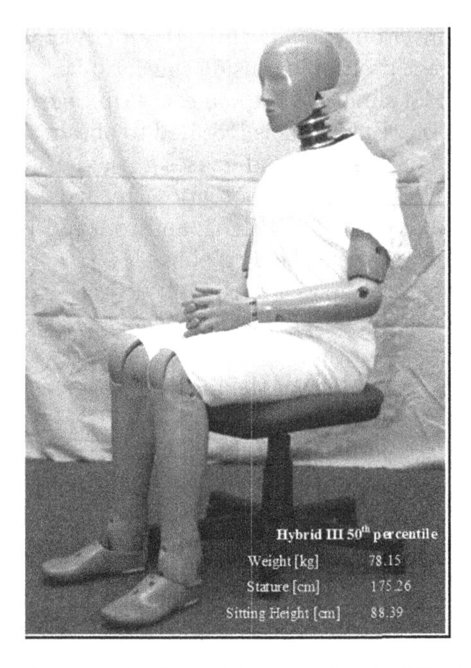

Fig. 13.64. Dummy for frontal impact: Hybrid III 50 percentile.

The correct anthropometry of the dummy is an important element to be considered: in fact, the *crash test* results are more or less realistic depending on the extent to which the dummy exhibits the typical anthropometric characteristics of humans. In particular the various shapes, locations of joints, centres of gravity and weights of the various components that comprise the dummy in the seated position represent the principal factors that determine its behaviour during a *crash test*: Hybrid III was designed to assume naturally the typical posture of a seated man inside a car.

The head is made of an aluminium skull onto which a skin is applied having a sufficient thickness to ensure the biomechanical fidelity and response repeatability of the head with respect to impact with hard surfaces. At the location of the centre of gravity of the head three accelerometers are mounted orthogonally to measure the triaxial acceleration (Fig. 14.17).

The neck is a flexible component that has stiffness and damping characteristics with good biofidelity both in bending and extension; it is made by three vertebral rigid discs in aluminium with elastomeric elements between them. A single steel cable runs along the neck centre to ensure high axial stiffness. The transversal section of neck is asymmetric to guarantee high stiffness for the head movement foward (bending), rather than backward (extension), as requested by biomechanical data.

Appropriate transducers measure shear stresses, axial stresses and the moments at the upper joint to the head (occipital condyle) and in the first thoracic vertebra: this last connection is made through an adjustable bracket to enable head leveling and correct positioning inside the cockpit.

The chest of Hybrid III simulates the spine and rib cage and is covered by a removable protective jacket. Ballasting is required to obtain the correct weight distribution. The thoracic part is rigid and houses a triaxial accelerometer mounted at the centre of gravity. A rotational potentiometer fixed to the extremity of a fork linked to the lumbar spine on one side and to a rod linked to the sternum on the other in order to measure chest crushing.

The model of the rib cage allows a maximum deflection of 90 mm. The curved lumbar spine enables the dummy to assume a not entirely erect position when placed on the vehicle seat, so that its posture similar to that of a human, permitting also a higher repeatability of the initial position as a consequence of the dummy not being forced into a specific position.

The abdomen, pelvis and legs of Hybrid III were essentially borrowed from the previous version of dummy (ATD 502) and subject to small alterations in order to obtain a better weight balance of the legs, an improvement of the knee covering and an increase in the reliability of the various components.

The foot-ankle system of Hybrid III is not instrumented: The ankle includes a spherical joint with range restrictors that limit the maximum foot excursions in the different directions similar to the human body; however Hybrid III does not allows the complete investigation of the types of damage to the lower parts of lower limbs.

Fig. 13.66 shows the measurement devices mounted on a dummy. The relatively large number of signals to be acquired (56) and the high speed of the impact phenomena require the use of complex data acquisition and analysis systems.

Side impact dummies

The EuroSID-2 (Fig. 13.65) is the dummy used in Europe to evaluate the biomechanical response of occupant in a side impact. The EuroSID-2 was born as an evolution of the previous EuroSID-1 after problems registered during experimental tests by american NHTSA in 1998.

The EuroSid-2 represents a 50° %ile male adult, without the lower part of arms. The head is the same as Hybrid III, while the legs are those of Hybrid II. The chest is made of three identical ribs that can be assembled on to the spine through several of elastic and damping elements positioned equally on the left and right sides, in order to enable crash tests with both impact configurations (and permit its use also with right hand drive cars and verify the behaviour for both the driver and the front passenger).

The abdomen is made of a metallic cast covered by a polyurethane foam mixed with rubber parts with well defined weight and curvature. The pelvis shape is typical of the human bone part, particularly as concerns the points relevant for side impact and those involved in the interaction with the seat and the belt.

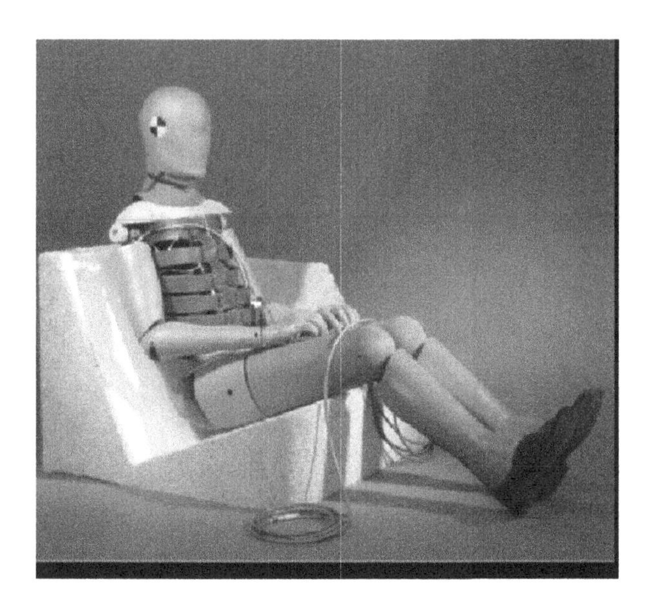

Fig. 13.65. EuroSID dummy for side impact.

The pelvis and the rib-damper units implement mechanisms to reduce the friction caused by the compression movement of the ribs. The two iliac crests are made with a special plyurethane plastic and are removable. The lumbar spine is erect, while a "jacket" of elastomeric material covers the chest, the two upper parts of arms and the lower region of the pelvis.

Fig. 13.66 shows the type and the number of the measurement devices installed on a dummy.

The WorldSID program for the research and the development of a single dummy to replaces, as a global *standard*, the different types of *side impact dummy* currently used is now nearing conclusion.

Other types of dummies

Other types of dummies have been developed for specific applications and types of impacts, including the following:

- BioRID: A dummy designed to evaluate the effects of a rear impact, the main objective being to study whiplash injuries and help designers to develop more effective head and neck control solutions. BioRID is more advanced in terms of spine construction than Hybrid: 24 vertebrae simulators enable BioRID to assume a seated posture which is more natural and demonstrate the correct neck movement and configuration associated with a rear impact.

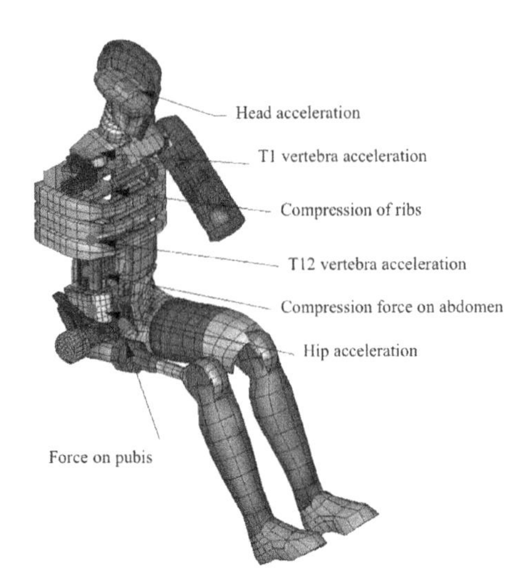

Fig. 13.66. Instrumentation and measurable quantities from EuroSID-2.

- CRABI is a child dummy used to evaluate the effectiveness of retaining devices for children including child seats, safety belts and *airbags*. There are three CRABI models that simulate children of 6, 12 and 18 months.
- THOR is an advanced male dummy, and is the successor of Hybrid III, with a more "human" spine and pelvis and a face equipped with a number of sensors that enables the analysis of facial impact with a degree of accuracy not attainable with other dummies. The number and sensitivity of the sensors in THOR are much higher than Hybrid III.

Virtual dummies

Bearing in mind the quality of a vehicle with respect to a *crash* is evaluated principally through the biomechanical response of dummies during experimental *crash tests*, the similar importance that virtual dummies models have in the CAE approach is also evident.

These virtual dummies are used in the structuring stage of vehicle when the impact absorption characteristics of the system need to attain the objective relative to the safety level of the occupants.

Although physical *crash test* dummies have provided highly valuable data on how human body reacts during impacts and have contributed to improve the design of the structure, a vehicle can be used once for a physical *crash test* which cannot be repeated exactly in the same way.

A second problem with physical dummies is that they only approximately represent humans; furthermore the sensors installed in a Hybrid III, for example,

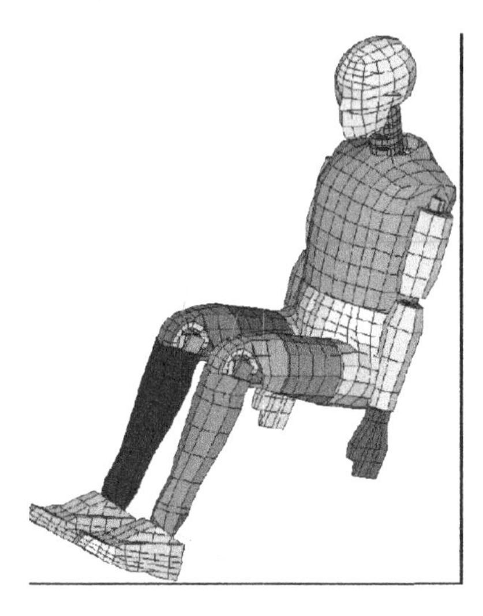

Fig. 13.67. Finite elements model for frontal impacts.

still only reproduce remotely the number of sensitive elements on a living human, and the simulation of internal organs is still at an extremely rudimentary level.

Today virtual models of dummies used during the experimental *test* can be ceated and refined to accurately reproduce the physical *crash*. In addition to simulating the complete occupant, studies of injuries to single parts can be made with a high level of detail; results have demonstrated reliability and are encouraging.

The potentially significant advantages of virtual impact simulation with respect to physical testing are evident: A virtual vehicle crashed once can be successively modified, eg. in terms of the configuration of the belts, and the crash repeated. Since each variable is under control and each event is repeatable, the need and cost of physical tests would be significantly reduced.

13.6 Impact Tests Equipment

The equipment used for impact tests are of two main types: those for tests on a whole vehicle and those for tests on parts of the vehicle or simple components or subsystems.

13.6.1 Equipment for Tests on a Whole Vehicle

The impact tests on the whole vehicle require for it to be launched into an appropriate obstacle in order to investigate its behaviour. The launch of the vehicle into the obstacle can be performed in two ways:

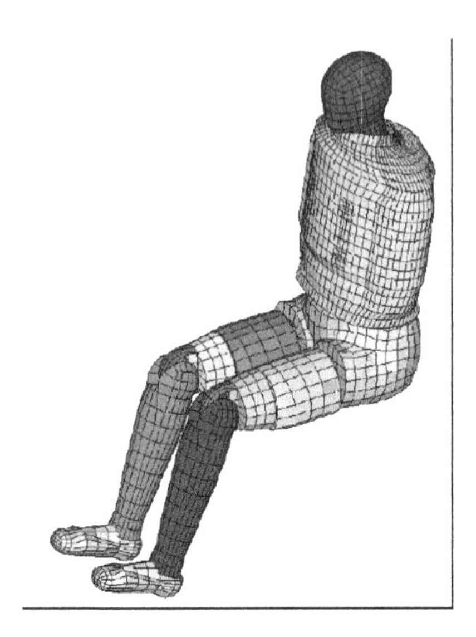

Fig. 13.68. Finite elements model for side impacts.

Fig. 13.69. THUMS (Total Human Model for Safety) developed by Toyota to improve the detail of responses obtained from virtual tests.

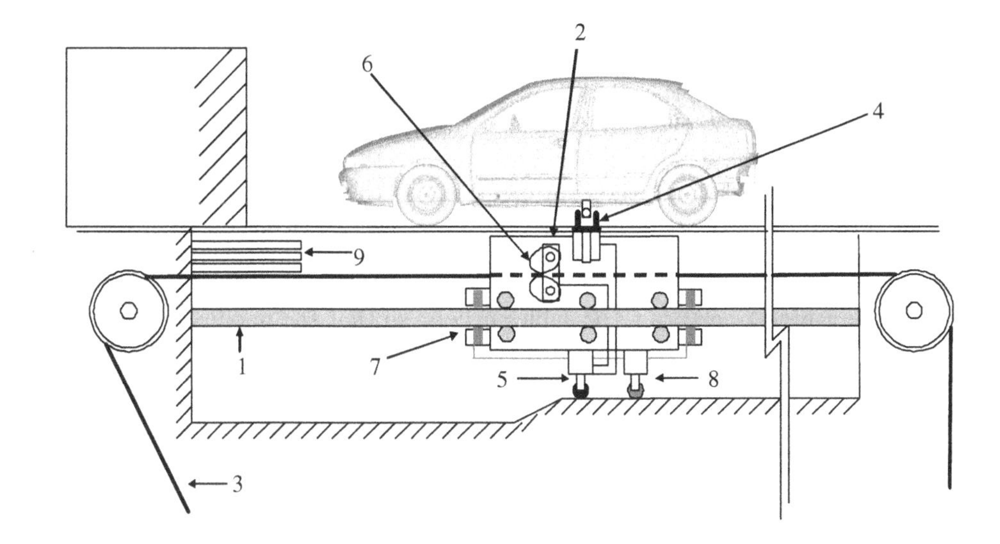

Fig. 13.70. Test equipment on vehicle: working scheme. 1) Rail; 2) bogie; 3) towline mounted on pulleys; 4) hook to drag the vehicle; 5) device for releasing the bogie to the towline; 6) coupling of bogie to the towline; 7) brakes of bogie; 8) brakes control; 9) emergency impact absorber.

- taking advantage of the engine of the same vehicle, using a device to remotely control the vehicle during the obstacle approaching phase;
- using an appropriate launching system that accelerates the vehicle (with engine off) into the obstacle;

The first approach has the advantage of not requiring expensive permanent facilities (it is sufficient to have a part of a road with appropriate length before the obstacle) and to be effectively closer to the real situation since the engine works at the moment of impact; however remote control of the vehicle is complex and there is the risk of a fire which may jeopardise the test. Though this system was used in the past (ex: for the AMUS impact done by the German magazine of the same name) and is still available, today has been largely superseded by independent cable launching systems, integrated into the test tracks at safety test centres (Fig. 13.70).

In practice, in the middle of the track and under the level of the road, a railroad (1) is housed in which a bogie runs (2) towed by a cable system (3) activated by an electric motor. The bogie has an appropriate upper hook (4) that protrudes from the road surface through a slot that runs along the middle of the entire track, via which the vehicle is towed into the obstacle.

The hook has an opening device (5) that, when activated at the right moment, releases the vehicle on its way towards the obstacle. Finally, the bogie hangs on the cable, that moves at the desired launching speed, through a clamping system

(6) similar to a cable car: clearly this system is disconnected before the bogie is braked (7), controlled by the system (8).

The activation first of the release mechanism and the uncoupling of the towline mechanism and then of the braking device, is controlled by special profiles of the bogie that activate appropriate switches as the obstacle is approached.

The vehicle can be connected to the hook of the bogie through a special transversal rigid bar welded to the platform (between the front and rear axles), or through cables or chains connected to the appropriate points (eg. the front suspension arms).

The launching track has to be sufficient long so that the maximum speed of impact into the obstacle (which is a function of the power of the electric motor and of the mass of the vehicle under test) is reached, to be maintained until the impact, from standing start, with an acceleration not too high in order to allow transitory conditions to disappear before impact.

13.6.2 Component Test: HYGE Slide

The *HYGE slide* is a test device designed to simulate the effects of a collision, both in acceleration and deceleration. It provides extremely repeatable and reproducible acceleration impulses, thus enabling accurate experimental simulation of *crash* conditions for cockpit elements (eg. the seats), without destroying a whole vehicle each time.

Systems operational for more than thirty years, with 15.000 tests, declare high levels repeatability with scatter of results in the order of 2%. The HYGE slide is used all over the world by car makers to test devices such as the occupant safety systems (*airbags*), children retaining systems, safety belts and components including seats, door locking mechanisms, windshields and fuel tanks

The HYGE slide receives a powerful, repeatable push from the action of two gases at different pressures applied to a piston in a cylinder. Using a runner system, the HYGE slide (Fig. 13.71) moves along two rails approx. 30m long. On the slide the entire car, the body or even just the seat with the belted dummy can be assembled. The diameter of the piston is calibrated depending on the mass to move.

At the start of the test, the slide in contact with the face of the piston being is lightly braked on the rails through the appropriate runners to avoid on-off movements. When the piston moves out of the slide, the test ends. From that moment the slide is decelerated through a braking system.

The HYGE slide is ideal to simulate the longitudinal deceleration conditions that passengers are subjected to in a vehicle involved in a frontal *crash*; such situations are simulated by moving the system under test in an opposite direction respect to reality.

Just before a real impact, the vehicle and its occupants move with constant speed. At the moment of impact they are stopped very quickly, suffering a deceleration in the opposite direction with respect to the direction motion. These conditions are simulated during the *test* on HYGE slide, starting from null speed

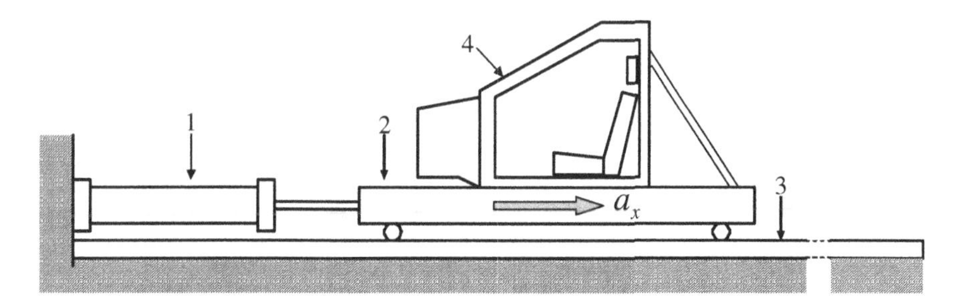

Fig. 13.71. HYGE slide scheme. 1) pneumatic hydraulic piston; 2) bogie; 3) rails; 4) element to test.

and accelerating the slide with the elements to be tested assembled in the opposite direction respect to the direction of motion.

The condition of null starting speed simulates the condition of constant speed before the impact. Successively, the slide acceleration moves the subsystem to be tested, providing a response comparable to that produced by a quick deceleration of a vehicle initially moving to a constant speed. For the simulation of the impact on the occupant, the dummy is placed backwards. The occupant has therefore the shoulders in the same direction of slide motion so that it suffers a deceleration; instead, in case of rear impact, the dummy facing the direction of motion.

Compared to the destructive tests of a whole vehicle into a barrier, the HYGE slide tests have several advantages:

- cost-effectiveness: once the system has been set up the test can be repeated many times without extra cost;
- repeatability: acceleration profiles are defined very accurately;
- modularity: since the whole car is not required, subsystems can be tested at a very early stage.

On the other hand, using this device it is difficult to repeat exactly the same acceleration profile and all the secondary movements the car is subjected to during a real impact.

13.7 Non Linear FEM Analysis

To describe a dynamic system, it is possible write a non linear general equation, where the equilibrium between internal F and external R forces is guaranteed by the presence of inertial forces. Inertial forces can be written as product between mass matrix and the acceleration vector $\ddot{q}$. This product is equal to the equilibrium between internal and external forces. [48], [49].

Fig. 13.72. HYGE slide.

Internal forces depend, generally, on the stiffness and damping characteristics of the material at that moment (then a function of time t) and on the values of displacements and speeds (vectors $\mathbf{q}$ and $\dot{\mathbf{q}}$):

$$\boldsymbol{M}\ddot{\boldsymbol{q}} = \boldsymbol{R}(\boldsymbol{q}, \dot{\boldsymbol{q}}, t) - \boldsymbol{F}_{internal}(\boldsymbol{q}, \dot{\boldsymbol{q}}, t). \tag{13.104}$$

Furthermore, functions that link the expression of internal forces to the variables q, $\dot{q}$, and t are, generally, non linear and even the mass matrix M can be function of the system time evolution.

Instead, when the behaviour of system can be considered to be linear, the Eq. 13.104 can be written as:

$$\boldsymbol{M}\ddot{\boldsymbol{q}} + \boldsymbol{C}\dot{\boldsymbol{q}} + \boldsymbol{K}\boldsymbol{q} = \boldsymbol{R}(t), \tag{13.105}$$

where the internal forces are represented by the sum of the elastic and damping forces; with much simpler functions, forces are linked to the values of displacements and speeds, through matrices C and K that, in the simplest case are constant over time, and represent linear functions (ie. not function of q).

13.7.1 Solution of Non Linear Static Problems

When the solution of a non linear problem where the variation of the load over time is very slow compared to the system dynamics, such a system can be defined as quasi static; inside the system equation the accelerations vector is set to zero:

$$\dot{q} = \ddot{q} = \mathbf{0}. \tag{13.106}$$

In addition, it is possible that some non linearities depend on how the load is applied, so it is generally necessary to treat the case where the load increases

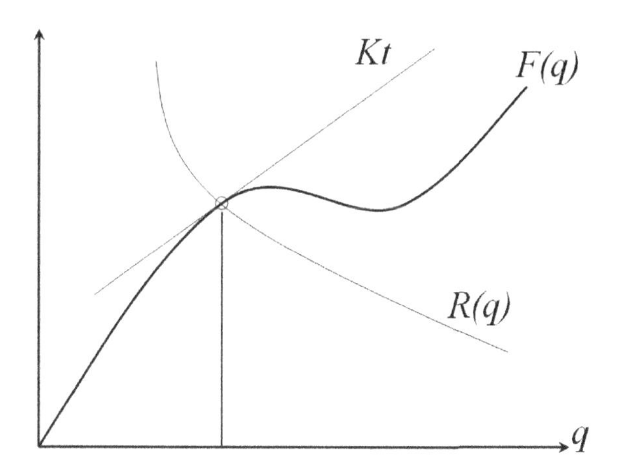

Fig. 13.73. Determination of the equilibrium point between internal F and external R forces in the non linear static problem.

up to the final value over time. Setting the accelerations vector to zero, the equilibrium condition of a system representing a body can be expressed as:

$$\boldsymbol{R}_t(\boldsymbol{q}, t) - \boldsymbol{F}_t(\boldsymbol{q}, t) = 0, \tag{13.107}$$

where the vector R represents the external forces applied, and the vector F represents the internal elastic forces.

Furthermore this relation represents the system equilibrium in the deformed conditions at the instant t, including all the non linearities. The Eq. 13.107 must be satisfied for each instant, with time t varying.

In the general case of a non linear equation the internal forces vector F_t is a function of time and displacement:

$$\boldsymbol{F}_t = \boldsymbol{F}_t(\boldsymbol{q}, t). \tag{13.108}$$

In a non linear problem, the material stiffness is not be constant with deformation. For a fixed instant of time, the stiffness characteristic of material is similar to that shown in Eq. 13.73 with $F(q)$, indicating with q the displacement.

If the stiffness matrix is always equal with the variation of material deformation, in each tension and deformation condition it is possible to treat the problem as a linear characteristic and obtain a linearization of the stiffness characteristic valid for a limited range of considered conditions.

For each operation of this type, a matrix of stiffness K_t, called matrix of tangent stiffness, is obtained. The K_t matrix in the non linear field has to be recalculated for each condition of tension and deformation of the material that generates modifications in the stiffness characteristics, and depends on the instant t considered.

Therefore to obtain the material behaviour at a certain point of the characteristic, one can choose to use the following expression:

$$F(q,t) = K_t \cdot q, \tag{13.109}$$

that, as already mentioned, is valid only for the t assigned.

The search of the solution to a problem of this type reduces to the determination of the deformation vector q that satisfies the equilibrium equation 13.107, ie. the determination of q corresponding to the intersection between R_t and F_t at the instant t .

A typical incremental approach assumes that the solution for certain instant t is known and that the objective is to calculate the solution for another instant $t + \Delta t$, where Δt is the time increment selected appropriately.

For instant $t + \Delta t$ one can write:

$$R_{t+\Delta t} - F_{t+\Delta t} = 0; \tag{13.110}$$

assuming that the external forces vector R is indipendent from deformations, since the solution at time t is known, one can write:

$$F_{t+\Delta t} = F_t + F, \tag{13.111}$$

where F is the increment in the nodal forces between t and $t+\Delta t$. This vector can be approximated with the tangent stiffness matrix, calculated with a linearization at time t:

$$F = K_t q, \tag{13.112}$$

where q is a nodal displacements vector and K_t is the tangent stiffness matrix

$$K_t = \frac{dF_t}{dq_t}. \tag{13.113}$$

substituting the last two equations in Eq. 13.107, it is possible to obtain:

$$K_t q = R_{t+\Delta t} - F_t \tag{13.114}$$

and making q explicit:

$$q_{t+\Delta t} = q_t + q, \tag{13.115}$$

which is an approximation since the tangent stiffness matrix was used.

Evaluating only an approximation of displacements at time $t+\Delta t$ it is possible to obtain only an approximation of stresses and corresponding nodal forces and then proceed to the next time interval. However, because of the approximations induced by the use of tangent mass matrix, the solution will be subject to errors and could give rise to numerical instability depending on the load type. To reduce the error is necessary to iterate until the equation 13.107 is satisfied with sufficient accuracy.

Different iteration methods exist; an example is the Newton-Raphson method that comes from the incremental technique. The method is based on the calculation of an increment in the nodal displacements to define a new vector of total displacements and repeat the incremental solution using the calculated displacements as starting point, and not anymore those at t time.

The equations used for the Newton-Raphson method are, for value of $i = 1, 2, 3$:

$$K_{t+\Delta t}^{(i-1)} \Delta q^{(i)} = R_{t+\Delta t} - F_{t+\Delta t}^{(i-1)}, \tag{13.116}$$

$$q_{t+\Delta t} = q_{t+\Delta t}^{(i-1)} + \Delta q^{(i)}, \tag{13.117}$$

with initial conditions

$$q_{t+\Delta t}^{(0)} = q_t, \tag{13.118}$$

$$K_{t+\Delta t}^{(0)} = K_t, \tag{13.119}$$

$$F_{t+\Delta t}^{(0)} = F_t. \tag{13.120}$$

For the first iteration, the system equations are:

$$Kq = R_t - F_t \tag{13.121}$$

and

$$q_{t+\Delta t} = q_t + q. \tag{13.122}$$

In successive iterations, each estimate of the displacement of nodal points is used as starting point for the next iteration in order to evaluate tensions and displacements and even for the linearization of stiffness matrix. The $R - F$ vector, called vector of unbalanced loads by elements tensions, is balanced in this way with an error that reduces itself at each iteration.

13.7.2 Characteristics of Non Linear Dynamic Problems

The solution to non linear dynamic problems necessarily requires the treatment of all model parameters as time dependent requiring the time integration of all the FEM model degrees of freedom. The speed which the parameters on the model vary is very important and distinguishes between the various types of models and integration time interval dimensions to be used.

As concerns structural components (treated here), Belitschko states that problems related to reflection and diffraction are not important: structural problems are called "inertials" because the time response is long compared with the time taken by pressure waves to cross the structure.

There are various approaches for the integration technique of equations depending on the method used to write the dynamic equations: Sometimes it is possible to express the speed and the position value in a time instant as a function of the previous instant, while in other cases it is not.

Integration techniques are therefore divided in into explicit and implicit.

Explicit code

Writing the equation for a dynamic system and neglecting the damping matrix effect, it is possible to obtain:

$$M\ddot{q}_t = R_t - F_t; \tag{13.123}$$

The solution of this equation requires the selection of the calculation technique for accelerations $\ddot{q}$; the explicit method involves writing the equilibrium at the instant $t + \Delta t$ to calculate displacements at instant t.

Usually used in this case is the central difference method. The derivatives of q and $\dot{q}$ are written as function of themselves at the previous instant, using an expression at the instant t, or rather as function of $q_{t+\Delta t}$ and $\dot{q}_{t+\Delta t}$ besides q_t and $\dot{q}_t$; using this technique, the acceleration $\ddot{q}_t$ results as:

$$\ddot{q}_t = \frac{\dot{q}_{t+\Delta t} - \dot{q}_t}{\Delta t}; \tag{13.124}$$

substituting in Eq. 13.123 and writing the velocity in the same way:

$$\begin{cases} M\dfrac{\dot{q}_{t+\Delta t} - \dot{q}_t}{\Delta t} = R_t - F_t \\ \dot{q}_t = \dfrac{q_{t+\Delta t} - q_t}{\Delta t} \end{cases}, \tag{13.125}$$

where $F = F(q_t, t)$ is a non linear expression.

In this system $\dot{q}_{t+\Delta t}$ and $q_{t+\Delta t}$ are the only unknowns and each can be calculated in an explicit way from its own equation; thus it is possible to calculate the system behaviour at the instant $t + \Delta t$ as function of the conditions at instant t.

Furthermore operating in a explicit manner it is possible to calculate the accelerations $\ddot{q}$ for each instant as function of the previous instant only if the mass matrix M is inverted. Fortunately this inversion is not usually particularly difficult from the numerical calculation point of view due to the fact it is possible to assume the mass to be concentrated (*lumped*), and the mass matrix M to be diagonal, making matrix inversion more straightforward.

The main disadvantage of using the explicit technique and the central difference method concerns the restrictions on the time interval dimensions (*time step*) since the time interval must be higher of the critical value to obtain a stable algorithm:

$$\Delta t_{cr} = \frac{T_n}{\pi}, \tag{13.126}$$

where T_n is the smallest period on the finite elements model.

This interval is often even defined as the one used by an elastic wave (of pressure) to cross the smallest element inside the model. Therefore the explicit techniques are only conditionally stable; their stability is subordinated to the dimensions of the integration *step* Δt, and the stability limits are expressed in terms of interval time dimensions.

In addition, in a linear analysis, the variation in the material or geometric conditions involve also the determination of the F_t vector; therefore the value of T_n is not constant over the length of the phenomenon, and the integration interval must be adapted to the material characteristics, in particular reduced, if the material becomes more rigid, ensuring the maintenance of the 13.126 condition for each time instant.

The explicit solution, imposing an integration interval within a fixed value, can require many, albeit not particularly complex, calculations; generally the procedure is best suited to relatively short phenomena characterized by fast time evolution.

Implicit code

The acceleration is written at time $t + \Delta t$ as:

$$\ddot{q}_{t+\Delta t} = \frac{\dot{q}_{t+\Delta t} - \dot{q}_t}{\Delta t}; \tag{13.127}$$

The dynamic equation is written at time $t + \Delta t$, to calculate q at instant $t + \Delta t$

$$\begin{cases} M\dfrac{\dot{q}_{t+\Delta t} - \dot{q}_t}{\Delta t} = R_{t+\Delta t} - F_{t+\Delta t} \\ \dot{q}_{t+\Delta t} = \dfrac{q_{t+\Delta t} - q_t}{\Delta t} \end{cases}, \tag{13.128}$$

where:

$$F_{t+\Delta t} = F(q_{t+\Delta t}, t + \Delta t) \tag{13.129}$$

is a non linear expression function of time and displacement q.

The system can be solved by substituting the expression of $\dot{q}_{t+\Delta t}$ in the first equation to obtain:

$$M\frac{\dfrac{q_{t+\Delta t} - q_t}{\Delta t} - \dot{q}_t}{\Delta t} = R_{t+\Delta t} - F_{t+\Delta t}. \tag{13.130}$$

Therefore, the problem reduces in this case to something similar to the non linear static solution, shown previously, with an equation similar to 13.107 where a contribution is added to the R function.

Inside Eq. 13.130, $F_{t+\Delta t}$ is also function of the K stiffness matrix; therefore, to calculate $q_{t+\Delta t}$ it is necessary to invert the stiffness matrix and use an iteration solution as for non linear static systems, using for example the method described in 13.116 and 13.117 (Newton-Raphson), but with the insertion of the product of mass matrix and acceleration vector $\ddot{q}$ in the first equation.

Furthermore, since in each non linear problem the solution is highly conditioned by the applied load history, it is very important to limit the error of the solution calculation for each time instant; the iteration in this case is more important than for the static solutions.

Implicit schemes remove the constraint of maximum dimension of time interval, calculating the dynamic quantities at time $t + \Delta t$, not only by their values at instant t, but also on their values at instant $t + \Delta t$.

In structural problems the implicit integration produces acceptable solutions with time interval values one or two order of magnitude higher than the stability limit of the explicit method (when compared with the simplest methods), but the forecast of the response deteriorates as the integration step increases.

Choice of the step size for the implicit method

Three factors need to be considered when selecting the maximum value of the *step size*:

- the rate of variation of the applied load,
- the complexity of the non linear of the stiffness and damping properties,
- the natural periodof oscillation of the structure.

Generally, a good rule to obtain reliable results is that a maximum time increment divided by the period must respect the following:

$$\frac{\Delta t}{T} < \frac{1}{10}. \tag{13.131}$$

Choice of best technique

Between explicit and implicit, the choice of the more appropriate integration technique depends on:

- the stability limits of explicit scheme (a very small time interval could be required);
- the relative ease with which the non linear equations can be solved with the implicit method;
- the relative dimension of time increments, that yields acceptable accuracy in the implicit scheme compared with the stability limit of the explicit one;
- the model size;
- rate of variation of the external loads;
- the length of the phenomenon.

Some finite element solvers offer the possibility to identify solutions of non linear dynamic problems, both with the explicit method and the implicit one (ABAQUS, RADIOSS), while others adopt only one of the two techniques (eg. explicit - MARC and LSDYNA ; implicit - ANSYS and PAMCRASH).

References

[1] Toffetti, A., Nodari, E., Zoldan, C., Rambaldini: Il carico mentale nell'interazione guidatore veicolo. In: Atti Del Convegno Nazionale Sulla Sicurezza Stradale, Torino (2002)

[2] Demontis, S., Giacoletto, M.: Prediction of car seat comfort from human-seat no. 2002-01-0781 interface pressure distribution. SAE Transactions (2002)

[3] Andreoni, G., Santambrogio, G.C., Rabuffetti, M., Pedotti, A.: Method for the analysis of posture and interface pressure of car drivers. Applied Ergonomics 33, 511–522 (2002)

[4] Andersson, G.: Loads on the Spine During Sitting. In: The Ergonomics of Working Posture, pp. 309–318. Taylor & Francis, London (1986)

[5] Akerblom, B.: Standing and Sitting Posture with Special Reference to the Construction of Chairs. PhD thesis, A. B. Nordiska Bokhandeln, Karolinska Institutet, Stockholm (1948)

[6] Reed, M.P., Schneider, L.W., Eby, B.A.H.: The effects of lumbar support prominence and vertical adjustability on driver postures. tech. rep., University of Michigan - Transportation Research Institute - Bioscience Division, 2901 Baxter Road, Ann Arbor, Mi, 48109-2150, March 31 (1995)

[7] Rebiffé, R.: The driving seat: Its adaption to functional and anthropometric requirements. In: Proceedings of a Symposium on Sitting Posture, pp. 132–147 (1969)

[8] Porter, J.M., Gyi, D.E.: Exploring the optimum posture for driver comfort. International Journal of Vehicle Design 19, 255–266 (1998)

[9] Grandjean, E., Hunting, W., Pidermann, M.: VDT workstation design: Preferred setting and their effects. Human factors 25, 16–175 (1983)

[10] Porter, J.M., Case, K., Marshall, R., Gyi, D., Olivier, R.S.N.: 'beyond jack and jill': Designing for individuals using HADRIAN. International Journal of Industrial Ergonomics 33, 249–264 (2004)

[11] Head contour. In: SAE Handbook, vol. 4, pp. 34.XXX–34.YYY. Society of Automotive Engineers, Inc., Warrendale (1994)
[12] Motor vehicle dimensions. In: SAE Handbook, vol. 4, pp. 34.85–34.103. Society of Automotive Engineers, Inc., Warrendale (1994)
[13] Hubbard, R.P., Haas, W.A., Boughner, R.L., Canole, R.A., Bush, N.J.: New biomechanical models for automobile seat design. SAE Technical Paper Series, vol. 930110 (1993)
[14] Park, S.J., Kim, C.B.: The evaluation of seating comfort by the objective measures. SAE Technical Paper Series, vol. 970595 (1997)
[15] Na, S., Lim, S., Choi, H.-S., Chung, M.K.: Evaluation of driver's discomfort and postural change using dynamic body pressure distribution. International Journal of Industrial Ergonomics 35, 1085–1096 (2005)
[16] Griffin, M.J.: Handbook of Human Vibration, pp. 404–408. Academic Press, London (1990)
[17] Describing and measuring the driver's field of view. In: SAE Handbook, vol. 4, pp. 34.157–34.167. Society of Automotive Engineers, Inc., Warrendale (1994)
[18] Vetri di sicurezza e materiali per vetri sui veicoli a motore e sui loro rimorchi, Tech. Rep. 92/22, Direttiva del Consiglio delle Comunità Europee (March 31, 1992)
[19] ASHRAE Handbook Fundamentals (1993)
[20] Alfano, G., D'Ambrosio, F.: La Valutazione Delle Condizioni Termoigrometriche Negli Ambienti Di Lavoro: Comfort e Sicurezza. Cuen (1987)
[21] McIntyre, D.: Indoor Climate. Architectural Science Series (1980)
[22] Fanger, P.: Thermal Comfort. McGraw-Hill Book Company, New York (1972)
[23] I. 9920, Ergonomics of the thermal environment - estimation of the thermal insulation and evaporative resistance of a clothing ensemble (1995)
[24] I. 8996, Ergonomics - determination of metabolic heat production (1990)
[25] Genta, G.: Vibrazioni Delle Strutture e Delle Macchine. Springer, Torino (1996)
[26] Meirovitch, L.: Dynamics and Control of Structures. Wiley-Interscience, Hoboken (1990)
[27] Crandall, S.H., Mark, W.D.: Random Vibration in Mechanical Systems. Academic Press, New York (1963)
[28] Crandall, S.H.: The role of damping in vibration theory. Journal of Sound and Vibration 11, 3–18 (1970)
[29] Pawlowski, J.: Vehicle Body Engineering. Business Books (1969)
[30] Brown, J.C., Robertson, A.J., Serpento, S.T.: Motor Vehicle Structures. Butterworth Heinemann, Oxford (2003)
[31] Lee, S., Hahn, C., Rhee, M., Oh, J.: Effect of triggering on the energy absorption capacity of axially compressed aluminum tubes. Materials and Design 20, 31–40 (1999)
[32] Mahmood, H.F., Paluszy, A.: Design of thin walled columns for crash energy management - their strength and mode of collapse. SAE Transactions, no. 811302, pp. 4039–4050 (1982)
[33] Wierzbicki, T., Abramowicz, W.: On the crushing mechanics of thin-walled structures. ASME Journal of Applied Mechanics 50, 727–734 (1983)
[34] Witterman, W.J.: Improved Vehicle Crashworthiness Design by Control of the Energy Absorption for Different Collision Situations. PhD thesis, Technische Universiteit Eindhoven (1999)
[35] Drazetic, P., Payen, F., Ducrocq, P., Markiewicz, E.: Calculation of the deep bending collapse response for complex thin-walled columns i. pre-collapse and collapse phases. Thin-Walled Structures 33, 155–176 (1999)

[36] Markiewicz, E., Payen, F., Cornette, D., Drazetic, P.: Calculation of the deep bending collapse response for complex thin-walled columns II. post-collapse phase. Thin-Walled Structures 33, 177–210 (1999)
[37] Automotive Steel Design Manual. American Iron and Steel Institute Auto/Steel Partnership, Southfield (1996)
[38] Alghamdi, A.A.A.: Collapsible impact energy absorbers: An overview. Thin-Walled Structures 39, 189–213 (2001)
[39] Alexander, J.M.: An approximate analysis of the collapse of thin cylindrical shells under axial load. Quarterly Journal of Mechanics Applied Methematics 13, 10–15 (1960)
[40] Avalle, M., Belingardi, G.: Experimental evaluation of the strain field history during plastic progressive folding of aluminium circular tubes. International Journal of Mechanical Science 39, 575–583 (1997)
[41] Belingardi, G., Avalle, M.: Investigation on the crushing of circular tubes: Theoretical models and experimental validation. In: Crashworthiness and Occupant Protection in Transportation Systems, Proceedings of ASME Int. Mech. Eng. Congress, San Francisco, USA, November 1995. AMD, vol. 210, pp. 129–141 (1995)
[42] Avalle, M., Belingardi, G., Vadori, R.: Analisi teorica e sperimentale del collasso plastico progressivo di tubi a sezione circolare in alluminio. In: Atti Del XXII Congresso Nazionale AIAS (Forlì), pp. 347–354 (1993)
[43] Abramowicz, W., Jones, N.: Dynamic axial crushing of circular tubes. International Journal of Impact Engineering 2, 263–281 (1984)
[44] Pugsley, A.G.: On the crumpling of thin tubular struts. Quarterly Journal of Mechanics and Applied Mathematics 32, 1–7 (1979)
[45] Gibson, L.J., Ashby, M.F.: Cellular Solids: Structures and Properties, 2nd edn. Cambridge University Press, Cambridge (1997)
[46] Avalle, M., Belingardi, G., Montanini, R.: Characterization of polymeric structural foams under compressive impact loading by means of energy-absorption diagram. International Journal of Impact Engineering 25, 455–472 (2001)
[47] Avalle, M., Belingardi, G., Ibba, A.: Mechanical models of cellular solids: Parameters identification from experimental tests. International Journal of Impact Engineering 34, 3–27 (2007)
[48] Belytschko, T., Liu, W.K., Moran, B.: Nonlinear Finite Elements for Continua and Structures. Wiley, New York (1997)
[49] Bathe, K.J.: Finite Element Procedures. Prentice Hall, Englewood Cliffs (1996)

Index